"Barry and Hall-McKim have produced the ultimate book
both as an excellent scientific/academic contribution and ?
the wider public concerned about the Polar environments, past history, and future p
tially serious impacts of the current climate warming on the entire world. The authors have
unsurpassed experience to undertake such an immense and important task. The book is
replete with highly relevant maps, diagrams, tables, and photographs. It also provides easy
access to the vast amount of research and data sets scattered through an immense
literature."

- Jack D. Ives,
Carleton University

"The book is well structured with a flow from chapter to chapter, which leads the
reader constructively through a set of themes and trends in the changing role that the
polar regions play in the planet's changing climate. The nine chapters provide this
flow of important understanding to a book that has the clear potential as a textbook
and a 'good' read for a broader audience. The development of and use of figures,
boxes, and graphics is well done and very usefully connected to the text. Further, the
references, chapter by chapter, are extensive and well thought through to give the
reader additional material that enhances the value of the book as a textbook and
provides the broader readership with particularly important added information. The
summaries at the end of each chapter are a clear composite of what is in the chapter:
just reading these can provide the reader with a substantive understanding of the polar
regions and climate change. Finally, the listing of important questions at the end of
each chapter is excellent.

It is, in my view, a book that will be an excellent textbook, but, of equal importance, a
'good solid' read for a broader audience. I highly recommend the book to Cambridge
University Press, with the hope that it will be published soon. It is deeply a moment of
sadness that Roger passed away so near to the end of the review period and close to – what
I hope will be – its publication. I hope that Dr. Eileen A. Hall-McKim will both carry the
effort to its completion and honor Roger by fostering its use as a textbook and as a book for
that 'good read' by a broader audience. I hope too, that she will follow his incredible
leadership in polar science and understanding by picking up his remarkable professional
mantle."

- Robert W. Corell,
Principal, Global Environment and Technology Foundation,
and its lead at its Center for Energy and Climate Solutions (US);
Adjunct Professor, University of Miami;
and Professor at the University of the Arctic (Norway)

"Roger Barry and Eileen Hall-McKim have produced a compelling, state-of-the-art synthesis of the current state of the physical components of the Arctic and Antarctic climate
systems. Complementary perspectives from the paleoclimate record, climate change

projections, and the ThirdPole expand upon the primary material. Extensive use of contemporary literature, chapter summaries, and chapter questions that probe understanding make this text ideal for a graduate course on polar environments."

- *David Bromwich,*
Byrd Polar Research Center, Ohio State University

"*Polar Environments and Global Change* by Barry and Hall-McKim focuses on the environmental change in the polar regions that comes as a consequence of global climate change. One of the very few currently available, this timely book could easily be the foundation textbook for an upper-level undergraduate or a graduate-level class on environmental change in the polar regions. It is comprehensive, providing a wealth of information, and giving much of the necessary scientific and historical background which allows the student to view and understand the contemporary change. Barry and Hall-McKim set the stage with a strong, historical component which touches on polar discovery and exploration. They also examine the polar climate over an arc of time linking the paleoclimatic to the contemporary, effectively placing present-day climate change within the context of the distant, past climate. Their detailing of how the environment developed, before moving to an exposition of how present-day practices induced change, fosters a fuller comprehension of the material.

As is normal in textbooks written/led by Barry, this book is very lucidly written. The authors managed quite successfully to provide a wealth of information in a very accessible way. It reads easily and is well illustrated. The summary pieces at the end of each chapter tie together the major ideas expressed in the chapter, allowing the reader to get a large-scale view of how these ideas are linked. It is also very well referenced; its comprehensive bibliography includes research very recently published.

I should also say that I am preparing a course on polar environmental change and this textbook is a prime candidate for use in this class."

- *Marilyn Raphael,*
University of California, Los Angeles

"Packed with a wealth of important and fascinating information, this is a timely, valuable, and highly readable synthesis of our current understanding of diverse aspects of polar environments and their significance and vulnerability – including the Third Pole. Though aimed particularly at tertiary students, this book will be of great interest to all those who care about our planet and its future."

- *Robert Massom,*
Antarctic Climate and Ecosystems Cooperative Research Centre,
University of Tasmania

"This is another excellent book from Roger Barry and Eileen Hall-McKim, discussing the three most important cold regions of the world: Arctic, Antarctic, and the Third Pole (encompassing the mountains of Central Asia, Hindukush-Himalayas system and Tibet). The book provides a comprehensive account of the interactions between climate, ocean, and the cryosphere, and an excellent review of the observational systems collecting data in these remote regions. The book is written in an engaging and very understandable way with plenty of useful references and self-check questions and will be indispensable in teaching undergraduate and master's courses in geography, glaciology, and climate science."

- Maria Shahgedanova,
University of Reading

Polar Environments and Global Change

The polar regions are the "canary in the coal mine" of climate change: They are likely to be hit the hardest and fastest. This comprehensive textbook provides an accessible introduction to the scientific study of polar environments against a backdrop of climate change and the wider global environment. The book assembles diverse information on polar environmental characteristics in terrestrial and oceanic domains, and describes the ongoing changes in climate, the oceans, and components of the cryosphere. Recent significant changes in the polar region caused by global warming are explored: shrinking Arctic sea ice, thawing permafrost, accelerating loss of mass from glaciers and ice sheets, and rising ocean temperatures. These rapidly changing conditions are discussed in the context of the paleoclimatic history of the polar regions from the Eocene to the Anthropocene. Future projections for these regions during the twenty-first century are discussed. The text is illustrated with many color figures and tables, and includes further reading lists, review questions for each chapter, and a glossary.

Roger G. Barry was Distinguished Professor Emeritus of Geography at University of Colorado at Boulder and was the former director of the National Snow and Ice Data Center. He published 29 textbooks and 260 research articles, and supervised 65 graduate students. He was a Guggenheim Fellow, a Fulbright Teaching Fellow, and a visiting professor in eight countries.

Eileen A. Hall-McKim is a climatologist receiving her doctorate from the University of Colorado at Boulder. Her interdisciplinary degrees include work in the geological sciences, paleoclimatology, meteorology, oceanography, and water resources. She is the co-author, with Roger G. Barry, of *Essentials of the Earth's Climate System* (Cambridge University Press, 2014).

Polar Environments and Global Change

Roger G. Barry

UNIVERSITY OF COLORADO AT BOULDER

Eileen A. Hall-McKim

UNIVERSITY OF COLORADO AT BOULDER

CAMBRIDGE UNIVERSITY PRESS

CAMBRIDGE
UNIVERSITY PRESS

University Printing House, Cambridge CB2 8BS, United Kingdom

One Liberty Plaza, 20th Floor, New York, NY 10006, USA

477 Williamstown Road, Port Melbourne, VIC 3207, Australia

314–321, 3rd Floor, Plot 3, Splendor Forum, Jasola District Centre, New Delhi – 110025, India

79 Anson Road, #06–04/06, Singapore 079906

Cambridge University Press is part of the University of Cambridge.

It furthers the University's mission by disseminating knowledge in the pursuit of education, learning, and research at the highest international levels of excellence.

www.cambridge.org
Information on this title: www.cambridge.org/9781108423168
DOI: 10.1017/9781108399708

© Roger G. Barry and Eileen A. Hall-McKim 2018

First published 2018

Printed and bound in Great Britain by Clays Ltd, Elcograf S.p.A.

A catalogue record for this publication is available from the British Library.

Library of Congress Cataloging-in-Publication Data
Names: Barry, Roger G. (Roger Graham), 1935– author. | Hall-McKim, Eileen A., author.
Title: Polar environments and global change / Roger G. Barry, Eileen A. Hall-McKim.
Description: First edition. | Cambridge, UK ; New York, NY : Cambridge University Press, 2019. |
    Includes bibliographical references and index.
Identifiers: LCCN 2018000613 | ISBN 9781108423168 (hardback) | ISBN 9781108436359 (pbk.)
Subjects: LCSH: Environmental monitoring–Polar regions. | Polar regions–Environmental
    conditions. | Arctic regions–Environmental conditions. | Antarctica–Environmental
    conditions. | LCGFT: Textbooks.
Classification: LCC QC903.2.P73 B37 2019 | DDC 551.6911–dc23
    LC record available at https://lccn.loc.gov/2018000613

ISBN 978-1-108-42316-8 Hardback
ISBN 978-1-108-43635-9 Paperback

Additional resources for this publication at www.cambridge.org/polarenvironments.

In memoriam

Roger G. Barry, Distinguished Professor Emeritus at the University of Colorado (CU) Geography Department, longtime CU faculty member, colleague, and friend to many, passed away on March 19, 2018, concluding a distinguished career in the study of the cryosphere and mountain climates. His work as a scientist and professor, and his dedication to the formation of centers for the study of the cryosphere, helped shape the evolution of climate science. He was the founder of the National Snow and Ice Data Center (NSIDC) and director there for more than 30 years.

Along with training and recruiting a dedicated staff at NSIDC, Dr. Barry also fostered international collaboration with many countries, and from 1971-2011, he supervised 67 graduate students, 36 of whom earned Ph.D. degrees. Over the course of his career, he authored hundreds of research papers and review articles, and was a prolific author of textbooks, some of which are: *Atmosphere, Weather and Climate (9th edition), Mountain Weather and Climate*; *The Global Cryosphere: Past, Present and Future*; *Essentials of the Earth's Climate System*; *Microclimate and Local Climate*; and his final text, *Polar Environments and Global Change*.

Roger Barry was recognized around the world for both his scientific accomplishments and service to the community, receiving many prestigious awards throughout his career.

Further information is available at the following websites:

References and Tributes

Barry, R. G. 2015: The shaping of climate science: half a century in personal perspective. *History of Geo- and Space Sciences* 6:87-105. doi:10.5194/hgss-6-87-2015. www.earthdata.nasa.gov/
Boulder Daily Camera: Roger Barry Obituary. http://www.legacy.com/obituaries/dailycamera/obituary.aspx?n = roger-barry&pid = 188547268
Cooperative Institute for Research in Environmental Sciences: Hats off to Barry. http://cires1.colorado.edu/science/spheres/snow-ice/barry.html
Geography Department, University of Colorado, Boulder: In Memoriam: Professor Emeritus Roger Barry. https://www.colorado.edu/geography/2018/03/21/memoriam-professor-emeritus-roger-barry
Global Cryosphere Watch: A giant has fallen: In Memoriam, Roger Barry http://globalcryospherewatch.org/news/rgb/roger_barry.html.
National Snow and Ice Data Center: First 25 years: the history of the WDC for Glaciology and NSIDC in Boulder, Colorado. https://nsidc.org/arc/history.html.
National Snow and Ice Data Center: In Memoriam: Roger Barry, NSIDC Founding Director. https://nsidc.org/research/bios/barry.html.

Media Contact

Natasha Vizcarra
National Snow and Ice Data Center, University of Colorado, Boulder.
press@nsidc.org

1. Roger Barry on the Boas Glacier on Baffin Island, Canadian Arctic, in August 1970
(photo: Ronald Weaver)

2. Roger Barry at his initiation as a Fellow of the American Geophysical Union, San Francisco, December 1999
(photo: Ronald Weaver)

Contents

Figures

Boxes

About the Authors

We believe it is appropriate to outline our scientific history as it relates to the polar regions.

Roger has been fascinated by cold environments since spending a year at the McGill Subarctic Research Station in Schefferville, Quebec, as a graduate student and weather observer during the International Geophysical Year, 1957–1958. Professor Jack Ives introduced him to the history of glaciation and the Laurentide ice sheet, a topic that became a part of his MSc thesis and subsequent PhD dissertation.

In summer 1963 and spring 1964, Roger took part as a meteorologist in Operation Tanquary, a Canadian Defense Research Board program on Ellesmere Island led by Dr. Geoff Hattersley-Smith. In 1966–1967, he worked at the Geographical Branch, Department of Energy Mines and Resources, in Ottawa studying the climate of Baffin Island. This included a two-week spell at the Branch's base camp in Inugsuin Fiord with a visit to the Barnes ice cap.

In October 1968, Roger accepted a faculty position at the University of Colorado, Boulder, in the Geography Department, rostered in the Institute of Arctic and Alpine Research. In 1969, along with John Andrews, he directed the work of a group of graduate students studying glaciology and glacial geomorphology in eastern Baffin Island. The "Boas" glacier, near Narpaing Fiord, was the object of glacier–meteorological research with John Jacobs and Ron Weaver. Subsequently, they shifted their attention to the landfast ice at Broughton Island. In May 1970, Roger and John Jacobs conducted aircraft measurements over the Davis Strait sea ice from a Queenair of the National Center for Atmospheric Research in Boulder.

A team-taught course on arctic and alpine environments at the Institute of Arctic and Alpine Research led to a multiauthored book, of the same title, that was edited by Jack Ives and Roger.

In 1981, Roger transferred his faculty position to the Cooperative Institute for Research in the Environmental Sciences (CIRES) in partnership with the World Data Center for Glaciology, of which he was director until 2008. This center was granted the title National Snow and Ice Data Center (NSIDC) by the National Oceanic and Atmospheric Administration (NOAA) in 1981. As part of the International Tundra Biome Programme, Roger was able to visit tundra sites at Toolik Lake, Eagle Summit, and Barrow, Alaska. The award of a J. S. Guggenheim Fellowship in 1982–1983 enabled him to spend time at the National Institute for Polar Research in Tokyo, the University of Bern, Switzerland, and the Scott Polar Research Institute in Cambridge, United Kingdom, researching snow and ice data.

Roger has also visited various parts of Central Asia. In 1979 and again in the 1990s, he visited the Abramov Glacier in the Pamir. In 1981, he traveled with Jack Ives and Gordon Young, together with Professor Shi Yafeng, Qiu Jiachi, and Kang Ersi, to No. 1 glacier in the Tien Shan. In 2005, he traveled by road from DunHuang in northwest China to Golmud and Lhasa. The railroad was under construction at the time, and the party inspected the embankments designed to prevent thawing of the permafrost. In 2009, he traveled to the terminus of No. 12 glacier at 4,250 meters in the Qilian Shan, following a meeting in Lanzhou.

In 1990, and again in 1997, Roger spent six months at the Institute of Geography, ETH, Zurich, where he lectured on mountain climates and cryospheric topics. In 1994, Roger spent four months at the Alfred Wegener Institute for Polar and Marine Research, Bremerhaven, Germany, writing a paper about snow and sea ice albedo. In spring 2000, he spent four months at Moscow State University lecturing on snow and ice. In 2004, he spent four months at the Laboratoire de Géophysique et Glaciologie in Grenoble, France, writing a paper on glacier research. In 2005, Mark Serreze and Roger published *The Arctic Climate System* through Cambridge University Press, with a second edition appearing in 2014.

Roger received a Humboldt Prize Fellowship in 2009–2010 at the Bavarian Academy of Sciences Commission on Glaciology while writing *The Global Cryosphere: Past, Present, and Future* with Thian Gan of the University of Alberta for Cambridge University Press.

In September 2005, Roger was a lecturer for a Summer School of the University of Alaska, Fairbanks, on the Russian icebreaker *Kapitan Dranitsin* as it sailed from Kirkenes, Norway, into the Laptev Sea, where it conducted research at an ice station. In February 2012, he had the opportunity to lecture on a cruise ship that sailed from Ushuaia, Argentina, across Drake Passage to the South Shetland Islands, with a landing on the Antarctic peninsula. This provided some impressions of Antarctica and the Southern Ocean. Roger continues to work on scientific papers at the NSIDC, gives invited lectures internationally, and consults on climate and cryospheric topics.

Eileen Hall-McKim is a PhD climatologist, receiving her doctorate from the University of Colorado, Boulder. Her interdisciplinary degrees include work in the geological sciences, paleoclimatology, meteorology, hydrology, oceanography, and water resource research.

Eileen's interest in weather began early. Growing up on the high plains of Kansas, she became fascinated by big sky horizons filled with supercell thunderstorms and cloud formations. Her interest in geology and paleoclimate was largely initiated by visits to Utah's Canyonlands in early 1970s. Her intrigue with the World of Ice was fueled while she engaged in an undergraduate astronomy class with astrophysicist Dr. Joe Romig, University of Colorado, as he discussed the Milankovitch cycles and the coming and going of ice ages. Emeritus Professor Gary Thomas, working in the field of astrophysics and planetary sciences, also greatly encouraged Eileen's interest in meteorology and climate and helped facilitate undergraduate research into high-altitude noctilucent clouds as possible harbingers of climate change in early 1990s.

In 1993, during a University of Colorado INSTARR/Geography program, Eileen did extensive fieldwork in glacial geology on an Arctic River Expedition, exploring the Mara and Burnside Rivers by inflatable canoe and following the flow of springtime ice breakup of the rivers, into the Arctic Ocean near Bathurst Inlet, in Canada's Northwest Territories. Throughout the expedition, she conducted research/data collection for paleoclimate indicators of the ice flow of the Laurentide Ice Sheet through mapping of glacial striations and other landform features on the northern Canadian Shield bedrock.

Other interesting fieldwork followed observing glacial landscapes on Colorado's Niwot Ridge and Sangre de Christo Mountain Range of the southern Rockies through the University of Colorado's Arctic and Alpine Mountain Research Station.

Eileen completed her MSc at the National Snow and Ice Data Center, Boulder, and her PhD while working as editor and writer for the Intermountain West Climate Summary of the NOAA/Western Water Assessment. Her honors include being elected a member of the Phi Beta Kappa National Honor Society; the Outstanding Women in Geosciences Student Award from the American Association of Women in Geosciences; the Graduate Research Fellowship Award from the Cooperative Institute for Research in Environmental Sciences (CIRES); and membership in the Magna Cum Laude National Honor Society, University of Colorado.

In 2008, Eileen traveled to the No. 12 glacier in the Qilian Shan in northwest China following a meeting on the World Glacier Inventory. In 2012, she sailed on the *Antarctic Dream* across Drake Passage to the South Shetland islands and Antarctic peninsula, followed by travels through Argentina and the Perito Moreno Glacier. Other travel includes visiting glaciers and periglacial landscapes in Iceland, Norway, Switzerland, Canada, and Alaska.

Eileen completed professional certification in Sustainable Community Management from the University of Colorado Sustainable Practices Program in 2011–2014 and has continuing interest in methods of carbon sequestration, with special interest in biomimetic techniques. She co-authored *Essentials of the Earth's Climate System* (with Roger G. Barry, Cambridge University Press, 2014).

Roger G. Barry and Eileen A. Hall-McKim

Preface

The polar world is undergoing unprecedented changes as a result of global warming and its amplification in high latitudes. Arctic sea ice is shrinking and thinning at an unprecedented rate, permafrost is thawing, glaciers and the polar ice sheets are losing mass at an accelerating rate, and ocean temperatures are rising. These changes are beginning to have large impacts on plants, animals, and human society that will increase in the future, whether or not greenhouse gas emissions are reduced.

The Third International Conference on Arctic Research Planning (ICARP 3) lists as the first of its three themes "The Arctic system and its transformation." The first key finding of the 2017 Arctic Monitoring and Assessment Programme (AMAP) report *Snow, Water, Ice, and Permafrost in the Arctic (SWIPA)* is that "the Arctic's climate is shifting to a new state." It seems timely, therefore, to assemble the diverse information on polar environmental change for the benefit of students in environmental sciences, geography, biology, and climate sciences, as well as planners and northern residents.

Following an introduction that outlines the setting and research history of the polar regions, Chapter 2 describes the broad outlines of climatic history from the Eocene to the Anthropocene. Chapter 3 gives an overview of in situ and remote sensing observations of polar regions. Chapter 4 surveys the atmospheric and oceanic circulations and climatic conditions. Chapter 5 details the characteristics of the terrestrial environments and the processes at work in them. Chapter 6 examines the ice sheets of Greenland and Antarctica and ice shelves. Chapter 7 treats oceanic and sea ice environments. Chapter 8 introduces the Third Pole in the Central Asian highlands. Finally, Chapter 9 discusses projections for the future environments of the polar regions during this century. Each chapter concludes with a summary of the main points and a mixture of review and discussion questions that encourage students to check their understanding and think critically.

Acknowledgments

We wish to thank the following colleagues for reviewing and commenting on drafts of the text:

Professor Ray Bradley, University of Massachusetts, Amherst, for Chapter 2
Dr. Matt Druckenmiller, NSIDC, for Section 9.10
Dr. William Emery, University of Colorado, for Section 4.2 and Chapter 6
Professor Jack Ives, Carleton University, Ottawa, Canada, for Chapter 5
Dr. Ersi Kang, CAREERI (now Northwest Institute of Eco-Environment and Resources (NIEER), Lanzhou, China, for Chapter 8
Dr. Lora Koening, National Snow and Ice Data Center, University of Colorado, Boulder, for Section 6.1
Dr. Ted Scambos, National Snow and Ice Data Center, University of Colorado, Boulder, for Chapter 3 and Sections 6.2–6.4
Dr. Neil Wells, Emeritus University Fellow, University of Southampton, United Kingdom, for Chapters 4B and 7
Dr. Qinghua Ye, Institute for Tibet Plateau Research, Beijing, China, for the glacier section of Chapter 8
Dr. Tingjun Zhang, Lanzhou University, Gansu, China, for the permafrost section of Chapter 8

We also wish to thank Jack Ives, Koni Steffen, D.A. "Skip" Walker, and Kevin Schaefer for their photographs. Special thanks to student assistants at NSIDC for illustration assistance, Florence Fetterer, NSIDC and Matt Lloyd of Cambridge University Press for his enthusiastic support of this project. We also thank the following publishers for permissions to reproduce the figures whose sources are identified in the text:

Academic Press
American Geophysical Union
American Meteorological Society
Arctic Institute of North America
Cambridge University Press
CRC Press
Elsevier
Geological Survey of Canada
Oxford University Press
Routledge/Taylor and Francis
Springer

The Setting, History of Studies, and the Climatic Role of the Cryosphere

1

1.1 Introduction

Polar environments experience the coldest conditions on Earth, including extensive areas of persistent snow cover, land ice, permafrost, and sea ice. They are, however, undergoing the most rapid changes in environmental conditions of any place on Earth. The process of polar amplification of global warming has led to temperature increases of 2–4 °C over the last five decades, with increases that are two to three times the global average occurring in most of the Arctic and around the Antarctic Peninsula. The area covered by summer sea ice in the Arctic has decreased by more than half since the 1980s, and Arctic ice caps and glaciers are shrinking at an unprecedented rate (Zemp et al. 2015). Since 2000 there has been accelerating loss of ice from the ice sheets of Greenland and Antarctica. A review of the global changes in the cryosphere is provided by Williams and Ferrigno (2013), while the Arctic Mapping and Assessment Program (AMAP 2011, 2017) has published reports on Arctic snow, water, ice, and permafrost, and Barry (2017) has surveyed changes in the Arctic cryosphere in this century. These changes are of vital importance for global sea level, plants, land and marine animal species, and indigenous populations. Arctic states (members of the Arctic Council: Canada, Denmark, Finland, Iceland, Norway, the Russian Federation, Sweden, and the United States, and six organizations representing indigenous peoples) are seeking to exploit mineral resources in the Arctic and make use of the diminishing cover of Arctic sea ice to facilitate and increase Arctic shipping, as well as to ensure environmental protection and sustainable use of the Arctic environment. Twelve other nations (France, Germany, India, Italy, Japan, Korea, the Netherlands, People's Republic of China, Poland, Singapore, Spain, and the United Kingdom) are now Official Observers at the Arctic Council.

The Antarctic Treaty entered into force in 1961 and now has fifty-three member states (see www.ats.aq/seleccion.htm). The Scientific Committee for Antarctic Research (SCAR) is part of the International Council for Science (ICSU) and has Science Groups on geosciences and physical sciences, as well as Scientific

Research Programs that include Antarctic climate change in the twenty-first century and past Antarctic ice sheet dynamics (www.scar.org).

A major survey of climate impact assessments for the Arctic was published by the Arctic Climate Impact Assessment (ACIA 2004). The Fifth Assessment Report of the Intergovernmental Panel on Climate Change (IPCC) provides an assessment of both polar regions (Larsen and Anisimov 2014). From 1964 to 2003 the American Geophysical Union published seventy-nine volumes in its Antarctic Research Series; these were put online in 2013.

Comprehensive encyclopedia volumes on each polar region have been published by Nuttall (2005) for the Arctic, and by Stonehouse (2002) and Riffenburgh (2007) for the Antarctic.

This book aims to present an up-to-date account of the various environments in the polar and subpolar regions, including the mountain ranges of Central Asia and the Tibetan Plateau – Himalaya–Karakorum, now commonly referred to as the "Third Pole." It also seeks to show how these environments have evolved over the last fifty million years, how they are responding to current global climate changes, and how they are projected to change over the rest of the twenty-first century.

1.2 Setting

The definition of "polar" depends very much on the viewpoint that is adopted. The Arctic and Antarctic circles at 66.3° N and S latitude define the regions where the sun does not set and does not rise for at least one day per year. Each of these polar regions encompasses approximately 4 percent of Earth's surface. Geographically, the Arctic and Antarctic can be defined by their climatic and biotic conditions. For example, Köppen (1923) determined that a July mean temperature below 10 °C approximately defined the limit of tree growth, with tundra (treeless land) found poleward of this limit. There are outliers such as Greenland, which extends to 60° N, and temperatures well above the zonal average value are found at 80° N around Svalbard. Antarctica is well defined by the southern continent, but ice caps cover many of the sub-Antarctic islands and sea ice extends seasonally to 40° S. In the northern hemisphere, sea ice is present to 40° N off East Asia in winter.

The two polar regions differ markedly in their geography. The Arctic is largely an ice-covered ocean ringed by the continents of Eurasia and North America/ Greenland (Figure 1.1). The Arctic Ocean is connected to the North Pacific Ocean by the narrow Bering Strait and to the North Atlantic Ocean by the Greenland–Icelandic–Norwegian (GIN) seas. These latter passageways transport warm water from the North Atlantic Current into the Arctic between Svalbard and northern Norway, and sea ice and cold polar water out of the Arctic via Fram Strait and the Canadian Arctic archipelago. By contrast, the Antarctic is almost totally covered by a massive ice sheet that rises to more than 4 km elevation, and is encircled by the cold Southern Ocean, which has an extensive seasonal sea ice cover. The Arctic mainland and islands are occupied by polar desert, tundra, glaciers, and ice caps, whereas only approximately 2 percent of Antarctica is ice-free polar desert. There are some seventeen Antarctic islands larger than 1,000 km^2 lying south of the oceanic Antarctic Convergence at about 60° S. This convergence is where cold, northward-flowing Antarctic waters meet the relatively warmer waters of the sub-Antarctic. Most of the island surfaces are ice covered or bare rock with moss and lichens.

The Tibetan Plateau and surrounding mountain ranges of Central Asia are often referred to as the "Third Pole" because this region contains the largest area of ice and frozen ground outside the polar regions proper. It has comparable climatic conditions as a result of its continentality (large annual range of temperature) and its high altitude, implying low average temperatures.

The southern hemisphere high latitudes are characterized by a high degree of zonality (i.e., latitudinal arrangement of their characteristic features). Only the Antarctic Peninsula markedly disrupts this pattern. The Arctic, by comparison, is highly azonal, with cold currents and winter sea ice extending far into mid-latitudes off the eastern coasts of Asia and North America and open water extending far into Arctic latitudes in the vicinity of the Svalbard archipelago (80° N). There are similar environmental contrasts between the eastern and western parts of the northern continents in terms of climates, plant cover, land ice, and permafrost.

1.3 History of Scientific Study

1.3.1 Arctic

The term "Arctic" is derived from the Greek word *arktikos,* which means "near the Great Bear" (constellation). Systematic scientific study of the polar regions began only in the late nineteenth century with the First International Polar Year (IPY, 1882–1883). Barr (1985) provides accounts of twelve national expeditions during this era. Only five years earlier George De Long had led the last US Navy Arctic expedition in the *USS Jeanette* seeking a passage northwest of Bering Strait to the mythical Open Polar Ocean, which was thought to be maintained by warm waters branching northward from the Kuro Shio Current (Sides 2014). During 1878–1879, the *Jeanette*, frozen into the sea ice, drifted for some 1,250 km from

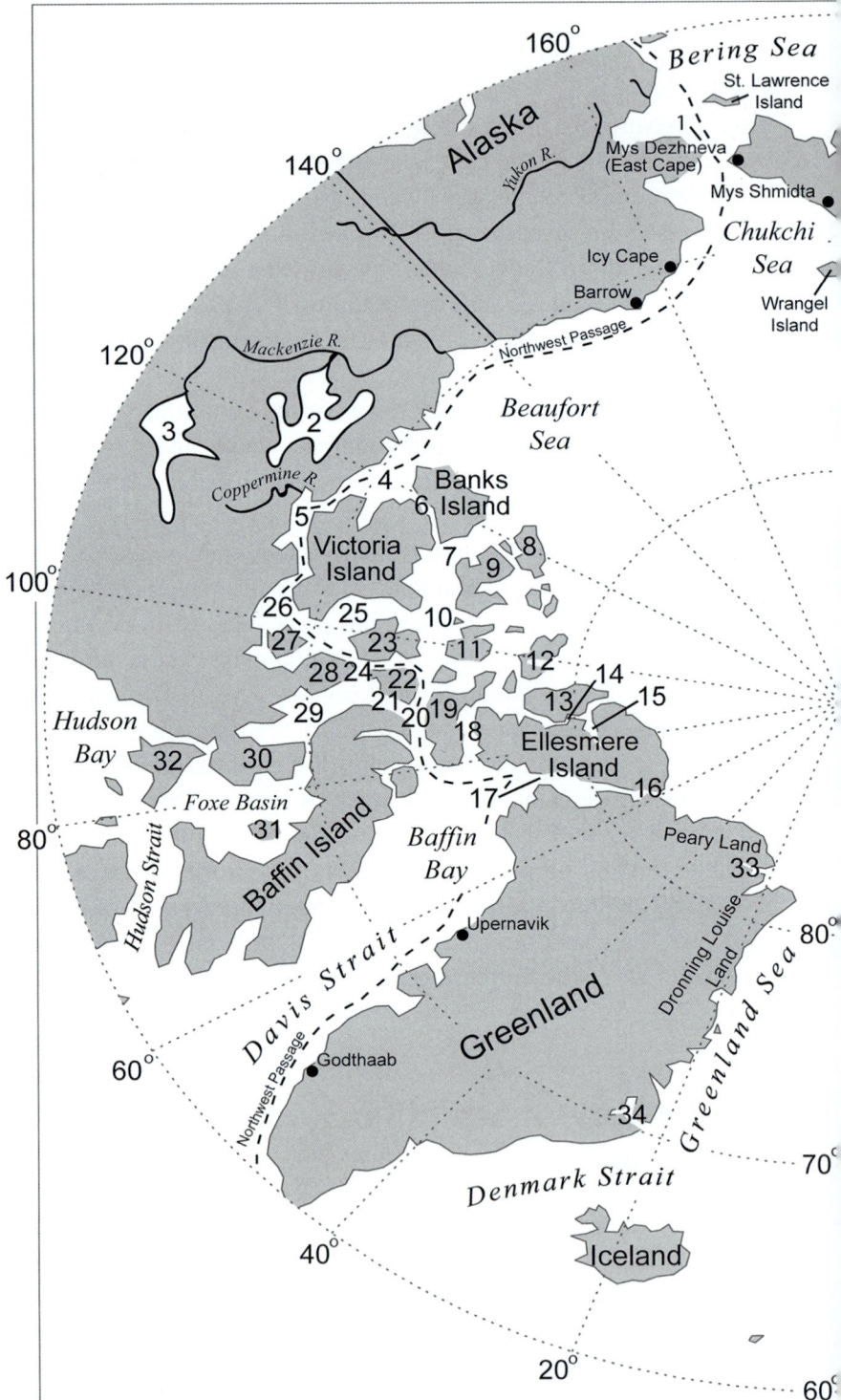

Figure 1.1 Map of the Arctic. 1. Bering Strait. 2. Great Bear Lake. 3. Great Slave Lake. 4. Amundse
9. Melville Island. 10. Melville Strait. 11. Bathurst Island. 12. Ellef and Admund Ringnes Islan
Sound. 18. Jones Sound. 19. Devon Island. 20. Lancaster Sound. 21. Prince Regent's Inle
26. Queen Maud Gulf. 27. King William Island. 28. Boothia Peninsula. 29. Gulf of Boothi.
Fiord. 34. Scoresby Sound.

Source: Courtesy M. Lavrakas, National Snow and Ice Data Center, University of Colorado, Boulder.

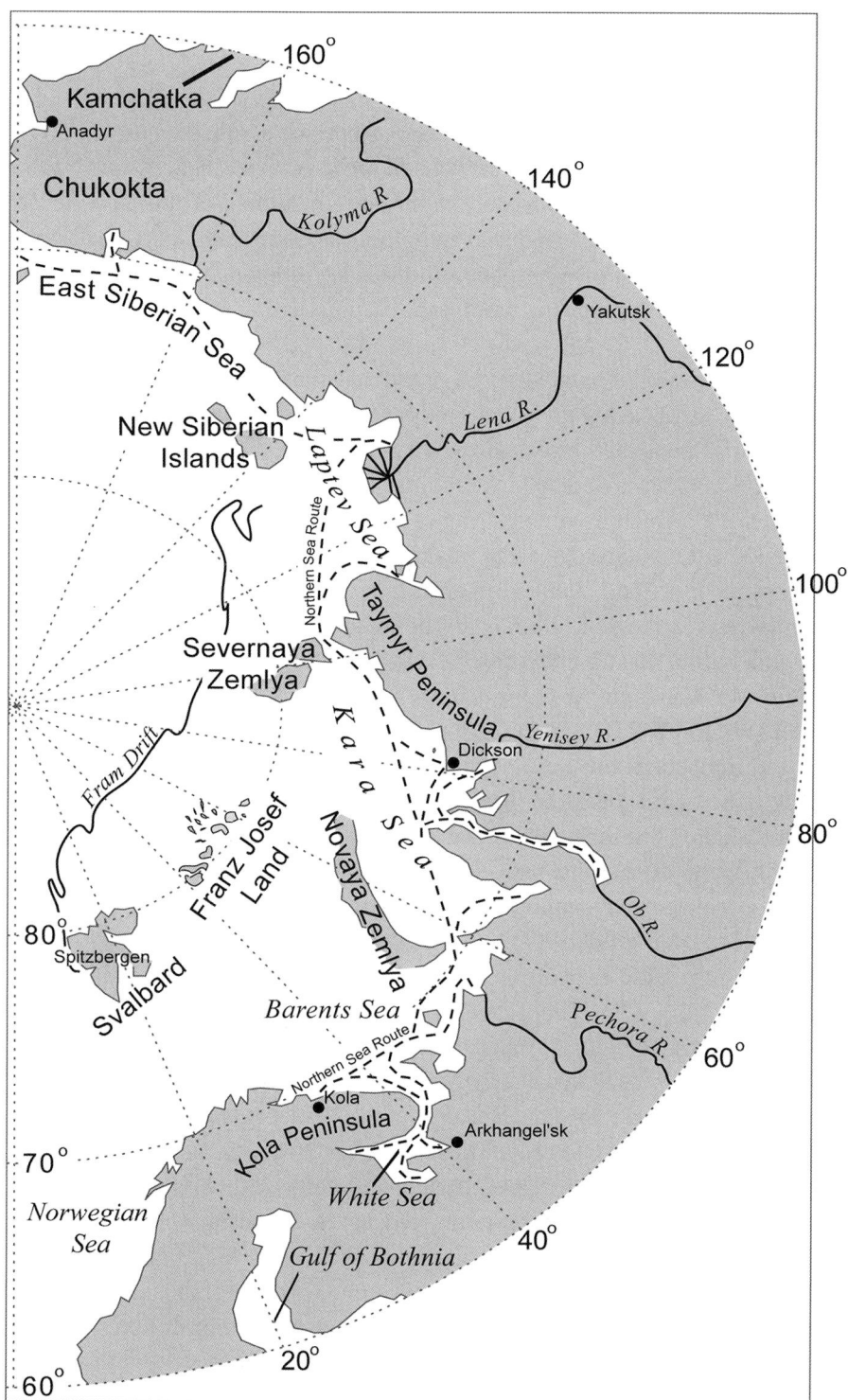

Gulf. 5. Coronation Gulf. 6. Prince of Wales Strait. 7. M'Clure Strait. 8. Prince Patrick Island.
13. Axel Heiberg Island. 14. Eureka Sound. 15. Greeley Fiord. 16. Lady Franklin Bay. 17. Smith
22.Somerset Island. 23. Prince of Wales Island. 24. Bellot Strait. 25. M'Clintock Channel.
30. Melville Peninsula. 31. Prince Charles Island. 32. Southampton Island. 33. Independence

north of Wrangel Island to northeast of the New Siberian Islands, where it was crushed by the ice and sank. The crew made their way by sledges and boats to the Lena delta, where De Long and twenty-one members of the crew of thirty-three men died from starvation and hypothermia. Wreckage from the *Jeanette* drifted to the east coast of Greenland, and this sighting encouraged the Norwegian explorer Fridtjof Nansen to mount the famous *Fram* expedition in 1893–1896.

The Austrian Carl Weyprecht, who had co-organized the Austro-Hungarian Polar Expedition of 1872–1874, led the planning for the First International Polar Year. The scientific and political context of the First IPY (and the International Geophysical Year, 1957–1958) is detailed by Launius et al. (2010). Eleven nations took part in the First IPY, which involved a ring of twelve stations set up around the Arctic that carried out meteorological, geomagnetic, and auroral observations (Corby 1982; Wood and Overland 2006). However, the stations were too far apart to enable synoptic meteorological studies to be made. The Second IPY in 1932–1933 saw ninety-four Arctic meteorological stations established (Laursen 1982), although World War II prevented much of the data from being published and analyzed. The Second IPY benefited from radio transmissions and airborne reconnaissance and transportation.

A review of climatological research programs undertaken in the Arctic is provided by Barry (2005). The first significant data for the central Arctic were collected by Henrik Mohn during the 1893–1896 drift of the *Fram* (Mohn 1905). Nansen's expedition carried out significant meteorological and oceanographic research (Barry 2016). During 1910–1915, the Russian government organized the Arctic Ocean Hydrographic Expedition (GESLO) to explore the Arctic coast of Siberia (Starokadomskii 1976). This scientific expedition comprised two ice-breakers and was commanded by B. A. Vilitsky. The expedition charted the entire Arctic coast of Siberia and in 1913 discovered Severnaya Zemlya, the last significant land discovery on the globe. The Norwegian A. Hoel organized annual expeditions to Svalbard from 1911 to the mid-1920s (Barry 2016). Roald Amundsen organized the *Maud* expedition, 1922–1925, to perform a drift similar to Nansen's in the Arctic. The scientific program, led by H. Sverdrup (1933), is noteworthy because the observations of Finn Malmgren laid the foundation for modern sea ice research. Another major pre-war milestone was the establishment by aircraft of the first of the Soviet Union's Arctic drifting stations, North Pole 1, on an ice floe in 1937. This floe eventually drifted to East Greenland, where the party was picked up by a Soviet icebreaker.

There were two major expeditions to Greenland in the 1930s. The British Arctic Air Route Expedition (Merles 1932) and Alfred Wegener's Greenland Expedition, which operated their stations – Watkins Ice Cap (67.1° N, 41.8° W, 2,440 m) and

Eismitte (70.9° N, 40.7° W, 3,000 m) – during 1930–1931, yielded the first detailed information on the ice sheet climate (Loewe 1936; Putnins 1969).

The Soviet North Pole (NP) drifting station program resumed in 1950 and continued until July 1991, with stations NP-2 to NP-31 (Romanov et al. 2000). Most of these stations were established on multiyear ice floes. The program was resumed by the Russian Federation in 2003 with NP-32, and continues to operate, with NP-2015 being established in 2015.

The United States had two stations on "ice islands" that had broken away from the Ward Hunt Ice Shelf (Hattersley-Smith et al. 1952; Sater 1968). They were T-3 (originally called Fletcher's ice island), established in 1952, and the Arctic Research Laboratory Ice Station (ARLIS) II, established in 1961; other stations were set up on multiyear sea ice. Sporadic work on T-3 lasted until 1974. ARLIS II had a more extensive and successful program, lasting from its occupation in 1961 in the southern Beaufort Sea until its evacuation in Denmark Strait in 1965. Many of the synoptic weather observations were reported by radio to the Global Telecommunications System (GTS) and so incorporated into operational weather maps. The presence of even two reporting stations from the central Arctic proved invaluable in detecting large-scale weather systems within the Arctic Basin. Meteorological data from the NP and US drifting stations, and other Arctic climate data for the period 1951–1990 have been assembled on digital media (Arctic Climatology Project 2000; National Snow and Ice Data Center 2000).

The establishment of the Norwegian Polar Institute in 1948 led to extensive research on the glaciers and snow cover in the Svalbard archipelago and to oceanographic and sea ice research in the Barents Sea and Arctic Ocean (Barry 2016).

The US Air Force began weather reconnaissance flights from Alaska to the North Pole in the 1940s (Anonymous 1950), which in turn led to the compilation of Arctic cloud statistics by Huschke (1969). In 1998, the First International Satellite Cloud Climatology Project (ISCCP) Regional Experiment (FIRE) Arctic Clouds Experiment was carried out in the Arctic (Curry 2001). Data on Arctic clouds remain uncertain in several respects – the seasonal cycle, thickness, and phase.

Following World War II, it was decided to mount an International Geophysical Year (IGY), lasting from July 1957 to December 1958. While the emphasis was on Antarctic observations (discussed later in this chapter), some specific programs were carried out in the Arctic. McGill University operated the first station in the interior of the Canadian Arctic archipelago at Lake Hazen, Ellesmere Island (Jackson 1959); the permanent weather stations in the Canadian Arctic archipelago, established during 1947–1950, were all at coastal sites. Barry and

Jackson (1969) published summer data for Tanquary Fiord, Ellesmere Island, for 1963–1967, while Atkinson et al. (2000) assembled climate data for the High Arctic from archives of the Canadian Polar Continental Shelf Project.

Ice Station Alpha in the Arctic Ocean was the first US drifting station with a large, multidisciplinary research program. Russian scientific expeditions were mounted to study the glacial meteorology of the ice caps of Franz Josef Land (Krenke 1961).

During 1972–1976, a US–Canadian–Japanese program – the Arctic Ice Dynamics Joint Experiment (AIDJEX) – was carried out in the Beaufort Sea. Apart from advances in modeling sea ice dynamics, an important outcome of this program was the improved understanding of the energy balance over sea ice. Data from the main experiment in summer 1975 are archived at the National Snow and Ice Data Center (http://nsidc.org/noaa/aidjex). A Marginal Ice Zone Experiment (MIZEX) was carried out in the Bering Sea (MIZEX-West) and the Greenland Sea (MIZEX-East) by the US Office of Naval Research in the 1980s (MIZEX '87 Group 1989). The Labrador Ice Margin Experiment was conducted in the Labrador Sea in 1987 (McNutt et al. 1988). These efforts were followed by the Coordinated Eastern Arctic Research Experiment, 1988–1989, in the northern Norwegian and Greenland seas. The data are archived at http://nsidc.org/data/docs/daac/nsidc0020_cearex.gd.html.

Drifting buoy technology for the Arctic was developed during AIDJEX, which led to new information on surface pressure, air temperature, and ice drift in the central Arctic Ocean. Beginning in 1979, the Arctic Buoy Program was initiated by the University of Washington. In 1991, it became the International Arctic Buoy Program (IABP), which now involves eight nations. Initially, approximately twenty buoys were deployed, mainly from airdrops, but in the 1990s the number rose to more than thirty operating at any one time. Data from these buoys are relayed via the Argos satellite system to the GTS. The quality of Arctic surface pressure analyses greatly improved as a result.

Greenland has been the site of numerous ice core drilling projects since the first at Camp Century in northwest Greenland led by W. Dansgaard in 1969. Data on Greenland Summit ice cores obtained in the 1970s were reported by Hammer et al. (1976). The deep cores of the European Greenland Ice Core Project (GRIP), 1989–1995, and the Greenland Ice Sheet Project (GISP2), 1988–1993, were both extracted from the central dome and spanned approximately 100 kiloyears (ka). The North Greenland Eemian Ice Drilling Project (NEEM) was conducted in 2009–2013 in northwest Greenland with the aim of collecting Eemian-age (115–125 ka) ice. A Greenland Climate network of eighteen automatic weather stations was established around the ice sheet in 1994 as part of the National Aeronautics and Space Administration's (NASA's) Program for Arctic Regional

Climate Assessment (PARCA) that aimed to improve knowledge of the ice sheet's mass balance (Thomas et al. 2001).

The International Biological Program (IBP) included a Tundra Biome component in the 1970s. Each site carried out detailed climatological measurements, including energy budget studies. There were comparative field measurement programs in 1972 at Barrow, Alaska; Truelove Lowland, Devon Island; and Abisko, Sweden (Barry et al. 1981). In the 1990s, the US National Science Foundation funded the Land–Atmosphere–Ice Interactions (LAII) project in the Arctic (Kane and Reeburgh 1998; LAII Science Steering Committee 1997).

The *USS Nautilus* made the first underwater traverse of the Arctic Ocean in August 1957; this feat was repeated by the *USS Queenfish* in August 1970. The sonar data of ice draft were analyzed by McLaren (1989), skipper of the *Queenfish*, for his doctoral dissertation (see Chapter 7). United States and British submarines performed numerous studies of Arctic sea ice draft (Wadhams 1997; Yu et al. 2004).

Surface vessels began sailing to the North Pole in 1977, when the Soviet nuclear icebreaker *NS Arktika* completed the first passage. Subsequently, many nations deployed icebreakers in both polar oceans, including Russian icebreakers, *Polarstern* (Germany), *Oden* (Sweden), *Aurora Australis* (Australia) and *Amundsen and Des Grosseiller* (Canada). The United States currently has only one heavy-duty icebreaker, the fifty-year old *Polar Star,* and a research icebreaker, the *Healy.*

Through the Arctic Climatology Project (2000) of the Russia–US Environmental Working Group, comprehensive atlases of Arctic oceanography, sea ice, and meteorology/climate during the second half of the twentieth century were prepared. Other projects with Arctic components include those of the World Climate Research Program (WCRP): the Global Energy and Water Experiment (GEWEX) projects for the Mackenzie GEWEX Study (MAGS) during 1992–1994 and the GEWEX Asian Monsoon Experiment (GAME) (see www.gewex.com). MAGS involved large-scale hydrological, atmospheric, and land–atmosphere studies within the Mackenzie Basin to improve understanding of cold-region, high-latitude hydrological and meteorological processes and their roles in global climate. GAME began in 1998 with hydroclimatological measurements along the Lena River in Yakutia, related to the Asian winter monsoon.

The Study of Environmental Arctic Change (SEARCH) was launched by the Arctic Research Consortium of the United States (ARCUS) in 2001 following workshops in 1997 and 1999 (SEARCH Science Steering Committee 2001). Currently it involves three Action Teams on sea ice loss, permafrost degradation, and land ice loss and sea level impacts.

A list of polar institutes may be found in Appendix A.

1.3.2 Antarctic

During the nineteenth century, numerous discoveries were made by expeditions in the Southern Ocean: Antarctic islands, the Antarctic Peninsula, and coastal ice shelves. However, the first overwintering expeditions were the Belgian one led by A. de Gerlache, whose vessel the *Belgica* was beset in the Bellingshausen Sea in 1897–1899, and the British Southern Cross Expedition of C. Borchgrevink, 1898–1900, which overwintered at Cape Adare on the continent and whose explorers reached 78.8° S. In 1904, Scottish explorer W. S. Bruce established a permanent weather station at 60° S at Orcadas in the South Orkney Islands that was later transferred to Argentina.

The early twentieth century saw numerous expeditions vying to reach the South Pole, a goal that was achieved by R. Amundsen in December 1911 and by R. F. Scott in January 1912. Scientific work in the Southern Ocean by the *Guass* during 1901–1903 was extensively reported by von Drygalski and Raraty (1991). In 1912, Australian D. Mawson established a base at Cape Denison, where an annual *mean* wind speed of $20\,\mathrm{m\,s^{-1}}$ was recorded during 1912–1913, making it the windiest place on Earth. Mawson was the first person to reach the South Magnetic Pole. Scientific expeditions to Antarctica essentially began with Richard Byrd's expedition of 1928–1930, when Little America was established on the Ross Ice Shelf and Byrd flew to the South Pole and back. The British–Australian–New Zealand Research Expedition (BANZARE) of 1929–1931 mapped parts of the coastline. However, despite the various efforts aimed at ground and satellite mapping, a complete coastline of Antarctica was not available until 2004, when Liu and Jezek (2004) produced one from SAR imagery.

In the 1940s, the number of Antarctic expeditions increased greatly with many nations taking part. The listing found at https://en.wikipedia.org/wiki/List_of_Antarctic_expeditions#20th_century includes these expeditions.

In 1946, Rear Admiral Richard Byrd organized Operation Highjump, a US Navy expedition that established Little America IV during August 1946 through February 1947. Also, soon after World War II (1949–1952), a Norwegian–British–Swedish expedition was conducted in Dronning Maud Land. It was the first in Antarctica involving an international team of scientists (Robin 1953). A base camp was established at a location named Maudheim (71.1° S, 10.9° W) on a floating ice shelf some 3 km from an inlet. The program addressed geology, glaciology, and meteorology, as well as photo-reconnaissance.

In 1954, the Australian Antarctic Division established Mawson station, located at 67.6° S, 62.5° E; it is the oldest continuously manned base in Antarctica. Most bases on the continent were established immediately prior to, or during, the International Geophysical Year (IGY), 1957–1958, which focused on Antarctica.

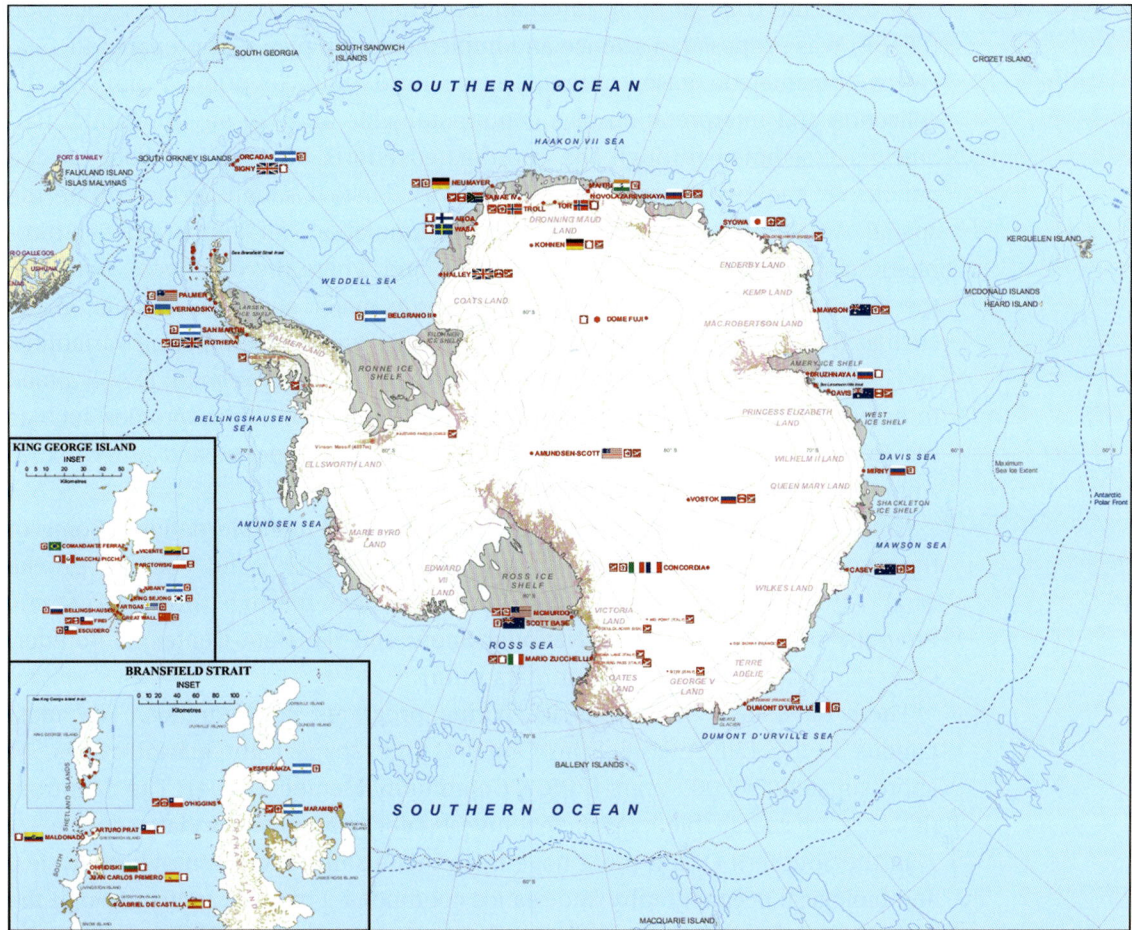

Figure 1.2 Permanent stations in the Antarctic.
Source: Wikipedia.

The US Navy's Operation Deep Freeze set up McMurdo and Little America V in 1956 (Belanger 2006). The United States installed South Pole station and McMurdo on the coast, while the USSR established the most remote station Vostok on the East Antarctic Plateau. The permanent stations are shown in Figure 1.2.

Between 1955 and 1992, there were thirty-six Soviet Antarctic expeditions. The third expedition during the IGY saw the establishment and operation of six stations: Mirny, Vostok, Sovetskaya, Oasis, Pionerskaya, and Komsomol'skaya.

The Antarctic Peninsula has been a focal point of research since the 1940s. The retreat of ice shelves along the peninsula has particularly attracted attention as area of investigation and research.

In 1990, the concept of an International Trans-Antarctic Scientific Expedition (ITASE) was proposed; a science and implementation plan was prepared in 1996 (www2.umaine.edu/itase/content/Science/intro.html). Its primary aim was the collection and interpretation of a continental-wide array of environmental parameters assembled through the coordinated efforts of scientists from several nations. It had two key scientific objectives. First, it sought to determine the spatial variability of Antarctic climate (accumulation, air temperature, atmospheric circulation) over the last 200 years, and where the data were available, the last 1,000 years. These variations include major atmospheric phenomena such as the El Niño Southern Oscillation (ENSO); snow accumulation variations; and extreme events such as volcanic eruptions and storms. Second, ITASE aimed to determine environmental variability based on environmental proxies such as sea ice variation, ocean productivity, anthropogenic impacts, and other, extra-Antarctic continental influences.

Numerous Antarctic traverses were performed by multiple nations. Some of the results are described by Kreutz and Mayewski (1999) and Van den Broeke et al. (1999). All glaciological data sets generated through the US Antarctic Program are archived at the National Snow and Ice Data Center in Boulder, Colorado (https://nsidc.org/agdc/data.html).

There have been numerous ice coring programs in Antarctica. The most famous is the EPICA core from Dome C that spans eight glacial cycles, or approximately 740,000 years (EPICA Community Members 2004). The data from this core are searchable by principal investigator, parameter, or data set title.

In the Southern Ocean there have been many ship surveys studying physical and biological oceanography and sea ice conditions. From the 1940s through the 1960s, icebreakers were primarily involved in missions to supply Antarctic bases, but new ones designed for ocean science research were built in the 1970s. These included the *US Polar Star* and *Polar Sea*, the German *Polarstern*, the British *RRS James Clark Ross*, the Australian *Aurora Australis*, and Soviet/Russian vessels.

1.3.3 Central Asia

Exploration of the extensive region known as Central Asia was long hindered by its remoteness and the high altitudes. Study of the mountain ranges of the Himalaya began with British explorers in the late nineteenth century, mainly related to climbing expeditions, and these continued from several countries into the 1950s. In Soviet Central Asia, hydrometeorological stations were installed from the 1930s. Glaciological surveys and mapping began in the mid-1950s in the

Tien Shan and Pamirs, organized by the Institute of Geography, Russian Academy of Sciences under the leadership of Professor Vladimir Kotlyakov. China began setting up a network of weather stations across the Tibetan Plateau in the 1950s and Chinese glaciers were mapped starting in the eastern Tien Shan in 1958. The Lanzhou Institute of Glaciology, Cryopedology, and Desert Research was established in 1965 to lead these efforts under the direction of Professor Shi Yafeng. Ice core research was conducted in Tibet by Lonnie Thompson of the Byrd Polar Research Center, Ohio State University, with Chinese collaborators, on Dunde ice cap in the mid-1980s, Guliya ice cap in the 1990s and in 2015, Dasuopu glacier in the Himalaya in the late 1990s, and Puruogangri ice cap in Tibet in 2000.

Institutions that address the environmental conditions of the region are as follows:

The Lanzhou Institute of Glaciology and Cryopedology was reorganized in 1999 to become part of the Cold and Arid Regions Environmental and Engineering Research Institute (CAREERI) of the Chinese Academy of Sciences (CAS). In 2016, it was further reorganized into the Northwest Institute of Eco-Environment and Resources (NIEER), CAS. NIEER comprises five former CAS institutes – namely, CAREERI, the Lanzhou Center for Oil and Gas Resources, the Lanzhou Branch of the National Science Library, the Northwest Institute of Plateau Biology, and the Qinghai Institute of Salt Lakes. The experimental and analytical unit of NIEERI is home to the State Key Laboratory of Frozen Soils Engineering and the State Key Laboratory of Cryospheric Sciences. (http://english.nieer.cas.cn/au/ct/)

The Institute of Tibetan Plateau Research, CAS, has campuses in Lhasa, Beijing, and Kunming. It operates five field stations on the Tibetan Plateau. (http://english.itpcas.cas.cn/)

The Institute of Mountain Hazards and Environment, CAS, is located in Chengdu. (http://english.imde.cas.cn/)

The International Centre for Integrated Mountain Development (ICIMOD) was set up in Kathmandu, Nepal, in 1984 as a regional intergovernmental learning and knowledge sharing center serving the eight member countries of the Hindu Kush Himalayas: Afghanistan, Bangladesh, Bhutan, China, India, Myanmar, Nepal, and Pakistan. Water, Hazards and Environmental Management was developed as a theme in the 2000s. Currently, the Cryosphere Initiative focuses on the monitoring of glaciers, snow, permafrost, glacial lakes, and glaciohydrology, with an emphasis on in situ measurements, remote sensing, and modeling. (www.icimod.org/?q = abt)

1.4 The Role of Polar Snow and Ice in the Climate System

Land ice covers 14.1 million km^2 of the Earth, most of it in the Antarctic. It is noteworthy that 97 percent of all fresh water is at present locked up in the Greenland and Antarctic ice sheets. If these ice sheets melted, global sea level would rise by approximately 63 m. Permafrost underlies approximately 17 million km^2 in the northern hemisphere. Seasonally, snow covers up to 48 million km^2 in the northern hemisphere, while sea ice covers 20 million km^2 in September around Antarctica and approximately 15 million km^2 in March in the Arctic and adjacent seas. More extensive is seasonally frozen ground that at maximum underlies 57 million km^2 of the northern hemisphere.

The cryosphere interacts with the climate system in numerous ways, as illustrated in Figure 1.3.

- Ice–albedo (positive) feedback, referring to changes in snow/ice cover that result in changes in radiative forcing, which then lead to changes in temperature. Reduced (increased) albedo results in warming (cooling).
- Lags (one to six months) introduced into the hydrologic cycle.
- Seasonal insulation of snow on land and sea ice on ocean, cutting off heat and moisture fluxes.
- Cryospheric margins affect the development/steering of synoptic systems.
- Sea ice role in ocean salinity and vertical circulation.
- Land ice melt contribution to global sea level.

Figure 1.3
Schematic diagram of the interaction of cryospheric components with land, atmosphere, and ocean.

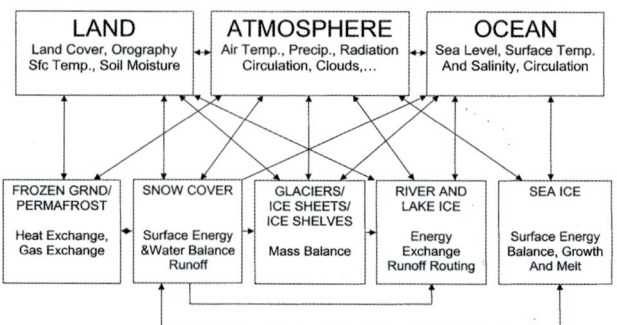

-List in upper boxes indicate important state variables
-Lists in lower boxes indicate important processes involved in interactions.
-Arrows indicate **direct** interactions

In more detail, we can consider the permanent and seasonal cryosphere components:

Ice sheets

- Ice sheet topography affects planetary waves and storm tracks, cloudiness, and precipitation.
- Freshwater input to the ocean from iceberg calving and melt runoff affect ocean salinity and the ocean thermohaline circulation.
- Ice volume affects global sea level (glacial/interglacial range is 140 m).
- Major iceberg calving – Heinrich and higher-frequency events.
- Surface inversion and katabatic winds; local cooling.
- Basal melt and surges.

Permafrost/frozen ground

- Affect energy and moisture fluxes, runoff, methane retention, and vegetation cover.

Processes involving the seasonal cryosphere

- Snow/ice albedo feedback.
- Lower atmospheric thickness over cold snow cover leads to changes in planetary waves, storm tracks, cloudiness, and precipitation.
- Insulation of land by snow pack and of ocean by sea ice leads to decoupling of the surface and the boundary layer, affecting heat, moisture, and gas fluxes.
- Time delay in hydrologic cycle due to water storage in snow and ice.
- Sea ice growth/decay and export affect ocean salinity.
- Seasonally frozen ground affects heat, moisture and gas fluxes, runoff, and plant growth.

SUMMARY

The climate of high northern and southern latitudes is primarily determined by the annual solar cycle and low temperatures. The resulting covers of ice and snow are predominant features, but they are undergoing rapid changes as a result of global warming and its amplification near the poles. There are various definitions of "polar," and the two polar regions differ markedly in their physical characteristics – an ice-covered ocean versus a continental ice sheet. The southern high latitudes have a high degree of zonality, in contrast to the Arctic. Polar regions include the Third Pole in Central Asia, where altitude replaces latitude as the overriding climatic factor.

Systematic study of high latitudes began with the twelve weather stations operated in the First International Polar Year, 1882–1883. Extensive data for the Arctic Ocean were obtained during the 1893–1896 *Fram* expedition of F. Nansen.

Norwegians carried out annual expeditions to Svalbard from 1911 to the mid-1920s. Two stations were operated on the Greenland ice sheet in 1930–1931. The Soviet Union deployed North Pole drifting stations in 1937–1938 and in 1950–1991, and these operations were resumed by the Russian Federation from 2003 to the present. The United States operated ice island stations in the Arctic from 1952 to the 1970s. Five weather stations were installed on Canadian Arctic islands during 1947–1950. The International Biological Program conducted a Tundra Biome project in the 1970s. The US Arctic Buoy program began in 1979 and became an international venture in 1991. Sea ice draft data collected by US submarines across the Arctic in 1957 and 1970 were analyzed and published in 1989 by A. McLaren. Icebreakers began sailing to the North Pole in 1977. In the 1990s the World Climate Research Program planned the Global Energy and Water Experiment (GEWEX) project for the Mackenzie GEWEX Study and the GEWEX Asian Monsoon Experiment to study cold-region hydrometeorology. In 2000 the Russia–US Environmental Working Group assembled and published atlases of Arctic oceanography and meteorology.

The first overwintering expeditions in Antarctica were made in the late 1890s. The South Pole was reached in 1911 and 1912.

Mawson established a base at Cape Denison, showing it to be the windiest place on Earth. Richard Byrd's 1928–1930 expedition established Little America on the Ross ice shelf. Little America IV was established by the US Navy in 1946. Major coordinated Antarctic science began with the International Geophysical Year, 1957–1958. Coastal bases were established by many nations; the United States set up its South Pole station, while the USSR installed a base at Vostok in East Antarctica. In the mid-1990s the International Trans-Antarctic Scientific Expedition conducted multinational traverses collecting data on Antarctic climate variables, snow parameters, and past (200–1,000 years) climates. Southern Ocean and sea ice conditions have been surveyed by ships since the *Gauss* in 1903. Research icebreakers have been widely utilized since the 1970s.

Science in Central Asia developed in the Soviet Union with the hydrometeorological station network set up in the 1930s. China installed weather stations in Tibet in the 1950s. Glacier mapping began in the USSR and in China in the later 1950s. From the mid-1980s to 2000s, ice cores were extracted from ice caps in Tibet by L. Thompson and Chinese collaborators. There are three major institutes in China and one in Nepal involved in environmental research in Central Asia.

Ice and snow are the dominant features of polar regions. The cryosphere interacts with the climate system in a multitude of ways, including ice–albedo feedback that amplifies small changes in radiation and temperature; time lags introduced into the hydrological cycle; insulation effects of snow and ice on the land and ocean, modifying energy and gas fluxes to the atmosphere; and the effects of land ice volume on global sea level.

QUESTIONS

1. Compare the geographical characteristics of the two polar regions.
2. Which circumstances limited the scientific value of the First and Second IPYs?
3. Summarize the technological advances that facilitated research on Arctic climate and sea ice in the twentieth century.
4. Examine the "big science" aspects of Antarctic research during the second half of the twentieth century.
5. Compare the climatic characteristics of the Third Pole region with those of the Arctic.

References

Anonymous. 1950. "U.S. Air Force Weather Reconnaissance Flights to the North Pole." *Polar Record* 6(42): 268.

Arctic Climate Impact Assessment (ACIA). 2004. *Arctic Climate Impact Assessment (ACIA)*. Cambridge, UK: Cambridge University Press.

Arctic Climatology Project. Environmental Working Group. 2000. *Arctic Meteorology and Climate Atlas* [CD-ROM], edited by F. Fetterer and V. Radionov. Boulder, CO: National Snow and Ice Data Center.

Arctic Monitoring and Assessment Programme (AMAP). 2011. *Snow, Water, Ice and Permafrost in the Arctic (SWIPA): Climate Change and the Cryosphere*. Oslo, Norway: Arctic Monitoring and Assessment Programme.

Arctic Monitoring and Assessment Programme (AMAP). 2017. *Snow, Water, Ice and Permafrost in the Arctic (SWIPA): Climate Change and the Cryosphere*. Oslo, Norway: Arctic Monitoring and Assessment Programme.

Atkinson, D. E., B. Alt, and K. Gajewski. 2000. "A New Database of High Arctic Climate Data from the Polar Continental Shelf Project Archives." *Bulletin of the American Meteorological Society* 81(11): 2621–9.

Barr, W. 1985. *The Expeditions of the First International Polar Year, 1982–83*. Tech. Paper No. 29. Arctic Institute of North America. Calgary, Alberta: University of Calgary.

Barry, R. G. 2005. "Climate: Research Programs." In *Encyclopedia of the Arctic*. Vol. 1, edited by M. Nuttall, 379–83. New York: Routledge.

Barry, R. G. 2016. "Norwegian Contributions to Arctic Environmental Sciences from the 1880s to the Third International Polar Year." *Advances in Polar Science* 27: 1–7.

Barry, R. G. 2017. "The Arctic Cryosphere in the 21st Century." *Geographical Review* 107: 69–88.

Barry, R. G., G. H. Courtin, and B. C. Labibe. 1981. "Tundra Climates." In *Tundra Ecosytems: A Comparative Analysis*, edited by L. C. Bliss, O. W. Heal, and J. J. Moore, 81–114. Cambridge: Cambridge University Press.

Barry, R. G., and C. I. Jackson. 1969. "Summer Weather Conditions at Tanquary Fiord, N.W.T. 1963–67." *Arctic and Alpine Research* 1: 169–80.

Belanger, D. O. 2006. *Deep Freeze*. Boulder, CO: University Press of Colorado.

Corby, G. A. 1982. "The First International Polar Year, 1882/83." *WMO Bulletin* 31(3): 197–214.

Curry, J. 2001. "Introduction to Special Section: FIRE Arctic Clouds Experiment." *Journal of Geophysical Research* 106(D14): 14985.

EPICA Community Members. 2004. "Eight Glacial Cycles from an Antarctic Ice Core." *Nature* 429: 623–8.

Hammer, C., P. A. Mayewski, D. Peel, and M. Stuiver. 1976. "Preface. (Special Issue, Greenland Summit Ice Cores)." *Journal of Geophysical Research* 102(C12): 26, 315–16.

Hattersley-Smith, G., et al. 1952. "Arctic Ice Islands." *Arctic* 5(2): 67–103.

Huschke, R. E. 1969. *Arctic Cloud Statistics from "Air Calibrated" Surface Weather Observations*. Rand Corporation Memo RM-6173. Santa Monica, CA: Rand Corporation.

Jackson, C. I. J. 1959. *Operation Hazen. The Meteorology of Lake Hazen, NWT Based on Observations Made during the International Geophysical Year. Part I: Analysis of the Observations*. Publications in Meteorology No. 15. Montreal: Arctic Meteorology Research Group, Geography Department, McGill University.

Kane, D. D. I., and W. S. Reeburgh. 1998. "Introduction to Special Section: Land–Air–Ice Interactions (LAII) Flux Study." *Journal of Geophysical Research* 103(D22): 28913–15.

Köppen, W. 1923. *Die Klimate der Erde: Gundriss der Klimakunde*. Berlin: Walter de Gruyter Co.

Krenke, A. N. 1961; translated 1997. "The Ice Dome with Firn Nourishment in Franz Josef Land." In *34 Selected Papers on Main Ideas in the Soviet Glaciology, 1940s–1980s*, edited by V. M. Kotlyakov, 132–44. Moscow: Institute of Geography, Russian Academy of Sciences.

Kreutz, K. J., and P. A. Mayewski. 1999. "Spatial Variability of Antarctic Surface Snow Glaciochemistry: Implications for Paleoatmospheric Circulation Reconstructions." *Science* 511: 105–18.

LAII Science Steering Committee. 1997. *Arctic System Science: Land–Atmosphere–Ice Interactions: A Plan for Action*. Fairbanks, AK: University of Alaska, Fairbanks.

Larsen, J. N., and O. A. Anisimov. 2014. "Polar Regions." In *Climate Change 2014: Impacts, Adaptation, and Vulnerability. Part B: Regional Aspects*. Contribution of Working Group II to the Fifth Assessment Report of the Intergovernmental Panel on Climate Change, 1567–617. Cambridge: Cambridge University Press.

Launius, R. D., J. R. Fleming, and D. H. Devorkin, eds. 2010. *Globalizing Polar Science*. New York: Palgrave Macmillan.

Laursen, V. 1982. "The Second International Polar Year (1932/33)." *WMO Bulletin* 31: 214–26.

Liu, H.-X., and K. C. Jezek. 2004. "A Complete High-Resolution Coastline of Antarctica Extracted from Orthorectified Radarsat SAR Imagery." *Photogrammetric Engineering & Remote Sensing* 70: 605–16.

Loewe, F. 1936. "The Greenland Ice Cap as Seen by a Meteorologist." *Quarterly Journal of the Royal Meteorological Society* 62: 359–77.

McLaren, A. S. 1989. "The Under-Ice Thickness Distribution of the Arctic Basin as Recorded in 1958 and 1970." *Journal of Geophysical Research: Oceans* 94: 4971–83.

McNutt, A. S., et al. 1988. "LIMEX '87: The Labrador Ice Margin Experiment 1987: A Pilot Experiment in Anticipation of Radar Sat and ERS 1 Data." *Eos* 69(23): 634–5, 643.

Merles, S. T. A. 1932. *Meteorological Results of the British Arctic Air Route Expedition: 1930–31*. Geophysical Memoir 7. London: Meteorological Office.

MIZEX '87 Group. 1989. "MIZEX East 1987." *Eos* 70(17): 545, 548–54, 554–5.

Mohn, H. 1905; reprinted 1969. "Meteorology, XVII." In *The Norwegian North Polar Expedition, 1893–1896. Scientific Results*. Vol. 6, edited by F. Nansen. New York: Greenwood Press.

National Snow and Ice Data Center. 2000. *Arctic Ocean Snow and Meteorological Observations from the North Pole Drifting Stations: 1937, 1950–1991* [CD-ROM]. Boulder, CO: National Snow and Ice Data Center, University of Colorado.

Nuttall, M., ed. 2005. *Encyclopedia of the Arctic*. 3 volumes. London: Routledge.

Putnins, P. 1969. "The Climate of Greenland." In *Climates of the Polar Regions. World Survey of Climatology*. Vol. 14, edited by S. Orvig, 3–128. Amsterdam: Elsevier.

Riffenburgh, B., ed. 2007. *Encyclopedia of the Antarctic*. London: Routledge.

Robin, G. de Q. 1953. "Norwegian–British–Swedish Antarctic Expedition, 1949–52." *Polar Record* 6(45): 608–14.

Romanov, I. P., Yu. B. Konstantinov, and N. A. Kornilov. 2000. "North Pole Drifting Stations (1937–1991)." In *Arctic Climatology Project, Environmental Working Group Arctic Meteorology and Climate Atlas*, edited by F. Fetterer and V. Radionov [CD-ROM]. Boulder, CO: National Snow and Ice Data Center.

Sater, J. E. (coordinator). 1968. *Arctic Drifting Stations: A Report on Activities Supported by the Office of Naval Research*. Washington, DC: Arctic Institute of North America.

SEARCH Science Steering Committee. 2001. *SEARCH: Study of Arctic Environmental Change. Science Plan*. Seattle: Polar Science Center, University of Washington.

Sides, H. 2014. *In the Kingdom of Ice*. New York: Doubleday.

Stamnes, K., et al. 1999. "Review of Science Issues and Deployment Strategy and Status for the ARM North Slope of Alaska–Adjacent Arctic Ocean Climate Research Site." *Journal of Climate* 12(1): 461–3.

Starokadomskii, L. M. 1976. *Charting the Russian Northern Sea Route: The Arctic Ocean Hydrographic Expedition 1910–1915*. Translated by W. Barr. Montreal: Arctic Institute of North America.

Stonehouse, B., ed. 2002. *Encyclopedia of the Antarctic and Southern Ocean*. Chichester, UK: John Wiley and Sons.

Sverdrup, H. U. 1933. *The Norwegian North Pole Expedition with the* Maud, *1918–1925, Scientific Results*. Bergen: Geofysisk Institutt.

Thomas, R. H., and investigators. 2001. "Program for Arctic Regional Climate Assessment (PARCA): Goals, Key Findings, and Future Directions." *Journal of Geophysical Research* 106(D24): 33691–706.

Van den Broeke, M., et al. 1999. "Climate Variables along a Traverse Line in Dronning Maud Land, East Antarctica." *Journal of Glaciology* 45(150): 295–302.

von Drygalsi, E., and M. Raraty. 1991. *The German South Polar Expedition, 1901–3*. Norwich, UK: Erskine Press.

Wadhams, P. 1997. "Variability of Arctic Sea Ice Thickness: Statistical Significance and Its Relationship to Heat Flux." In *Operational Oceanography: The Challenge for European Co-Operation*, 368–84. Amsterdam: Elsevier.

Williams, R. S., Jr., and J. G. Ferrigno, eds. 2013. *State of the Earth's Cryosphere at the Beginning of the 21st Century: Glaciers, Global Snow Cover, Floating Ice, and Permafrost and Periglacial Environments*. US Geological Survey Professional Paper 1386-A, A425–96.

Wood, K. R., and J. E. Overland. 2006. "Climate Lessons from the First International Polar Year." *Bulletin of the American Meteorological Society* 87: 1685–97.

Yu, Y., G. A. Maykut, and D. A. Rothrock. 2004. "Changes in the Thickness Distribution of Arctic Sea Ice between 1958–1970 and 1993–1997." *Journal of Geophysical Research: Oceans* 109: C08004. doi: 10.1029/2003JC001982.

Zemp, M., et al. 2015. "Historically Unprecedented Global Glacier Decline in the Early 21st Century." *Journal of Glaciology* 61(228): 745–6.

Paleoclimatic History

2

This chapter traces the paleoclimatic history of the polar regions by focusing on a series of time snapshots of environmental conditions. Beginning with the Eocene Thermal Maximum, we then proceed to the Oligocene–Miocene growth of the Antarctic ice sheet, followed by an account of Pliocene–Pleistocene glaciations, the post-glacial Holocene, and the current Anthropocene.

2.1 Temperatures over Geologic Time

The overall setting of the Earth's temperature changes is summarized in Figure 2.1. This figure demonstrates that there were very large fluctuations in the first 500 to 50 MY of the planet's history, but these need not concern us here, as terrestrial geography was so radically different as a result of plate tectonics and continental drift. Since the Eocene Thermal Maximum there has been an irregular cooling and fluctuations about a much lower average temperature than in the Earth's early history. These phases are discussed in this chapter.

Before reviewing paleoclimate history, however, it is appropriate to introduce the main sources of evidence for long-term climate change and the primary drivers.

2.2 Geological Records

2.2.1 Ocean Sediment Cores

Beginning in the 1950s, ocean sediment cores were collected and analyzed to determine past climatic conditions over millions of years. There were two main approaches. Cold and warm assemblages of planktonic (near surface) and benthic (deep water) foraminfera were identified and used to determine water masses and their changes over time. These microorganisms have calcium carbonate shells. In the 1960s, the use of oxygen isotope ratios (see Box 2.1) in ocean

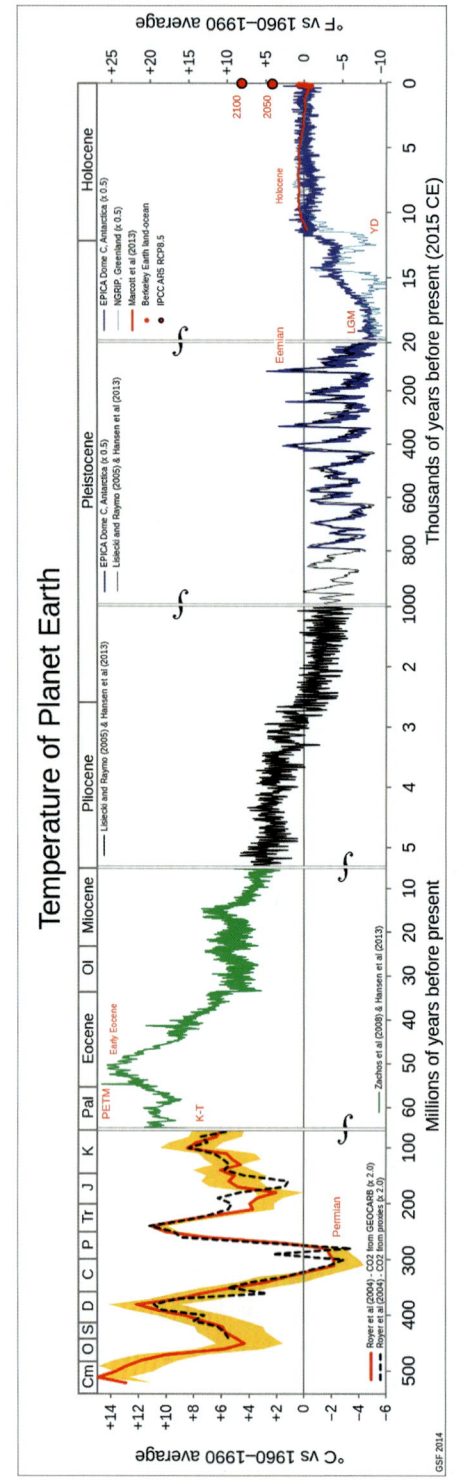

Figure 2.1 Temperatures of planet Earth during geological time.
Note that the time scale is divided into different segments.
Source: Wikipedia, https://upload.wikimedia.org/wikipedia/commons/thumb/5/5f/All_palaeotemps.svg/1760px-All_palaeotemps.svg.png

Box 2.1 **Oxygen Isotope Analysis**

Oxygen has two main isotopes: the heavier ^{18}O and the lighter ^{16}O. In isotope analysis, only the ratio of ^{18}O to ^{16}O present in the calcium carbonate of the sample is considered. The calculated ratio is then compared to that for standard mean ocean water (SMOW) to yield information about the temperature at which the sample was formed. The ratio varies slightly depending on the temperature of the water, as well as other factors such as the water's salinity and the volume of water locked up in ice sheets. $\delta^{18}O$ is measured in parts per mil (per thousand, or ‰). Epstein et al. (1951) estimated that an increase of $\delta^{18}O$ by 0.22‰ was equivalent to a cooling of 1 °C if only temperature change was involved.

^{16}O is preferentially evaporated from the surface because it is lighter. Consequently, the surface ocean contains greater amounts of ^{18}O in the subtropics and tropics, where there is more evaporation, and lesser amounts of ^{18}O in the mid-latitudes, where it rains more. For oceanic $\delta^{18}O$, the range of values between glacial and interglacial conditions is approximately 2–3‰.

sediments gained prominence. Warm and cold periods in the Earth's past climate were deduced from oxygen isotope records derived from foraminiferal shells in deep-sea sediment core samples collected through the long-term Deep Sea Drilling Project (DSDP) that operated from 1968 to 1983.

Numerous ocean sediment cores have been collected in both polar regions. There is a long history of drilling on the Lomonosov, Gakkel, Morris Jessup, Alpha, and Northwind ridges. In the Antarctic, cores were collected under the Ross Ice Shelf through the ANtarctic geological DRILling (ANDRILL) program in 2006 and 2007. A core spanning almost 20 million years was recovered.

In a reinterpretation of marine core records, Shackleton (1967, 1977) showed that changes in oxygen isotope composition from benthic samples correspond not to ocean temperature variations, which are slight in the deep ocean, but rather to the extraction of large amounts of water from the oceans during glacial periods and their deposition in ice sheets.

Alternating cold (even numbers) and warm (odd numbers) periods in the ocean records are known as marine isotope stages (MIS). The MIS system was devised by Emiliani (1961). MIS 104 began 2.614 million years ago (MYA). We are currently in MIS 1 – the Holocene – and the last interglacial period was MIS 2.

2.2.2 Ice Cores

Ice cores were first extracted from Camp Century in northern Greenland in 1969 by Dansgaard et al. (1971). The oxygen isotope record spanned almost

100 ka and showed marked fluctuations in past temperatures. In contrast to ocean records, the departure of $\delta^{18}O$ can be in the range of -40 to -50 ppm, because the heavier isotope readily precipitates out as air moves poleward.

Numerous deep ice cores have been extracted in Greenland, Antarctica, Tibet, and high mountain ice caps around the world over the last forty-five years. Their analysis has vastly expanded to include measurement of atmospheric gas concentrations, chemical species, volcanic and terrestrial aerosol deposition, snow stratigraphy, pollen, sea salts, and forest fires. Raynaud et al. (2000) describe the variations of greenhouse gases identified in ice cores for the Holocene, Younger Dryas, last glaciation, and the last four glacial cycles over 0–400 ka. The oldest core obtained to date is 800 ka from the European Project for Ice Coring in Antarctica (EPICA) site on Dome C, Antarctica (EPICA Community Members 2004).

2.3 Climate Drivers

A major driver of long-term climate change is the variation of solar radiation (insolation) as a result of changes in the orbital geometry of the Earth with respect to the Sun. These orbital cycles are detailed in Box 2.2.

Orbital variations represent a primary external driver of global climate. An equally important internal driver is the atmospheric composition, especially the concentration of greenhouse gases – notably, carbon dioxide (CO_2) and methane (CH_4). CO_2 concentrations decreased from approximately 760 parts per million

Box 2.2 Orbital Cycles

In the nineteenth century, Joseph Adhémar (1842) and James Croll (1864) proposed links between orbital variations and global climate changes, although the relationships could not be tested (Berger 2012). The quantitative basis for these relationships was provided by the Serbian astronomer Milankovitch (1930), who made detailed calculations of the three orbital signatures. The significance of these signatures for glacial cycles was first recognized by Hays, Imbrie, and Shackleton (1976) in a landmark paper (Hodell 2016). The 23,000- and 19,000-year precession cycles and the 41,000-year obliquity cycle in solar radiation were found to be linearly related to spectral peaks in the climatic record in 450,00 years of sub-Antarctic ocean sediments (Figure 2.2). The approximately 100,000- and 400,000-year cycles of eccentricity, however, produced only a small change in solar radiation, necessitating the assumption of some nonlinear forcing. For this reason, Hays et al. (1976) used the analogy of a pacemaker that sets the rhythm of the ice ages but does not account for the amplitude of the signal. Even today, it remains unclear which aspects of the climate system are most sensitive to the changes in seasonal insolation anomalies induced by orbital geometry.

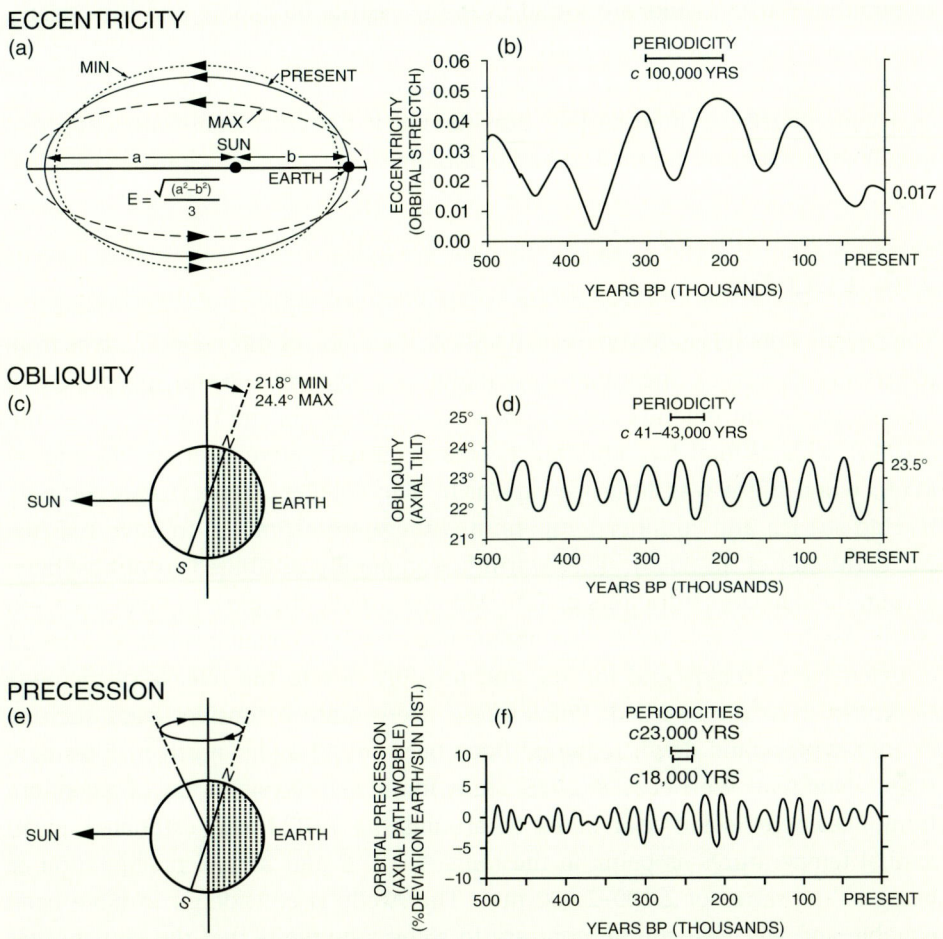

Figure 2.2 Summary of astronomical (orbital) effects on solar irradiance and their relevant time scales over the past 500,000 years. (A and B) Eccentricity of the orbit, (C and D) obliquity or axial tilt, (E and F) precession or axial path wobble.
Sources: Partly after Broecker and Van Donk (1970) and Henderson-Sellers and McGuffie (1984). B, D, and F from *Review of Geophysics and Space Physics* 8, 1970. Reproduced by kind permission of the American Geophysical Union. Barry and Chorley 2010, 437, figure 13.3.

(ppm) around 34 MYA to 180 ppm during glacial cycles of the last million years, with major effects on atmospheric absorption of emitted terrestrial radiation and global temperatures (discussed later in this chapter). The long-term carbon cycle involves the movement of carbon between the atmosphere, oceans, soil, rocks, and volcanism. Volcanic outgassing is a major natural source of carbon dioxide.

Other such sources include combustion of organic matter by wildfires and plant respiration. Carbon sinks are found in the terrestrial biosphere and the oceans, where carbonate rock is eventually formed. Methane levels have hovered around 500 ppm over the last 30 million years. Methane is twenty-five times more potent as a greenhouse gas than carbon dioxide and it currently contributes approximately 20 percent of the total radiative forcing.

2.4 Eocene

The Eocene Epoch lasted from 56 to 34 MYA; the name of this epoch derives from the Greek *eos kainos*, meaning "new dawn" – a reference to the appearance of modern fauna. During the Eocene Epoch, a Paleocene–Eocene Thermal Maximum (PETM) with several hypothermal events occurred between about 55 and 49 MYA, when there was little ice on Earth. The smaller Arctic Ocean certainly had no sea ice and summer temperatures there are estimated to have reached 24 °C (Moran et al. 2005). This warmth is generally attributed to atmospheric carbon dioxide concentrations of 700–900 ppm, twice the present level (Pearson 2010). Methane levels were also about twice today's concentration as a result of extensive wetlands and forests, and possibly due to the release of *methane clathrates* from the sea bed. Polar forests were quite extensive; trees such as swamp cypress and dawn redwood have been found as far north as Ellesmere Island, Nunavut. West et al.'s (2015) study found leaf physiognomy of paleoflora from Ellesmere Island that dates to around the PETM and estimated mean annual temperatures as being in the range 9–12 °C and annual precipitation as being in the range of 2,100–2,400 mm. The Arctic is considered to have been equable and wet year-round, contrary to some arguments that the climate was monsoonal.

A major unresolved issue is the existence of warm equable climatic conditions over the globe in the early Eocene. Model experiments indicate that the higher greenhouse gas concentrations and poleward ocean heat transport are insufficient to account for this state (Huber and Caballero 2011). However, Alley (2016) suggests that igneous outpourings in the North Atlantic led to a major carbon dioxide increase over a few millennia that triggered the Thermal Maximum (PETM) about 55.9 MYA; the PETM persisted about 150,000 years and had major impacts on biota.

Anagnostou et al. (2016) generated a new high-fidelity record of CO_2 concentrations using the boron isotope composition of well-preserved planktonic foraminifera from the Tanzania Drilling Project. During the PETM, CO_2 concentrations were approximately 1,400 ppm. The relative decline in CO_2

concentration through the Eocene was more robustly constrained at about 50 percent, with a further decline occurring into the Oligocene. Provided that the latitudinal dependence of sea surface temperature (SST) change for a given climate forcing in the Eocene was similar to that of the late Quaternary period, this CO_2 decline was sufficient to drive the well-documented high- and low-latitude cooling that occurred through the Eocene. Anagnostou et al. (2016) also demonstrated that both the PETM and the late Eocene exhibited an equilibrium climate sensitivity relative to the preindustrial period of 2.1–4.6 °C per CO_2 doubling, which is similar to the canonical range (1.5–4.5 °C) identified in most general circulation model (GCM) calculations. This indicates that a large fraction of the warmth of the early Eocene greenhouse was driven by increased CO_2 concentrations.

Moran et al. (2005) have found evidence from the Lomonosov Ridge in the Arctic Ocean for the first occurrence of *ice-rafted debris* (IRD) in the middle Eocene Epoch (approximately 45 MYA), some 35 million years earlier than previously thought. Fresh surface waters were present at approximately 49 MYA, before the onset of ice-rafted debris from icebergs. Also, the temperatures of surface waters during the Paleocene/Eocene Thermal Maximum (approximately 55 MYA) appear to have been substantially higher than those previously estimated. Wetter conditions and strong rainfall seasonality suggest an enhanced hydrological cycle in high latitudes of the northern hemisphere. The transition from a warm "greenhouse" world during the late Paleocene and early Eocene Epochs, to a colder "icehouse" world was characterized by sea ice and icebergs from the middle Eocene Epoch onward.

In the early Eocene in Antarctica, moist, cool temperate rainforests were present. These forests were dominated by *Nothofagus* and conifers (Francis et al. 2009). On King George Island in the South Shetlands, *Nothofagus*-dominated forests were characteristic of the middle to late Eocene, with ferns and tree ferns becoming increasingly important. Estimated mean annual temperatures were 5–8 °C.

Around 34 MYA, continental drift separated Australia from Antarctica, which led to the routing of warm equatorial water away from the southern continent, cooling the Southern Ocean around Antarctica and allowing the seas to freeze. The opening of the Tasman–Antarctic and Drake passages to deep-water flow occurred around 34 and 31 MYA, respectively. However, Huber and Nof (2006) argue from model simulations that the resulting changes in ocean heat transport were insufficient to account for glacial onset; instead, these authors favor the role of decreasing greenhouse gas concentrations, as also proposed by DeConto and Pollard (2003). Regardless, the Eocene/Oligocene transition appears to mark the initiation of the Antarctic Circumpolar Current (see Sections 4.1 and 6.2) and thus the onset of thermal isolation of Antarctica, with the first major growth in ice volume on east Antarctica. Antarctic summer temperatures appear to be critical in allowing winter snowfall to persist.

Earlier studies suggested unipolar glaciation during the late Eocene and Oligocene, but the Lomonosov cores analyzed by Moran et al. (2005) instead indicate that the northern hemisphere cooling occurred around 45 Ma. This finding suggests contemporaneous coevolution of ice at both poles, and thus symmetry in cooling and a bipolar transition from the early Eocene "greenhouse" to an "icehouse" world.

2.5 Oligocene–Miocene–Antarctic Glaciation

Galeoti et al. (2016) argue that about 34 MYA, global cooling caused an ice sheet to form on East Antarctica as atmospheric CO_2 concentrations fell below approximately 750 ppm (DeConto and Pollard 2003). Sedimentary cycles from a core in the western Ross Sea demonstrated orbitally controlled glacial cycles between 34 and 31 MYA. At first, under atmospheric CO_2 levels of 600 ppm or greater, the ice sheet was restricted to the terrestrial continent. It was highly responsive to local insolation forcing. A more stable, continental-scale ice sheet then developed, reaching its maximum extent between 33.6 and 33.2 MYA. Calving at the coastline did not begin until approximately 32.8 MYA, when atmospheric CO_2 levels fell below 600 ppm for the first time. Galeoti et al. (2016) conclude that a CO_2 threshold for the continental-scale Antarctic ice sheet occurred at approximately 600 ppm with much increased sensitivity to radiation forcing above this level. Based on two subtropical Pacific cores, Holbourn et al. (2005) found that relatively constant, low summer insolation over Antarctica coincided with declining atmospheric CO_2 levels at the time of Antarctic ice sheet expansion and global cooling (14.7–12.7 MYA). They also showed that Antarctic glaciation was rapid, taking place within two obliquity cycles, and coincided with a striking transition from obliquity to eccentricity as the driver of change.

Global temperatures were depressed between about 32.5 and 25.5 MYA, a period during which global ice volume increased. According to Francis et al. (2009), a stable, cold, dry climate was not established in the western Ross Sea until the Eocene/Oligocene boundary, with major ice sheet growth occurring at the Early/Late Oligocene boundary. A temperate glacier regime is suggested for the Early Oligocene, following by cooling into the Miocene, typified by polythermal glaciers. The Early Oligocene landscape was characterized by temperate glaciers flowing from the early East Antarctic ice sheet. According to Francis et al. (2009), glacier ice first reached the continental shelf edge in Prydz Bay, East Antarctica, in earliest Oligocene time, while the Victoria Land coast was influenced by iceberg rafting. The Late Oligocene saw repeated ice expansions. Nevertheless, the cold, frigid regime of today only began much later, at the end of Pliocene time.

At the Eocene/Oligocene transition around 34 MYA, the oxygen isotopic composition ($\delta^{18}O$) of benthic and planktonic foraminifera in the Southern Ocean increased by more than 1‰. This shift is thought to represent a combination of global cooling and the growth of a large ice sheet on the Antarctic continent. Petersen and Schrag (2015) have demonstrated that the long-term change in $\delta^{18}O$ seen in planktonic foraminifera on the Maud Rise (64 °S) in the Weddell Sea is predominantly due to changes in ice volume, with only a modest cooling contribution. They estimate a global ice growth equivalent to approximately 110–120 percent of the volume of the modern Antarctic ice sheet or approximately 80–90 m of eustatic sea level change. The relations between atmospheric CO_2 concentrations, ocean temperatures, atmospheric moisture transport, and ice buildup have not yet been precisely determined, however.

An Oligocene (33.9–23 MYA) data set of benthic foraminfera from the equatorial Pacific has been analyzed by Pälike et al. (2006). These researchers identified two major glaciations: one immediately after the Eocene/Oligocene boundary and the other at the Oligocene/Miocene boundary. They also found a persistent 405-ka eccentricity pacing of the carbon cycle recognized in the $\delta^{13}C$ record. Naish et al. (2001) showed that sedimentary records from shallow marine cores in the western Ross Sea exhibited well-dated cyclic variations during the Oligocene/Miocene transition (24.1–23.7 MYA), linking the extent of the East Antarctic ice sheet directly to orbital cycles of obliquity and eccentricity.

It remains unclear how much ice persisted in Antarctica during the Oligocene, but global climate and sea-level data suggest that a smaller Antarctic ice sheet was predominant in the Oligocene to mid-Miocene (Aitken et al. 2016). A major buildup took place during the Middle Miocene soon after 15 Ma. Flowers and Kennett (1994) showed that from 16 to 14.8 MYA, poleward transport of warm, saline deep water inhibited polar cooling. This flow apparently then ended, with a major buildup of the East Antarctic ice sheet ensuing from 14.8 to 12.9 MYA. This gave rise to the modern-scale ice sheet, including the first significant West Antarctic ice sheet (WAIS) (Mackensen 2004). The full WAIS developed approximately 6 MYA, subsequently disappeared at times of warming, and then re-formed (Pollard and DeConto 2009).

In the Arctic, mid-Miocene cooling, evidenced by sea ice and iceberg rafting of debris, was synchronous with the expansion of East Antarctic ice (approximately 14.5 MYA) and of Greenland ice (approximately 3.2 MYA). However, there were intervals around 8 and 9.2 MYA when the sea ice could have been seasonal, rather than perennial (Moran et al. 2005).

Based on their analysis of ocean sediments off northeast and southern Greenland, Thiede et al. (2011) propose that the Greenland ice sheet formed in the middle Miocene by coalescence of ice caps and glaciers. There has been a more or less continuous presence of ice covering significant parts of Greenland for the last 18 million years. There was an intensification of glaciation during the Late Pliocene from about 11 to 10 MYA, but then the Greenland ice sheet was greatly reduced in size.

Teschner et al. (2016) report that the first significant continental glaciation of the northern hemisphere occurred during the late Miocene with the development of an ice sheet on southern Greenland (Jansen and Sjøholm 1991). At 3.3 MYA, distinct ice-rafted debris (IRD) peaks have been documented for the Nordic Seas, indicating a pronounced glacial expansion with the Greenland ice sheet as the most important source (Kleiven et al. 2002). During this period reduced sea-surface temperatures in the North Atlantic Ocean of about 2–3 °C have been documented. Major intensification of the northern hemisphere glaciation occurred at 2.7 MYA.

2.6 Plio-Pleistocene Glacial Cycles

Understanding of the behavior of the marine-based West Antarctic ice sheet during the "warmer-than-present" Early Pliocene Epoch (5–3 MYA) has been advanced by Naish et al. (2009). They presented a marine glacial record from a sediment core recovered from beneath the Ross ice shelf by the ANDRILL program. It shows approximately 40-ka cyclic variations in ice sheet extent linked to obliquity cycles in insolation. The WAIS periodically collapsed during the Pliocene, resulting in a switch from grounded ice, or ice shelves, to open waters in the Ross embayment when planetary temperatures were up to 3 °C higher than today and atmospheric CO_2 concentration was up to 400 ppm.

The start of the Pleistocene (meaning "newest") Epoch, which marks the beginning of the Quaternary period, is dated to 2.6 MYA. Ehlers and Gibbard (2004a, 2004b) provide detailed regional accounts of Quaternary glaciations. The climate was characterized by repeated glacial cycles, with the buildup of a major ice sheet in Greenland during the Pliocene Epoch (3 MYA) and later in North America and northwestern Europe. Lunt et al. (2008) used a fully coupled atmosphere–ocean GCM and an ice sheet model to assess the impact of proposed driving mechanisms for glaciation and the influence of orbital variations on the development of the Greenland ice sheet. They found that Greenland glaciation has been mainly controlled by an approximately 200 ppm decrease in atmospheric carbon dioxide during the Late Pliocene. By contrast, the model results suggest that climatic shifts associated with the tectonically driven closure of the

Panama seaway, with the termination of a permanent El Niño state, or with tectonic uplift, are not large enough to contribute significantly to the growth of the Greenland ice sheet. Koenig et al. (2011) used a coupled climate–vegetation–ice sheet model to investigate the climatic sensitivity of Greenland to external forcings and internal feedbacks. They showed that tundra expansion at the expense of boreal forest, in response to the decline of CO_2 concentration from 400 to 200 ppm and decreasing summer insolation, impacted surface energy budgets, amplifying summer cooling and expanding snow cover over Greenland, and allowing the growth of small ice caps.

Global average surface temperatures for the last 2 million years have recently been compiled by Snyder (2016) from a multi-proxy database containing more than 20,000 sea surface temperature point reconstructions. He demonstrated a gradual reduction in global temperature by about 6 °C until roughly 1.2 MYA, when cooling stalled, until the present, except for glacial/interglacial oscillations of approximately ±5 °C. Over the past 800,000 years, polar amplification has been largely stable, and global temperature and atmospheric greenhouse gas concentrations have been closely coupled across glacial cycles. Polar amplification can be estimated as the change in Antarctic temperature for every 1 °C change in global average surface temperature, which is estimated as 1.6 °C per 1 °C unit change. A comparison of the new temperature reconstruction with greenhouse gas (GHG) radiative forcing estimates the total global climate response as a 2.5 °C (1.8–3.6 °C) change in global average surface temperature per 1 W m^{-2} change in GHG radiative forcing. This relationship has not changed significantly over the past 800 ka.

From 2.6 MYA until 0.8 MYA, global climate was characterized by cycles of 41,000-year duration that were driven by the axial tilt of the Earth (the obliquity) varying between 22.1° and 24.5°. The obliquity effect on solar radiation receipts is most marked in high latitudes, giving a 10 percent range between obliquity extremes at latitudes 80–90°. Gibbard and Lewin (2016) note that the start of the Quaternary occurred in the context of progressive Cenozoic cooling. Between 4.5 and 3.1 MYA, a pronounced long-term minimum in the amplitude of the 41-ka obliquity cycle provided sufficiently warm summers to prevent winter ice buildup; in contrast, between 3.1 and 2.5 MYA, increased amplitude in the obliquity cycle appears to have led to repeated cold summers in the northern hemisphere, enabling ice to build up.

Around 1.4 MYA, there was a progressive increase in the amplitude of climate oscillations, increasing long-term average global ice volume, and the establishment of strong asymmetry in global ice-volume cycles. Then, for poorly understood reasons, the cyclicity switched to approximately 100,000 years after 0.8 MYA (see Box 2.1). This periodicity roughly coincides with the timing of the eccentricity

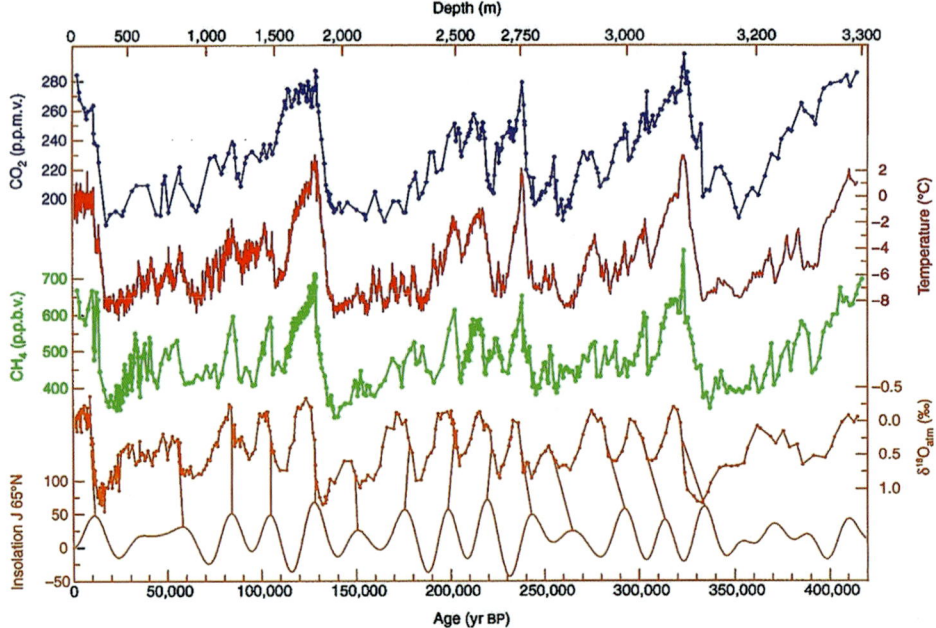

Figure 2.3 The 420-ka ice core record from Vostok, East Antarctica.
Left side: carbon dioxide and methane concentrations and solar insolation at 65° N; right side: $\delta^{18}O$ and temperature.
Source: Wikipedia Commons, https://en.wikipedia.org/wiki/Milankovitch_cycles

of the Earth's orbit around the Sun, but its effect on solar radiation was too small by itself to cause glacial/interglacial cycles. Other processes, particularly ice–albedo feedback, must amplify the small fluctuations in solar energy. Between 0.8 and 0.43 MYA, interglacial episodes have been shown by the isotopic record in the EPICA ice core from Dome C in Antarctica to have been longer and cooler than subsequent interglacials. The interglacial from 428 to 397 ka known as MIS 11 was the longest and gave rise to the disappearance of the Greenland ice sheet and the West Antarctic ice sheet. Figure 2.3 shows the 420-ka record from Vostok.

Interglacial periods before about 430 ka, a time identified as the Mid-Brunhes (geomagnetic) event, featured larger continental ice sheets, lower sea level, cooler conditions in Antarctica, and lower atmospheric CO_2 concentrations, relative to the subsequent interglacials, according to Yin and Berger (2010). Modeling suggests that the more recent interglacials were mainly warmer due to higher global mean temperatures during boreal winters. This warmth was a result of increased insolation in winter relative to the preceding interglacials. When Yin and Berger (2012) sought to determine the contributions of insolation and greenhouse gases to the interglacial climates of the past 800,000 years, they found that greenhouse gases played the dominant role in variations of the annual mean temperature globally and in southern high latitudes, whereas insolation dominated the variations of tree fraction, precipitation, and of northern high latitude temperature and sea ice. The relative importance of the two factors varied between interglacials. For example, the effects reinforced one another during the warmest (MIS 9 and MIS 5) and coolest (MIS 17 and MIS 13) interglacials. The researchers also showed that obliquity is highly linearly correlated with both global and Antarctic temperatures, while precession is important for Arctic temperatures.

Regional-scale glaciations (Cordilleran and continental) started in northwestern Canada and east-central Alaska between 2.9 and 2.6 MYA (Dud-Rodkin et al. 2004, 2010), creating one of the oldest known continental glacial records (Late Pliocene) preserved in stratigraphic sections. Hidy et al. (2013) date the Klondike gravel – a widespread glaciofluvial gravel marking the earliest and most extensive Cordilleran ice sheet (CIS) – to 2.64 MYA based on terrestrial cosmogenic nuclide (TCN) burial ages. This age coincides with the onset of 41-ka obliquity (or axial tilt) cycles. This angle is currently about 23.4°. When the Earth's tilt is large (small), the contrast between the seasons is enhanced (reduced).

According to Bailey et al. (2013), the onset of abundant IRD deposition (see Box 2.3) in the Nordic Seas and subpolar North Atlantic Ocean around 2.72 MYA

Box 2.3 Ice-Rafted Debris (IRD) and Heinrich Events

Ice-rafted debris (IRD) layers were first identified in the northern Labrador Sea in a 1928 US Coast Guard expedition by the *Marion*; subsequently, in 1940, glacial sediments and erratics were traced across the North Atlantic by the US Geological Survey (Kuipers et al. 2014). Ruddiman (1977) followed up on this work to study paths of ice dispersal in the subpolar North Atlantic. A major milestone was an analysis by Henrich (1988) of cyclical ice rafting. The events occurred approximately 7,000 years apart, and each is thought to have lasted about 500 years.

The mechanism responsible for these layers has not been firmly established. It has been proposed that the ice sheet underwent "binge and purge" cycles of ice sheet buildup and collapse, that massive floods occurred from a glacially dammed lake in Hudson Bay, or that an ice shelf fringing the ice sheet suddenly collapsed as in the Antarctic Peninsula in 2002 (Andrews 2000). Bassis et al. (2017) propose that the magnitude and timing of Heinrich events can be explained by the same processes that drive the retreat of modern marine-terminating glaciers. Subsurface ocean warming associated with variations in the meridional overturning circulation increases underwater melt along the calving face, triggering rapid margin retreat and increased iceberg discharge. On millennial time scales, isostatic adjustment causes the bed to uplift, isolating the terminus from subsurface warming and allowing the ice sheet to advance again.

records the Pliocene onset of major northern hemisphere glaciation in Greenland and Scandinavia. However, the Laurentide ice sheet in North America did not extend to 39° N until approximately 2.4 MYA and the grounding line of continental ice on northeastern North America may not have extended onto the continental shelf during glacial intervals until 2.64 MYA.

During the Early Pleistocene, three Cordilleran glaciations occurred in the Yukon–Alaska, while one to five continental glaciations (Keewatin ice sheet and Horton ice cap) are inferred from the Banks Island stratigraphic record (Dud-Rodkin et al. 2004). Three Middle Pleistocene glaciations are recorded for the Cordillera as well as three continental (Keewatin ice sheet and Horton ice cap) events. During the Late Pleistocene, a well-defined, extensive Keewatin ice sheet covered western and northwestern Canada, while in the Yukon Cordillera and Yukon–Tanana Uplands, two glaciations are recognized. Successive Cordilleran glaciations diminished in size, while continental glaciations increased.

In Missouri, the first recorded advance of the Laurentide ice sheet reached 39° N, near the extreme southern limit of North American glaciation at 2.4 MYA, thus postdating the Cordilleran ice sheet. The next advance to this latitude took place near the beginning of the mid-Pleistocene transition, 1.3 MYA, and three more advances took place from 0.75 to 0.2 MYA (Balco and Rovey 2010).

The glacial cycles of the late Pliocene to early Pleistocene (from about 3 to 1 MYA) were marked by a regular 41,000-year cycle that reflects the obliquity (or axial tilt) in the Earth's orbit (see Box 2.2). Precession of the equinoxes, which occurs mainly at 23,000- and 19,000-year intervals, is the orbital component that most influences summer insolation intensity (Raymo and Huybers 2008). It is clearly identified in ice volume and sea level records for the last 700 ka, but is strangely absent in the late Pliocene and early Pleistocene. Raymo and Huybers (2008) point out that the effect of precession on summer insolation intensity is out of phase between hemispheres, whereas obliquity is in phase. Hence, precession-paced changes in ice volume in each hemisphere would cancel out in globally integrated proxies such as ocean ^{18}O or sea level. Field evidence is needed to test this hypothesis.

The mid-Pleistocene transition from 41-ka obliquity cycles to 100-ka orbital cycles, for unknown reasons, occurred around 0.8 MYA. The eccentricity of the Earth's orbit actually varies at periods of 413,000 and approximately 100,000 years, but the latter appears to regulate the last eight glacial cycles. Increased eccentricity leads to increased changes in the seasons. When the eccentricity is at its maximum (the orbit is most elliptical), the amount of incoming solar radiation at perihelion (when the Earth is closest to the Sun) will be about 23 percent more than at aphelion (when the Earth is farthest from the Sun). The timing of perihelion and aphelion is determined by the precession of the vernal equinox, which has an approximately 23,500-year period. Currently, perihelion occurs on January 3 and aphelion is July 4.

The phase relation of the 100,000-year cycle in ice volume to orbital eccentricity, temperature, and carbon dioxide concentration was investigated by Shackleton (2000). The ice volume component of the $\delta^{18}O$ signal in deep-sea sediments was separated from the deep-water temperature signal by using the record of $\delta^{18}O$ in atmospheric oxygen trapped in Antarctic ice at Vostok. At the 100,000-year period, atmospheric carbon dioxide, Vostok air temperature, and deep-water temperature are in phase with orbital eccentricity, whereas ice volume lags these three variables. Hence, the 100,000-year cycle does not arise from ice sheet dynamics, but rather the response of the global carbon cycle likely generates the eccentricity signal by causing changes in atmospheric carbon dioxide concentration.

Raymo (1997) first suggested that the excess ice that is characteristic of late Quaternary "100-ka" climate cycles typically accumulates when July insolation at 65° N has been unusually low for more than a full precessional cycle, or more than 21 ka. Once established, however, it does not last beyond the next increase in summer insolation. Thus, the timing of the growth and decay of large 100-ka ice sheets is strongly influenced by eccentricity – namely, through its modulation of the orbital precession component of northern hemisphere summer insolation.

The paradox of the weak variation of insolation associated with eccentricity, despite the presence of the approximately 100-ka variation in ice extent, has recently been addressed by Cheng et al. (2016) using a 640,000-year CO_2 record from stalagmites in central China. They show that the seven terminations (TI–TVII) in the record occurred during the rising limbs of northern hemisphere summer insolation, separated by 4, 5, 5, 4, 5, and 5 precession cycles, respectively. The durations between successive terminations were about 93, 105, 92, 92, 113 and 115 ka, respectively, rather than strict 100-ka cycles. Thus, the "100-ka cycle" is an approximate average of intervals that are generally a little longer or shorter than 100 ka.

Lee et al. (2017) add a further suggestion for the transition from 40- to 100-ka cycles. They demonstrate, using climate model simulations, that the pace of the glacial cycle depends on the asymmetric pattern of hemispheric sea ice growth. In a cold climate, the sea ice grows asymmetrically between two hemispheres under changes to Earth's orbital precession, because sea ice growth potential outside of the Arctic Circle is limited. In an environment cooler than today, Antarctic sea ice can expand at all longitudes, whereas only about one sixth to one twelfth of Arctic longitudes are open to growing sea ice. This difference in hemispheric sea ice growth leads to an asymmetry in absorbed solar energy for the two hemispheres, particularly when eccentricity is high. In the past 1 MYA, the dominant glacial cycle has a 100-ka periodicity because sea ice grows asymmetrically between the two hemispheres in response to the precessional cycle, particularly when eccentricity is high, modulating absorbed solar energy. When the southern hemisphere receives less solar energy in winter, increased sea ice should trigger positive feedbacks, amplifying the initial cooling.

Tzedakis et al. (2017) have developed a rule to determine which insolation peaks led to an interglacial in the Quaternary period. They first distinguish between interglacials and interstadials. The former are intervals when there is little ice in the northern hemisphere outside Greenland. They also identify "continued interglacials," in which an isotopic minimum contains two intergla-cials. The insolation metric is the "caloric summer half-year insolation" calculated by Milankovitch – that is, the amount of energy integrated over the time interval defined as any day of the summer half that receives more insolation than any day of the winter half. In high latitudes, the insolation metric combines nearly equally the effects of precession and obliquity. Before 1 MYA, 36 caloric peaks out of 67 were interglacials, 7 continued interglacials received greater than $5.945\ GJ\ m^{-2}$, and 21 associated with interstadials were below that threshold. After 1 MYA, no interglacials occurred below a higher threshold of $5.979\ GJ\ m^{-2}$. Tzedakis et al. propose that the threshold decreases with the amount of elapsed time since the onset of the previous interglacial; in other words, an ice sheet becomes more sensitive with time to insolation received. They conclude that a gradual rise in the deglaciation threshold started about 1.55 MYA, which led to the skipping of 12 out of 25 caloric summer insolation peaks of above-average obliquity after 1 MYA.

Recent work by Jakobsson et al. (2016) has revived suggestions made by Mercer (1970) and Hughes et al. (1977) that the central Arctic Ocean (approxi-mately 4 million km^2) was occupied by a 1,000-m-thick ice shelf based on sea floor evidence of grooving and lineations on the Lomonosov Ridge. This ice shelf is now dated to the penultimate glaciation of 160–140 ka (MIS 6) rather than the last (Würm or Weichselian) glaciation. The shelf was fed by ice streams originat-ing from the Laurentide and Kara–Barents ice sheets.

The last interglacial event (the Eemian) is dated to 130–118 ka (MIS 5e) and is notable because global temperatures rose by approximately 1 °C and sea level was 5.5–9 m above the present level, implying significant loss of land ice (Dutton and Lambeck 2012). They consider that a 2–4 m contribution came from Green-land and a maximum contribution of 3.3 m from West Antarctica. Hence, the upper estimate would require additional ice loss from East Antarctica. Proxy data summarized by Otto-Bliesner et al. (2006) and model simulations indicate summer temperature increases of 2–4 °C occurred in northeastern Canada, the seas adjacent to northern and eastern Greenland, northern Europe, and Siberia. Forests moved northward, reaching North Cape, Norway (71.1° N), and St. Lawrence Island in the Bering Strait, and trees grew in southern Baffin Island.

Conditions of the Laurentide ice sheet in the James Bay and Hudson Bay lowlands (HBL) since 130 ka were analyzed by Andrews et al. (1983). Using amino acid dating of shells from tills and marine and fluvial sediments, they identified ice-free periods about 35, 75, and 100 ka between the last interglacial Missinaibi (at 130 ka) and the Tyrell Sea (at 8 ka), rather than a stable ice sheet. These conclusions have recently been reviewed by Dalton et al. (2016). Data from the Missinaibi formation suggest that an ice-free HBL may have existed during parts of MIS 7; circa 243,000–190,000 yr BP), MIS 5 (circa 130,000–71,000 yr BP), and MIS 3 (circa 57,000–29,000 yr BP). While MIS 7 and MIS 5 are well-documented interglacial periods, the development of peat, forest bed, and fluvial deposits dating to MIS 3 suggests that the Laurentide ice sheet retreated and remained beyond, or somewhere within, the boundaries of the HBL during this interstadial.

At the Last Glacial Maximum (LGM), ice covered approximately 65 percent more area than at present (Clark et al. 2009). The Laurentide ice sheet at the LGM in North America has been described by Dyke et al. (2002). Ice advanced to its Late Wisconsinan (MIS 2) limit in the northwest, south, and northeast about 23–24 ^{14}C ka BP and in the southwest and far north about 20–21 ^{14}C ka BP. Figure 2.4 shows the margins and thickness contours. They state that major centers of outflow were located over Quebec–Labrador, Keewatin, and Foxe Basin, forming a system surrounding Hudson Bay, where the ice surface was generally lower. The period of maximum ice extent generally spanned the interval from approximately 24/21 to 14 ^{14}C ka BP.

Lambeck et al. (2017) use observations of sea level and crustal response to glacial loading cycles to provide constraints on the mantle rheology as well as on ice loading for the late Wisconsin ice sheet over North America. They model two principal domes, located over southern Nunavut (the Keewatin dome) and over Quebec–Labrador, both of 3,500 m thickness, separated by an ice ridge that was 1,500 m lower.

In the Canadian Arctic during the Last Glacial Maximum (25–18 ka), the Innuitian ice sheet covered the Queen Elizabeth Islands (QEI), north of the massive Laurentide ice sheet (England et al. 2006). While there is still some debate, sea level evidence suggests that it was relatively thick and extensive. It comprised an alpine sector in the northeast and a lowland sector in the central and western QEI. Most of the ice sheet remained on the continental shelf during the Younger Dryas. The delayed buildup of the ice sheet until after 20 ka BP was out of phase with the growth of the Laurentide ice sheet, which had attained its maximum extent between 24 and 20 ka BP. It is hypothesized that growth of the Laurentide ice sheet culminated in a split jet stream that temporarily favored augmented precipitation and growth of the Innuitian ice sheet to its north.

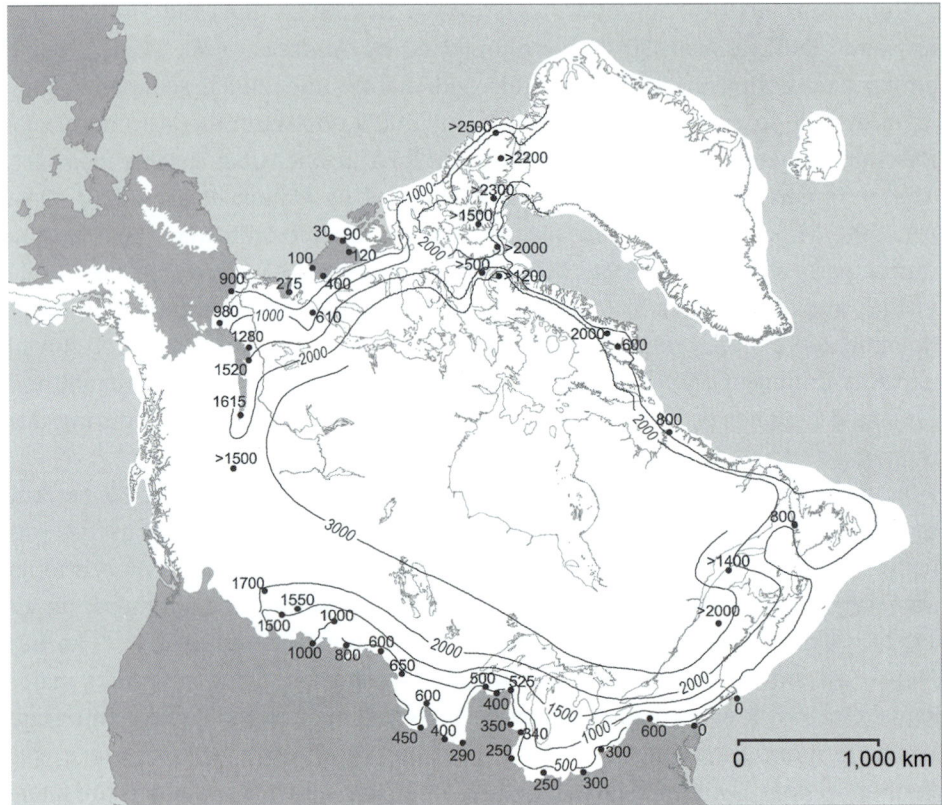

Figure 2.4 Ice surface contours based on elevations along the Last Glacial Maximum ice margin of the Laurentide ice sheet and topographic high points overridden by ice.
Source: Dyke et al. 2002, 21, figure 4. Courtesy of Elsevier.

Dyke (2004) has reviewed the deglaciation of central and northern Canada during the late Wisconsin, providing maps at 500-year intervals from 18 to 5 ^{14}C ka BP (21.4 to 5.75 cal ka BP). Radiocarbon dates increase in frequency from 14 to 10 ka BP, reflecting the gradual increase in the deglaciated area. A corridor between the Cordilleran and Laurentide ice sheets is considered to have opened between 12.5 and 12 ka BP.

The Younger Dryas cold interval (11–9.6 ka BP) was associated with extensive moraine formation in many sectors of the Laurentide ice sheet, including the northwest margin on the Arctic mainland (68.5° N, 119.7° W). There was also a readvance of Labrador ice across the mouth of Hudson Strait and outer Frobisher Bay on Baffin Island. Most of the Arctic archipelago was deglaciated after the Younger Dryas.

Ice recession accelerated after 10 ka BP, such that the Laurentide and Innuitian ice sheets were separated by 9 ka BP. By 8.6 ka BP, the latter had fragmented into island ice caps in the eastern archipelago, but the ice remained confluent with the Greenland ice sheet until about 7.75 ka BP. The youngest moraines of Keewatin ice on the mainland are from readvances dated to 8.2–8.1 ka BP, but the ice's recession is undated (Dyke 2004). Nevertheless, 8.2 cal ka BP was a notable cold event. Baffin sector ice, which spread radially out of Foxe Basin, had a near-maximum configuration until 10 ka BP. Extensive end moraines were constructed around much of the Baffin sector ice margin between 8.5 and 7 ka BP. During 7–6 ka BP, a calving bay moved northward from Hudson Bay, leading to the breakup of the Foxe dome. This left residual ice caps in Arctic Quebec, Southampton Island, Melville Peninsula, and Baffin Island.

The major glaciations of Scandinavia and the Barents Sea–Svalbard area started around 2.5–2.8 MYA, according to Mangerud et al. (1996). For most of the time until 0.9 MYA, the ice sheets were of intermediate size. Recent work by Pope et al. (2016) has documented the Barents Sea ice sheet over the last 140 ka. Ice advanced to the shelf edge during four distinct periods. By far the largest sediment volumes were delivered during the oldest Saalian advance, more than 128,000 years ago. Later Weichselian advances occurred from 68,000 to 60,000, 39,400 to 36,000, and 26,000 to 20,900 BP. The configuration of the ice sheet was also different between the two glacial periods, implying that the ice feeding the Bear Island ice stream came predominantly from Scandinavia during the Saalian, while it drained more ice from east of Svalbard during the Weichselian. The last glacial cycle is the best documented. In Eurasia, there was a marine-based ice sheet in the Barents–Kara Sea during the mid-Weichselian (last glaciation), designated as MIS 4/3, 74–25 ka. Svendsen et al. (1999) have shown that the maximum Weichselian ice sheet extent in the Barents and Kara Sea region occurred during the Early/Middle Weichselian (Figure 2.5). In the Yenisei basin, the ice extended 800 km south of the coastline. The extensive mid-Weichselian ice sheet blocked the Arctic-draining rivers, forming massive proglacial lakes over northwest Russia. This ice sheet collapsed before 40 ka. During the Late Weichselian (25,000–10,000 yr BP), much of the Russian Arctic remained ice-free, with localized ice caps on the Arctic islands, polar Urals, and Putorana Plateau (Astakhov 2004). A shelf-based ice sheet covered the Barents Sea and coalesced with the Scandinavian ice sheet, which reached its maximum extent at that time.

The Eurasian ice sheet complex (EISC) during the LGM approximately 23–19 ka BP comprised domes over the British Isles, Fennoscandinavia, and the Barents–Kara Seas, with a volume equivalent to 20 m of sea level (Patton et al. 2017) (see Figure 2.6). Patton et al. provided maps of ice extent at intervals from

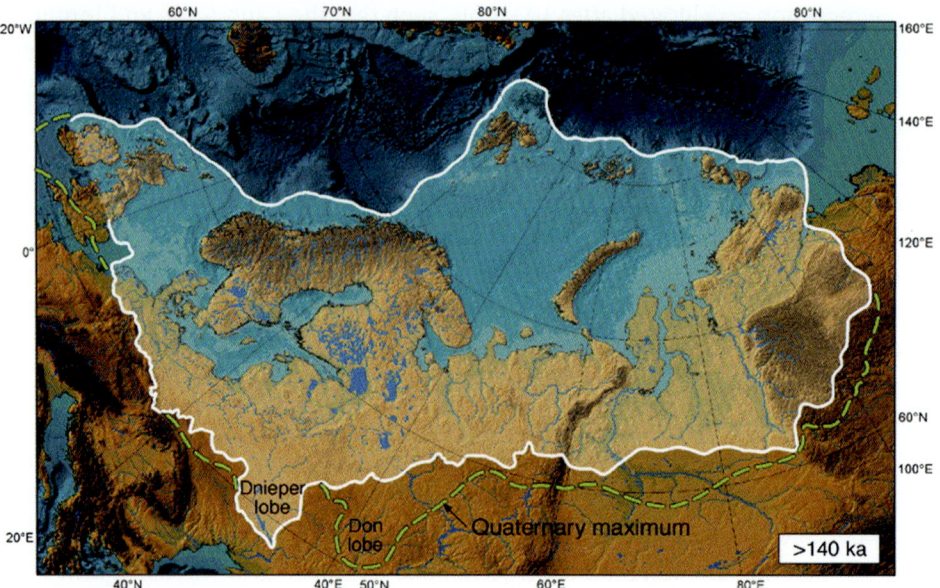

Figure 2.5 A reconstruction of the maximum ice sheet extent in Eurasia during the Late Saalian (circa 160–140 ka), based on review of published material. The approximate maximum extent of the Quaternary glaciations (drift limit) is indicated by a dotted line. Source: Svendsen et al. 2004, 1252, figure 13. Courtesy of Elsevier.

20.4 to 10 ka BP. They also showed that during glacial times, the Channel/Fleuve Manche palaeoriver routed substantial volumes of water to the North Atlantic from a catchment of 2.56 million km^2, half of it ice covered, incorporating drainage from the Seine, Rhine, Thames, Elbe, and Vistula.

The deglaciation of Fennoscandinavia has recently been documented between 22 and 13 cal ka BP, and between 11.6 and 9.7 cal ka BP, by Stroeven et al. (2016), based on 335 radiocarbon, 794 cosmogenic nuclide (CN), and 138 optically simulated luminescence (OSL) dates, as well as late- and post-glacial varve chronologies. Here, the changes are discussed for the western, eastern, southern, and central sectors of the ice sheet.

The Norwegian shelf was deglaciated between the local LGM and 14–15 cal ka BP. The more maritime setting of the western margin, and its proximity to an Atlantic moisture supply, may explain why Younger Dryas readvances were more extensive on the western margin than elsewhere. The local LGM was attained as early as 19 cal ka BP in the southern part of the eastern sector. From its local LGM extent, the ice margin retreated into a depression spanning the entire length from northern Germany to Lake Onega in western Russia. This led to the formation of a number of ice-dammed lakes in Poland, Lithuania, Latvia, Estonia, and northwest Russia,

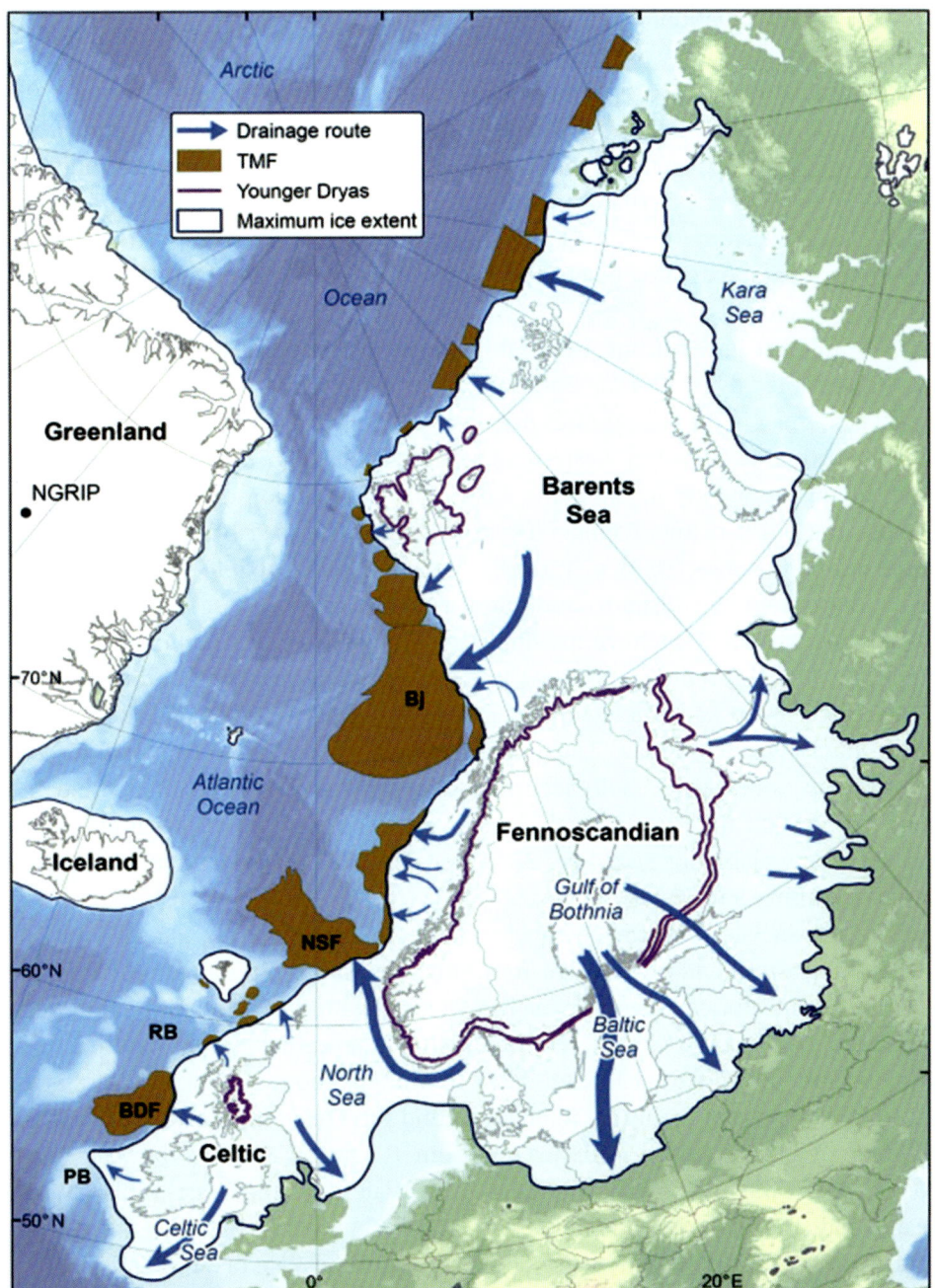

Figure 2.6 Major drainage routes of the Eurasian ice sheet complex. PB: Porcupine Bank; BDF: Barra and Donegal Fans; RB: Rosemary Bank; NSF: North Sea Fan; Bj: Bjørnøyrenna Fan. TMF denotes trough mouth fans.
Source: Patton et al. 2017, 150, figure 1. Courtesy of Elsevier.

and later the formation of the Baltic Ice Lake that persisted until the end of the Younger Dryas, at approximately 11,620 cal yr BP. The LGM extent of the ice sheet in northwestern Russia was located farther east and peaked later than the rest of the ice sheet perimeter, at around 17–15 cal ka BP. There were three major lobes located in the Dvina, Vologda, and Rybinsk basins. Fast-flowing, low-gradient ice lobes expanded into the exceptionally wide valleys and basins of the northwestern Russian Plain. These three major ice lobes at the eastern ice sheet margin extended beyond the major basins (the White Sea, Lake Onega, and Lake Ladoga).

The uniform retreat from the local LGM configuration in the southern sector was interrupted by two Young Baltic advances in Denmark. There is a remarkable spatial contrast in the retreat rates between 16 and 14 cal ka BP: A slow terrestrial margin retreat occurred in western and southern Sweden, and a rapid retreat of the calving ice margin in the Baltic Basin. The deglaciation of southern Sweden before the onset of the Younger Dryas was characterized by a predominantly terrestrial and slow (less than 150 m yr^{-1}) ice marginal retreat that was repeatedly interrupted by stillstands and minor readvances.

Hughes et al. (2016) have assembled a time-slice reconstruction of British–Irish, Svalbard–Barents–Kara Seas, and Scandinavian ice sheets during the last glaciation (40–10 ka). Their work demonstrates that the advances and retreats were highly asynchronous. The westernmost limit along the British–Irish and Norwegian continental shelf occurred at approximately 27–26 ka, while the eastern limit on the Russian Plain was at approximately 20–19 ka. Maps of ice extent are available for 38–34 ka, 32–30 ka, 29–28 ka, 27 ka, and at 10,000-year intervals from 25 to 10 ka.

The British Isles experience severe arctic conditions during glacial intervals when the eastern end of the polar front migrates south to the Iberian Peninsula. During such historical intervals, sea level lowering linked Britain to the European continent, enhancing climatic continentality. The earliest evidence from East Anglia is the Baventian glaciation dated to MIS 68, about 1.86 MYA (Clark et al. 2004). There was probably ice in Scotland and the Welsh mountains at this time. Numerous glacial episodes followed – from 1.7 to 0.9 MYA to the Anglian at 0.45 MYA (MIS 12), based on erratics from Wales and the West Midlands and sand and gravel deposits in the River Thames terraces (Bowen et al. 2002). During the Anglian glaciation, ice advanced from the Vale of York and Lincolnshire into west and central East Anglia. Ice from centers in northern Britain moved south along the lowlands of eastern and western England. Glaciation of south Wales occurred from local ice sheets, and ice from Scandinavia reached eastern England from County Durham to East Anglia. The maximum limits of the LGM Devensian ice (22–18 ka BP) in Britain remain uncertain at locations in

Wales and Scotland. In part, this reflects the fact that the maxima occurred at different times in different places. Recent studies (Bowen et al. 2002) indicate that the most extensive Devensian glacial advance occurred about 40 ka (Henrich event 4) when all of Scotland and Ireland were covered and British ice was confluent with Scandinavian ice in the North Sea. The British ice sheet reached its maximum extent around 22 cal ka, just after Heinrich event 2 (see Box 2.3). This was followed by extensive deglaciation at 17.4 cal ka. The map developed by Bowen et al. (Figure 2.7) is a closer match to the limited extent model of Boulton et al. (1991).

In the Russian Far East, Wennrich et al. (2016) report periods of exceptional warmth in the Pleistocene record of Lake El'gygytgyn, with dense boreal forests around, and peaks of primary production in, the lake. These are assigned to so-called super-interglacial periods dated to circa 2.38 Ma, 2.34 Ma, 2.03 Ma, 1.60 Ma, 1.07 Ma, and 420 ka, respectively. For two of these intervals, MIS 31 and 11c, the warmest and wettest conditions in the entire Pleistocene record of Lake El'gygytgyn have been inferred, reaching 4–5 °C higher in summer and with approximately 300 mm higher annual precipitation compared with the Holocene thermal maximum. The occurrence of these super-interglacials corresponds well to collapses of the West Antarctic ice sheet recorded in ice-free periods in the ANDRILL core.

The retreat phases of the Laurentide glaciation were marked by a sequence of sedimentation events in the North Atlantic that appear to have originated at a calving front in Hudson Strait. Heinrich (1988) originally identified six layers in ocean sediment cores with extremely high proportions of rocks of continental origin. These were interpreted as having been carried by icebergs or sea ice, which broke off ice streams or ice shelves and then dumped the rocks on the sea floor as the icebergs melted. The origin of these debris layers was mostly the Laurentide ice sheet then covering North America for Heinrich events 1 (16.8 ka), 2 (24 ka), 4(38 ka), and 5 (45 ka), and probably the European ice sheets for the minor events 3 (31 ka) and 6 (60 ka) (Hemming 2004).

During the last glacial cycle (115–15 ka), the climates of the two polar regions were slightly out of phase. Arctic temperatures led Antarctic ones by a quarter of a period, leading to a bipolar seesaw on a millennial time scale. The Antarctic warms while Greenland is cold; the Antarctic then cools as Greenland warms rapidly. This linkage of the two polar regions is attributed to the inter-hemispheric ocean thermohaline circulation. Greenland ice cores reveal that there were twenty-five major climatic oscillations (Dansgaard–Oeschger oscillations) during the last glacial cycle that were correlated with iceberg discharge events in the North Atlantic (Bond and Lotti 1995; Dansgaard et al. 1982). The oscillations were characterized by rapid warming followed by a gradual cooling, with each event over the last 50,000 years averaging approximately

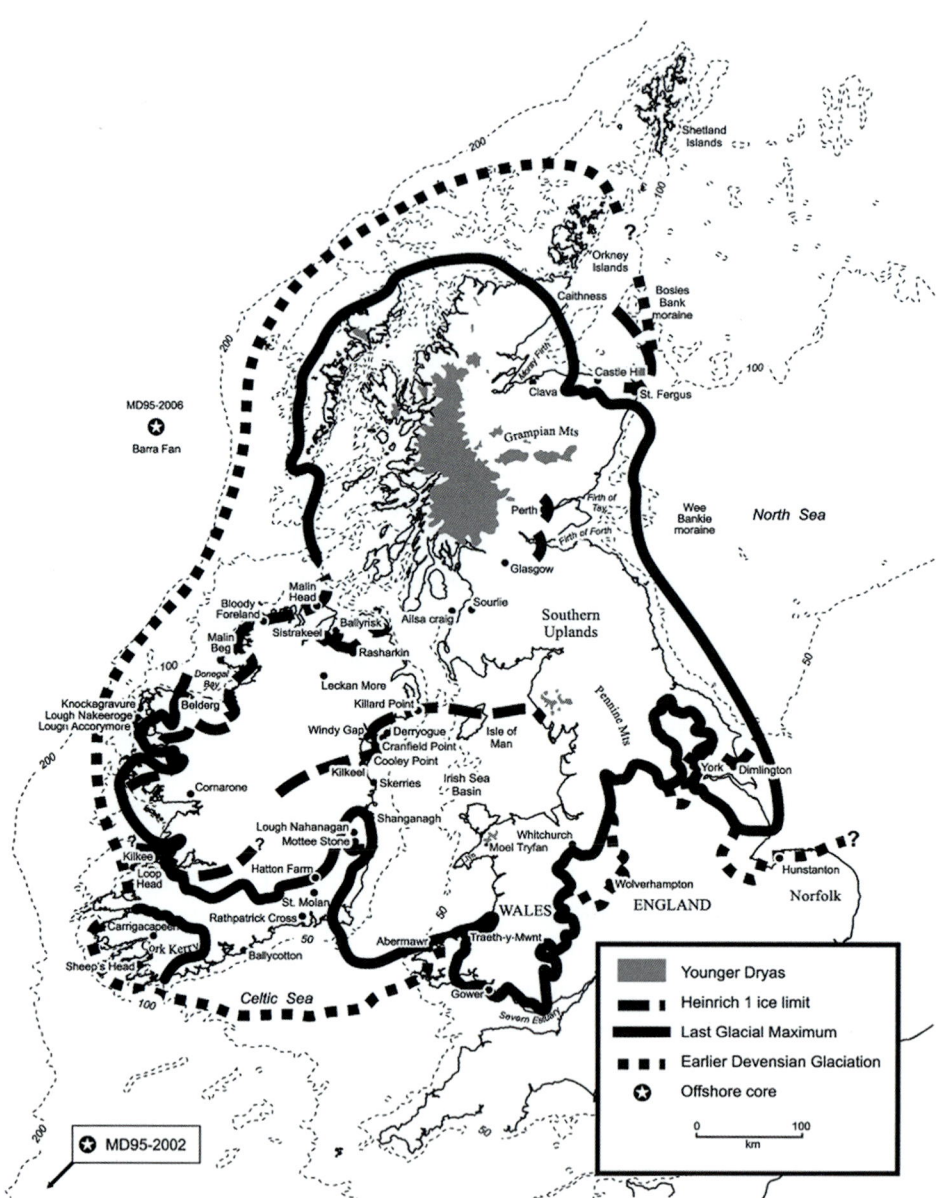

Figure 2.7 The Devensian ice limits in the British Isles.
Source: Bowen et al. 2002, 91, figure 1. Courtesy of Elsevier.

1,470 years ± 20 percent, according to Schultz (2002). However, Ditlevsen et al. (2007) demonstrated using the well-dated NGRIP core that the oscillations were essentially random. They were probably caused by influxes of fresh water into the North Atlantic related to "binge–purge" cycles in the mass of the Laurentide ice sheet over North America. Henry et al. (2016) have shown that changes in the Atlantic Meridional Overturning Circulation (AMOC) were associated with each of thirteen stadial/interstadial oscillations during MIS 3 between 60 and 25 ka. Moreover, both SST and Greenland temperature proxies lag the ocean circulation in a consistent fashion by approximately 200 years and, in turn, these northern changes have been demonstrated to lead Antarctic temperatures by a similar amount (WAIS Divide Project Members 2015).

Gregoire et al. (2015) have modeled what drove northern hemisphere ice sheet melt during the last deglaciation (21–7 ka). Their work shows that by 9 ka, orbital forcing caused 50 percent of the deglaciation, greenhouse gases 30 percent, and the interaction between the two 20 percent.

The retreat of the Greenland ice sheet has recently been analyzed using 673 [10]Be and 791 ^{14}C dates (Sinclair et al. 2016). This work indicates that following initial retreat of the marine margins from the continental shelf, most land-based deglaciation occurred following the end of the Younger Dryas cold period (12.9–11.7 ka). However, deglaciation in east Greenland peaked significantly earlier (13.0–11.5 ka) than that in south Greenland (11.0–10 ka) or west Greenland (10.5–7.0 ka). The terrestrial deglaciation of east and south Greenland coincided with adjacent ocean warming; peak ocean warmth occurred between approximately 12 and 9.5 ka.

Nevertheless, Jennings et al. (2017) have shown that in central West Greenland, the ice sheet first retreated from the continental shelf margin beginning around 17.1 cal ka BP, associated with subsurface warm Atlantic water. Calving retreat was delayed there until 15.3 cal ka BP, again linked to warm ocean water. The GrIS reached a smaller-than-present extent in the early to middle Holocene.

In the Antarctic, considerable attention has been given to the West Antarctic ice sheet (WAIS) in view of its potential for runaway retreat and the fact that its ice volume contains the equivalent of about 5 m of global sea level. It is currently buttressed by fringing ice shelves, but grounding lines are several hundred meters below sea level and the bed deepens toward the interior. Pollard and DeConto (2009) employed a combined ice sheet/ice shelf model, capable of high resolution, nested with a new treatment of grounding-line dynamics and ice-shelf buttressing to simulate Antarctic ice sheet variations over the past five million years. During this interval, there were essentially no changes in the East Antarctic ice sheet (EAIS), but the WAIS showed substantial variations. These range from full glacial extent with grounding lines near the continental shelf break, intermediate states similar to modern states, and brief but dramatic retreats, leaving only small, isolated ice caps on West Antarctic islands. Transitions between glacial, intermediate and collapsed states are relatively rapid, taking one to several thousand years. The simulation is in good agreement with a sediment record (ANDRILL AND-1B) from beneath the Ross Ice Shelf in the western Ross Sea (Naish et al. 2009). It indicates a long-term trend from more frequently collapsed states to more glaciated states, dominant 40-kyr obliquity cycles in the Pliocene (5–3 Ma), and major retreats at MIS 31 (approximately 1.07 Ma) and during other super-interglacials.

Ice sheet retreat phases can have important effects on sea level and ice sheet stability. Jakobsson et al. (2014) point out that mass loss of ice leads to weakening of the gravitational force on the adjacent ocean surface, causing a lowering of sea level and potentially stabilizing the ice sheet margin. This effect can also be enhanced by the sea level lowering due to glacio-isostatic rebound.

Sediment core evidence consisting of diatoms indicates that during the LGM, Antarctic sea ice was twice its current extent in winter and also increased in extent in summer in at least some ocean sectors (Turner et al. 2014). In the Scotia Sea, there is evidence that winter sea ice reached its maximum extent about 24,000 yr BP, and summer sea ice reached its peak between 30,800 and 23,500 yr BP (Collins et al. 2012). The polar front and the sub-Antarctic front both shifted to the north during the LGM by between 2° and 10° latitude from their present locations.

The temperature rise that accompanied the deglaciation of West Antarctica is surveyed by Cuffey et al. (2016). Based on a deep bore hole and ice core data, they infer that the deglacial warming was 11.3 ± 1.8 °C, approximately two to three times the global average (see the later discussion of polar amplification). It was mostly completed by 15 ka BP, several millennia earlier than in the northern hemisphere.

2.7 Holocene

The start of the post-glacial Holocene (meaning "wholly recent") period is dated to 11.7 ka. Glaciers worldwide retreated from 11 to 5 ka as a result of precession-based regional warmth. In the Arctic, the summer temperature anomaly during the Holocene Thermal Maximum around 8 ka is estimated to have been 1.7 ± 0.8 °C due to polar amplification (Serreze and Barry 2011), while for the northern hemisphere the rise was only 0.5 ± 0.3 °C compared with today (Miller et al. 2010a). Miller et al. (2010b) demonstrated that for both cold and warm climates – the Last Glacial Maximum (approximately 20 ka), Holocene Thermal Maximum (approximately 8 ka), Last Interglaciation (130–125 ka), and the Middle Pliocene (3.5 MYA) – summer temperature changes were amplified three to four times in the Arctic above the global average.

In the northern hemisphere, the massive ice sheets delayed the Holocene warming. The Fennoscandian ice sheet disappeared approximately 9 ka and the

Laurentide ice sheet finally melted away approximately 6.5 ka in northern Labrador–Ungava. Ulman et al. (2016) use [10]Be dating of surface exposure ages to show that the final Labrador dome deposited moraines during North Atlantic cold events at approximately 10.3, 9.3, and 8.2 ka, suggesting that these regional climate events helped stabilize the retreating Labrador dome in the early Holocene. After Hudson Bay became seasonally ice free at 8.2 ka, the majority of the Laurentide ice sheet melted abruptly within a few centuries. Its last surviving remnant is the present-day Barnes ice cap in Baffin Island.

In Arctic Canada and Greenland, temperature-sensitive records indicate more consistent and earlier Holocene warmth in the north and east, and a more diffuse and later Holocene thermal maximum in the south and west (Briner et al. 2016). From a core in the Agassiz ice cap, Ellesmere Island, Lecavalier et al. (2017) determined a peak temperature of 6.1 °C at 10 ka and values that regularly exceeded present ones for around 3,000 years. The temperature decreases from the warmest to the coolest periods of the Holocene were 3.0 ± 1.0 °C on average. The Greenland ice sheet retreated to its minimum extent between 5 and 3 ka, consistent with many sites from around Greenland depicting a switch from warm to cool conditions around that time. The spatial pattern of temperature change through the Holocene was likely driven by the decrease in northern latitude summer insolation through the Holocene, the varied influence of waning ice sheets in the early Holocene, and the variable influx of Atlantic water into the study region.

Beginning around 5 ka, orbitally forced cooling led to glacier readvances, which occurred during five or six time intervals with a general similarity in millennial trends in high and middle latitudes of the northern hemisphere (Solomina et al. 2015). The intermittent readvances, despite smooth orbital forcing, point to local thresholds being crossed as a result of other factors (solar variations, volcanic eruptions, ice–albedo feedbacks, ocean cooling from ice sheet meltwater inputs, and ocean–atmosphere variability). Eight out of nine advances of the last 4.5 ka closely correspond to cooling episodes in the North Atlantic forced by negative anomalies of total solar irradiance of multi-decadal duration (Renssen et al. 2006).

In Alaska, advances are documented about 4.5–4.0 ka, 3.3–2.9 ka, 2.2–2.0 ka, 550–720 CE, and in the last millennium spanning 1180–1320, 1540–1710, and 1810–1880 CE. The last two intervals represent the later part of the Little Ice Age (LIA), considered to span 1300–1850 CE. In Baffin Island two distinct intervals of ice-margin variability were identified: (1) minor glacier advances during the early Holocene that punctuated overall glacier retreat and (2) a series of glacier advances in the Neoglacial (3.5–2.5 ka), culminating in the abrupt onset of the LIA about 1275–1300 CE and intensification and readvance about 1430–1455 CE

Box 2.4 **Lichenometry**

The dating of moraines by measuring lichen thalli was pioneered by Roland Beschel in West Greenland (Beschel 1961). The method was taken up by Andrews and Webber (1964) near the northwest margin of the Barnes ice cap. Thousands of measurements were made of lichen and the percentage of lichen cover during the 1963 field season of the Geographical Branch, Department of Energy, Mines and Resources, Canada (Ives 2016, 188–9). The main species used were the slow-growing, yellow-green, crustose *Rhizocarpon geographicum* (0.064 mm a^{-1} up to approximately 50 mm) and the black, foliose-fructose *Pseudephebe (Alectoria) miniscula* (0.4 mm a^{-1} up to ~150 mm). Later work in the same area supported those results (Andrews and Barnett 1979). A manual on the technique was published by Locke et al. (1979).

(Miller et al. 2012). Records of ice-cap growth from Arctic Canada and Iceland show that cold summers and ice growth began abruptly between 1275 and 1300 CE and this was followed by an intensification around 1430–1455 CE. These intervals coincide with two of the most volcanically active half-centuries of the past millennium. A climate model simulation showed that explosive volcanism produces abrupt summer cooling, and that cold summers can be maintained by sea-ice/ocean feedbacks after the volcanic aerosols have been removed.

On forty-three outlet and cirque glacier moraines in Cumberland Peninsula, Baffin Island, Miller (1973) licheno-metrically (see Box 2.4) dated the earliest Neoglacial advance to 3200 ± 600 BP. Subsequent advances ended just prior to 1650, 780, 350, and 65 yr BP. The most recent of these marked the maximum Neoglacial ice coverage. It appears that the Barnes ice cap persisted through the entire Holocene, which is surprising in view of the summer warmth in the mid-Holocene (Thomas et al. 2010).

Around the Greenland ice sheet, local glacier advances that occurred between 1200 and 1940 CE were the most extensive since the early Holocene deglaciation (Kelly and Lowell 2009). In western Greenland, the Jakobshavn Isbræ retreated rapidly after 8 ka, most likely due to warmer summers. The glacier remained behind its present margin for about 7.0 ka, but between 1500 and 1800 CE it advanced at least 2–4 km as a response to cooling during the LIA. In western Spitsbergen, there were no glaciers from 11.3 to 5.0 ka, but then glaciers started to form, with advances at 3.0, 2.4–2.5, and 1.4–1.5 ka, and in the LIA; the maximum extent of advances occurred in the nineteenth century (Svendsen and Mangerud 1997). In the Russian Arctic (Franz Josef Land and Severnaya Zemlya), glaciers were smaller in the early Holocene than they are now (Lubinsky et al. 1999). In Franz Josef Land, the glaciers remained small until at least 5 ka. Ice advances occurred before 5.6–5.2 ka, at 2.1–2.0 ka, at 1.0 ka, after 0.8 ka, as well as at 1400 CE, at 1600 CE, and in the early twentieth century.

Holocene conditions in Antarctica differed from those in the northern hemisphere, according to Ingólfsson et al. (1998). The initial deglaciation of the shelf areas surrounding Antarctica took place before 10,000 ^{14}C yr BP, and was controlled by rising global sea level. This was followed by the deglaciation of some presently ice-free inner shelf and land areas between 10,000 and 8,000 yr BP. Continued deglaciation occurred gradually between 8,000 and 5,000 yr BP. For the Antarctic Peninsula, there are a few scattered glacier records. Ice retreated from LGM limits after 18 ka and continued with stillstands and minor readvances during the early Holocene (O'Cofaigh et al. 2014). At James Ross Island, there was a local advance after 6.1 ka until after 4 ka, and another is inferred to have occurred between 1 and 0.7 ka. On King George Island, there was a glacial advance about 1.56–1.33 ka (Solomina et al. 2015). Some readvances appear related to increased accumulation, and others to cooling. A number of Antarctic Peninsula ice shelves also experienced major fluctuations during the Holocene. The George VI Ice Shelf was absent between 9.6 and 7.7 ka (Smith et al. 2007), The Larsen A Ice Shelf was absent during 3.8–1.4 ka (Balco and Schaefer 2013). In contrast, the Larsen B and Larsen C ice shelves existed throughout the Holocene (Curry and Pudsey 2007). The best dated records from the Antarctic Peninsula and coastal Victoria Land suggest climatic optima occurred there from 4,000–3,000 and 3,600–2,600 yr BP, respectively. Thereafter, Neoglacial readvances are recorded. Relatively limited glacial expansions in Antarctica during the past few hundred years are correlated with the Little Ice Age in the northern hemisphere.

2.8 Anthropocene

The Anthropocene (from the Greek *anthropos*, meaning "human") is the name that has recently been applied to the period when human activity began to dominate global environmental processes. It was first proposed by Crutzen and Stoermer (2000). Box 2.5 outlines its recent history.

During this period of time, lake sediments show unprecedented combinations of plastics, fly ash, radionuclides, metals, pesticides, reactive nitrogen, and consequences of increasing greenhouse gas concentrations. Important drivers involve changes in air chemistry and in land cover. Land clearance began at least 6,000 years ago, and less than 40 percent of the original forest cover on Earth still remains today. As Syvitski (2012) points out, major, long-lasting changes have been made during the twentieth century to the hydrological cycle as a result of

Box 2.5 **Definition of the Anthropocene**

The Anthropocene has been dated loosely to the beginning of the Industrial Revolution about 1800 CE. Lewis and Maslin (2015) define it as starting either in 1610 with the colonization of North America, or with the atomic bomb tests in 1964. Waters et al. (2016) define it as starting in the 1950s. According to Steffen et al. (2016), "the Earth has been pushed out of the Holocene Epoch by human activities, with the mid-20th century a strong candidate for the start date of the Anthropocene." Voosen (2016) states that the start of the Anthropocene has been proposed as the end of World War II by the Anthropocene Working Group. Before formal submission of this terminology to the International Commission on Stratigraphy, however, researchers must identify some stratigraphic section rich in geochemical markers of this postwar transition. One of the most promising proxies comes from seventy-one lakebed sedimentary records worldwide, which display a 1950s spike in fly ash residue from the high-temperature combustion of coal and oil.

dam and reservoir building, irrigation of semi-arid environments, and the effects of sediment redistribution leading to subsidence of major deltas. These changes in sedimentation will be identifiable in geological deposits in the future.

Edwards (2016) notes that the term Anthropocene is already well established, although it is neither defined nor a formal part of the Geological Time Scale. This time scale identifies rock records that relate to specific segments of geological time. A key concern is whether the Anthropocene has a starting point that is the same everywhere, or whether it started at different times at different places. Another question is its rank – era, period, epoch, or age? The last of these would place it as an age of the Holocene Epoch. These issues are being addressed currently by the International Commission on Stratigraphy of the International Geological Union.

2.8.1 Historical Climatology

Historical climatology refers to the period when written documentary evidence of weather and climate became available in the form of diaries, chronicles, ships' logs, and rudimentary observations. Most of these records are from middle latitudes, where they span about five centuries. For the Arctic, there are sparse weather records collected by eighteenth- and nineteenth-century expeditions. The first systematic records were kept during the First IPY. Wood and Overland (2006) have demonstrated that the surface air temperature (SAT) and sea level pressure (SLP) observed during 1882–1883 were within the limits of recent climatology. However, there was a slight skew toward lower temperatures, and

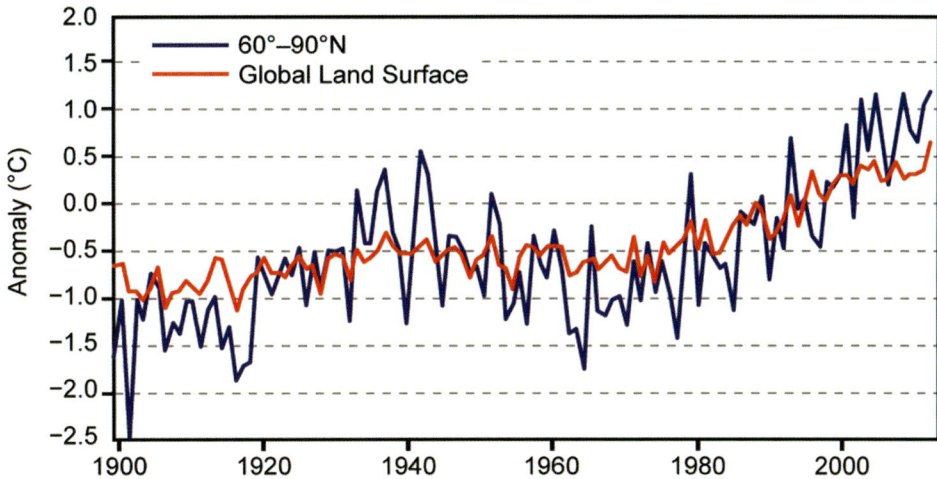

Figure 2.8 Arctic (land stations north of 60° N) and global mean annual land surface air temperature (SAT) anomalies (°C) for the period 1900–2015 relative to the 1981–2010 mean value. Note that there were few stations in the Arctic, particularly in northern Canada, before 1940 (based on CRUTEM4, Climatic Research Unit, University of East Anglia). Source: Richter-Menge and Mathis 2016, S 132, figure 5.1. Courtesy of American Meteorological Society.
© Copyright 2016 American Meteorological Society (AMS).

these authors showed a wide range of variability from place to place over the course of the year, which is a feature typical of the Arctic climate today. Monthly SAT, SLP, and associated phenological anomalies were regionally coherent and consistent with patterns of variability in the atmospheric circulation, such as the North Atlantic Oscillation (NAO) in mean SLP between the Azores anticyclone and the Icelandic low. Evidence of a strong NAO signature in the observed SAT anomalies during the First IPY highlights the impact of large-scale atmospheric circulation patterns on regional climate variability in the Arctic.

Since approximately 1980, the mean global temperature has risen due to increasing greenhouse gas concentrations by 0.5 °C, with the three warmest years on record (since 1850) being 2013–2016 (Blunden and Arndt 2016). This pattern differs from that of the high northern latitude warming in the early twentieth century in that it is a global phenomenon. In the Arctic, temperatures have increased at almost twice the global average rate in the past 100 years as a result of "polar amplification" involving the effects of the ice–albedo positive feedback of sea ice and snow cover (Serreze and Barry 2011). This effect, together with the advection (horizontal transport) of warm water and warm air from middle latitudes, has given rise to the amplified Arctic warming (Figure 2.8). In the central Arctic Ocean, temperatures from 1979 to 1995 rose by 0.5 °C or more

Table 2.1 Statistically significant decadal temperature trends in Antarctica, 1958–2012

	East	West	Peninsula
Annual		0.22 ± 0.12	0.33 ± 0.17
DJF			0.17 ± 0.13
MAM			0.32 ± 0.21
JJA		0.28 ± 0.27	0.58 ± 0.36
SON	0.14 ± 0.13	0.39 ± 0.21	0.28 ± 0.22

Source: Nicholas and Bromwich 2014, 8080, table 5.

in April–June and January and fell slightly during October–December. For locations north of 64° N, there have been Arctic-wide warm conditions since the 1990s in spring and, less strongly, in summer that are unique in the instrumental record. Before this, there was generally interdecadal negative covariability between northern Europe and Baffin Bay in winter. This pattern of temperature anomalies is associated with the NAO (van Loon and Rogers 1978). The Arctic-wide warming pattern of the 1990s indicates a change in atmospheric circulation.

Graversen and Burtu (2016) analyzed the role of the atmospheric circulation in polar amplification. They used a Fourier decomposition to distinguish the transports with respect to zonal wave numbers. Reanalysis and model data reveal that the planetary waves (n = 1–5) impact Arctic temperatures much more than the synoptic-scale waves (n = 6–10) do. In addition, the latent heat transport by these waves affects the Arctic climate more than does the dry-static part. The EC-Earth model suggests that changes in energy transport over the twenty-first century will contribute to Arctic warming, despite the model's assumption that the total energy transport to the Arctic will decrease. This apparent contradiction is due to the cooling induced by a decrease of the dry-static transport by planetary waves being more than compensated for by a warming caused by the latent heat counterpart.

For 1958–2012, Nicholas and Bromwich (2014) provide a reconstruction of Antarctic monthly mean near-surface temperatures. This demonstrates statistically significant annual warming in the Antarctic Peninsula and virtually all of West Antarctica, but no significant temperature changes in East Antarctica (Table 2.1). Importantly, the warming is of comparable magnitude in central West Antarctica and in most of the peninsula, rather than being concentrated either in one or the other region as previous reconstructions have suggested. The Transantarctic Mountains act for the temperature trends as a clear dividing line between East and West Antarctica, reflecting the topographic constraint on warm air advection from the Amundsen Sea basin (Turner and Marshall 2011).

Ice-core reconstructions indicate that Antarctica has warmed approximately 0.2 °C since the late nineteenth century (Steig et al. 2013). Over this period, temperatures were broadly in phase with those in the southern hemisphere, but temperatures on the continent and in the Antarctic Peninsula were basically anti-phase. Temperatures over most of Antarctica showed slight changes in the twentieth century. The exception is the Antarctic Peninsula, where a 2.5 °C warming has occurred since the 1950s (when observational records began), mainly in winter and spring. Analysis of records, including those from automatic weather stations, since 1957 shows warming of 0.10 °C decade^{-1} in East Antarctica and 0.17 °C decade^{-1} in West Antarctica.

Jones et al. (2016) have surveyed satellite proxy records of surface temperature since 1979 as well as historical surface observations. Their work shows strong surface warming over the West Antarctic ice sheet and Antarctic Peninsula regions and surface cooling over Adélie Land in East Antarctica, but with considerable interannual variability. Accompanying changes in snow cover, sea ice, lake ice, and permafrost are discussed in Chapter 5.

2.8.2 Tropical–Polar Links

A variety of recent studies have suggested links between the tropical atmosphere–ocean system and climatic and cryospheric responses in both polar regions.

The Tropically Excited Arctic Warming Mechanism (TEAM) ascribes warming of the Arctic surface to tropical convection, which excites poleward-propagating Rossby wave trains that transport water vapor and heat into the Arctic (Flourney et al. 2016). A crucial component of the TEAM mechanism is the increase in downward infrared radiation (IR) that precedes the Arctic warming. Previous studies have examined the downward IR associated with the TEAM mechanism using reanalysis data. To corroborate previous findings, the most recent study examines the linkage between tropical convection, Rossby wave trains, and downward IR with Baseline Surface Radiation Network (BSRN) downward IR station data. Both the Barrow and the Ny-Ålesund station downward IR anomalies are preceded by anomalous tropical convection and poleward-propagating Rossby wave trains. The wave train associated with Barrow resembles the Pacific–North America teleconnection pattern, while that for Ny-Ålesund corresponds to a northwestern Atlantic wave train. Both wave trains promote warm and moist advection from the mid-latitudes into the Arctic.

The interference between transient eddies and climatological stationary eddies in the northern hemisphere has been investigated by Goss et al. (2016). The amplitude and sign of the interference is represented by the stationary wave index (SWI), which is calculated by projecting the daily 300-hPa stream function anomaly field onto the 300-hPa climatological stationary wave. ERA-Interim data for the years 1979–2013 are used. The amplitude of the interference peaks during boreal winter. The evolution of outgoing longwave radiation, Arctic temperature, 300-hPa stream function, 10-hPa zonal wind, Arctic sea ice concentration, and the Arctic Oscillation (AO) index are examined for days of large SWI values during the winter. Constructive interference during winter tends to occur about one week after enhanced warm pool convection; it is followed by an increase in Arctic surface air temperature, along with a reduction of sea ice in the Barents and Kara seas. The warming of the Arctic does occur without prior warm pool convection, but it is enhanced and prolonged when constructive interference occurs in concert with enhanced warm pool convection.

Park et al. (2016) have extended the analysis of tropical–Arctic coupling in terms of winter sea ice. The northward flux of moisture into the Arctic during the winter strengthens the downward IR by 30–40 W m^{-2} over one to two weeks. This is followed by a decline of up to 10 percent in sea ice concentration over the Greenland, Barents, and Kara seas. A climate model simulation indicates that the wind-induced sea ice drift leads a decline of sea ice thickness during the early stage of the strong downward IR events, but that within one week the cumulative downward IR effect appears to be dominant. Further analysis indicates that strong downward IR events are preceded several days earlier by enhanced convection over the tropical Indian and western Pacific oceans.

Relations between the Madden Julian Oscillation (MJO) and northern hemisphere spring snow depth have been investigated by Barrett al. (2014). A physical pathway is proposed for intraseasonal variability of spring snow depth changes: Poleward-propagating Rossby waves in response to tropical MJO convection interact with northern hemisphere background flow, leading to anomalous troughing and ridging. These anomalous circulation centers then impact daily snow depth change via precipitation processes and anomalies in surface air temperature.

Intraseasonal variability in springtime northern hemisphere daily snow depth change (ΔSD) appears to be related to the phase of the MJO. Statistically significant regions of lagged ΔSD anomalies for multiple phases of the MJO have been found in March, April, and May in both North America and Eurasia. In each month, lagged ΔSD anomalies are physically supported by corresponding lagged anomalies of 500-hPa height (Z500) and surface air temperature (SAT). There is a moderate to strong relationship between both Z500 and ΔSD and SAT and ΔSD in both Eurasia and North America for phases 5 and 7 in March. Phases 5–6 of the MJO occur when the main center of convection is over the central-eastern tropical Pacific Ocean. In April, a moderately strong relationship between Z500 and ΔSD has been found over Eurasia for phase 5, but the relationship between

SAT and ΔSD is weak. In May, correlations between ΔSD and both Z500 and SAT over a hemisphere-wide latitude band from 60° to 75° N are close to -0.5 and -0.4, respectively.

Several southern hemisphere studies of tropical linkages have examined relationships between high latitude circulation and El Niño–Southern Oscillation (ENSO). Welhouse et al. (2016) have analyzed ERA-Interim data for 1979–2013 to investigate the relationship between ENSO and Antarctica for each season. For nine composites of each mode 2-m temperature (T_{2m}), sea level pressure (SLP), 500-hPa geopotential height, sea surface temperatures (SST), and 300-hPa geopotential height anomalies were calculated separately for El Niño minus neutral and La Niña minus neutral conditions. These anomaly patterns can differ in important ways from El Niño minus La Niña composites, which may be expected from the geographical shift in tropical deep convection and associated pattern of planetary wave propagation into the southern hemisphere. The researchers have found a robust signal, during La Niña, of cooling over East Antarctica from December to August. Both El Niño and La Niña experience the weakest signal during austral autumn. The peak signal for La Niña occurs during austral summer, while El Niño is found to peak during austral spring.

Li et al. (2016) have examined the role of the different tropical ocean basins in winter circulation in the Amundsen Sea low. Their work shows that Atlantic Ocean warming, Indian Ocean warming, and eastern Pacific cooling are all able to deepen the Amundsen Sea low located adjacent to West Antarctica, while western Pacific warming increases the pressure to the west of the international date line, encompassing the Ross Sea and regions south of the Tasman Sea. In austral winter, these tropical ocean basins work together linearly to modulate the atmospheric circulation around West Antarctica. The Amundsen Sea low has also been studied by Fogt and Wovrosh (2016). Tropical SST forcing alone explains much of the climatological variability and extreme intensities of the low (both strong and weak central pressures). The role of radiative forcing is best observed in the trends of the low. The simulation leads to a marked deepening of the low and pressures across the southern hemisphere that is consistent with atmospheric reanalysis in austral summer.

2.8.3 Polar–Mid-latitude Links

The relationship between the state of the Arctic climate system and mid-latitude weather has been an active area of research in the last decade, in view of the rapid changes in the Arctic sea ice and snow cover. Francis and Vavrus (2012) examined the links between Arctic amplification and mid-latitude tropospheric Rossby (planetary) waves. They found weaker zonal flow and increased wave amplitude, especially in autumn and winter in association with loss of sea ice, but also in summer possibly related to decreased snow cover. They proposed that the slower eastward movement of Rossby waves would give rise to more persistent circulation regimes, perhaps enhancing the frequency of weather extremes.

SUMMARY

Evidence of long-term climate change is derived primarily from ocean sediment cores that span tens of millions of years. These contain planktonic and benthic foraminifera that reflect ocean temperatures in upper and bottom waters, respectively. Temperature conditions were initially reconstructed from faunal assemblages for different water masses and then from the 1960s by analysis of oxygen isotope ratios ($\delta^{18}O$ and $\delta^{16}O$). Changes in benthic ^{18}O were later shown to reflect land ice volume rather than ocean temperatures. Ice core records that span 100,000–800,000 years provide annual to decadal records of temperature from $\delta^{18}O$, as well as atmospheric greenhouse gases, aerosols, chemical species, pollen, and snow stratigraphy.

A major external driver of long-term climate change is the effect of orbital geometry on solar radiation and temperature, first detailed by Milankovitch. Obliquity, or axial tilt (41 ka period), and precession of the vernal equinox (19- and 23-ka cycles) have opposite effects on seasonal radiation amounts in each hemisphere, whereas orbital eccentricity (100- and 400-ka periods) has small global effects. A major internal driver of climate is the atmospheric concentration of greenhouse gases – carbon dioxide and methane – that affect the absorption of terrestrial infrared radiation.

The Eocene Epoch (55–34 MYA) experienced major warmth in the Paleocene–Eocene Thermal Maximum (PETM), perhaps associated with high CO_2 levels. Forest cover was extensive in polar regions and global temperatures were equable, a state not yet understood. Cooling occurred during the Eocene, such that the Arctic Ocean recorded ice-rafted debris around 45 MYA. Antarctica had cool temperate rainforest in the Early Eocene, but ice appeared on East Antarctica around the Eocene/Oligocene transition.

The decrease of CO_2 levels below 750 pp led to this ice growth, which reached continental scale between 33.6 and 33.2 MYA. Sea level data imply that there was a smaller ice sheet during most of the Oligocene and Early Miocene, but a major buildup occurred in the Middle Miocene soon after 15 MYA. This was also the first appearance of West Antarctic ice. Mid-Miocene cooling in the Arctic is evidenced by sea ice and iceberg rafting as well as by some ice on Greenland. Ice rafting in the Nordic seas around 3.3 MYA indicates glacial expansion on Greenland. In the Russian Far East, the climate in the mid-Pliocene was significantly warmer and wetter than it is now, but there were also sharp interruptions and Late Pliocene cooling.

The Plio-Pleistocene is characterized by pronounced glacial cycles. Notably, 40-ka obliquity cycles dominate the West Antarctic ice sheet (WAIS) record from 5 to 3 MYA, and the WAIS periodically collapsed. Global average surface temperatures decreased by approximately 6 °C from 2 MYA and then stalled around 1.2 MYA, except for glacial/interglacial oscillations. Global temperatures and atmospheric GHG concentrations are closely coupled across glacial cycles. From 2.8 to 0.8 MYA, global climate showed 41-ka oscillations that abruptly switched to 100-ka oscillations. The cause of these variations has been controversial, as the eccentricity effect is weak. Interglacials from 0.8 to 0.43 MYA were longer and cooler than those that occurred subsequently.

Regional glaciation began 2.9–2.6 MYA in northwest Canada and east central Alaska. Its onset in Greenland and Scandinavia is dated to 2.72 MYA. The Laurentide ice sheet reached the continental shelf about 2.64 MYA. There were three Cordilleran and three Keewatin ice sheets in the Middle Pleistocene.

Various explanations for the 100-ka phasing of glacial cycles after 0.8 MYA have been proposed. Recent work notes that before 1 MYA, 36 caloric peaks out of 67 were interglacials and 21 interstadials received solar radiation below a given threshold. This threshold has decreased with time elapsed since the onset of the prior interglacial. During 160–140 ka, a thick ice shelf covered the Arctic Ocean.

The last Eemian interglacial saw sea level rise by 5.5–9 m, implying loss of WAIS and contributions from Greenland. Temperatures from northeast Canada to northern Eurasia were 2–4 °C higher than they are now.

During the Last Glacial Maximum (LGM), the Laurentide ice sheet had centers over Quebec–Labrador, Keewatin, and Foxe Basin. The Innuitian ice sheet covered the Queen Elizabeth Islands. Deglaciation was interrupted by the Younger Dryas cold, but then ice recession accelerated.

Glaciation of Scandinavia and the Barents–Kara Sea began approximately 2.5–2.8 MYA. The Barents Sea ice sheet reached the shelf four times over the last 140 ka, with the maximum being over 128 ka. During the last glaciation, 74–25 ka, the Barents–Kara Sea ice sheet blocked Arctic-draining rivers, leading to the formation of massive proglacial lakes in northwest Russia. During the LGM, there were domes over the British Isles, Fennoscandinavia, and the Barents–Kara Sea. The deglaciation of Fennoscandinavia shows large regional differences. There were ice-dammed lakes in Poland, the Baltic states, and northwest Russia, and the Baltic Ice Lake later formed.

The British Isles experienced Arctic climate during glacials as the polar front shifted southward. Continentality was enhanced by sea level lowering. The earliest glaciation (approximately 1.86 MYA) in East Anglia was followed by several other glaciations in the Anglian at 0.45 MYA. British ice was confluent with Scandinavian at 40 ka, but reached its maximum extent at 22 ka.

In the North Atlantic, Heinrich identified layers of ice-rafted debris (IRD) related to iceberg calving events from Hudson Bay and Europe. They occurred between 60 and 16.8 ka, though the precise mechanism is still debated. During the last glacial cycle, twenty-five (Dansgaard–Oeschger) climatic oscillations are attributed to fresh water inputs from the ice sheets affecting the Atlantic Meridional Overturning Circulation (AMOC). The AMOC results in a bipolar seesaw: The Antarctica warms while Greenland cools, and vice versa.

The WAIS has undergone dramatic changes over the last 5 million years from full glaciation, to the present state, to a collapsed state during super-interglacials. Deglacial warming of West Antarctica was two to three times the global average and was mostly complete by 15 ka.

The post-glacial Holocene started in 11.7 ka. Glaciers retreated worldwide due to precession-based warming until 5 ka. The Holocene Thermal Maximum was around 8 ka, except where delayed by northern hemisphere ice. Intermittent readvances after 5 ka point to regional thresholds due to multiple factors affecting smooth orbital forcing. These Neoglacials culminated in the Little Ice Age, 1300–1850 CE. Maximum ice cover in eastern Baffin Island is dated by lichenometry to 65 yr BP; the Barnes ice cap – a Laurentide remnant – persisted through the Holocene.

Ice returned to West Spitzbergen after 5 ka and reached its maximum in the nineteenth century. The maximum of local Greenland glaciers occurred in the twentieth century. Antarctic deglaciation continued from 8 to 5 ka, followed by irregular readvances. There were optima in the Antarctic Peninsula from 4 to 3 ka and coastal Victoria Land from 3 to 2.6 ka, but the Larsen B and C ice shelves were present through the Holocene.

The starting point of the period when human activity began to dominate world climate – the Anthropocene – is not yet firmly established. Since the Industrial Revolution, atmospheric CO_2 concentration has risen by 40 percent and methane (CH_4) concentration by 2.5 times. Arctic temperatures rose to a peak in the 1930–1940s, apparently due to internal variability and then from circa 1980 due to greenhouse gas concentrations. Polar amplification caused the warming in the Arctic and the Antarctic Peninsula to be double the global average.

Tropical convection excites Rossby wave trains that transport heat and moisture into the Arctic. Relations have been shown between the Madden Julian Oscillation and northern hemisphere spring snow depth. For East Antarctica, there is cooling from December to August when La Niña is present. Tropical SST forcing explains much of the variability of the Amundsen Sea low. Loss of Arctic sea ice in autumn is linked to increased planetary wave amplitude.

QUESTIONS

1. What are the major sources of evidence for long-term climate change?
2. What are the main drivers of long-term changes in global climate?
3. Account for the Eocene/Oligocene climate shift from greenhouse to icehouse conditions.
4. Describe the Earth's orbital variations and their significance for global climate. What is problematic about the 100-ka eccentricity effect?
5. Summarize the Quaternary glaciation for *one* of its main centers.
6. What is the significance of IRD, and what mechanisms may explain it?
7. Compare the Holocene and the LGM in any one region.
8. Do you consider the Anthropocene a useful concept? Why or why not?
9. Discuss Arctic climate change since the late nineteenth century.

References

Adéhmar, J. A. 1842. *Révolutions de la mer*. Paris: Carilian-Goeury et V. Dalmont.

Aitken, A. R. A., et al. 2016. "Repeated Large-Scale Retreat and Advance of Totten Glacier Indicated by Inland Bed Erosion." *Nature* 533: 385–9.

Alley, R. B. 2016. "A Heated Mirror of Future Climate." *Science* 352: 351–2.

Anagnostou, E., et al. 2016. "Changing Atmospheric CO_2 Concentration Was the Primary Driver of Early Cenozoic Climate." *Nature*. doi: 10.1038/nature17423.

Andrews, J. T. 2000. "Icebergs and Iceberg Rafted Detritus (IRD) in the North Atlantic: Facts and Assumptions." *Oceanography* 13: 100–10.

Andrews, J. T., and D. M. Barnett. 1979. "Holocene (Neoglacial) Moraines and Proglacial Lake Chronology: Barnes Ice Cap, Canada." *Boreas* 8: 342–58.

Andrews, J. T., W. W. Shilts, and G. J. Miller. 1983. "Multiple Deglaciations of the Hudson Bay Lowlands, Canada, since Deposition of the Missinaibi (Last-Integlacial?) Formation." *Arctic and Alpine Research* 19: 18–37.

Andrews, J. T., and P. J. Webber. 1964. "Lichenometrical Study on the Northwestern Margins of the Barnes Ice Cap: A Geomorphological Technique." *Geographical Bulletin* 22: 80–104.

Astakhov, V. 2004. "Pleistocene Ice Limits in the Russian Northern Lowlands." In *Quaternary Glaciations: Extent and Chronology. Part 1. Europe*, edited by J. Ehlers and P. J. Gibbard, 309–19. Amsterdam: Elsevier.

Bailey, I., G. M. Hole, G. L. Foster, P. A. Wilson, C. D. Storey, C. N. Trueman, and M. E. Raymo. 2013. "An Alternative Suggestion for the Pliocene Onset of Major Northern Hemisphere Glaciation Based on the Geochemical Provenance of North Atlantic Ocean Ice-Rafted Debris." *Quaternary Science Reviews* 75: 181–94.

Balco, G., and C. W. Rovey II. 2010. "Absolute Chronology for Major Pleistocene Advances of the Laurentide Ice Sheet."*Geological Society of America Bulletin* 38: 795–8.

Balco, G., and J. M. Schaefer; LARISSA Group. 2013. "Exposure-Age Record of Holocene Ice Sheet and Ice Shelf Change in the Northeast Antarctic Peninsula." *Quaternary Science Reviews* 59: 101–11.

Barrett, B. S., et al. 2014. "The Influence of the MJO on the Intraseasonal Variability of Northern Hemisphere Spring Snow Depth." *Journal of Climate* 28(18): 7250–62.

Barry, R. G., and R. J. Chorley. 2010. *Atmosphere, Weather and Climate*, 9th ed. London: Routledge.

Bassis, J. N., S. V. Petersen, and L. MacCathles. 2017. "Heinrich Events Triggered by Ocean Forcing and Modulated by Isostatic Adjustment." *Nature* 542: 332–4.

Berger, A. 2012. "A Brief History of the Astronomical Theories of Paleoclimate." In *Climate Change*, edited by A. Berger et al., 107–29. Vienna: Springer.

Beschel, R. 1961. "Dating Rock Surfaces by Lichen Growth and Its Application to Glaciology and Physiography (Lichenometry)." In *Geology of the Arctic*, Vol. 2, edited by G. A. Rasch, 1044–62. Toronto, Ontario: University of Toronto Press.

Blunden, J., and D. S. Arndt, eds. 2016. "State of the Climate in 2015." *Bulletin of the American Meteorological Society* 97(8): S1–275.

Bond, G. C., and R. Lotti. 1995. "Iceberg Discharges into the North Atlantic on Millennial Time Scales during the Last Glaciation." *Science* 267: 1006–10.

Boulton, G. S., J. D. Peacock, and D. G. Sutherland. 1991. "Quaternary." In *Geology of Scotland*, 3rd ed., edited by G. V. Craig, 503–41. London: Geological Society.

Bowen, D. Q., et al. 2002. "New Data for the Last Glacial Maximum in Great Britain and Northern Ireland." *Quaternary Science Reviews* 21: 89–101.

Brigham-Grete, J., et al. 2013. "Pliocene Warmth, Polar Amplification, and Stepped Pleistocene Cooling Recorded in NE Arctic Russia." *Science* 340(6139): 1421–7.

Briner, J. P., et al. 2016. "Holocene Climate Change in Arctic Canada and Greenland." *Quaternary Science Reviews* 147: 340–64.

Broecker, W. S., and J. Van Donk. 1970. "Insolation Changes, Ice Volumes, and the O^{18} Record in Deep-Sea Cores." *Reviews of Geophysics* 8: 169–98.

Cheng, H., et al. 2016. "The Asian Monsoon over the Past 640,000 Years and Ice Age Terminations." *Nature* 534: 640–6.

Clark, C. D., P. L. Gibbard, and J. Rose. 2004. "Pleistocene Glaciation Limits in England, Scotland and Wales." In *Quaternary Glaciations: Extent and Chronology. Part 1. Europe*, edited by J. Ehlers and P. J. Gibbard, 47–82. Amsterdam: Elsevier.

Clark, P. U., et al. 2009. "The Last Glacial Maximum." *Science* 325: 710–14.

Collins, L. G., et al. 2012. "High-Resolution Reconstruction of Southwest Atlantic Sea-Ice and Its Role in the Carbon Cycle during Marine Isotope Stages 3 and 2." *Paleoceanography* 27: PA3217.

Croll, J. 1864. "On the Physical Cause of the Change of Climate during Geological Epochs." *Philosophical Magazine (London)* 28: 121–37.

Crutzen, P. J., and Stoermer, E. F. 2000. "The Anthropocene." *Global Change Newsletter* 41: 17–18.

Cuffey, K. M., et al. 2016. "Deglacial Temperature History of West Antarctica." *Proceedings of the National Academy of Sciences* 113(14): 249–54.

Curry, P. J., and C. J. Pudsey. 2007. "New Quaternary Sedimentary Records from Near the Larsen C and Former Larsen B Ice Shelves: Evidence for Holocene Stability." *Antarctic Science* 19: 355–64.

Dalton, A. S., et al. 2016. "Constraining the Late Pleistocene History of the Laurentide Ice Sheet by Dating the Missinaibi Formation, Hudson Bay Lowlands, Canada." *Quaternary Science Reviews* 146: 288–99.

Dansgaard, W., et al. 1971. "Climate Record Revealed by the Camp Century Ice Core." In *Late Cenozoic Glacial Ages: Symposium 1969*, edited by K. K. Turekian, 37–56. New Haven, CT: Yale University Press.

Dansgaard, W., et al. 1982. "A New Greenland Deep Ice Core." *Science* 218: 1273–7.

DeConto, R. M., and D. Pollard. 2003. "Rapid Cenozoic Glaciation of Antarctica Induced by Declining Atmospheric CO_2." *Nature* 421: 245–9.

Ditlevsen, P. D., K. K. Andersen, and A. Svensson. 2007. "The DO-Climate Events Are Probably Noise Induced: Statistical Investigation of the Claimed 1470 Years Cycle." *Climate of the Past* 3: 129–34.

Dud-Rodkin, A., et al. 2004. "Timing and Extent of Plio-Pleistocene Glaciations in North-Western Canada and East-Central Alaska." *Developments in Quaternary Science* 2 (Part B): 313–45.

Dud-Rodkin, A., R. W. Barendregt, and J. M. White. 2010. "An Extensive Late Cenozoic Terrestrial Record of Multiple Glaciations Preserved in the Tintina Trench of West-Central Yukon: Stratigraphy, Paleomagnetism, Paleosols, and Pollen." *Canadian Journal of Earth Sciences* 47: 1003–28.

Dutton, A., and K. Lambeck. 2012. "Ice Volume and Sea Level during the Last Interglacial." *Science* 337(6091): 216–19.

Dyke, A. S. 2004. "An Outline of North American Deglaciation with Emphasis on Central and Northern Canada." In *Quaternary Glaciations: Extent and Chronology. Part 2: North America*, edited by J. Ehlers and P. J. Gibbard, 373–424. Amsterdam: Elsevier.

Dyke, A. S., et al. 2002. "The Laurentide and Innuitian Ice Sheets during the Last Glacial Maximum." *Quaternary Science Reviews* 21: 9–31.

Edwards, L. P. 2016. "What Is the Anthropocene?" *Eos* 97(2): 6–7.

Ehlers, J., and P. J. Gibbard, eds. 2004a. *Quaternary Glaciations: Extent and Chronology. Part 1: Europe*. Amsterdam: Elsevier.

Ehlers, J., and P. J. Gibbard, eds. 2004b. *Quaternary Glaciations: Extent and Chronology. Part 2: North America*. Amsterdam: Elsevier.

Emiliani, C. 1961. "Cenozoic Climatic Changes As Indicated by the Stratigraphy and Chronology of Deep-Sea Cores of Globigerina-Ooze Facies." *Annals of the New York Academy of Sciences* 95: 521–36.

England, J., et al. 2006. "The Innuitian Ice Sheet: Configuration, Dynamics and Chronology." *Quaternary Science Reviews* 25: 689–703.

EPICA Community Members. 2004. "Eight Glacial Cycles from an Antarctic Ice Core." *Nature* 429: 623–8.

Epstein, S., et al. 1951. "Carbonate Water Isotopic Temperature Scale." *Geological Society of America Bulletin* 62: 417–26.

Flournoy, M. D., et al. 2016. "Exploring the Tropically Excited Arctic Warming Mechanism with Station Data: Links between Tropical Convection and Arctic Downward Infrared Radiation." *Journal of the Atmospheric Sciences* 73: 1143–58.

Flowers, B. P., and J. P. Kennett. 1994. "The Middle Miocene Climatic Transition: East Antarctic Ice Sheet Development, Deep Ocean Circulation and Global Carbon Cycling." *Palaeogeography, Palaeoclimatology, and Palaeoecology* 108: 537–55.

Fogt, R. L., and A. J. Wovrosh. 2016. "The Relative Influence of Tropical Sea Surface Temperatures and Radiative Forcing on the Amundsen Sea Low." *Journal of Climate* 28(21): 8540–55.

Francis, J. A., and S. J. Vavrus. 2012. "Evidence Linking Arctic Amplification to Extreme Weather in Mid-latitudes." *Geophysical Research Letters* 39(6): L06801.

Francis, J. E., et al. 2009. "From Greenhouse to Icehouse: The Eocene/Oligocene in Antarctica." In *Developments in Earth & Environmental Sciences*, 8th ed., edited by F. Florindo and M. Siegert, 311–72. Amsterdam: Elsevier.

Galeoti, S., et al. 2016. "Antarctic Ice Sheet Variability across the Eocene–Oligocene Boundary Climate Transition." *Science* 352: 76–80.

Gibbard, P. L., and J. Lewin. 2016. "Partitioning the Quaternary." *Quaternary Science Reviews* 151: 127–39.

Goss, M., S. B. Feldstein, and S.-Y. Lee. 2016. "Stationary Wave Interference and Its Relation to Tropical Convection and Arctic Warming." *Journal of Climate* 29: 1369–89.

Graversen, R. G., and M. Burtu. 2016. "Arctic Amplification Enhanced by Latent Energy Transport of Atmospheric Planetary Waves." *Quarterly Journal of the Royal Meteorological Society* 142: 2046–54.

Gregoire, L. J., P. J. Valdes, and A. J. Payne. 2015. "The Relative Contribution of Orbital Forcing and Greenhouse Gases to the North American Deglaciation." *Geophysical Research Letters*. doi: 10.1002/2015GL066005.

Hays, J. D., J. Imbrie, and N. J. Shackleton. 1976. "Variations in the Earth's Orbit: Pacemaker of the Ice Ages." *Science* 194: 1121–32.

Heinrich, H. 1988. "Origin and Consequences of Cyclic Ice Rafting in the Northeast Atlantic Ocean during the Past 130,000 Years." *Quaternary Research* 29: 142–52.

Hemming, S. R. 2004. "Heinrich Events: Massive Late Pleistocene Detritus Layers of the North Atlantic and Their Global Climate Imprint." *Reviews of Geophysics* 42: 1005H.

Henderson-Sellers, A., and K. McGuffie. 1984. *A Climate Modelling Primer*. Hoboken, NJ: John Wiley and Sons.

Henry, L. G., et al. 2016. "North Atlantic Ocean Circulation and Abrupt Climate Change during the Last Glaciation." *Science* 353: 470–4.

Hidy, A. J., J. C. Gosse, D. G. Froese, J. D. Bond, and D. H. Rood. 2013. "A Latest Pliocene Age for the Earliest and Most Extensive Cordilleran Ice Sheet in Northwestern Canada." *Quaternary Science Reviews* 61: 77–84.

Hodell, D. A. 2016. "The Smoking Gun of the Ice Ages." *Science* 354: 1235–6.

Holbourn, A., et al. 2005. "Impacts of Orbital Forcing and Atmospheric Carbon Dioxide on Miocene Ice-Sheet Expansion." *Nature* 438: 483–7.

Huber, M., and R. Caballero. 2011. "The early Eocene Equable Climate Problem Revisited." *Climate of the Past Discussion Papers* 6: 241–304.

Huber, M., and D. Nof. 2006. "The Ocean Circulation in the Southern Hemisphere and Its Climatic Impacts in the Eocene." *Palaeogeography, Palaeoclimatology, and Palaeocology* 231: 9–28.

Hughes, A. L. C., et al. 2016. "The Last Eurasian Ice Sheets: A Chronological Database and Time-Slice Reconstruction, DATED-1." *Boreas* 45: 1–45.

Hughes, T. J., G. H. Denton, and M. G. Grosswald. 1977. "Was There a Late-Würm Arctic Ice Sheet?" *Nature* 266: 596–602.

Ingólfsson, O., et al. 1998. "Antarctic Glacial History since the Last Glacial Maximum: An Overview of the Record on Land." *Antarctic Science* 10: 326–44.

Ives, J. D. 2016. *Baffin Island: Field Research and High Arctic Adventure 1961–1967.* Calgary, Alberta: University of Calgary Press.

Jakobsson, M., et al. 2014. "The Dynamic Arctic." *Quaternary Science Reviews* 92: 1–8.

Jakobsson, M., et al. 2016. "Evidence for an Ice Shelf Covering the Central Arctic Ocean during the Penultimate Glaciation." *Nature Communications* 7. doi: 10.1038/ncomms10365.

Jansen, E., and J. Sjøholm. 1991. "Reconstruction of Glaciation over the Past 6 Myr from Ice-Borne Deposits in the Norwegian Sea." *Nature* 349: 600–3.

Jennings, A. E., et al. 2017. "Ocean Forcing of Ice Sheet Retreat in Central West Greenland from LGM to the Early Holocene." *Earth and Planetary Sciences Letters* 472: 1–15.

Jones, J. A., et al. 2016. "Assessing Recent Trends in High-Latitude Southern Hemisphere Surface Climate." *Nature Climate Change* 6: 917–26.

Kelly, M. A., and T. W. Lowell. 2009. "Fluctuations of Local Glaciers in Greenland during Latest Pleistocene and Holocene Time." *Quaternary Science Reviews* 28: 2088–106.

Kleiven, H. F., et al. 2002. "Intensification of Northern Hemisphere Glaciations in the Circum Atlantic Region (3.5–2.4): Ice-Rafted Detritus Evidence." *Palaeogeography, Palaeoclimatology, and Palaeocology* 184: 213–23.

Koenig, S. J., R. M. DeConto, and D. Pollard. 2011. "Late Pliocene to Pleistocene Sensitivity of the Greenland Ice Sheet in Response to External Forcing and Internal Feedbacks." *Climate Dynamics* 37: 1247–68.

Kuijpers, A., P. Knutz, and M. Moros. 2014. "Ice-Rafted Debris (IRD)." In *Encyclopedia of Marine Geosciences.* Dordrecht, Netherlands: Springer Media. doi: 10.1007/978-94-007-6.

Lambeck, K., A. Purcell, and S. Zhao. 2017. "The North American Late Wisconsin Ice Sheet and Mantle Viscosity from Glacial Rebound Analyses." *Quaternary Science Reviews* 158: 174–210.

Lecavalier, B. S., et al. 2017. "High Arctic Holocene Temperature Record from the Agassiz Ice Cap and Greenland Ice Sheet Evolution." *Proceedings of the National Academy of Sciences* 114(23): 5952–7.

Lee, J.-E., et al. 2017. "Hemispheric Sea Ice Distribution Sets the Glacial Tempo." *Journal of Geophysical Research* 44: 1008–14.

Lewis, S. L., and M. A. Maslin. 2015. "Defining the Anthropocene." *Nature* 519: 171–80.

Li, X.-Ch., et al. 2016. "Rossby Waves Mediate Impacts of Tropical Oceans on West Antarctic Atmospheric Circulation in Austral Winter." *Journal of Climate* 28(20): 8151–64.

Locke, W. W. III, J. T. Andrews, and P. W. Webber. 1979. *A Manual for Lichenometry.* London: British Geomorphological Research Group.

Lubinsky, D. J., S. L., Forman, and G. H. Miller. 1999. "Holocene Glacier and Climate Fluctuations on Franz Josef Land, Arctic Russia, 80°N." *Quaternary Science Reviews* 18: 85–108.

Lunt, D. J., G. K. Foster, A. M. Haywood, and E. J. Stone. 2008. "Late Pliocene Greenland glaciation Controlled by a Decline in Atmospheric CO_2 Levels." *Nature* 454: 1102–5.

Mackensen, A. 2004. "Changing Southern Ocean Palaeocirculation and Effects on Global Climate." *Antarctic Science* 16: 369–86.

Mangerud, J., E. Jansen, and J. Y. Landvik. 1996. "Late Cenozoic History of the Scandinavian and Barents Sea Ice Sheet." *Global and Planetary Change* 12: 11–26.

Mercer, J. H. 1970. "A Former Ice Sheet in the Arctic Ocean?" *Palaeogeography, Palaeoclimatology, and Palaeecology* 8: 19–27.

Milankovitch, M. 1930. "Mathematische Klimalehre und Astronomische Theorie der Klimaschwankungen." In *Handbuch der Klimatologie, Band I, Teil A*, edited by W. Köppen and R. Geiger. Berlin: Gebrüder Borntraeger.

Miller, G. H. 1973. "Late Quaternary Glacial and Climatic History of Northern Cumberland Peninsula, Baffin Island, N.W.T., Canada." *Quaternary Research* 3: 561–83.

Miller, G. H., et al. 2010a. "Temperature and Precipitation History of the Arctic." *Quaternary Science Reviews* 29: 1679–715.

Miller, G. H., et al. 2010b. "Arctic Amplification: Can the Past Constrain the Future?" *Quaternary Science Reviews* 29: 1779–90.

Miller, G. H., et al. 2012. "Abrupt Onset of the Little Ice Age Triggered by Volcanism and Sustained by Sea-Ice/Ocean Feedbacks." *Geophysical Research Letters* 39: L02708.

Moran, K., et al. 2005. "The Cenozoic Palaeoenvironment of the Arctic Ocean." *Nature* 441: 601–5.

Naish, T. R., et al. 2001. "Orbitally Induced Oscillations in the East Antarctic Ice Sheet at the Oligocene/Miocene Boundary." *Nature* 413: 719–23.

Naish, T. R., et al. 2009. "Obliquity-Paced Pliocene West Antarctic Ice Sheet Oscillations." *Nature* 458: 322–8.

Nicholas, J. P., and D. H. Bromwich. 2014. "New Reconstruction of Antarctic Near-Surface Temperatures: Multidecadal Trends and Reliability of Global Reanalyses." *Journal of Climate* 27: 8070–93.

O'Cofaigh, C., et al. 2014. "Reconstruction of Ice Sheet Changes in the Antarctic Peninsula since the Last Glacial Maximum." *Quaternary Science Reviews* 100: 87–110.

Otto-Bliesner, B. L., et al. 2006. "Simulating Arctic Climate Warmth and Icefield Retreat in the Last Interglaciation." *Science* 311: 1751–3.

Pälike, H., et al. 2006. "The Heartbeat of the Oligocene Climate System." *Science* 314: 1894–8.

Park, H.-S., et al. 2016. "The Impact of Poleward Moisture and Sensible Heat Flux on Arctic Winter Sea Ice Variability." *Journal of Climate* 28(13): 5030–40.

Patton, H., et al. 2017. "Deglaciation of the Eurasian Ice Sheet Complex." *Quaternary Science Reviews* 169: 148–72.

Pearson, P. N. 2010. "Increased Atmospheric CO_2 during the Middle Eocene." *Science* 330: 763–4.

Petersen, S. V., and D. P. Schrag. 2015. "Antarctic Ice Growth Before and After the Eocene–Oligocene Transition: New Estimates from Clumped Isotope Paleothermometry." *Paleoceanography*. doi: 10.1002/2014PA002769.

Pollard, D., and R. M. DeConto. 2009. "Modelling West Antarctic Ice Sheet Growth and Collapse through the Past Five Million Years." *Nature* 458: 329–32.

Pope, E. L., et al. 2016. "Long-Term Record of Barents Sea Ice Sheet Advance to the Shelf Edge from a 140,000-Year Record." *Quaternary Science Reviews* 150: 55–66.

Raymo, M. E. 1997. "The Timing of Major Climate Terminations." *Paleoceanography* 12: 577–85.

Raymo, M. E., and P. Huybers. 2008. "Unlocking the Mysteries of the Ice Ages." *Nature* 451: 284–5.

Raynaud, D., et al. 2000. "The Ice Core Record of Greenhouse Gases: A View in the Context of Future Changes." *Quaternary Science Reviews* 19: 9–17.

Renssen, H., H. Goosse, and R. Muscheler. 2006. "Coupled Climate Model Simulation of Holocene Cooling Events: Oceanic Feedback Amplifies Solar Forcing." *Climate of the Past* 2: 79–90.

Richter-Menge, J., and J. Mathis, eds. 2016. "The Arctic." In *State of the Climate in 2015*, edited by J. Blunden and D.S. Arndt. *Bulletin of the American Meteorological Society* 97(8): S131–53.

Ruddiman, W. F. 1977. "Late Quaternary Deposition of Ice-Rafted Sand in the Sub-Polar North Atlantic (40–60°N)." *Geological Society of America Bulletin* 88: 1813–27.

Schultz, M. 2002. "On the 1470-Year Pacing of Dansgaard-Oeschger Warm Events." *Paleoceanography* 17: 4.1–4.9.

Serreze, M. C., and R. G. Barry. 2011. "Processes and Impacts of Arctic Amplification: A Research Synthesis." *Global and Planetary Change* 77: 85–91.

Shackleton, N. J. 1967. "Oxygen Isotope Analyses and Pleistocene Temperatures Re-Assessed." *Nature* 215: 15–17.

Shackleton, N. J. 1977. "The Oxygen Isotope Stratigraphic Record of the Pleistocene." *Philosophical Transactions of the Royal Society (London) B* 280: 169–82.

Shackleton, N. J. 2000. "The 100,000-Year Ice-Age Cycle Identified and Found to Lag Temperature, Carbon Dioxide, and Orbital Eccentricity." *Science* 289: 1897–902.

Sinclair, G., et al. 2016. "Diachronous Retreat of the Greenland Ice Sheet during the Last Deglaciation." *Quaternary Science Reviews* 1345: 243–58.

Smith, J. A., M. J. Bentley, D. A. Hodgson, and A. J. Cook. 2007. "George VI Ice Shelf: Past History, Present Behaviour and Potential Mechanisms for Future Collapse." *Antarctic Science* 19: 131–42.

Snyder, C. W. 2016. "Evolution of Global Temperature over the Past Two Million Years." *Nature* 583: 226–8.

Solomina, O., et al. 2015. "Holocene Glacier Fluctuations." *Quaternary Science Reviews* 111: 9–34.

Steffen, W., et al. 2016. "Stratigraphic and Earth System Approaches to Defining the Anthropocene." *Earth Future*. doi: 10.1002/2016EF000379.

Steig, E. J., et al. 2013. "Recent Climate and Ice-Sheet Changes in West Antarctica Compared with the Past 2,000 Years." *Nature Geoscience* 6: 372–5.

Stroeven, A. P., et al. 2016. "Deglaciation of Fennoscandinavia." *Quaternary Science Reviews* 147: 91–121.

Svendsen, J. I., et al. 1999. "Maximum Extent of the Eurasian Ice Sheets in the Barents and Kara Sea Region during the Weichselian." *Boreas* 28: 234–42.

Svendsen, J. I., et al. 2004. "Late Quaternary Ice Sheet History of Northern Eurasia." *Quaternary Science Reviews* 23: 1229–71.

Svendsen, J. I., and J. Mangerud. 1997. "Holocene Glacial and Climatic Variations on Spitsbergen, Svalbard." *Holocene* 7: 45–57.

Syvitski, J. 2012. "An Epoch of Our Making." *Global Change* 8(78): 12–15.

Teschner, C., et al. 2016. "Plio-Pleistocene Evolution of Water Mass Exchange and Erosional Input at the Atlantic–Arctic Gateway." *Paleoceanography* 31. doi: 10.1002/2015PA002843.

Thiede, J. C., et al. 2011. "Millions of Years of Greenland Ice Sheet History Recorded in Ocean Sediments. *Polarforschung* 80(3): 141–59.

Thomas, E. K., J. S. Szymanksi, and J. P. Briner. 2010. "The Evolution of Holocene Alpine Glaciation Inferred from Lacustrine Sediments on Northeastern Baffin Island, Arctic Canada." *Journal of Quaternary Science* 25: 146–61.

Turner, J., et al. 2014. "Antarctic Climate Change and the Environment: An Update." *Polar Record* 50: 237–59.

Turner, J., and G. J. Marshall. 2011. *Climate Change in the Polar Regions*. Cambridge: Cambridge University Press.

Tzedakis, P. C., et al. 2017. "A Simple Rule to Determine Which Insolation Cycles Lead to Interglacials." *Nature* 542: 427–32.

Ulman, D. J., et al. 2016. "Final Laurentide Ice-Sheet Deglaciation and Holocene Climate-Sea Level Change." *Quaternary Science Reviews* 152: 49–59.

van Loon, H., and J. C. Rogers. 1978. "Seesaw in Winter Temperatures between Greenland and Northern Europe. 1. General Description." *Monthly Weather Review* 108: 296–310.

Voosen, P. 2016. "Anthropocene Pinned to Postwar Period." *Science* 353: 852–3.

WAIS Divide Project Members. 2015. "Precise Interpolar Phasing of Abrupt Climate Change during the Last Ice Age." *Nature* 520: 661–5.

Waters, C., et al. 2016. "The Anthropocene Is Functionally and Stratigraphically Distinct from the Holocene." *Science* 351(6269). doi: 10.1126/science.aad2622.

Welhouse, L. J., et al. 2016. "Composite Analysis of the Effects of ENSO Events on Antarctica." *Journal of Climate* 29: 1797–808.

Wennrich, V., et al. 2016. "Impact Processes, Permafrost Dynamics, and Climate and Environmental Variability in the Terrestrial Arctic As Inferred from the Unique 3.6 Myr Record of Lake El'gygytgyn, Far East Russia: A Review." *Quaternary Science Reviews* 147: 221–44.

West, C., D. R.Greenwood, and J. Basinger. 2015. "Was the Arctic Eocene "Rainforest" Monsoonal? Estimates of Seasonal Precipitation from Early Eocene Megafloras from Ellesmere Island, Nunavut." *Earth and Planetary Science Letters* 427: 18–30.

Wood, K. R., and J. E. Overland. 2006. "Climate Lessons from the First International Polar Year." *Bulletin of the American Meteorological Society* 87: 1685–97.

Yin, Q. Z., and A. Berger. 2010. "Insolation and CO_2 Contribution to the Interglacial Climate Before and After the Mid-Brunhes Event." *Nature Geoscience* 3: 243–6.

Yin, Q. Z., and A. Berger. 2012. "Individual Contribution of Insolation and CO_2 to the Interglacial Climates of the Past 800,000 Years." *Climate Dynamics* 38: 709–24.

Observing Polar Environments

3

This chapter first describes the major observing networks that have operated in the polar regions. It then summarizes various in situ measurements of meteorological and oceanographic variables, followed by an account of remote sensing observing systems that are applied in polar environments. It concludes with a discussion of reanalysis products.

3.1 Observing Networks

The first observational network to be established was that of surface weather observations, which became organized nationally, and soon thereafter internationally, in the 1860s–1880s following the invention of the telegraph and the standardization of observing practices. Arctic stations remained few and far between, however, until the early twentieth century. A limited network consisting of twelve geomagnetic and weather stations and thirteen auxiliary stations was established in the Arctic during the First International Polar Year in 1882–1883, but it is only recently that those data have been used in climate change assessments (Wood and Overland 2006). For example, negative departures of surface air temperature were more frequent in the past relative to the current climatology (1968–1997), especially in the summer of 1883. Similar efforts during the Second International Polar Year, 1932–1933, were largely overshadowed by World War II.

Historical marine climate and oceanographic data in high latitudes are equally scarce, with the exception of particular expedition records. These sources include the 1807–1822 Scoresby expeditions in the Greenland Sea, the 1819–1825 Parry expedition in the Canadian Arctic, Nansen's 1893–1896 *Fram* expedition (Mohn 1905), the second *Fram* voyage of 1898–1902 (Mohn 1907), the Norwegian *Maud* expedition of 1918–1825 (Sverdrup 1930, 1933), the 1914–1916 Shackleton expedition in the Southern Ocean, and US Arctic logbooks for the Bering Sea dating from 1870–1846. These data are incorporated in the International

Comprehensive Ocean–Atmosphere Data Set (ICOADS) Release 3.0, which includes data from 1662 to 2014 (Freeman et al. 2017).

Methods of observing polar environments have evolved rapidly over the last fifty years with the advent of airborne and satellite remote sensing, automatic weather stations, ocean data buoys, and the miniaturization of instruments. The International Geophysical Year (IGY), 1957–1958, can be regarded as the first major step in "Big Data" collection in the Antarctic, which also witnessed the designation of World Data Centers (WDCs) by the International Council on Scientific Unions (ICSU) for glaciology, meteorology, and oceanography in the Soviet Union and United States. However, this particular research activity was not guided by "Big Science" theories (Aronova et al. 2010).

Biological projects were performed in Antarctica and elsewhere during the IGY, which laid the foundation for the observing sites of the International Biological Program (IBP) during 1964–1974. These projects included the Tundra Biome studies at Point Barrow, Alaska, and Truelove Lowland on Devon Island (Barry et al. 1981). The IBP was initially led by Roger Revelle, who also developed programs in oceanography. A major weakness of the IBP was the failure to establish data centers and standardize data formats. The National Science Foundation set up the Long-Term Ecological Research Program (LTER) in 1980 as a successor to the IBP's ecosystem studies; the LTER included careful data management and archival. These studies included data collected at the Arctic site at Toolik Lake on Alaska's North Slope as well as the Palmer station and the McMurdo Dry Valleys in Antarctica. Starting in late 2017 are the Northern Gulf of Alaska and Beaufort Lagoon Ecosystem LTERs (https://lternet.edu/lter-sites).

Systematic glacier observations began in the 1880s. Data on glacier fluctuations, dating back to 1881, were first published for the Alps in 1895. These records have been maintained in the series *Fluctuations of Glaciers*, whose latest issue covers 2005–2010.

In 1986, the Permanent Service for Fluctuations of Glaciers and the World Glacier Inventory, both established in the 1960s, merged in the World Glacier Monitoring Service (WGMS) (http://wgms.ch/). The WGMS collects and disseminates mass balance data for worldwide glaciers. Long-term Arctic glacier mass balance records – dating back to the 1960s – are available for the Devon ice cap NW, Meighen ice cap, Melville South ice cap, and White Glacier in the Canadian Arctic as well as for two glaciers on Svalbard. In Antarctica, glacier mass balances have been derived from satellite radar measurements that began in 1992, except for field observations in the Dry Valleys that began in 1993.

Networks of permafrost observations were established in the 1990s. They are now organized through the Global Terrestrial Network on Permafrost (GTN-P 2015). There are about 1,070 bore holes where ground temperatures are monitored (Figure 3.1). Approximately 27 percent of the bore holes have a depth greater than 25 m.

The Circumpolar Active Layer Monitoring (CALM) observational network, established in 1991, observes the long-term response of the active layer and near-surface permafrost to changes and variations in climate at more than 240 sites in both hemispheres (Figure 3.2). Almost 4 percent of Thermal State of Permafrost (TSP) and CALM sites are in polar desert, 27 percent in tundra, and 11 and 13 percent, respectively, in shrub tundra.

The 1990s witnessed the advent of the Earth Observation System (EOS) program organized by the National Aeronautics and Space Administration (NASA).

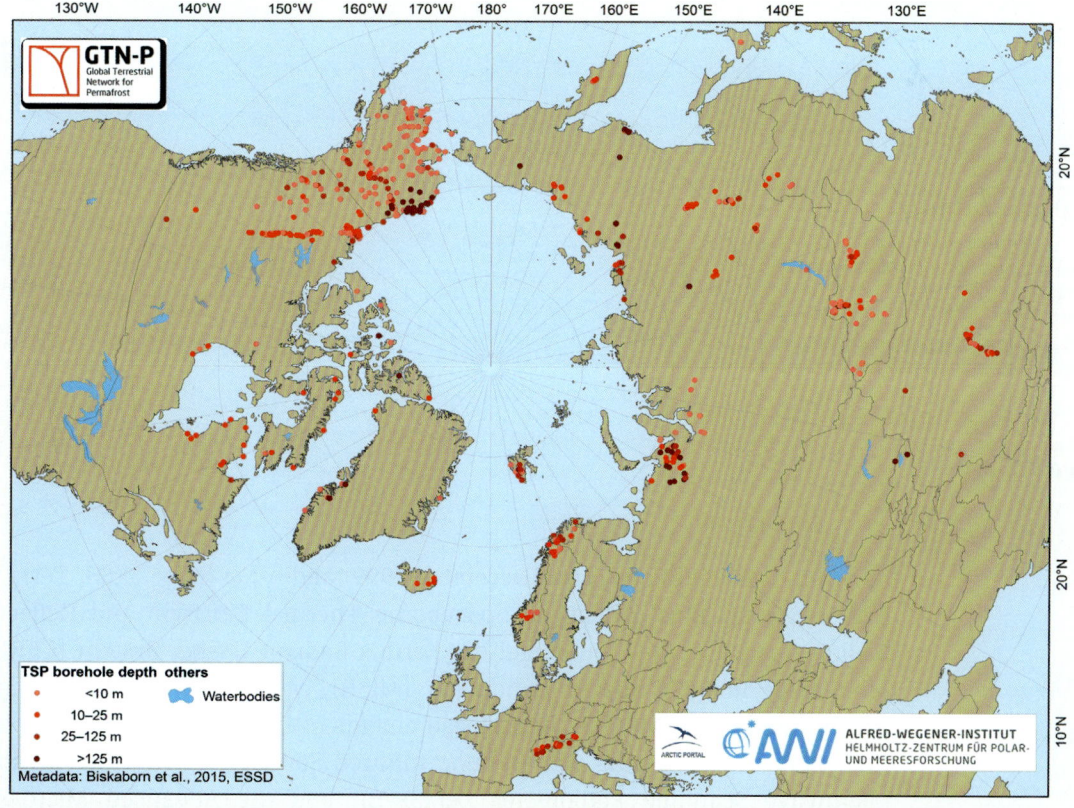

Figure 3.1 Distribution of permafrost bore holes in the Global Terrestrial Network of Permafrost (GTN-P): (a) Arctic, (b) Antarctic. (Biskaborn et al. 2015).
Source: http://gtnp.arcticportal.org/resources/maps/12-resources/37-maps-boreholes

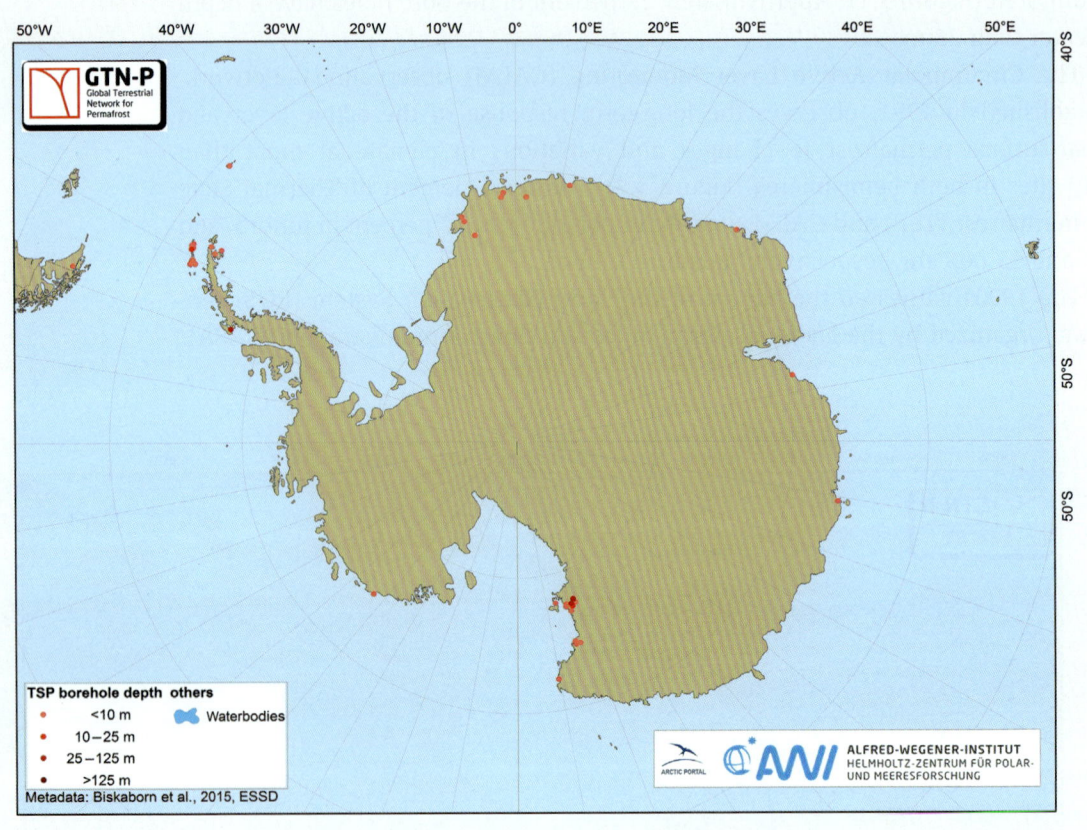

Figure 3.1 (cont.)

Its primary satellites are Terra (December 1999) and Aqua (May 2002). Terra has five sensor systems: Advanced Spaceborne Thermal Emission and Reflection Radiometer (ASTER), Clouds and the Earth's Radiant Energy System (CERES), Multi-angle Imaging SpectroRadiometer (MISR), Moderate Resolution Imaging Spectroradiometer (MODIS), and Measurements of Pollution in the Troposphere (MOPITT). Aqua has the Atmospheric Infrared Sounder (AIRS), the Advances Microwave Scanning Radiometer (AMSR-E), and the Advanced Microwave Sounding Unit (AMSU), in addition to CERES and MODIS. The flood of data and data products that resulted from the EOS program led to the establishment of discipline-specific Data Active Archive Centers (DAACs) to archive and distribute the various data products: Alaska Synthetic Aperture Radar (SAR) Facility (ASF) for SAR data and polar processes, USGS EROS Data Center (EDC) for Landsat data and land processes, Goddard Space Flight Center (GSFC) for earth science, Jet Propulsion Laboratory (JPL) for physical oceanography, Langley Research Center (LaRC) for atmospheric science, Marshall Space Flight Center (MSFC) for global hydrology, National Snow and Ice Data Center (NSIDC) for cryospheric science, and Department of Energy (DoE) Oak Ridge National Laboratory (ORNL) for biogeochemistry (Ramapriyan 2009).

The fourth International Polar Year (2007–2008) was a major multinational science program in both polar regions. In the Arctic, there were studies on sea ice, glaciers and the Greenland ice sheet, and frozen ground. An overview of the major achievements is provided by the Polar Research Board (2012).

Building on a series of workshops that began in 2006, the Arctic Council initiated the Sustaining Arctic Observing Networks (SAON) in 2012. Working through several task teams, it is designed to improve coordination and to promote sustained, integrated Arctic observing networks that provide free, open, and timely access to high-quality data.

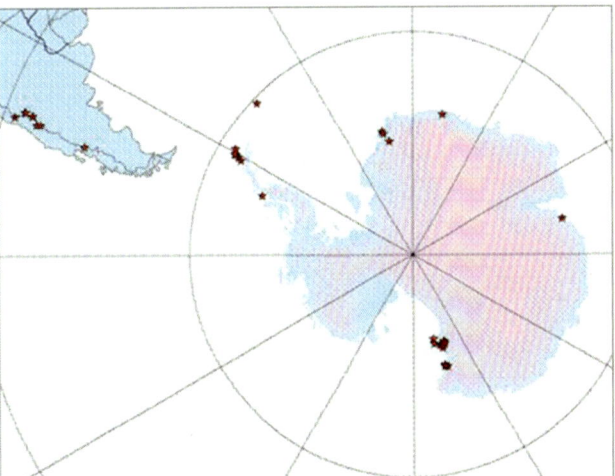

Figure 3.2 Distribution of CALM sites: (a) Arctic, (b) Antarctic (Biskaborn et al. 2015).
Source: www2.gwu.edu/~calm/data/data-links.html

3.2 In Situ Measurements

3.2.1 Meteorological Observations

Weather stations typically measure air temperature, barometric pressure, humidity, wind speed, wind direction, and precipitation amounts at synoptic times (00h00, 06h00, 12h00, and 18h00 UTC) and at intermediate synoptic hours (03h00, 09h00, 15h00, and 21h00 UTC). Station networks are sparse away from the coasts in both polar regions and have changed considerably over time, with transitions to automatic stations or closure of some remote locations.

Between 1947 and 1950, five sites in the Canadian Arctic archipelago were selected and weather stations were built jointly by Canada and the United States at Eureka and Alert on Ellesmere Island, Isachsen on Ellef Ringnes Island, Mould Bay on Prince Patrick Island, and Resolute on Cornwallis Island. The United States withdrew from Isachsen and Mould Bay in 1971 and from Eureka and Resolute in 1972. Budgetary constraints and reductions caused the decommissioning of Isachsen in September 1978, but an automatic weather station (AWS) installed there in 1989 is routinely interrogated by satellite. Mould Bay also has had an AWS with satellite uplink since 1993. Alert became a location for monitoring atmospheric trace gases in 1975. The World Meteorological Organization (WMO) Global Atmosphere Watch (GAW) observatory at Alert is the most northerly site in the network. In 1993, the Arctic Stratospheric Observatory (ASTRO) with two lidars began operations on a mountain ridge 15 km west of Eureka.

Permanent manned weather stations were established in Antarctica during the IGY in 1957–1958. All except two – at the South Pole and at Vostok on the East Antarctic Plateau – were located on the coast. In Antarctica, there is a well-established AWS network that was developed and has been maintained by the University of Wisconsin since 1980 (Figure 3.3). An overview is presented by Lazzara et al. (2012). These AWS collect data on the basic weather variables using sensors hardened for polar conditions. Some stations also now collect information on snow accumulation using acoustic depth gauges.

3.2.2 North Pole Drifting Stations

Romanov et al. (2000) describe the thirty-one North Pole Drifting Stations of the Soviet Union that operated in 1937–1938 and 1950–1991. The Russian Federation reinstituted the program in 2003, and it continues to the present (Figure 3.4).

Data from the program through the 1990s are available from Fetterer and Radionov (2000). Their atlas provides gridded fields over the Arctic Ocean of sea level pressure (decadal monthly means for the 1950s–1990s); precipitation (monthly mean fields for 1951–1990); cloud (decadal monthly mean fields of total and low cloud cover for 1952–1995); observed snow depths (monthly mean snow depth fields on land for 1966–1982); monthly mean snow depth fields for the Arctic Ocean for 1954–1991; observed monthly mean snow water equivalent fields for the Arctic Ocean for 1954–1991; global solar radiation (climatological monthly means); and a climatology of direct, global and net

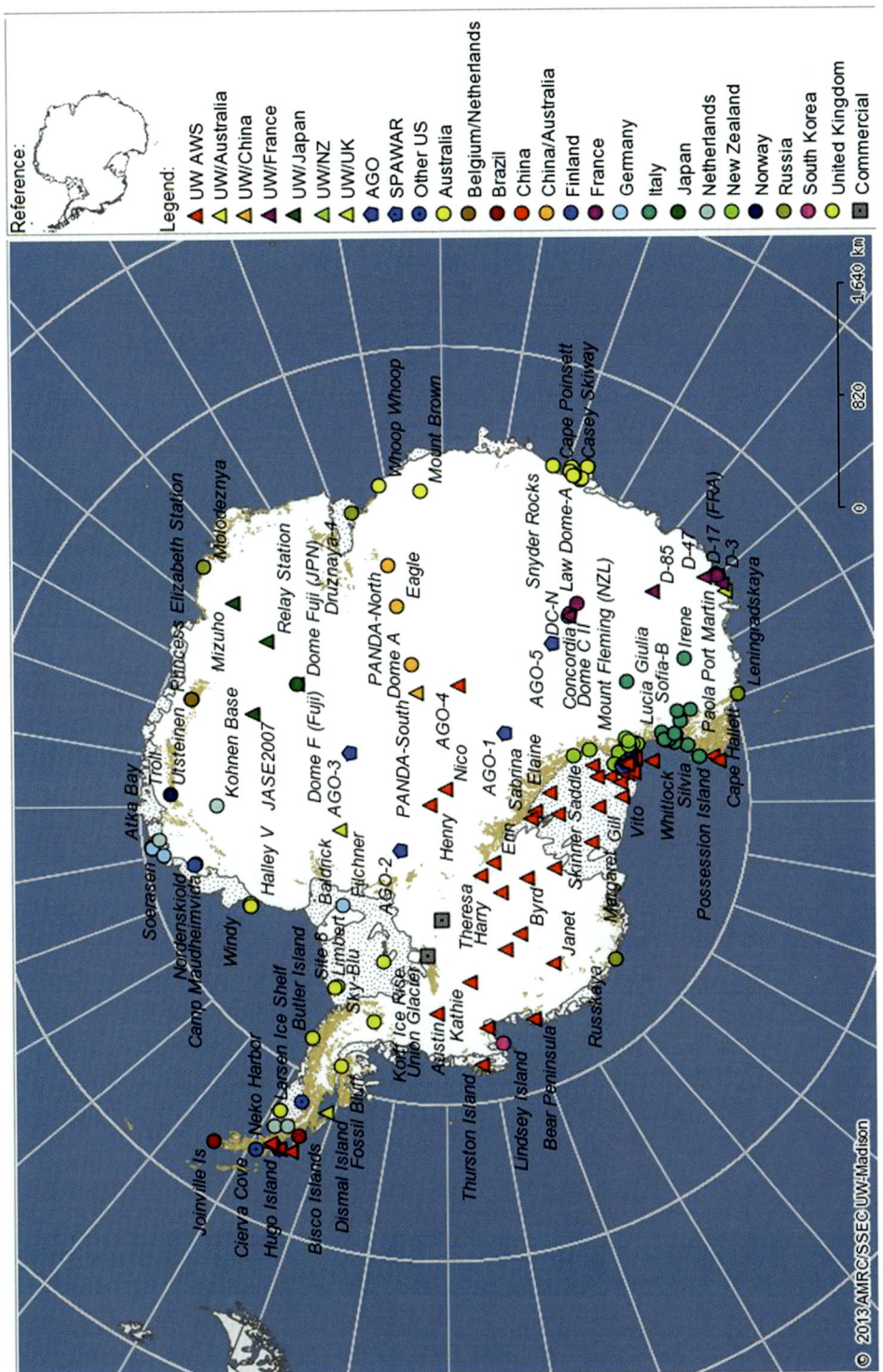

Figure 3.3 Antarctic automatic weather stations, 2016.
Courtesy of AMRC, Space Science and Engineering Center, University of Wisconsin, Madison, WI.

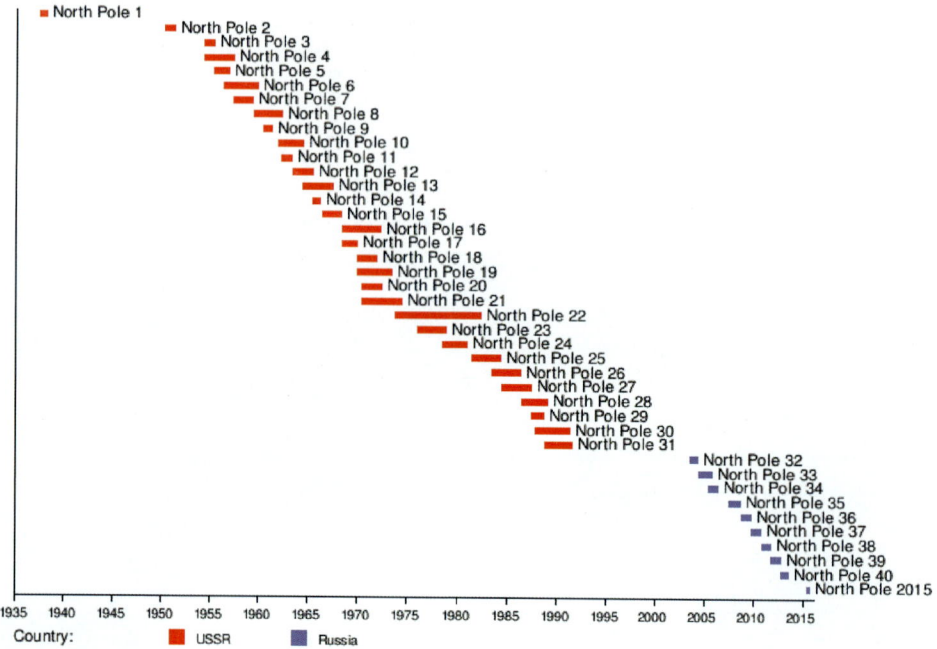

Figure 3.4 Timeline of North Pole drifting stations.
Source: Wikipedia, https://en.wikipedia.org/wiki/Drifting_ice_station.

radiation. Russian and Western drifting station data are given as 3- or 6-hour synoptic data and monthly means. Western data begin with the *Fram* expedition in 1893 and extend to AIDJEX in 1975. Russian Ice Patrol ships provided wind speed and direction, air pressure, air temperature, sea surface temperature, total cloud amount, low cloud amount, and relative humidity data for voyages from 1952 through 1982. Data from Russian Drifting Automatic Radio-meteorological Stations (DARMS) include once-daily measurements of wind speed and direction, surface air pressure, and air temperature from 1958 through 1975. Monthly means of meteorological observation data from sixty-five Russian and twenty-four Western coastal and island stations are also available for a period that spans the early 1950s through 1990 (nsidc.org/data/docs/noaa/g01938_ewg_arctic_met_atlas/).

Oceanographic observations are greatly hindered by the presence of sea ice in both polar regions. However, the ice may also serve as a platform for deploying ocean instruments.

An Arctic Ocean Buoy Program was initiated by the University of Washington's Applied Physics Laboratory in 1979 and provided the basis for the International Arctic Ocean Buoy Program (IABP) established in 1991 by eight nations.

Commonly, twenty-five to forty buoys operate at any given time and provide real-time position, pressure, temperature, and interpolated ice velocity. A description of the IABP data products can be found at http://iabp.apl.washington.edu/data.html.

Since 2000, the National Science Foundation's (NSF) Division of Polar Programs has supported the operation of the North Pole Environmental Observatory (NPEO). This program involves the deployment of buoys close to the North Pole each April, with the buoys then drifting over the next six months or so toward Fram Strait (http://psc.apl.washington.edu/northpole/Buoys.html).

In their report on the sea ice mass balance studies from the NPEO during 2000–2013, Perovich et al. (2014) failed to find any definitive trends in ice mass balance near the North Pole. There has been large interannual variability, however, with surface melt ranging from 0.02 to 0.50 m and bottom melt ranging from 0.10 to 0.57 m. The largest amounts of bottom melt have occurred in the past few years. For all nine years covered by the Perovich et al. report, the ice at the end of the melt season was at least 1.2 m thick.

3.2.3 Icebreakers

Dedicated scientific icebreaker expeditions, starting with IB *Ymer* in 1980 and RV *Polarstern* in 1983, increased observational activity in the Fram Strait area significantly during the 1980s. These expeditions have contributed data to large international programs like the Marginal Ice Zone Experiment (MIZEX), the Greenland Sea Project (GSP), and the Fram Strait Project (FSP) (Rudels 2015).

Regular cruises in both polar regions have been made since 1982 by the German icebreaker *Polarstern*. Reports are provided in the journal *Berichte zur Polar- und Meeresforschung*. The most recent is PS89 from the Weddell Sea in 2014–2015. Data are archived in the Pangaea system at the Alfred Wegener Institute in Bremerhaven (Diepenbroek et al. 2002). König-Langlo et al. (2006) provide a summary of twenty-five years of polar meteorological observations performed on the forty-four expeditions carried out between 1982 and 2006. Other national icebreaker cruises include those of the Swedish *Oden*, which undertook Antarctic research expeditions in 2007–2011 as well as earlier Arctic voyages.

Increasing numbers of icebreakers have entered the inner Arctic since the 1990s. The Soviet icebreaker *Rossiya* found warming of Atlantic water in the Eurasian Basin in 1990; the CCGS *Henry Larsen* also reported this phenomenon in the Makarov Basin in 1993.

Data for a section across the Arctic Ocean from the Chukchi Sea to the Nansen Basin were obtained by the Canadian CCGS *Louis S. St. Laurent* during the Arctic Ocean Section expedition of 1994. This section showed a significant increase in Atlantic water temperature on the Amundsen Basin side of the Lomonosov Ridge as well as at the North Pole compared to the 1991 temperatures (Carmack et al. 1997; Swift et al. 1997). The Canadian Coast Guard icebreaker *Sir John Franklin* operated in the Gulf of St. Lawrence and the Canadian Arctic from 1980 to 1996; it then became inactive but was recommissioned as a research icebreaker in 2003. The *Amundsen* has been frozen into the ice of the southern Beaufort Sea during several winters. Under a National Science Foundation project, the Canadian icebreaker *Des Groseilliers* was frozen into the Beaufort Sea during the Surface Heat Budget of the Arctic (SHEBA) experiment, which spanned October 1997 to October 1998. The data collected during SHEBA made major contributions to Arctic meteorology, oceanography, and sea ice research.

The *Nathaniel B. Palmer* is an icebreaker operated by the National Science Foundation for research around Antarctica. The US Coast Guard has several heavy-duty icebreakers. The *Polar Sea* has operated in the Arctic from 1979; in August 1994, it became the first US surface ship to reach the North Pole. It is currently inactive. The *Polar Star* primarily resupplies McMurdo. It was commissioned in 1976, was in "caretaker" status during 2006–2013, and then was reactivated. The UCGS *Healy* was commissioned in 2000 and operates in the western Arctic.

3.2.4 Oceanographic Observations

Surface drifters were developed for the First Global Atmospheric Research Program (GARP) Experiment (FGGE) of 1978–1979 and were deployed largely in the Southern Ocean to measure sea level pressure. These systems were both large and expensive (Clarke 2016). By comparison, the Tropical Ocean Global Atmosphere program (1985–1994)/World Ocean Circulation Experiment (1990–1998) (TOGA/WOCE) drifters used in the Surface Velocity Programme (SVP) were smaller and cheaper and had a very low freeboard, such that they were subject to minimal windage. A global array of these drifters, currently numbering 1,372 devices, is still maintained. Those in the Southern Ocean measure both SST and SLP.

The Argo program is a major component of the Global Ocean Observing System designed to monitor the temperature and salinity fields of the upper ocean. The profiling floats used in Argo are 2-m-long, freely drifting robotic devices that adjust their depth in the ocean by changing their buoyancy. The ARGO float starts at the surface and then dives to a depth of 1,000 m (the parking depth), where it rests for nine to ten days. After nine days at rest, it dives to a depth of 2,000 m, turns on its sampling equipment, and measures ocean properties as it rises to the surface, where it rests for sufficient time to transmit the data collected to satellite systems. It then returns to the parking depth to start another cycle. The typical duration of a complete cycle is ten days (Riser et al. 2016). The satellite Système Argos transmitters allow the position of a platform be determined up to fourteen times per day to an accuracy of a few hundred meters. Currently, an Argos or Iridium satellite link can be used to collect the data from the floats. The Argos program began in 1999 and reached a total of some 3,000 floats in 2007. There are currently about 3,900 floats worldwide. Figure 3.5 shows a map of their locations in September 2015.

The original Argo target called for temperature and salinity accuracies of 0.005 °C and 0.01 salinity units, respectively, with a pressure accuracy of 2.5 dbar (equivalent to a depth error of about 2.5 m). About 80 percent of the raw profile data transmitted from the floats meet these standards, with little or no correction required. The other 20 percent of the data are corrected, and nearly all of these profiles eventually meet the accuracy goals.

The initial coverage of the Argo program was devoted to the open ocean from 60° N to 60° S, but recent work (Wong and Riser 2011) has demonstrated that operation in seasonally ice-covered waters is possible. Algorithms have been developed that greatly increase a float's chances of surviving the winter in the ice zone by inferring the presence of ice from the near-surface temperature structure. If the float determines that ice is present through an onboard analysis of the stratification, it avoids the surface (and, therefore, the ice), stores its profile, and descends for another cycle. In the spring, when the ice retreats and the float can reach the surface, all of its winter data can be transmitted.

Beginning in 2004, ice-tethered profilers (ITPs) have been deployed in the Arctic Ocean. Toole et al. (2011) describe the instrumentation and early results from ITPs. With this type of device, a surface instrument package is installed on an ice floe with a line to about 700–800 m depth, where temperature and salinity profiles are collected. In most cases the ITPs operate for one to three years. The data are relayed via satellite to Woods Hole Oceanographic Institute in near-real time and processed and archived there (www.whoi.edu/page.do? pid = 26316). As of April 2017, there had been ninety-three ITPs. The data were applied by Rabe et al. (2011) to calculate freshwater budgets for the Arctic.

Figure 3.5 The location of 3,918 ARGO floats in September 2015.
Source: Joint Commission on Maritime Meteorology, http://argo.jcommops.org/maps.html.

As described by Rudels (2015), many long-term mooring deployments were undertaken in the first decade of the twenty-first century. The Nansen Amundsen Basin Observation System (NABOS) was established at the Laptev Sea continental slope in 2002. The North Pole Environmental Observatory (NPEO), consisting of one deep mooring in the Amundsen Basin close to the North Pole, was established in 2001 (Morison et al. 2002). The Beaufort Gyre Exploration Program (BGEP) started with four deep moorings in the Beaufort Sea in 2003 (Proshutinsky et al. 2009).

Instrumented elephant seals have been used in the Southern Ocean to measure oceanographic parameters (conductivity, temperature, and depth [CTD]) to average depths of approximately 500 m. The data are relayed by satellite through the ARGOS system. Roquet et al. (2009) provide data sets from eight instrumented seals that moved south from the Kerguelen islands to the coast of Antarctica between about 50° and 95° E. These measurements are invaluable in remote, inaccessible locations.

3.2.5 Upward-Looking Sonar

Sound navigation and ranging, commonly called sonar, was installed on submarines during World War II and began to be used to detect the underside of sea ice in the Arctic in the 1950s. The USS *Nautilus* was the first submarine to sail from Barrow and transit the North Pole in August 1958. The ice draft characteristics were analyzed from the upward-looking sonar (ULS) by McLaren (1986) and later compared with those from an identical transit in August 1970 by the USS *Queenfish*, commanded by McLaren (1989). Rothrock et al. (1999) analyzed measurements from submarine cruises between 1993 and 1997 and compared them with similar data acquired between 1958 and 1976. These comparisons show that the mean ice draft at the end of the melt season had decreased by about 1.3 m in most of the deep-water portion of the Arctic Ocean, from 3.1 m in 1958–1976 to 1.8 m in the 1990s.

The Office of Polar Programs at the National Science Foundation has supported a program named SCience ICe EXcercise (SCICEX) to acquire sea ice data from submarines since 1994. Edwards and Coakley (2003) provide an overview of activities and results during the first decade. ULS data for the Arctic Ocean collected by US and Royal Navy submarines between 1960 and 2005 are available at http://nsidc.org/data/g01360 (NSIDC 2006).

ULS systems can also be moored to the ocean floor. Two types of ULS observations are possible: Ice profiling sonar (IPS) is used to obtain ice draft data and an acoustic Doppler current profiler (ADCP) is used to obtain ice velocity data. NSIDC archives moored ULS data for the continental shelf of the east Beaufort Sea, the Greenland Sea and Fram Strait, and the Weddell Sea. Strass (1998) has reported on two years of moored ULS measurements at four locations in the Weddell Sea.

3.2.6 Traditional Knowledge and Community Science

Beginning during the Third IPY, community-based observations by Arctic residents and indigenous peoples started to be organized (Johnson et al. 2015). The National Snow and Ice Data Center at the University of Colorado received NSF funding to establish the Exchange for Local Observations and Knowledge of the Arctic (ELOKA) project. This project provides ongoing data management services and support to Arctic communities and others who are working with local and traditional knowledge (LTK) or who are gathering community-based monitoring data and information (https://eloka-arctic.org/). Data products include seasonal sea ice observations in northern and western Alaska since 2006, as well as near-real time weather observations in eastern Baffin Island by Inuit (Gearheard et al. 2013). Local sea ice observations at Barrow, Clyde River (Baffin Island), and Qaanaaq, Greenland, were organized by Shari Gearheard and Andrew Mahoney (Mahoney et al. 2009) and a handbook for community-based sea ice monitoring has been published (Mahoney and Gearheard 2008). For coastal areas around the Bering Sea, Williams et al. (2017) used survey data from a community-based observing network. They compared those observations with instrument-derived data for air temperature, sea ice breakup and freeze-up, and vegetation changes. Except for air temperature, where there are problems due to different spatial scales, there was good correspondence.

A recent addition to community science is being made by passengers on cruise ships sailing from Murmansk to the North Pole and back (Farmer et al. 2016). Guides developed an observation program for teams of observers on the ship's bridge, at 2-hour intervals, using a standardized procedure to observe ice conditions en route and at the North Pole. The data collected are made rapidly available.

3.3 Remote Sensing

In this section, we first trace the history of the remote sensing of polar regions. Sea ice in the Eurasian Arctic seas was routinely mapped in the Soviet Union using visual reconnaissance from aircraft flights starting in July 1933 and continuing until 1992 (Borodachev and Shilnikov 2003). From the 1950s onward, thirty to forty aircraft made 500–700 flights annually (Johannessen et al. 2007). Initially, coverage spanned only late summer, but from 1950 onward it was continuous throughout the year. These and additional satellite data acquired between 1933 and 2006 were analyzed by Mahoney et al. (2008).

A massive aerial campaign to photograph the Canadian Arctic was undertaken in the late 1940s through 1950s, and the vertical and oblique photographs obtained covered all land areas. The US Geological Survey (USGS) photographed numerous glaciers in western North America and Alaska beginning in the 1950s. Aerial photography transitioned to satellite photography and imagery with the launch of the Landsat series in July 1972, using a return beam Videocon (RBV) camera and multispectral scanner (MSS).

Pope et al. (2014) review open access sources of aerial and spaceborne imagery of the polar regions and for the cryosphere. They also discuss data products and tools. Lubin and Massom (2006) and Massom and Lubin (2006) published a two-volume study of polar remote sensing of the atmosphere, oceans, and ice sheets. Tedesco (2015) has edited a text that addresses the remote sensing of all components of the cryosphere. Various chapters are referenced in the following subsections.

3.3.1 Satellite Photography

Beginning in 1978, a major effort was undertaken by Williams and Ferrigno (1988) to compile a *Satellite Image Atlas of Glaciers of the World*. The authors used maps; aerial photographs; Landsat 1, 2, and 3 MSS images; and Landsat 2 and 3 RBV images to inventory the areal distribution of glacier ice between about 82° N and 82° S. Some later contributors also used Landsat 4 and 5 MSS and Thematic Mapper (TM), Landsat 7 Enhanced Thematic Mapper-Plus (ETM+), and other satellite images. Polar volumes include those covering Antarctica (B), Greenland (C), and Alaska (K). The series was completed in 2012.

A NASA-funded project for Global Land Ice Measurements from Space (GLIMS) is continuing at the NSIDC (www.glims.org/) and has provided outlines for almost 200,000 glaciers. GLIMS uses data collected primarily by the Advanced Spaceborne Thermal Emission and Reflection Radiometer instrument, aboard the Terra satellite, and the Landsat ETM+, along with historical observations. The processing of ASTER data for the GLIMS project is fully documented by Ramachandran et al. (2014). Raup et al. (2015) document the remote sensing of glaciers and provide details of the optical instruments suitable for glacier mapping.

Northern hemisphere snow cover extent has been mapped by the National Oceanographic and Atmospheric Administration (NOAA) since 1966, first with Very High Resolution Radiometer (VHRR) visible imagery, and since 1972 with Advanced Very High Resolution Radiometer (AVHRR) 1.1-km resolution data. Global snow cover maps are now available from 500-m resolution MODIS devices on Terra (February 2000 to present) and Aqua (July 2002 to present) (Hall et al. 2015).

The theoretical basis for monitoring snow grain size and albedo/pollution of a snow surface is presented by Khokanovsky (2015). A series of digital image maps of surface morphology and optical snow grain size that cover the Antarctic continent and its surrounding islands have been prepared by NSIDC and the University of New Hampshire (Haran et al. 2005; Scambos et al. 2007). The MODIS Mosaic of Antarctica (MOA) image maps are derived from composites of 260 MODIS orbit swaths spanning the austral summer seasons of 2003–2004 and 2008–2009; a 2013–2014 image set is in preparation. The USGS with the British Antarctic Survey, NASA, and the NSF have produced a Landsat Image Mosaic of Antarctica (LIMA) from more than 1,000 scenes of ETM+ to latitude 82.5° S at 15-m resolution (http://lima.usgs.gov/view_lima.php).

Photoclinometric methods of improving digital elevation models (DEMs) began with the work of Bindschadler and Vornberger (1994), who used Landsat Thematic Mapper data to derive a detailed elevation map of Ice Stream C, West Antarctica, within a grid of airborne laser altimetry data. They derived the photometric function from the laser elevation data and spatially filtered image data. Scambos and Fahnestock (1998) have shown how DEMs over ice sheets can be enhanced by using AVHRR-based photoclinometry. A DEM is used to establish the photometric relationship for two AVHRR images of a snow surface. Slopes from the DEM are compared with AVHRR data that are filtered (i.e., blurred) to the resolution of the DEM to give an empirical photometric determination. This information is then used to convert unfiltered AVHRR data into quantitative slope measurements of the surface in the along-sun direction in each image, resolving features not present (or poorly represented) in the DEM. Co-registration of the images is based on the assumption that the two slope fields from the images describe one continuous smooth surface, and the combined slopes are converted to topography. The improved topographic map reveals a ten-fold increase in local surface relief over the northeast Greenland ice stream feature. Scambos and Haran (2002) assembled an elevation grid for the Greenland ice sheet using a combination of the best current DEM and forty-four AVHRR satellite images acquired in spring 1997.

The images are used to quantitatively enhance the representation of surface undulations through photoclinometry.

Panchromatic imagery from Worldview 1 (launched in 2007), 2 (launched in 2009), and 3 (launched in August 2014) is now providing DEMs globally with spatial resolutions of approximately 0.5 m and vertical resolutions of 35–28 cm. An Arctic DEM is being compiled through the NSF and the National Geospatial-Intelligence Agency (NGA). As of September 2017, 97 percent of the Arctic was available (free to federal grantees; http://pgc.umn.edu/arcticdem).

3.3.2 Passive Microwave

Passive microwave data have been used to map sea ice extent and concentration since the 1970s (Carsey 1992). In December 1972, NASA launched the Electrically Scanning Microwave Radiometer (ESMR) on Nimbus 5. Until May 1977, it provided single-channel, horizontally polarized radiation at a frequency of 19 GHz. Sea ice can be discriminated in the microwave region through differences in the emissivity characteristics between ice and ocean; in general, sea ice is more emissive than the open ocean. The ESMR's ability to operate in darkness and through cloud cover yielded the first comprehensive maps of polar sea ice extent for 1973–1976 (Parkinson et al. 1987; Zwally et al. 1983). In October 1978, the Scanning Multichannel Microwave Radiometer (SMMR) was launched on Nimbus 7; it operated until August 1987. Frequencies of 18 and 37 GHz were used in various algorithms to derive sea ice concentrations for first-year and multiyear ice based on the SMMR data (Gloersen et al. 1992). The records continued with the Special Sensor Microwave Imager (SSM/I) and Sounder (SSMIS) on Defense Meteorological Satellite Program (DMSP) satellites. These instruments had five frequencies, including 19 and 37 GHz.

The DMSP provides global visible and infrared cloud data and other specialized meteorological, oceanographic, and solar–geophysical data in support of world-wide Department of Defense (DoD) operations. The satellites that have been used and are currently in use by NSIDC are as follows: the DMSP F8 satellite (June 1987–December 1991), F11 (September 1991–December 1995), F13 (March 1995–February 2015), F15 (launched in December 1999), F17 (launched in November 2006), and F19 (launched in 2014). All of these satellites were placed in a near-circular, sun-synchronous, polar orbit. Slight differences in the data obtained by the different satellites in periods of overlap are used to standardize the sea ice concentration time series. Details are provided at https://nsidc.org/data/docs/daac/nsidc0051_gsfc_seaice.gd.html#version.

The Advanced Microwave Scanning Radiometer-2 (AMSR-2) was launched on the Japan Aerospace Exploration Agency GCOM-W1satellite in 2012. It continues to operate. The AMSR-2 instrument is a follow-on to the AMSR-E instrument onboard the NASA Aqua satellite that ceased operation in 2011 http://www .remss.com/missions/amsr/.

An improved-resolution passive microwave data set, known as the Calibrated Passive Microwave Daily EASE–Grid 2.0 (CETB) gridded data (3.125–25 km), was produced under the NASA Making Earth System Data Records for Use in Research Environments (MEaSUREs) program and released in 2016 (Brodzik et al. 2016). It combines data from SMMR, SSM/I, SSMIS, and AMSR-E and is available at http://nsidc.org/data/nsidc-0630.

Sea ice can be discriminated in the microwave data through differences in the emissive characteristics between ice and ocean; in general, sea ice is more emissive than the open ocean. Use of combinations of frequencies allows more accurate discrimination between ice and ocean as well as the ability to estimate fractional ice cover within regions of mixed ice and water (Carsey 1992). The NASA Team algorithm (Cavalieri et al. 1984) uses a polarization ratio (PR) and a gradient ratio (GR). The polarization ratio is

$$PR_{19V/H} = \frac{T_B[19V] - T_B[19H]}{T_B[19V] + T_B[19H]} \tag{3.1}$$

The gradient ratio is

$$GR_{37V/19V} = \frac{T_B 37V - T_B 19V}{T_B 37V + T_B 19V} \tag{3.2}$$

The PR is small for ice and large for water, while the GR is small for first-year ice but large for multiyear ice (Cavalieri et al. 1984). Combinations of PR and GR enable the T_B signatures to be interpreted as ice type; there are some eight or so algorithms in use for this purpose.

Another approach is used in the Bootstrap algorithm (Comiso 1986), which employs linear combinations of 19- and 37-GHz frequencies at both horizontal and vertical polarizations to estimate fraction ice coverage. The NASA Team and Bootstrap algorithms, as well as other algorithms, require empirically derived "tie points," or coefficients for pure surface types (100 percent ice and 100 percent water).

Meier and Stroeve (2005) have examined ice-extent fields from SSM/I passive microwave and QuikSCAT scatterometer data for 1999–2004. They report that there is general agreement between fields derived from the two sources, though scatterometer estimates yield substantially lower ice extents during winter and larger ice extents in late summer. Comparisons with ice-edge locations estimated from AVHRR imagery indicate that enhanced scatterometer data can sometimes provide an improved edge location, but there is substantial variation in the results, depending on the local conditions.

Meier et al. (2015) have analyzed sea ice concentrations derived from the operational daily Multisensor Analyzed Sea Ice Extent (MASIE). This work is based on the Interactive Multisensor Snow and Ice Mapping System, which uses all available imagery. The 4-km spatial resolution, along with multiple input data and manual analysis, can potentially provide more precise mapping of the ice edge than passive microwave estimates gridded to 25 km in the NSIDC Sea Ice Index. Comparison of the results for 2006–2014 shows that MASIE identifies somewhat higher ice concentrations except during May–June and late September–October. The mean differences are about $\pm 0.25 \times 10^6$ km^2.

Sea ice motion mapping using AVHRR imagery was demonstrated by Emery et al. (1991) for Fram Strait. These researchers extracted ice velocity vectors from sequential AVHRR imagery by combining the maximum cross-correlation method with a spatial filtering technique on the image-inferred ice motion vectors. They computed the cross-correlations between images directly from the image brightness values. Agnew et al. (1997) applied the same basic method to SSM/I horizontally polarized 85-GHz data for December 1993 to January 1994 over the area from the Canadian Arctic archipelago to the coast of Siberia. Emery et al. (1997) extended cross-correlation image processing methods applied to 12.5-km, 85.5-GHz SSM/I data to provide details about sea ice motion in both the Arctic and Southern Ocean for 1988–1994. Updated records are provided by Tschudi et al. (2015) who gridded the SMMR, SSM/I, SSMIS and AMSR-E data to 25-km resolution. Figure 3.6 shows the mean motions in each hemisphere for 2000–2016. In the Antarctic, the motion vectors demonstrate a nearly continuous westward transport along the Antarctic coast, with well-defined regions of exchange between this East Wind Drift and the Antarctic Circumpolar Current.

Global snow depth and water equivalent can be estimated from passive microwave data. Maps of snow water equivalent gridded to 25 km from SMMR and SSM/I have been produced from 1978 to 2007 (with updates) by Armstrong et al. (2005) (http://nsidc.org/data/nsidc-0271.html). There are problems where topography, coastlines, or forests are present, and biases are introduced when there is snow melt or wet snow as well as early in the season when the snowpack is shallow.

Arthern et al. (2006) use the polarization ratio (P) at 6.9 GHz from AMSR-E to estimate snow accumulation over Antarctica. The ratio P is defined as follows:

$$\frac{T_B(V) - T_B(H)}{T_B(V) + T_B(H)}$$

where brightness temperature $T_B(V)$ is at vertical polarization and brightness temperature $T_B(H)$ is at horizontal orientation. In the dry snow region, the polarization ratio has been shown to be higher for sites with a low accumulation rate. Using density profiles from snow pits, Winebrenner et al. (2001) have shown that

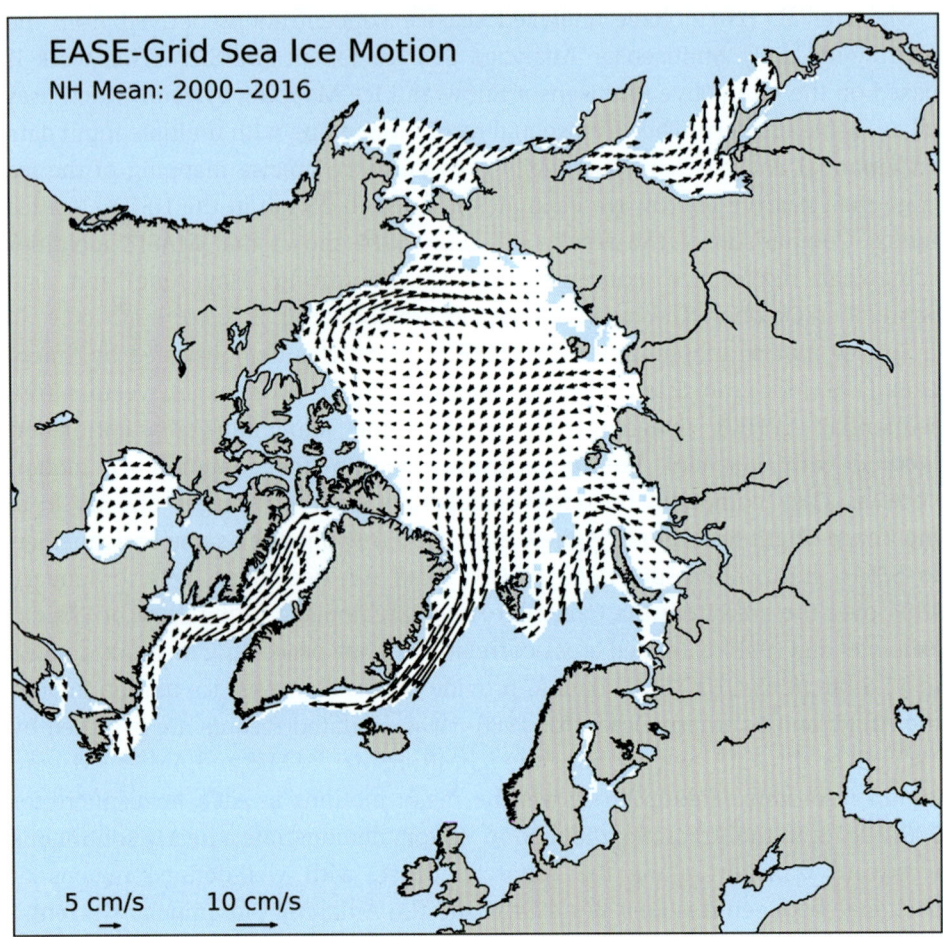

Figure 3.6 Mean sea ice velocity vectors of (a) the Arctic Ocean, (b) the Antarctic for 2000–2016.
Source: Tschudi et al. 2016. http://dx.doi.org/10.5067/O57VAIT2AYYY.

greater variation in layer densities (on the wavelength scale) increases polarization for the low-accumulation-rate sites above that of high-accumulation-rate sites. Koenig et al. (2015) note that passive microwave radiation (PMR) approaches are restricted to the dry snow zone and require ground truth measurements, which are not always readily available.

Between 2009 and 2011, approximately 610 12.5-km satellite grid cells were covered by seasonal sea ice where both satellite SD retrievals and Operation Ice Bridge (OIB) data were available. Using all the available data, Brucker and Markus (2013) calculated the difference between the AMSR-E product and the averaged OIB snow radar-derived standard deviation (SD) to be 0.00 ± 0.07 m. The satellite-derived SD was accurate in the Beaufort Sea and the Canadian archipelago, but

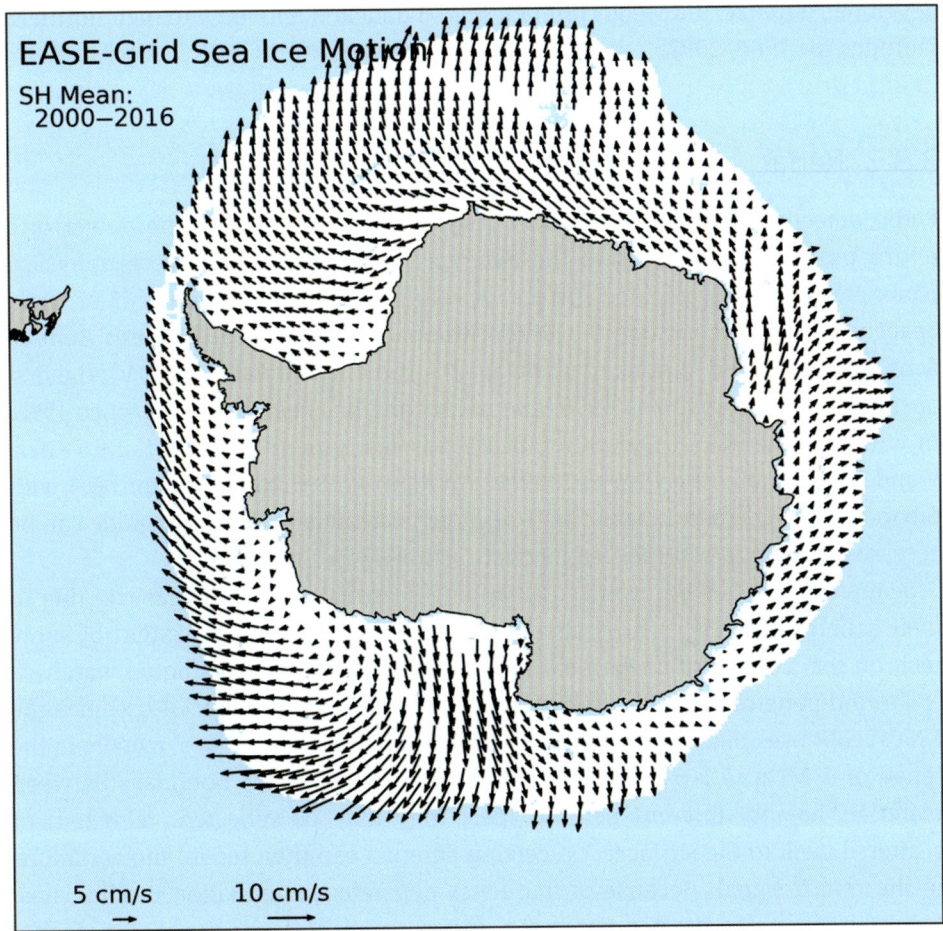

Figure 3.6 (cont.)

underestimated the sea ice thickness (by approximately 0.07 m) in the Nares Strait. The RMSE between the two products ranged between 0.03 and 0.15 m.

Snow melt over the Greenland ice sheet has been mapped using SSM/I data by Abdalati and Steffen (1996) and using SMMR and SSM/I data by Mote and Anderson (1995). The latter used a threshold value of the 37-GHz brightness temperature, while the former used a cross-polarized gradient ratio (XPGR), which is a normalized difference between the 19-GHz horizontally polarized and 37-GHz vertically polarized brightness temperatures. Tedesco et al. (2015) provide an overview of mapping melting snow over Greenland.

Passive microwave data have also been used to map seasonally frozen ground for 1978–2003 over the northern hemisphere (Zhang and Armstrong 2001, 2003). Frozen soils relative to unfrozen soils exhibit (1) lower thermal temperatures, (2) higher emissivity, and (3) lower brightness temperatures. Smith et al. (2004)

developed a freeze/thaw algorithm for SSM/I data and applied it to high northern latitudes for 1988–2002.

3.3.3 Radar

Radio detection and ranging (radar) instruments that measure the power of a return pulse scattered back to the antenna can be used to derive geophysical parameters of the illuminated surface, or volume, with a resolution of 15 or 30 m. Spaceborne sensors include the European Space Agency's (ESA) Earth Remote Sensing (ERS)-1 and -2 Active Microwave Instrument (C band, 3 GHz, V); the first operated between 1992 and 1996, and the second has been operating since 1996. In addition, the NASA QuikSCAT SeaWinds instrument (Ku band, 13.6 GHz, V and H) operated from 1999 to 2009. In April 2014, ESA launched Sentinel 1 with C band (5.4 GHz) radar. Level 1 and 2 products are available. A user guide can be accessed at https://sentinel.esa.int/web/sentinel/user-guides.

Scatterometry can be useful for mapping seasonal and multiyear ice, due to their salinity differences. It is also beneficial for measuring the extent of snow melt on sea ice, due to its extreme sensitivity to the presence of liquid water.

Ground-penetrating radar (GPR) has been widely used in glaciology since the 1950s. GPR uses high-frequency (generally polarized) radio waves, usually in the range of 2 MHz to 2.6 GHz. When the waves encounter a boundary between materials having different permittivities, they may be reflected, refracted, or scattered back to the surface. A receiving antenna can then record the variations in the return signal. Because of frequency-dependent attenuation mechanisms, higher frequencies do not penetrate as far as lower frequencies, but higher frequencies may provide improved resolution. Applications include ice thickness measurements in glaciers and ice sheets, snow depth, and permafrost thickness. Gogineni and Yan (2015) provide a detailed overview of the principles and applications to ice thickness measurements. The great difference in permittivity of liquid water compared to other materials makes radar an effective tool on glaciers and permafrost, where it is necessary to distinguish between water (permittivity 80) and ice (permittivity 3–4) (Evans 1963; Gruber and Ludwig 1996). Ice that is at or close to its pressure melting point is called temperate ice, whereas ice that is below this temperature is called cold ice. Temperate ice contains various amounts of water in liquid form in cavities, in channels, and between grain (crystal) boundaries. High polar glaciers and ice sheets usually contain only cold ice. In contrast, temperate ice is found in temperate glaciers and polythermal glaciers, the latter of which also contain cold ice. Those glaciers are situated in mountain areas and in subpolar regions.

In general, radar waves of higher frequencies are used to detect internal structures of glaciers due to their higher resolution, while lower frequencies are more suitable to penetrate the ice to view the bottom topography. GPR has been used at high frequencies (500 MHz) for mass balance measurements on temperate glaciers to detect the last summer's surface under the winter snow at the end of the winter season. Polar ice sheets usually contain only cold ice, so the internal reflections are not caused by changes in permittivity, which is mainly a function of depth due to increasing density. Instead, the reflections are caused by different conductivities, which are due to changes in acidity and other chemical impurities due to volcanic eruptions and changes in climatic conditions (Moore 1988). For example, the ice of the Greenland ice sheet that was deposited during the Wisconsin glaciation is alkaline due to higher dust content, whereas the Holocene ice in the same ice sheet is acidic.

The application of GPR in permafrost investigations uses the strong dielectric contrast between liquid water and ice to distinguish between frozen and unfrozen materials. GPR has been successfully used to detect ground ice bodies and to map the seasonal depth of thaw in permafrost areas (Judge et al. 1991).

Another active microwave sensor useful for cryospheric studies is synthetic aperture radar (SAR). In contrast to scatterometers, it is an imaging radar that synthesizes images from multiple looks during the satellite's motion in orbit to effectively create a large antenna and thereby attain much higher spatial resolution. In 1997, the Canadian RADARSAT-1 satellite was rotated in orbit so that its SAR antenna looked south toward Antarctica. This permitted the first high-resolution mapping of the entire Antarctic continent (Jezek et al. 2002). Calibrated radar backscatter data have been assembled into ninety tiles at 25-m resolution and an image mosaic available at 125-m to 1-km resolutions. The RADARSAT Image Map of Antarctica is the product of the RADARSAT-1 Antarctic Mapping Project (RAMP). This mosaic provides a detailed look at ice sheet morphology, rock outcrops, the coastline, and other features. Accompanying this mosaic is the high-resolution RAMP digital elevation model, which combines topographic data from a variety of non-SAR sources to provide consistent coverage of all of Antarctica.

Interferometric SAR (InSAR) (Gogineni and Yan 2015; Goldstein et al. 1993; Joughin et al. 2000) is used to map ice stream velocity. InSAR employs pairs of high-resolution SAR images to generate high-quality terrain elevation maps using phase interferometry methods. This technology provides an all-weather, day/night capability to generate measurements of terrain elevation on a dense grid of sample points with accuracies of a few meters. Both spaceborne and airborne systems are in use. Whereas SAR makes use of the amplitude and the absolute phase of the return signal data, interferometry uses the differential phase of the reflected radiation. These data may be obtained either from multiple passes along the same trajectory and/or from multiple displaced phase centers (antennas) on a single pass. Joughin et al. (2010) used interferometry to map the flow velocity over much of the Greenland ice sheet for the winters of 2000–2001 and 2005–2006.

Ice velocity mapping of glaciers and ice streams using image-to-image cross-correlation software has been described by Bindschadler and Scambos (1991) and Scambos et al. (1992). It is applied to pairs of digital satellite images to map the velocity field of moving ice in Ice Streams D and E, West Antarctica. This technique uses small-scale glacial surface features, such as crevasse scars and snow dunes, as markers on the surface of the moving ice.

3.3.4 Altimetry and Gravity Measurement

Radar altimeters and light detection and ranging (lidar) altimeters are used to estimate sea ice thickness by measuring the freeboard – that is, the height of the ice above the ocean surface, which is detected in leads and polynyas. The ERS radar altimetry has been used to estimate sea ice thickness in the Arctic (Laxon et al. 2003) and the Antarctic (Giles et al. 2008) but its orbit limits its application. Cryosat 2 with a Synthetic Aperture Radar (SAR)/Interferometric Radar Altimeter (SIRAL) was launched by ESA in April 2010 and is providing sea ice thickness and volume estimates for the Arctic (Laxon et al. 2012). The altimeter's along-track footprint is divided into more than sixty separate beams with a resolution of around 250 m each, sufficient to differentiate ice floes from open water and often the leads between them. Lidar applications from aircraft and satellites began around 1989 but their applications to the cryosphere surged from 2008 onward (Bardwaj et al. 2016). Bardwaj et al. summarize studies of polar sea ice, ice sheets, ice caps, glaciers, and permafrost since 2001.

The Geoscience Laser Altimeter System (GLAS) instrument on the Ice, Cloud, and Land Elevation Satellite (ICESat) was launched by NASA in January 2003 and operated until October 2009. Using frequencies of 532 and 1064 nm, it had a 70-m footprint and a 170-m along-path spacing. The GLAS data have been used to determine sea ice freeboard in the Arctic (Kwok et al. 2008). The main objective of the GLAS sensor was to measure ice sheet elevations and changes in elevation (Schutz et al. 2005). GLAS provided global coverage between 86° N and 86° S.

Airborne or ground surface measurement of snow depth is now being widely applied through the use of laser altimetry or lidar systems. Deems et al. (2013) provide a review of the procedures. Sensors allow mapping of vegetation height and snow at ground surface elevations below a forest canopy. Airborne data have vertical accuracies on a decimeter scale with 1-m point spacing, while ground systems have millimeter-scale accuracy.

Changes in mass balance of the two major ice sheets have been derived from a variety of satellite measurements, including the Gravity Recovery and Climate Experiment (GRACE) (Wingham et al. 2006), and from altimetry methods, including both radar-based and laser altimetry (Cryosat-2) (MacMillan et al. 2016; Martin Espanol et al. 2016) and a mass budget approach (Rignot et al. 2011). GRACE makes detailed measurements of Earth's gravity field anomalies. The pair of satellites, which were launched in March 2002, are separated by approximately 200 km along their orbit track. Cryosat-2 was launched in April 2010.

Shepherd et al. (2012) have provided an assessment of the various estimates of ice sheet imbalance in the 1990s through early 2000s. They suggest that East Antarctica was gaining some 25 Gt year^{-1} over this period, while West Antarctica was losing about 50 Gt year^{-1} and the Greenland ice sheet was losing about 100 Gt year^{-1}. Their estimate of the combined imbalance of Greenland and Antarctica is about 125 Gt year^{-1} of ice, enough to raise sea level by 0.35 mm year^{-1}. This is only a modest contribution to the present rate of sea-level rise of 3.0 mm year^{-1}. However, much of the loss from Antarctica and Greenland is the result of the flow of ice to the ocean from ice streams and glaciers, which has accelerated over the past decade. According to Velicogna et al. (2014), Greenland lost 280 ± 58 Gt of ice per year between 2003 and 2013, while Antarctica lost 67 ± 44 Gt year^{-1} in the same period. These losses equate to a total of 0.9 mm year^{-1} of sea-level rise. Tedesco et al. (2013) have plotted the cumulative mass anomaly for Greenland from GRACE data, showing that it declined to -1500 Gt in 2012.

3.3.5 Electromagnetic Sounding of Sea Ice

The use of electromagnetic induction (EMI) sounding to determine sea ice thickness was pioneered in the Arctic in the 1970s and 1980s using both airborne and ground-based surveys (Haas et al. 1997; Kovacs and Holladay 1990). The transmitter coil, excited by a sinusoidal current, produces a primary magnetic field that induces currents in a nearby conductive material – in this case, seawater. The currents then produce a secondary magnetic field that is sensed by the receiver coil. The received signal is converted into a measurement of apparent conductivity that represents the integrated electrical conductivity of the halfspace beneath the instrument. The results of EMI sounding have been most consistently reliable over level, undeformed ice, and are usually within 10 percent of drilled measurements. Worby et al. (1999) later applied EMI sounding in the Antarctic.

3.3.6 Global Positioning Systems

A novel aspect of satellite meteorology is the use of the effect of the atmosphere on the propagation of the radio signals of the Global Positioning System (GPS) to monitor the atmospheric water content. The GPS technique is especially valuable because it measures absolute water vapor content (or partial pressure) integrated from the surface to the top of the atmosphere. The propagation path delay is proportional to $(\cos z)^{-1} \, 1\cos z \frac{1}{\cos z}$, where z is the elevation angle of the satellite.

In December 2016, NASA launched the Cyclone Global Navigation Satellite System (CYGNSS). This mission uses eight micro-satellites to measure wind speeds over Earth's oceans (Marsik et al. 2016).

3.3.7 Unmanned Aerial Vehicles

Unmanned aerial vehicles (UAVs), or drones, are increasingly being used to image remote locations in the Arctic and Antarctic. James Maslanik at the University of Colorado deployed a fixed-wing aerosonde UAV off northern Alaska in 1999–2003; Curry et al. (2004) described the atmospheric boundary layer and remote sensing observations of the sea ice that were obtained during this project. Air space for UAV operation was designated over the Beaufort Sea north of Oliktok Point, Alaska, that is approximately 1,300 km by 75 wide and divided into two altitude ranges (0–600 and 600–3,000 m). Inoue et al. (2008) reported on melt pond studies using aerosondes in this area. Maslanik led the Marginal Ice Zone Ocean and Ice Observations and Processes EXperiment (MIZOPEX) that studied ice–ocean conditions in this space in the summers of 2012–2013. In Antarctica, Cassano et al. (2010) used a UAV to study atmospheric conditions over the Terra Nova Bay polynya. In high latitudes, UAVs can be lost due to icing build up, as reported by Curry et al. (2004).

3.4 Reanalysis Products

Reanalysis involves the production of multiyear, global, state-of-the-art, gridded representations of atmospheric states, generated by a constant numerical weather prediction model and constant data assimilation system. The first major reanalysis was undertaken by the National Centers for Environmental Prediction (NCEP) and National Center for Atmospheric Research (NCAR) in the United States by Kalnay et al. (1996). The data used in this product begin in 1948 and have been continually updated. Some errors were detected, however, and a corrected version of the reanalysis from 1979 to present was published by Kanamitsu et al. (2002).

The European Centre for Medium-Range Weather Forecasting (ECMWF) has generated several reanalysis products. The first, ERA-15, spanned 1979–1993; it was subsequently superseded by ERA-40 covering the period 1957–2002 (Uppala et al. 2005). ERA-Interim begins in 1979 and is continuously updated.

Others reanalysis products include the Modern-Era Retrospective Analysis for Research and Applications (MERRA), spanning 1980 to the present. MERRA-2 is a NASA reanalysis for the satellite era using a major new version of the Goddard Earth Observing System Data Assimilation System Version 5 (GEOS-5) produced by the NASA GSFC Global Modeling and Assimilation Office (GMAO).

The NCEP Climate Forecast System Reanalysis (CFSR) spans 1979 to the present; it is a global, high-resolution, coupled atmosphere/ocean/land surface/sea ice system. The T382 resolution atmospheric data span 1979 to 2010. The current T574 analysis is an extension of the CFSR as an operational, real-time product from 2011 into the future.

Seven reanalyses (NCEP-R1, NCEP-R2, CFSR, 20CR, MERRA, ERA-Interim, and JRA-25) for the Arctic over 1981–2010 have been evaluated by Lindsay et al. (2014). Three models stand out as being more consistent with independent observations: CFSR, MERRA, and ERA-Interim. The trend in the total ice volume in September is greatest with MERRA (-4.1×10^3 km^3 decade^{-1}) and least with CFSR (-2.7×10^3 km^3 decade^{-1}).

The Polar Meteorology Group at the Byrd Polar and Climate Research Center, Ohio State University, produces the Arctic System Reanalysis (ASR). This high-resolution regional assimilation of model output, observations, and satellite data across the mid- and high latitudes of the northern hemisphere covers the period 2000–2012. It has been performed at 30-km (ASRv1) and 15-km (ASRv2) horizontal resolution using the polar version of the Weather Research and Forecasting (WRF) model and the WRF Data Assimilation (WRFDA) system.

A new data product for the Arctic, known as the Arctic Observation and Reanalysis Integrated System (ArORIS), is described by Christensen et al. (2016). It merges state-of-the-art satellite reanalysis and in situ data sets. ArORIS incorporates the following geophysical quantities: Atmospheric Infrared Sounder (AIRS), Cloud and the Earth Radiative Energy System (CERES), CloudSat Earth Radiation Budget (ERB), Moderate Resolution Imaging Spectrometer (MODIS), Advanced Along-Track Scanning Radiometer (ATSR), Global Energy and Water Exchanges Project (GEWEX) Surface Radiation Budget, Gravity Recovery and Climate Experiment (GRACE), in situ data on snow and ice, precipitation, radiation, aerosols, cloud, water budget, and climate indices. The complete data span 1979 to the present for latitudes 60–90° N.

SUMMARY

Polar observing networks were first established in 1882–1883 during the First International Polar Year (IPY). Nineteenth- and early twentieth-century expedition data are incorporated in the International Comprehensive Ocean Atmosphere Data Set (ICOADS). The International Geophysical Year, 1957–1958, marked the installation of a network of national stations in Antarctica and the establishment of the World Data Center (WDC) system for the geophysical sciences. Such a system was lacking for the Tundra Biome project of the International Biological Program (IBP), 1964–1974, but has been adopted for the Long-Term Ecological Research Program sites operated by the National Science Foundation.

Glacier monitoring in the Arctic began only in the 1960s. Permafrost networks were established in the 1990s. At this time NASA launched the Earth Observing System (EOS) satellite program. The Fourth IPY in 2007–2008 witnessed a major expansion in polar science.

In situ meteorological observations began in the Canadian Arctic in the late1940s and in the Antarctic in 1957. An extensive network of automatic weather stations is currently found in the Antarctic. Soviet North Pole drifting stations operated in 1937–1938 and 1951–1991; they have been maintained by Russia since 2003. Gridded meteorological data from Russian and Western sources from the 1950s to circa 1990 over the Arctic Ocean are available from the National Snow and Ice Data Center (NSIDC).

The Arctic Ocean Buoy Program started in 1979 and expanded internationally in 1991. Scientific icebreaker cruises began in the Arctic in the 1980s and greatly expanded in the 1990s. During 1997–1998, an icebreaker was frozen into the Beaufort Sea for the Surface Heat Budget of the Arctic (SHEBA) experiment.

The ARGO float program began in 1999, and by 2007 had 3,900 profiling floats measuring salinity and temperature in the world ocean from 60° N to 60° S; it has recently been extended to the seasonal sea ice zone. Ice-tethered profilers have been deployed in the Arctic since 2004.

Upward-looking sonar (ULS) was deployed on navy submarines to measure sea ice draft in the Arctic in the 1950s. In 1994, a US Science Ice Exercise (SCICEX) program began to collect under-ice data. Moored ULS data are also collected in the Beaufort Sea, Fram Strait, and Weddell Sea.

Community science became organized in the 2000s, with residents and indigenous peoples collecting environmental observations. Starting in the Third IPY, the Exchange for Local Observations and Knowledge of the Arctic (ELOKA) project was launched.

Polar remote sensing began with aircraft reconnaissance of sea ice in the Eurasian Arctic by the USSR from 1933 to 1992. Land areas of the Canadian Arctic were photographed from the air in the 1940s and 1950s; the US Geological Survey has performed aerial photography of many Alaskan glaciers since the 1950s. Satellite photography began in 1972 with the Landsat series. Those data were used to compile the *Satellite Image Atlas of Glaciers of the World* from 1988 to 2012. The Global Land Ice Measurement from Space (GLIMS) project at the NSIDC has assembled glacier outlines for almost all of the world's 200,000 glaciers based mainly on Advanced Spaceborne Thermal Emission and Reflection Radiometer (ASTER) data.

Snow cover extent has been mapped with visible-band data for the northern hemisphere since 1966 and globally since 2002. Panchromatic Worldview images are being used to create digital elevation models of the Arctic with 0.5-m horizontal and 0.3-m vertical resolution.

Passive microwave single-channel data were employed to map global sea ice concentration regardless of cloud cover and darkness from December 1972 to May 1977. Since 1978, 19- and 37-GHz polarized microwave data have been used to map global sea ice and snow water equivalent at 25-km resolution. A variety of algorithms in use. Sea ice motion has been mapped from both visible and passive microwave data. Snow melt onset and frozen/unfrozen state of the ground surface have also been mapped from passive microwave data.

Airborne and spaceborne active microwave (or radar) has been widely used in mapping sea ice at high resolution (15 or 30 m). Ground-penetrating radar has been used since the 1950s by glaciologists to map ice thickness and snow depth. More recently, it has been applied to measure permafrost depth. Synthetic Aperture Radar (SAR) enabled high-resolution mapping of the Antarctic ice sheet in 2002. Interferometric SAR (InSAR) is used to map ice stream motion.

The use of radar and lidar altimeters to measure surface elevation more precisely began around 1989. It rapidly expanded in the 2000s, enabling estimation of sea ice freeboard and volume. Airborne electromagnetic induction (EMI) soundings of sea ice thickness began in the Arctic in the 1970s.

The Geoscience Laser Altimeter (GLAS), 2003–2009, collected extensive ice sheet elevation data over Greenland and Antarctica. Calculations of changes in ice sheet mass balance have been made possible by the availability data from the Gravity Recovery and Climate Experiment (GRACE) satellite since 2002.

Reanalysis products are generated by a constant numerical weather prediction model and a constant data assimilation system. They were first developed in 1996 in the United States for 1948–1990 and then in Europe for 1979–1993, and later 1957–2002. The models have been continually updated and additional data sources incorporated; likewise, new groups have released improved products. Comparisons of reanalyses for the Arctic have been published.

QUESTIONS

1. Discuss the role of international polar years in the development of polar science.
2. What have been important advances in technology that have provided significant in situ data for the polar regions?
3. Review the contributions and limitations of optical remote sensing from satellites.
4. Compare the utility of active and passive microwave remote sensing snow and ice.
5. What contributions has InSAR made to glaciology?
6. Discuss the applications of satellite altimetry in cryospheric science.
7. How has reanalysis contributed to polar meteorology?

References

Abdalati, W., and K. Steffen. 1996. "Passive Microwave-Derived Snow Melt Regions on the Greenland Ice Sheet." *Geophysical Research Letters* 22: 787–90.

Agnew, T., H. Le, and T. Hiros. 1997. "Estimation of Large Scale Sea Ice Motion from SSM/I 85.5 GHz Imagery." *Annals of Glaciology* 25: 305–11.

Armstrong, R. L., et al. 2005. *Global Monthly EASE-Grid Snow Water Equivalent Climatology* [Digital Media]. Boulder, CO: National Snow and Ice Data Center.

Aronova, E., K. S. Baker, and N. Oreskes. 2010. "Big Science and Big Data in Biology: From the International Geophysical Year through the International Biological Program to the Long Term Ecological Research (LTER) Network, 1957–Present." *Historical Studies in the Natural Sciences* 40: 183–224. doi: 10.1525/hsns.2010.40.2.183.

Arthern, R. J., D. P. Winebrenner, and D. G. Vaughan. 2006. "Antarctic Snow Accumulation Mapped Using Polarization of 4.3-cm Wavelength Microwave Emission." *Journal of Geophysical Research* 111: D06107.

Bardwaj, A., et al. 2016. "LiDAR Remote Sensing of the Cryosphere: Present Applications and Future Prospects." *Remote Sensing of Environment* 177: 125–43.

Barry, R. G., G. M. Courtin, and C. Labine. 1981. "Tundra Climates." In *Tundra Ecosystems: A Comparative Analysis*, edited by L. C. Bliss et al., 81–114. Cambridge: Cambridge University Press.

Bindschadler, R. A., and T. A. Scambos. 1991. "Satellite-Image-Derived Velocity Field of an Antarctic Ice Stream." *Science* 252: 242–6.

Bindschadler, R. A., and P. L. Vornberger. 1994. "Detailed Elevation Map of Ice Stream C, Antarctica, Using Satellite Imagery and Airborne Radar." *Annals of Glaciology* 20: 327–35.

Biskaborn, B. K., et al. 2015. "The New Database of the Global Terrestrial Network for Permafrost (GTN-P)." *Earth System Science Data* 7(2): 245–59.

Borodachev, B. E., and V. I. Shilnikov. 2003. *Istoriya L'dovoi Aviatsionnoi Razedki v Arktikei na Zamerzayushchikh Moryakh Rossii (1924–1993). (The History of Aerial Ice Reconnaissance in the Arctic and Ice-Covered Seas of Russia, 1924–1993).* St. Petersburg: Gidrometeoizdat.

Brodzik, M. J., et al. 2016. *MEaSUREs Calibrated Enhanced-Resolution Passive Microwave Daily EASE-Grid 2.0 Brightness Temperature ESDR, Version 1.* [Digital media]. Boulder, CO: National Snow and Ice Data Center.

Brucker, L., and T. Markus. 2013. "Arctic-Scale Assessment of Satellite Passive Microwave-Derived Snow Depth on Sea Ice Using Operation Icebridge Airborne Data." *Journal of Geophysical Research: Oceans* 118: 2892–905.

Carmack, E. C., et al. 1997. "Changes in Temperature and Tracer Distributions within the Arctic Ocean: Results from the 1994 Arctic Ocean Section." *Studies in Oceanography,* 44: 1487–93, 1495–1502.

Carsey, F., ed. 1992. *Microwave Remote Sensing of Sea Ice.* Geophysical Monograph 68. Washington, DC: American Geophysical Union.

Cassano, J. J., et al. 2010. "Observations of Antarctic Polynya with Unmanned Aircraft Systems." *Eos* 91: 245–326.

Cavalieri, D. J., P. Gloersen, and W. J. Campbell. 1984. "Determination of Sea Ice Parameters with Nimbus 7 SMMR." *Journal of Geophysical Research* 89(D4): 5355–69.

Christensen, M. W., et al. 2016. "Arctic Observation and Reanalysis Integrated System." *Bulletin of the American Meteorological Society* 97(6): 907–15.

Clarke, R. A. 2016. "The Revolution and Evolution in Ocean Observation: CLIVAR 20 Years of Progress: Special Issue." *CLIVAR Exchanges* 70: 9–11.

Comiso, J. C. 1986. "Characteristics of Arctic Winter Sea Ice from Satellite Multispectral Microwave Observation." *Journal of Geophysical Research* 91(C1): 957–94.

Curry, J. A., et al. 2004. "Applications of Aerosondes in the Arctic." *Bulletin of the American Meteorological Society* 85: 1855–61.

Deems, J. S., T. H. Painter, and D. C. Finnegan. 2013. "LIDAR Measurement of Snow Depth: A Review." *Journal of Glaciology* 59(215): 467–79.

Diepenbroek, M., et al. 2002. "PANGAEA: An Information System for Environmental Sciences." *Computers & Geosciences* 28: 1201–10.

Edwards, M. H., and B. J. Coakley. 2003. "SCICEX Investigations of the Arctic Ocean System." *Geochemistry* 63: 281–328.

Emery, W. J., et al. 1991. "Fram Strait Satellite Image-Derived Ice Motions." *Journal of Geophysical Research* 96(C5): 8917–20.

Emery, W. J., C. W. Fowler, and J. A. Maslanik. 1997. "Satellite-Derived Maps of Arctic and Antarctic Sea Ice Motions: 1988–1994." *Geophysical Research Letters* 24: 897–900.

Evans, S. 1963. "Dielectric Properties of Ice and Snow: A Review." *Journal of Glaciology* 5: 773–87.

Farmer, L., et al. 2016. "Citizen Scientists Train a Thousand Eyes on the North Pole." *Eos* 97. doi: 10.1029/2016EO054989

Fetterer, F., and V. Radionov, eds. 2000. *Arctic Meteorology and Climate Atlas.* [Digital media]. Boulder, CO: Arctic Climatology Project, Environmental Working Group, National Snow and Ice Data Center. nsidc.org/data/docs/noaa/g01938_ewg_arctic_met_atlas/

Freeman, E., et al. 2017. "ICOADS Release 3.0: A Major Update to the Historical Marine Climate Record." *International Journal of Climatology* 37: 2211–32.

Gearheard, S., et al., eds. 2013. *The Meaning of Ice: People and Sea Ice in Three Arctic Communities*. Hanover, NH: International Polar Institute.

Giles, K. A., S. W. Laxon, and A. P. Worby. 2008. "Antarctic Sea Ice Elevation from Satellite Radar Altimetry." *Geophysical Research Letters* 35: L03503.

Gloersen, P., et al. 1992. *Arctic and Antarctic Sea Ice, 1978–1987: Satellite Passive-Microwave Observations and Analysis*. Washington, DC: National Aeronautics and Space Administration.

Gogineni, P., and J.-B. Yan. 2015. "Remote Sensing of Ice Thickness and Surface Velocity." In *Remote Sensing of the Cryosphere*, edited by M. Tedesco, 187–230. Chichester, UK: Wiley and Sons.

Goldstein, R. M., et al. 1993. "Satellite Radar Interferometry for Monitoring Ice Sheet Motion: Application to an Antarctic Ice Streams." *Science* 262(5139): 1525–30.

Gruber, S., and F. Ludwig. 1996. *Application of Ground Penetrating Radar in Glaciology and Permafrost Prospecting*. [Study paper]. Rovaniemi, Finland: Arctic Studies Programme, Arctic Centre.

GTN-P. 2015. "Global Terrestrial Network for Permafrost Metadata for Permafrost Boreholes (TSP) and Active Layer Monitoring (CALM) Sites." doi: 10.1594/ PANGAEA.842821. Supplement to Biskaborn, B. K., et al. 2015. "The New Database of the Global Terrestrial Network for Permafrost (GTN-P)." *Earth System Science Data* 7(2): 245–59.

Haas, C., et al. 1997. "Comparison of Sea-Ice Thickness Measurements under Summer and Winter Conditions in the Arctic Using a Small Electromagnetic Induction Device." *Geophysics* 62: 749–57.

Hall, D. K., A. Frei, and S. J. Déry. 2015. "Remote Sensing of Snow." In *Remote Sensing of the Cryosphere*, edited by M. Tedesco, 31–47. Chichester, UK: Wiley and Sons.

Haran, T., et al. (Compilers). 2005, updated 2006. *MODIS Mosaic of Antarctica (MOA) Image Map*. [Digital media]. Boulder, CO: National Snow and Ice Data Center.

Inoue, J., J. A. Curry, and J. A. Maslanik. 2008. "Application of Aerosondes to Melt-Pond Observations over Arctic Sea Ice." *Journal of Atmospheric and Ocean Technology* 25: 327–34.

Jezek, K., and RAMP Product Team. 2002. *RAMP AMM-1 SAR Image Mosaic of Antarctica*. [Digital media]. Fairbanks, AK: Alaska Satellite Facility, in association with Boulder, CO: National Snow and Ice Data Center.

Johannessen, O. M., et al. 2007. *Remote Sensing of Sea Ice in the Northern Sea Route: Studies and Applications*. Chichester, UK: Springer, Praxis Publishing.

Johnson, N., et al. 2015. "The Contributions of Community-Based Monitoring and Traditional Knowledge to Arctic Observing Networks: Reflections on the State of the Field." *Arctic* 68: 28–40.

Joughin, I. R., M. A. Fahnestock, and J. L. Bamber. 2000. "Ice Flow in the Northeast Greenland Ice Stream." *Annals of Glaciology* 31: 141–6.

Joughin, L. R., et al. 2010. "Greenland Flow Variability from Ice-Sheet-Wide Velocity Mapping." *Journal of Glaciology* 56(197): 415–30.

Judge, A. S., et al. 1991. "Remote Sensing of Permafrost by Ground-Penetrating Radar at Two Airports in Arctic Canada." *Arctic* 44(suppl 1): 40–8.

Kalnay, E., et al. 1996. "The NCEP/NCAR 40-Year Reanalysis Project." *Bulletin of the American Meteorological Society* 77: 437–71.

Kanamitsu, M., et al. 2002. "NCEP-DOE-AMIP II Reanalysis (R-2)." *Bulletin of the American Meteorological Society* 83: 1631–43.

Khokanovsky, A. A. 2015. "Remote Sensing of Snow Albedo, Grain Size, and Pollution from Space." In *Remote Sensing of the Cryosphere*, edited by M. Tedesco, 48–72. Chichester, UK: Wiley and Sons.

Koenig, L., et al. 2015. "Remote Sensing of Accumulation over the Greenland and Antarctic Ice Sheets." In *Remote Sensing of the Cryosphere*, edited by M. Tedesco, 157–86. Chichester, UK: Wiley and Sons.

König-Langlo, G., B. Loose, and B. Bräuer. 2006. *25 Years of Polarstern Meteorology*. WDC-Mare Reports 0004. Bremerhaven: Alfred Wegener Institute for Polar and Marine Research.

Kovacs, A., and J. S. Holladay. 1990. "Sea-Ice Thickness Measurement Using a Small Airborne Electromagnetic Sounding System." *Geophysics* 55: 1327–37.

Kwok, R., et al. 2008. "Ice, Cloud, and Land Elevation Satellite (ICESat) over Arctic Sea Ice: Retrieval of Freeboard." *Journal of Geophysical Research* 112(C12013): 1–19.

Laxon, S. W., N. Peacock, and D. Smith. 2003. "High Interannual Variability of Sea Ice Thickness in the Arctic Region." *Nature* 425: 947–50.

Laxon, S. W., et al. 2012. "CryoSat 2 Estimates of Arctic Ice Thickness and Volume." *Geophysical Research Letters* 40: 732–7.

Lazzara, M. A., et al. 2012. "Antarctic Automatic Weather Station Program: 30 Years of Polar Observations." *Bulletin of the American Meteorological Society* 93: 1519–37.

Lindsay, R., et al. 2014. "Evaluation of Seven Different Atmospheric Reanalysis Products in the Arctic." *Journal of Climate*. doi: 10.1175/JCLI-D-13-00014.1.

Lubin, D., and R. A. Massom. 2006. *Polar Remote Sensing. Vol. 1. Atmosphere and Oceans*. Berlin: Springer.

MacMillan, M., et al. 2016. "A High-Resolution Record of Greenland Mass Balance." *Geophysical Research Letters*. doi: 10.1002/2016GL069666.

Mahoney, A., and S. Gearheard. 2008. *Handbook for Community-Based Sea Ice Monitoring*. NSIDC Special Report 14. Boulder, CO: National Snow and Ice Data Center. http://nsidc.org/pubs/special/nsidc_special_report_14.pdf

Mahoney, A. R., et al. 2008. "Observed Sea Ice Extent in the Russian Arctic, 1933–2006." *Journal of Geophysical Research* 113: C110015.

Mahoney, A. R., et al. 2009. "Sea Ice Thickness Measurements from a Community-Based Observing Network." *Bulletin of the American Meteorological Society* 90: 370–7.

Marsik, F., et al. 2016. "Eight Microsatellites, One Mission: CYGNSS." *Earth Observer* 28(6): 4–13.

Martin-Espanol, A., et al. 2016. "Spatial and Temporal Antarctic Ice Sheet Mass Trends, Glacio-Isostatic Adjustment, and Surface Processes from a Joint Inversion of Satellite Altimeter, Gravity, and GPS Data." *Journal of Geophysical Research: Earth Surface*. doi: 10.1002/2015JF003550.

Massom, R. A., and D. Lubin. 2006. *Polar Remote Sensing. Vol. 2: Ice Sheets*. Berlin: Springer.

McLaren, A. S. 1986. "Analysis of the Under-Ice Topography in the Arctic Basin as Recorded by the USS *Nautilus* during August 1958." *Arctic* 41: 117–26.

McLaren, A. S. 1989. "The Under-Ice Thickness Distribution of the Arctic Basin as Recorded in 1958 and 1970." *Journal of Geophysical Research: Oceans* 94(C4): 4971–83.

Meier, W. N., and J. Stroeve. 2005. "Comparison of Passive Microwave Ice Concentration Algorithm Retrievals with AVHRR Imagery in Arctic Peripheral Seas." *IEEE Transactions on Geoscience and Remote Sensing* 43(6): 1324–37.

Meier, W. N., et al. 2015. "How Do Sea-Ice Concentrations from Operational Data Compare with Passive Microwave Estimates? Implications for Improved Model Evaluations and Forecasting." *Annals of Glaciology* 56(69): 332–40.

Mohn, H. 1905. *The Norwegian North Polar Expedition, 1893–1896, Scientific Results. Vol. VI: Nansen F. Meteorology*. Christiania: J. Dybwad.

Mohn, H. 1907. *Report of the Second Norwegian Arctic Expedition in the "Fram" 1898–1902. Vol. 1: Meteorology*. Kristiana: Norske Videnkaps-akad.

Moore, J. C. 1988. "Dielectric Variability of a 130 m Antarctic Ice Core: Implications for Radar Sounding." *Annals of Glaciology* 11: 95–9.

Morison, J., et al. 2002. "North Pole Environmental Observatory Delivers Early Results." *Eos* 33: 357.

Mote, T. L., and M. R. Anderson. 1995. "Variations in Snowpack Melt on the Greenland Ice Sheet Based on Passive-Microwave Measurements." *Journal of Glaciology* 41(137): 51–60.

National Snow and Ice Data Center. 2006. *Submarine Upward Looking Sonar Ice Draft Profile Data and Statistics, Version 1*. Boulder, CO: National Snow and Ice Data Center. doi: 10.7265/N54Q7RWK.

Parkinson, C. L., et al. 1987. *Antarctic Sea Ice, 1973–1976: Satellite Passive-Microwave Observations*. SP 489. Washington, DC: National Aeronautics and Space Administration.

Perovich, D., et al. 2014. "Sea Ice Mass Balance Observations from the North Pole Environmental Observatory." *Geophysical Research Letters* 41: 2019–25.

Polar Research Board. 2012. *Lessons and Legacies of the Polar Year 2007–2008*. Washington, DC: National Academies Press.

Pope, A., et al. 2014. "Open Access Data in Polar and Cryospheric Remote Sensing." *Remote Sensing* 6: 6183–220.

Proshutinsky, A., et al. 2009. "Beaufort Gyre Freshwater Reservoir: State and Variability from Observations." *Journal of Geophysical Research: Oceans* 114(C1): C00A10.

Rabe, B., et al. 2011. "An Assessment of Arctic Ocean Freshwater Content Changes from the 1990s to the 2006–2008 Period." *Deep-Sea Research, Part I* 58: 173–85.

Ramachandran, B., et al. 2014. "ASTER Datasets and Derived Products for Global Glacier Monitoring." In *Global Land Ice Monitoring from Space*, edited by J. S. Kargel et al., 145–62. Berlin: Springer-Verlag.

Ramapriyan, H. K. 2009. "EOS Data and Information System (EOSDIS): Where We Were and Where We Are." Part 1: *Earth Observer* 21(4): 4–10. Part 2: *Earth Observer* 21(5): 8–14.

Raup, B. H., et al. 2015. "Remote Sensing of Glaciers." In *Remote Sensing of the Cryosphere*, edited by M. Tedesco, 123–56. Chichester, UK: Wiley and Sons.

Rignot, E., et al. 2011. "Acceleration of the Contribution of the Greenland and Antarctic Ice Sheets to Sea Level Rise." *Geophysical Research Letters* 38: L05503.

Riser, S. C., et al. 2016. "Fifteen Years of Ocean Observations with the Global Argo Array." *Nature Climate Change* 6: 145–53.

Romanov, I. P., Yu. B. Konstantinov, and N. A. Kornilov. 2000. "North Pole Drifting Stations (1937–1991)." In *Arctic Meteorology and Climate Atlas*, edited by F. Fetterer and V. Radionov. [Digital media]. Boulder, CO: Arctic Climatology Project, Environmental Working Group, National Snow and Ice Data Center.

Roquet, F., et al. 2009. "Observations of the Fawn Trough Current over the Kerguelen Plateau from Instrumented Elephant Seals." *Journal of Marine Systems* 78: 377–93.

Rothrock, D. A., Y. Yu, and G. A. Maykut. 1999. "Thinning of the Arctic Sea-Ice Cover." *Geophysical Research Letters* 26: 3469–72.

Rudels, B. 2015. "Arctic Ocean Circulation, Processes and Water Masses: A Description of Observations and Ideas with Focus on the Period Prior to the International Polar Year 2007–2009." *Progress in Oceanography* 132: 22–67.

Scambos, T. A., and M. A. Fahnestock. 1998. "Improving Digital Elevation Models over Ice Sheets Using AVHRR-Based Photoclinometry." *Journal of Glaciology* 44(146): 97–103.

Scambos, T. A., and T. Haran. 2002. "An Image-Enhanced DEM of the Greenland Ice Sheet." *Annals of Glaciology* 34: 291–8.

Scambos, T. A., et al. 1992. "Application of Image Cross-Correlation to the Measurement of Glacier Velocity Using Satellite Image Data." *Remote Sensing of the Environment* 42: 177–86.

Scambos, T. A., et al. 2007. "MODIS-Based Mosaic of Antarctica (MOA) Data Sets: Continent-Wide Surface Morphology and Snow Grain Size." *Remote Sensing of the Environment* 111(2–3): 242–57. doi: 10.1016/j.rse.2006.12.020.

Schutz, B., et al. 2005. "ICESat Mission Overview and History." *Geophysical Research Letters* 32: L21S01.

Shepherd, A., et al. 2012. "A Reconciled Estimate of Ice-Sheet Mass Balance." *Science* 338(6111): 1183–9.

Smith, N. V., S. S. Saatchi, and J. T. Randerson. 2004. "Trends in High Northern Latitudes Soil Freeze and Thaw Cycles from 1988 to 2002." *Journal of Geophysical Research* 109: D12101.

Strass, V. I. 1998. "Measuring Sea Ice Draft and Coverage with Moored Upward Looking Sonars." *Deep-Sea Research, Part 1. Oceanographic Research Papers* 45: 795–818.

Sverdrup, H. U. 1930. *The Norwegian North Polar Expedition with the "Maud" 1918–1925: Scientific Results I.* Bergen, Norway: A. S. J. Griegs Boktrykkeri.

Sverdrup, H. U. 1933. *The Norwegian North Polar Expedition with the "Maud" 1918–1925, Scientific Results. Vol. II: Meteorology.* Bergen, Norway: Geofys Institute.

Swift, J. H., et al. 1997. "Waters of the Makarov and Canada Basins." *Deep-Sea Research, Part II* 44(8): 1503–29.

Tedesco, M., ed. 2015. *Remote Sensing of the Cryosphere.* Chichester, UK: Wiley and Sons.

Tedesco, M., et al. 2013. "Evidence and Analysis of 2012 Greenland Records from Spaceborne Observations, a Regional Climate Model and Reanalysis Data." *Cryosphere* 7: 615–30.

Tedesco, M., et al. 2015. "Remote Sensing of Melting Snow and Ice." In *Remote Sensing of the Cryosphere*, edited by M. Tedesco, 99–122. Chichester, UK: Wiley and Sons.

Toole, J. M., et al. 2011. "The Ice-Tethered Profiler: Argo of the Arctic." *Oceanography* 24(3): 126–35.

Tschudi, M., C. Fowler, and J. Maslanik. 2015. *EASE-Grid Sea Ice Age Version 2*. Boulder, CO: National Snow and Ice Data Center, Distributed Active Archive Center. doi: 10.5067/1UQ.JWCYPVX61.

Uppala, S. M., et al. 2005. "The ERA-40 Reanalysis." *Quarterly Journal of the Royal Meteorological Society* 111: 2961–3012.

Velicogna, I., T. C. Sutterly, and M. R. van den Broeke. 2014. "Regional Acceleration in Ice Mass Loss from Greenland and Antarctica Using GRACE Time-Variable Gravity Data." *Journal of Geophysical Research: Space Physics* 119: 8130–7.

Williams, P., et al. 2017. "Community-Based Observing Networks and Systems in the Arctic: Human Perceptions of Environmental Change and Instrument-Derived Data." *Regional Environmental Change*. doi: 10.1007/s10113-017-1220-7

Williams, R. S. Jr., and J. G. Ferrigno, eds. 1989–2012. *Satellite Image Atlas of Glaciers of the World*. US Geological Survey Professional Paper 1386 A-K.

Winebrenner, D. P., R. J. Arthern, and C. A. Shumanm. 2001. "Mapping Greenland Accumulation Rates Using Observations of Thermal Emission at 4.5 cm Wavelength." *Journal of Geophysical Research* 106(D24): 33919–34.

Wingham, D. J., et al. 2006. "Mass Balance of the Antarctic Ice Sheet." *Philosophical Transactions of the Royal Society, Series A* 364: 1627–35.

Wong, A. P. S., and S. C. Riser. 2011. "Profiling Float Observations of the Upper Ocean under Sea Ice off the Wilkes Land Coast of Antarctica." *Journal of Physical Oceanography* 41: 1102–15.

Wood, K. R., and J. E. Overland. 2006. "Climate Lessons from the First International Polar Year." *Bulletin of the American Meteorological Society* 87: 1685–97.

Worby, A. P., et al. 1999. "On the Use of Electromagnetic Induction Sounding to Determine Winter and Spring Sea Ice Thickness in the Antarctic." *Cold Regions Science and Technology* 29: 49–58.

Zhang, T.-J., and R. L. Armstrong. 2001. "Soil Freeze/Thaw Cycles over Snow-Free Land Detected by Passive Microwave Remote Sensing." *Geophysical Research Letters* 28: 763–6.

Zhang, T., and R. Armstrong. 2003, updated 2005. *Arctic Soil Freeze/Thaw Status from SMMR and SSM/I, Version 2*. [Digital media.] Boulder, CO: National Snow and Ice Data Center/World Data Center for Glaciology.

Zwally, H. J., et al. 1983. *Antarctic Sea Ice, 1973–1976: Satellite Passive-Microwave Observations*. SP 459. Washington, DC: National Aeronautics and Space Administration.

Atmospheric and Oceanic Circulation and Climate

4

This chapter describes the atmospheric and oceanic circulations of the polar regions and their current climatic characteristics and recent changes to provide a broad setting for the subsequent chapters on terrestrial and oceanic environments. The two polar regions are discussed separately following a general introduction. Previous book-length studies of the two regions include the studies of King and Turner (1997) on the meteorology and climatology of the Antarctic, Karoly and Vincent (1998) on the meteorology of the southern hemisphere, Turner and Marshall (2011) on climate change in the polar regions, and Serreze and Barry (2014) on the Arctic climate system.

4.1 Atmospheric Circulation

The two polar regions have analogous atmospheric circulation characteristics in the middle and upper troposphere, above approximately 5 km in height, but very different surface features. In the upper air, each polar region displays a vast circumpolar vortex of low pressure (known as the polar vortex) that is surrounded in high and middle latitudes (40–60°) by a belt of strong westerly winds with jet streams. The vortex continues upward into the stratosphere.

The Arctic vortex is centered near the North Pole and has two or three major troughs in the northern hemisphere winter located over eastern North America and eastern Asia, and a less pronounced trough over eastern Europe. In summer, there is a single weaker trough over eastern North America. These features are illustrated in Figure 4.1A showing the geopotential heights of the 500-hPa surface that is approximately at 5.5 km of altitude. These mean troughs are formed by the interaction of the air flow with the major north–south mountain ranges, and of air–sea interaction with sea surface temperature (SST) patterns set up by ocean currents (Barry and Carleton 2001, 278–300).

Over the last three decades, the Arctic stratospheric polar vortex in winter has been displaced toward the Eurasian continent and away from North America in

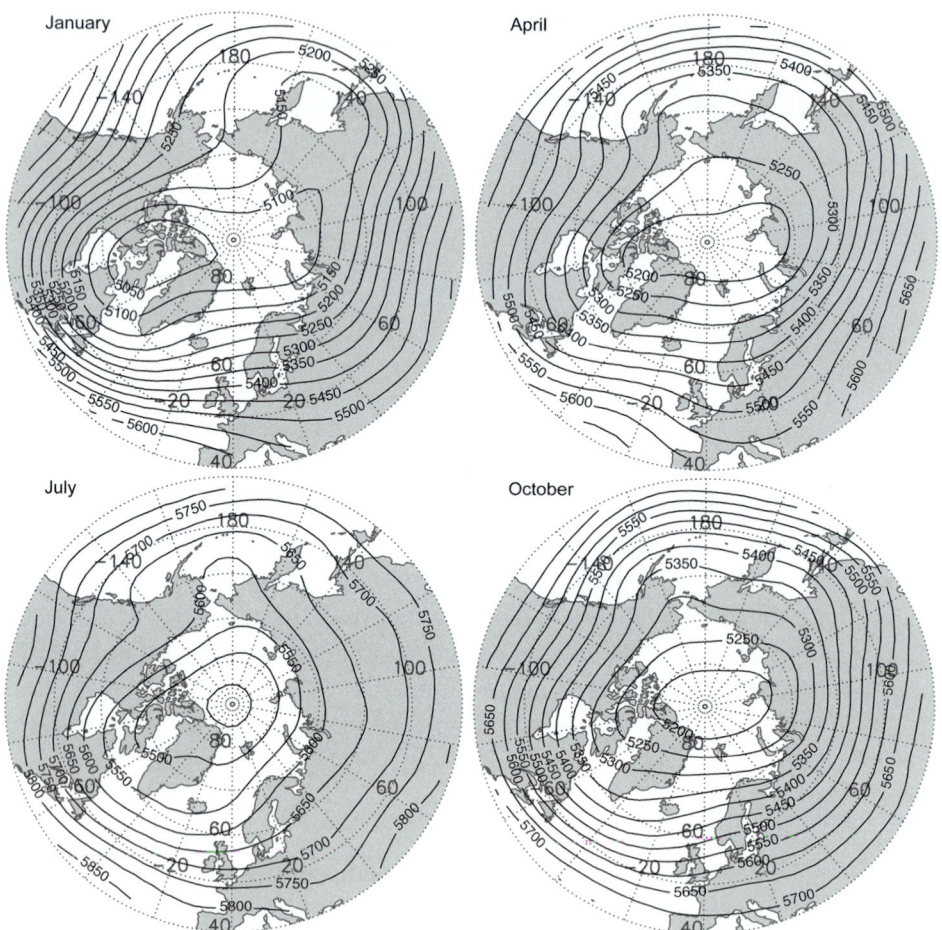

Figure 4.1A The pattern of 500-hPa geopotential heights in the northern hemisphere: (a) boreal winter, (b) summer.
Source: Serreze and Barry 2014, 102, figure 4.7.

February. This shift is found to be closely related to the enhanced zonal wave number 1 wave in response to Arctic sea-ice loss, particularly over the Barents–Kara Sea (Zhang et al. 2016). Increased snow cover over the Eurasian continent may also have contributed to the shift. The Antarctic vortex, by comparison, is more nearly circular with stronger, more consistently zonal westerly winds (Figure 4.1B).

According to James (1989), Ekman pumping by the boundary layer leads to the formation of the upper tropospheric cyclonic vortex above the Antarctic ice sheet. The strength and distribution of upper-level vorticity are determined by the shape of the underlying ice sheet. There is a basic wave number 1 pattern because

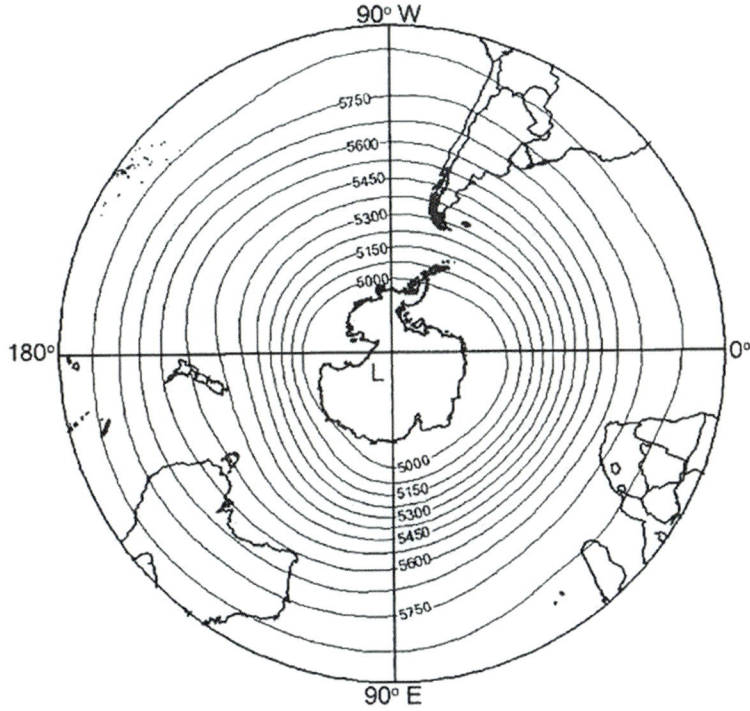

Figure 4.1B

The pattern of 500-hPa geopotential heights in austral summer and winter in the southern hemisphere, 1970–1999: (c) January, (d) July.

Source: NCEP/NCAR Reanalysis, data from the NOAA-CIRES Climate Diagnostics Center

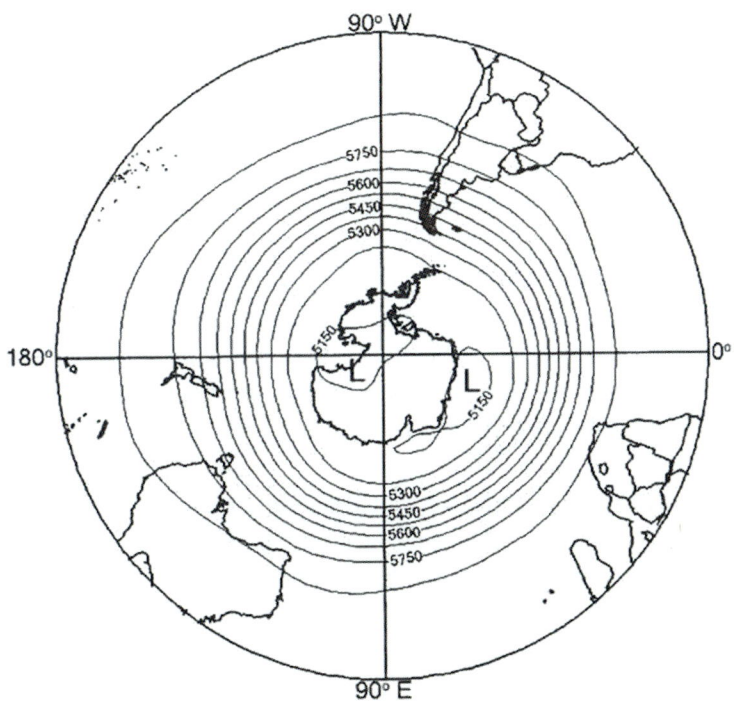

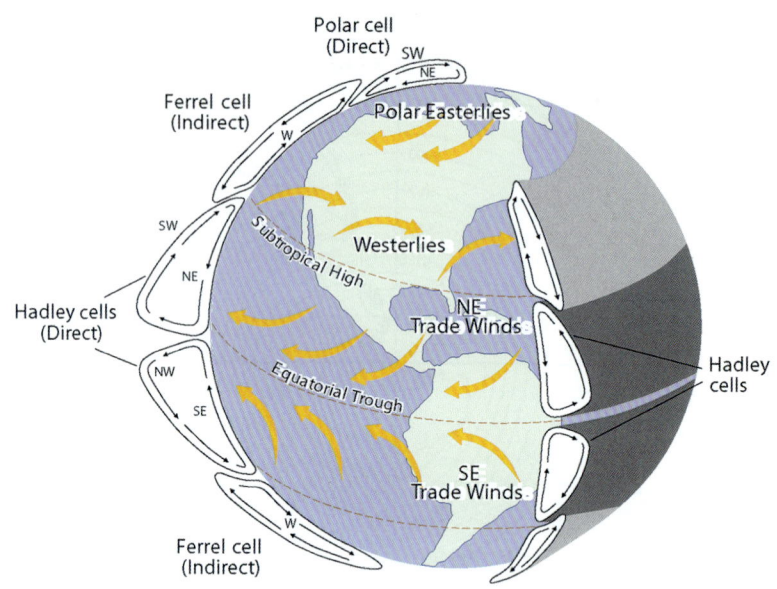

Figure 4.2

The global wind belts (westerlies and trade winds) and the vertical structure of the meridional cells: Hadley (thermally direct) and Ferrel (indirect).
Source: National Aeronautics and Space Administration.

the vortex is displaced toward the Indian Ocean sector. This is related to the fact that the most remote sector of the ice sheet (the "pole of inaccessibility" at 83° S, 55° E, and the highest part of the ice sheet) is displaced away from the pole toward the Indian Ocean and southwest South Pacific. Various mechanisms that could lead to the depletion of upper-level cyclonic vorticity, and hence to the retention of a substantial drainage flow, are discussed by James (1989). Disruption of the polar vortex by decaying mid-latitude cyclones, and the consequent export of cyclonic vorticity to lower latitudes, is the most probable mechanism.

In the meridional (north–south) plane, the classical view has been that there are three vertical cells in each hemisphere: the dominant thermally direct circulation tropical Hadley cells (where warm air rises and cold air sinks), the much weaker indirect Ferrel cells in mid-latitudes, and a weak Polar direct cell in high latitudes. This pattern is shown schematically in Figure 4.2. Notice that the polar cell in the Arctic is a minor component. Effectively, the circulation in middle and high latitudes is horizontal – westerly in middle latitudes, with polar easterlies in high latitudes. The latter are best developed around the Antarctic continent (not shown in Figure 4.2).

Recently, Qian et al. (2015) have shown that there are also weak Arctic and Antarctic cells in the troposphere of extreme northern and southern latitudes. Analyses of meridional-vertical section streamline, meridional mass stream function, and climatic vertical velocity for 1981–2010 from reanalyses of the National Centers for Environmental Prediction (NCEP) provide evidence that supports the existence of these cells. The mean central locations of the cells are listed in

Table 4.1. As the data show, the upper tropospheric Antarctic cell is almost nonexistent,
being only 500 km wide.

Table 4.1 Mean central locations of polar cells

	Boreal winter	Boreal summer
Arctic	75° N, 925 hPa	77.5° N, 850 hPa
	Austral summer	**Austral winter**
Antarctic	87.5° S, 300 hPa	87.5° S, 300 hPa

The surface climates of the two polar regions
are radically different from one another as a
result of the largely ice-covered, approximately
14 million km^2 Arctic Ocean, which is nearly
surrounded by continental land masses, and
the massive ice sheet that covers Antarctica, 3–4 km high and also 14 million km^2
in area. For this reason, the two hemispheres are treated separately. The only
common feature is the polar night and day, but the radiative conditions at the
surface are determined by the cloud regimes and surface properties, which differ
greatly.

Both hemispheres experience large-scale, long-lasting circulation oscillations
that were first identified by Sir Gilbert Walker in the 1920s. He noted
the Southern Oscillation in air pressure between the eastern and western equatorial Pacific, the North Atlantic Oscillation between the Azores high pressure
and the Icelandic low, and the corresponding North Pacific Oscillation. The
distant relationships in pressure and atmospheric circulation were named
teleconnections. These circulation modes and teleconnections (see Barry and
Hall-McKim 2014, 104–15) are discussed in this chapter and in Section 7.1.

The polar stratosphere merits special attention because of the presence of an
ozone maximum in the lower and middle stratosphere between about 10 and 50 km
of altitude. Ozone (O_3) is critical because it intercepts ultraviolet radiation that could
harm the lives of plants, animals, and humans. In the 1970s, it was recognized that
catalytic reactions involving manufactured chlorofluorocarbons (CFCs) were des-
troying ozone in the Antarctic stratosphere. Detailed monitoring of ozone concentrations and numerous chemical compounds in the atmosphere began worldwide,
and an "ozone hole" was identified (see Box 9.2). Figure 4.3 shows that October
values in southern high latitudes fell from approximately 350 Dobson units (DU) in
the 1970s to approximately 200 DU in the 1990s (DU = 0.01-mm thickness at
standard temperature and pressure [STP]). This led to the internationally agreed
Montreal Protocol in 1987 to phase out the production of CFCs by 2000 at the latest.
Kondratyev (2003) notes that the decline in total ozone (90 percent of it in the
stratosphere) from 1979 to 1997 was 20 percent decade^{-1} for September–November
in the Antarctic, compared with 7.7 percent in the Arctic for March–May and only
2 percent in September–November at 35–60° S. The fifth Intergovernmental Panel
on Climate Change (IPCC) assessment concluded that global stratospheric ozone
has declined from pre-1980 values, with most of the decline occurring prior to the
mid-1990s (Hartmann et al. 2013). Since then ozone has remained nearly constant at

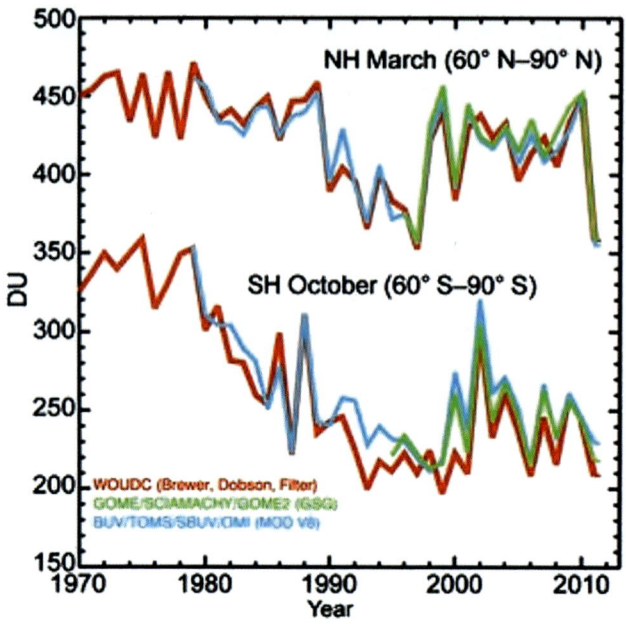

Figure 4.3 March and October polar total ozone (Dobson units, DU) in the northern and southern hemispheres, respectively. WOUDC ground-based measurements combining Brewer, Dobson, and filter spectrometer data (red: Fioletov 2008; Fioletov et al. 2002), the merged BUV/S BUV/TOM S/OMIMOD V8 (blue: Stolarski and Frith 2006), and GOME/SCIAMAC HY/GOME-2 "GSG" (green: Weber et al. 2007).

Source: Weber et al. 2012, S46, figure 2.49. Courtesy of American Meteorological Society.
© Copyright 2016 American Meteorological Society (AMS).

about 3.5 percent below the 1964–1980 level. The Antarctic ozone hole reached a minimum value of 73 DU in September 1994. Current minima are about 100–120 DU. The maximum daily area of the hole is currently in the range of 21–28 million km². Geographically, total ozone was lowest at approximately 200 DU in the sector 0–70° W between 70° and 80° S during spring 1979–2005 (Grytsal et al. 2007). This can be compared with levels of 340–370 DU over Antarctica in the 1970s.

4.1.1 Northern Hemisphere

A principal feature of the high-latitude circulation is the Arctic Oscillation (AO), also called the Northern Annular Mode (NAM). This was first described by Lorenz (1951), but documented in detail by Thompson and Wallace (1998). It features pressure and height anomalies of one sign over the Arctic and opposite anomalies over mid-latitudes centered at about 37–45° N. The AO extends throughout the troposphere. In its positive (negative) phase, pressure is low (high) over the pole and high (low) in middle latitudes. Correspondingly, the zonal westerlies are strong (weak) when the AO index is positive (negative). The AO has major effects on weather anomalies in mid-latitudes and is also linked to anomalies of Arctic sea ice. A high index features westerly surface winds along 55° N and transpolar flow from Russia toward Canada, whereas low-index conditions are marked by cold anticyclones centered over central Canada and Russia and an anticyclonic surface circulation throughout the Arctic basin.

During most of the twentieth century, the AO alternated between its positive and negative phases (Figure 4.4). Starting in the 1970s, the oscillation tended to be more positive, though it has tended to have a more neutral state since 2008. The North Atlantic Oscillation (NAO), which occurs between pressures in the Azores subtropical anticyclone and the Icelandic low pressure, is related to it, but

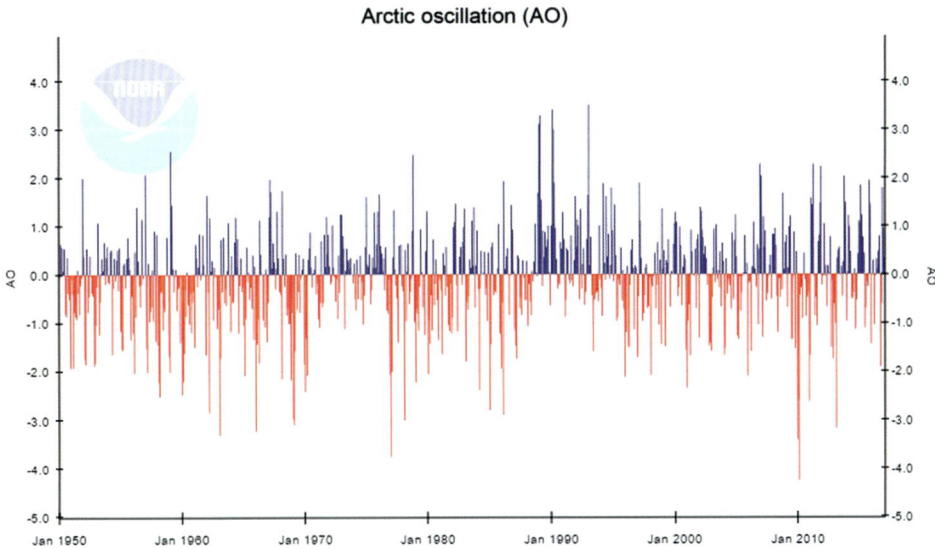

Figure 4.4 Arctic Oscillation index variations, 1950–2016. The AO index is obtained by projecting the AO loading pattern onto the daily anomaly 1000-hPa height field over 20–90° N. The AO loading pattern has been chosen as the first mode of EOF analysis using monthly mean 1000-hPa height anomaly data from 1979 to 2000 over 20–90° N. Source: www.ncdc.noaa.gov/teleconnections/ao/

can operate independently. The NAO is the dominant mode of winter climate variability in the northern hemisphere.

In winter, the Beaufort Sea region is influenced by a pressure ridge at sea level extending from the Siberian high to the Yukon high over northwestern Canada (Overland 2009; Serreze and Barrett 2011). Mapped at a 2-hPa contour interval, the Beaufort Sea high appears as a closed anticyclone in the long-term annual mean sea-level pressure field. A trough of low pressure extends over the eastern Arctic northeastward from the subpolar Icelandic low (Figure 4.5). While the Beaufort Sea ridge is relatively static and persistent, the mean trough represents the frequent passage of cyclone systems tracking across Iceland into the Barents Sea, where they usually decay. In spring, a mean anticyclone occurs north of the Canadian Arctic archipelago over the western Arctic Ocean; this is the only season when there is a discrete Arctic high pressure system. A strong closed high can be interpreted as the surface expression of an amplified western North American ridge at 500 hPa. There is some suggestion of a split flow, in which the ridge linked to the surface high becomes separated from the ridge to the south that lies within the main belt of westerlies. According to Serreze and Barrett (2011), a strong Beaufort Sea high is a feature of the positive phase of the Pacific–North American

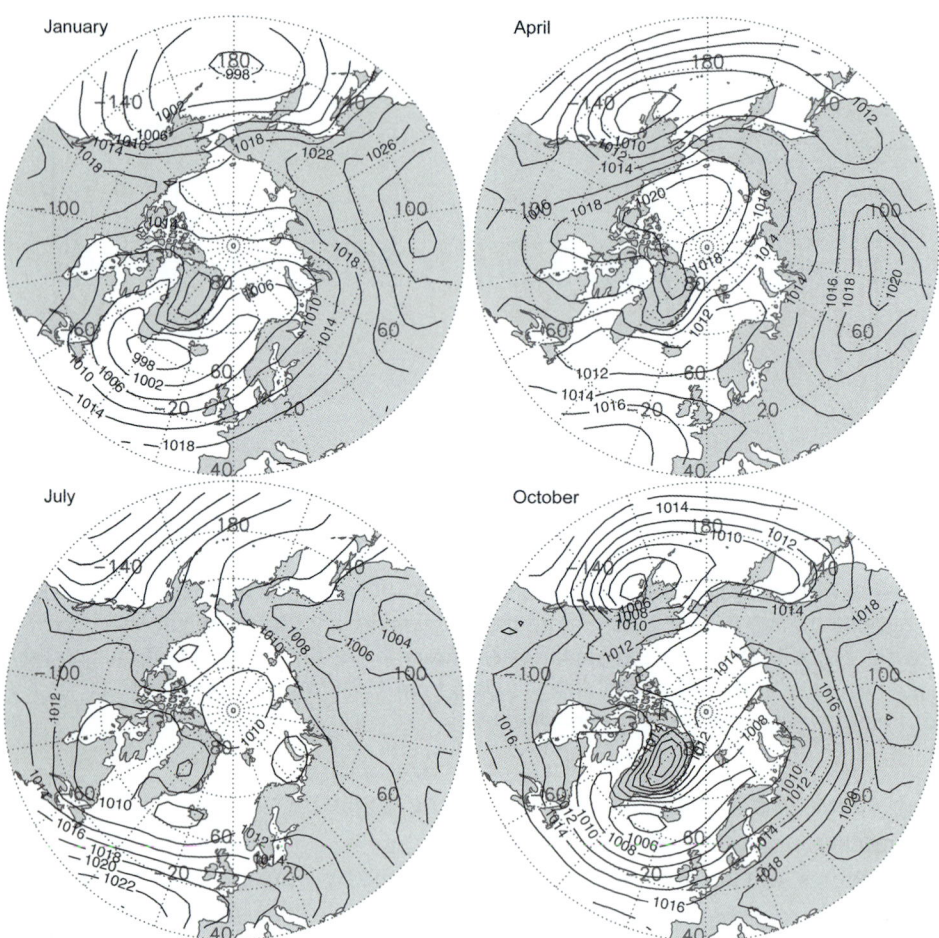

Figure 4.5 Mean sea-level pressure (hPa) in the Arctic for 1970–1999 based on NCEP/ NCAR data for the four mid-season months.
Source: Serreze and Barry 2014, figure 4.8.

wave train and, to a lesser degree, the positive phase of the Arctic dipole anomaly and Pacific Decadal Oscillation (PDO), discussed later.

The Arctic dipole anomaly is a pressure pattern characterized by two poles of opposite sign: one over the Canadian Arctic archipelago and northern Greenland, and the other over the Kara and Laptev seas. Anomalous winds are generally directed either toward the Greenland and Barents seas (positive Arctic dipole anomaly) or toward the Bering Strait (negative Arctic dipole anomaly). This pattern, which was first described by Wu et al. (2006), is defined as the spatial distribution of the second leading Empirical Orthogonal Function (EOF) mode of monthly mean sea-level pressure north of 70° N, where the first leading mode corresponds to the Arctic Oscillation. When defined for the winter season (October through March), the first leading mode (the AO) accounts for 61 percent of the total variance, while the second leading mode (Arctic dipole anomaly) accounts for 13 percent. Both oscillations play an important role in the export of sea ice out of the Arctic.

In summer, there is a weak mean low over the central Arctic. In individual summer months, a persistent deep low may occur as a result of lows moving in from Eurasia and stagnating and/or local cyclogenesis associated with baroclinicity over the Arctic Ocean. There is a prominent summer maximum in cyclone activity over the Arctic Ocean, centered near the North Pole in the long-term mean (Serreze and Barrett 2008). The seasonal onset of this pattern is associated with an eastward shift in the Urals trough, migration of the 500-hPa vortex core to near the pole, and development of a separate region of high-latitude baroclinicity close to the Arctic coastline. The latter two features are consistent with differential atmospheric heating between the mostly ice-covered Arctic Ocean and snow-free land. Variability in the strength of the cyclone pattern can be broadly linked to the phase of the summer Northern Annular Mode. When the cyclone pattern is well developed, the 500-hPa vortex is especially strong and symmetric about the pole, with negative sea-level pressure (SLP) anomalies over the pole and positive anomalies over middle latitudes. The opposite occurs when the mode is weak.

The Arctic frontal zone in summer is located near the Arctic coastline in Eurasia and Alaska due to the strong thermal gradient between the cold Arctic Ocean and the bare continental land surface (Crawford and Serreze 2015; Serreze et al. 2001) (Figure 4.6). Studies of this pattern have led to revision of the earlier postulate that the median frontal location is related to the contrast in energy budgets across the forest–tundra ecotone (Krebs and Barry 1970). The frontal zone involves cold maritime Arctic air and continental polar air that is relatively moist, as a result of evapotranspiration from the numerous small lakes and water-logged tundra vegetation. The summer frontal zone extends through the troposphere, and at 250 hPa is marked by an Arctic jet stream. Spatiotemporal variability of the frontal zone is primarily dependent on factors affecting temperature over land, especially variability in cloud cover, surface wind direction, and the timing of the annual snow cover retreat. In autumn, the circulation moves back toward its winter pattern.

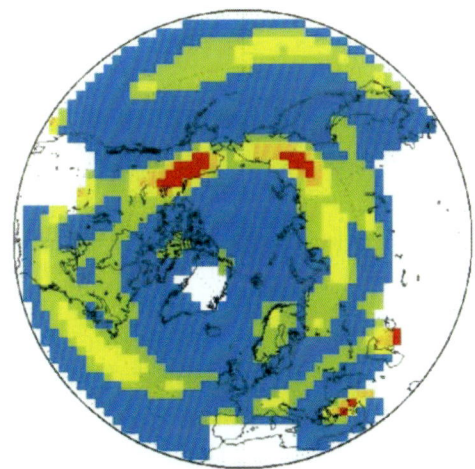

Figure 4.6

The Arctic frontal zone in summer defined by frontal frequencies (fronts day^{-1}) for 1979–1998 from the NCEP-NCAR reanalysis.

Source: Serreze et al. 2001, 1557, figure 4b. Courtesy of American Meteorological Society. © Copyright 2001 American Meteorological Society (AMS).

Tanaka et al. (2012) studied the characteristics of the Arctic cyclone, finding many differences in structure and behavior with it compared with the mid-latitude cyclone. The Arctic cyclone has a barotropic structure in the vertical from the surface to the stratosphere. The Arctic cyclone detected in the sea-level pressure field is connected with the polar vortex at the 500-hPa level and above. Importantly, it has a cold core in the troposphere and a warm core around the 200-hPa level, as a result of subsiding motion in the stratosphere. There are no frontal structures involved, but the vortex is surrounded by spiral cloud bands like a tropical cyclone. Arctic cyclones move rather randomly in direction over the Arctic Ocean and may last up to about a month.

Crawford and Serreze (2016) have examined the potential relations between the summer Arctic low-pressure systems and the Arctic frontal zone (AFZ). Their work shows that the Arctic coast (and therefore the AFZ) is not a region of cyclogenesis. Rather, the AFZ acts as an area of intensification for systems that form over Eurasia. As these systems migrate toward the Arctic Ocean, they experience greater deepening in situations when the AFZ is strong at midtropospheric levels. On a broader scale, intensity of the summer AFZ at midtropospheric levels has a positive correlation with cyclone intensity in the Arctic Ocean during summer, even when controlling for variability in the NAM.

4.1.2 Southern Hemisphere

The surface circulation in the southern hemisphere shows much less seasonal variability than its northern counterpart (Bromwich and Parish 1998; Simmonds 1998). Close to the Antarctic continent, south of about 65° S, the mean airflow is easterly, as a result of the prevailing high pressure over the ice sheet. There is a circum-Antarctic trough around the continent that has 1,414 centers near the major embayments of the Ross Sea and Weddell Sea, and off Wilkes Land in East Antarctica (Figure 4.7). Extratropical cyclones spiral in toward these centers over the Southern Ocean (Figure 4.8). The circumpolar trough actually consists of numerous disturbances that travel southeastward from middle latitudes toward Antarctica, or that develop in the vicinity of the continent (King and Turner 1997).

There are also numerous mesoscale polar low systems – predominantly comma clouds – that tend to move equatorward. In winter, the polar vortex expands equatorward but, in contrast to the situation in the northern hemisphere, the southern westerlies show little seasonal variation in strength.

Planetary waves during 1979–2013 are described by Turner et al. (2016) using the European Centre for Medium-Range Weather Forecasts (ECMWF) interim reanalyses. In addition, these authors identify the effects of tropical and extratropical forcing factors on the phase and amplitude of the planetary waves. The amplitudes

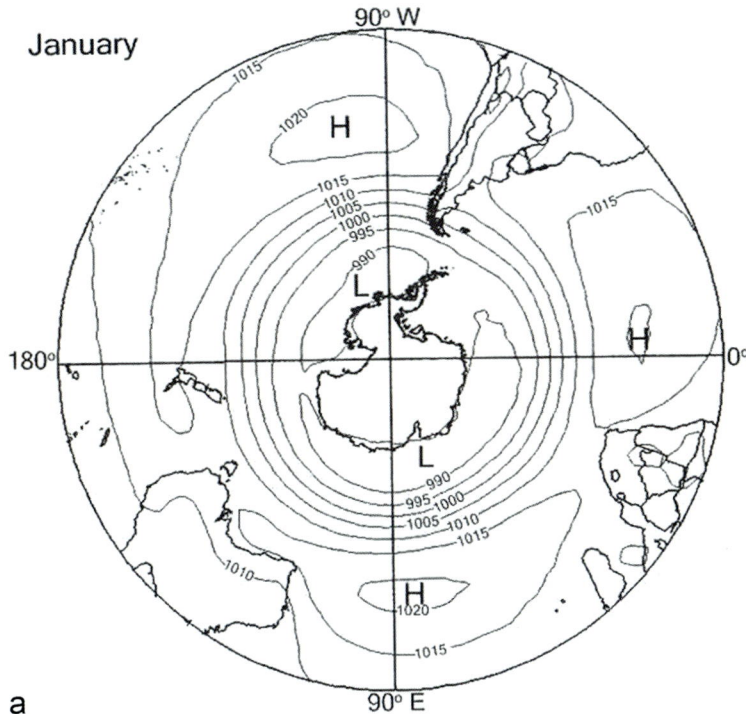

January

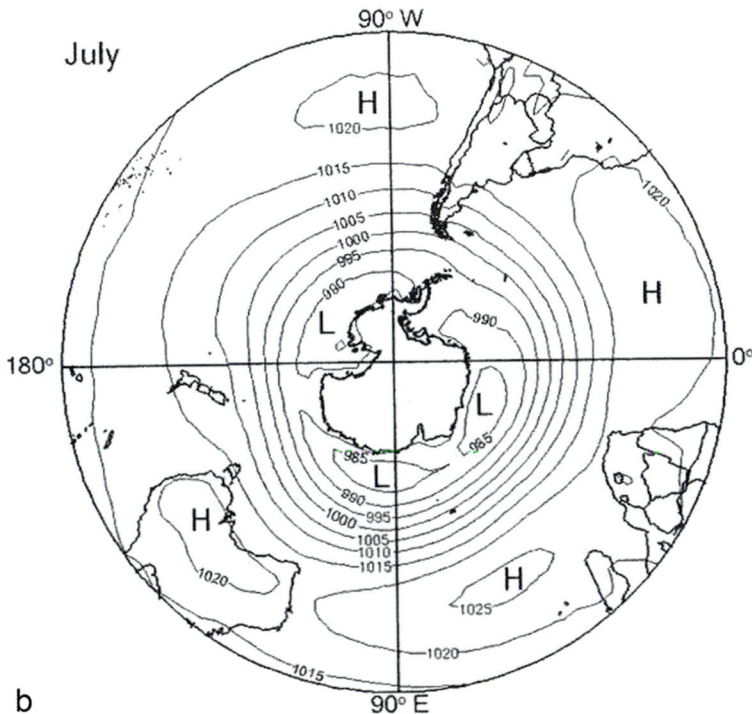

July

Figure 4.7
Mean sea-level pressure (hPa) in the southern hemisphere, 1970–1999: (a) January, (b) July.
Source: NCEP-NCAR Reanalysis, data from the NOAA-CIRES Climate Diagnostics Center.

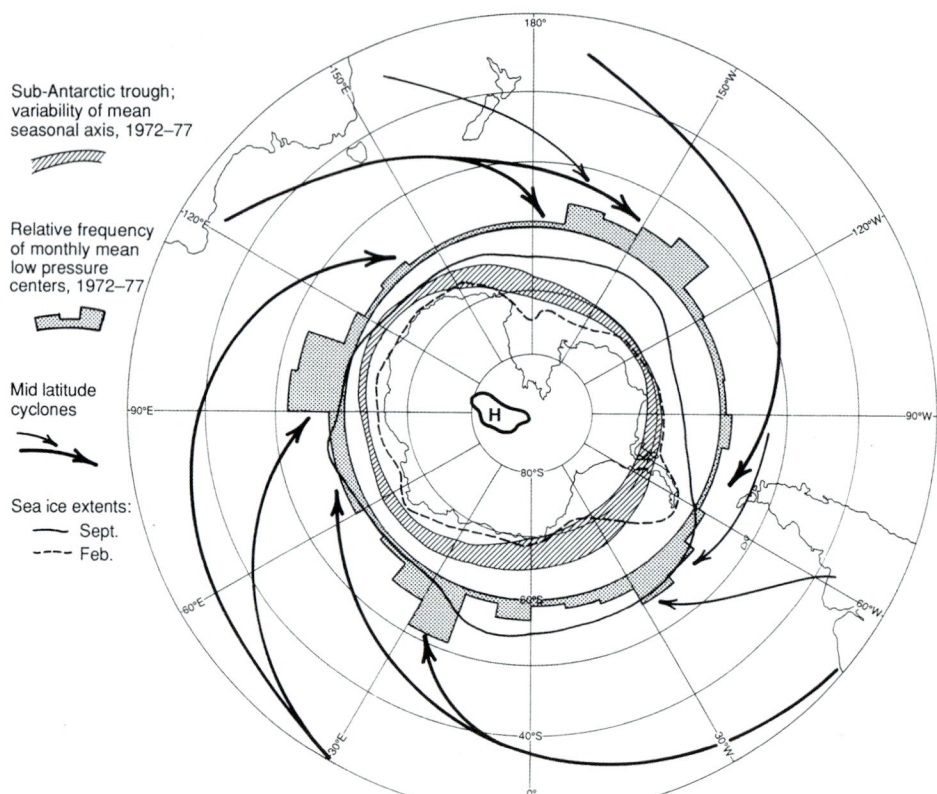

Figure 4.8 The sub-Antarctic trough, cyclone paths, and seasonal sea ice extent in the Southern Ocean around Antarctica.
Source: Carleton 2002, 133, figure 2. With permission of the author.

of wave numbers 1–3 exhibit an annual cycle, with a minimum in summer and a maximum over the extended austral winter period. The phase of wave number 1 has a semiannual cycle, moving east in austral spring/autumn and west in summer/winter, as a result of differences in the phase of the semiannual oscillation across the Pacific sector of the Southern Ocean. The phase of wave number 3 has an annual cycle, being more eastward (westward) in summer (winter). Year-to-year variability of the amplitude of wave number 1 is strongly associated with the Amundsen Sea low, which in turn is strongly influenced by the El Niño–Southern Oscillation (ENSO; Box 4.1), with the consequence that the amplitude of wave number 1 is larger during the El Niño phase of the cycle.

In austral winter (May–September), 500-hPa geopotential heights over Africa exhibit positive correlations with heights over South America and the central South Pacific near New Zealand, and negative correlations with heights over the Southern Ocean. A wave number 3 pattern is apparent – with three troughs and three ridges – showing the correlation of 500-hPa geopotential height anomalies in the subtropics with 500-hPa anomalies at 50° S, 95° E during winter. The wave number 3 pattern is commonly associated with blocking patterns in the southern hemisphere. During summer (November–March), anomalies over the three continents occur out of phase with anomalies over the subtropical oceans and in a wave number 3 pattern over the Southern Ocean near 55° S.

The cyclone climatology of the southern hemisphere reveals that there are typical mid-latitude baroclinic systems as well as mesocyclones (mainly comma clouds and less commonly spiral forms) (Box 4.2). A winter climatology for

Box 4.1 El Niño–Southern Oscillation (ENSO)

The Southern Oscillation is a fluctuation in surface air pressure of about ±2 hPa between the equatorial eastern Pacific Ocean and the western Pacific Ocean (Tahiti and Darwin, respectively). It has a time scale of about two to three years and is closely coupled to variations in tropical Pacific SSTs. Typically, when there is strong high pressure in the eastern tropical Pacific, the eastern ocean is cool with strong oceanic upwelling, and the western Pacific is warm. The upwelling in the eastern Pacific is caused by persistent easterly trade winds, which push the water away from the coast of South America. At three-to seven-year intervals, the trade winds weaken and the eastern Pacific warms, often around Christmas, and the pattern of vertical zonal circulation cells over the Pacific Ocean shifts eastward as pressure rises over Indonesia. The warm southward-flowing ocean current off Peru is called El Niño, Spanish for "the Christ child." The opposite cool mode is known as La Niña, "the girl," and is associated with the cool north-flowing Humboldt (or Peru) Current.

These two modes are the source of the greatest climatic anomalies on Earth. El Niño attained unprecedented amplitudes in 1982 and 1997 and was unusually prolonged in 1992.

Box 4.2 Mesocyclones

Mesocyclones are a class of small-scale (less than 1,000-km diameter) low-pressure systems that last only one to two days (Rasmussen and Turner 2003). In high latitudes they are often referred to as polar lows that form poleward of the main mid-latitude storm tracks. They develop most frequently in the cold season over ice-free ocean areas, where heat and moisture transfer from the ocean provides convective energy. They are found especially in the Norwegian and Barents seas, the Gulf of Alaska, and the Labrador Sea in the northern hemisphere and over the Southern Ocean. Their structure is most commonly either a comma cloud, poleward of the polar frontal zone, or a spiraliform system with cloud bands wrapped around the low center, analogous to a tropical hurricane. Strong spiraliform systems are sometimes referred to as Arctic hurricanes since deep cumulonimbus clouds may encircle a cloud-free eye. Figure 4.9 shows satellite imagery of polar lows.

Box 4.2 (continued)

Figure 4.9
Polar low in the
Greenland Sea as
viewed by MODIS,
March 19, 2010.
Source: Dr. Claude
Chantal, LMD,
Polytechnique, Paris.

1977–1983 was prepared from DMSP infrared images by Carlton and Carpenter (1990). In most winters, there is a positive relationship between the regional extent of the Antarctic sea ice, the longitudes of preferred occurrence of cold air outbreaks, and the incidence of polar lows. For the transition and winter seasons of 1988, Fitch and Carleton (1991) analyzed mesocyclones in the half-hemisphere sector centered on the Ross Sea. During active outbreak periods, a negative thickness anomaly ("cold pool") is located northeast of the Ross Sea, and mesoscale vortices tend to occur on the poleward side of that anomaly. In addition, an enhanced trough–ridge pattern is evident for the Ross Sea sector. Most systems are located between 60° and 70° S. For winter 1988 for 100° E to 120° W, and for winter 1989 for 160–60° W, Carleton and Fitch (1993) calculated a mean life of 15–28 hours and a motion speed of 10–16 m s^{-1}.

In a study of mesocyclones around Antarctica for June–September 2004, using high-resolution infrared (IR) mosaics, Verezemskaya et al. (2017) counted 1,735 systems, of which three fourths were polar lows, mainly comma clouds. They had a mean diameter of 300 km and a lifetime of approximately 10 hours. Maxima in track density were observed over the Bellingshausen Sea and around East Antarctica.

Uotila et al. (2011) used the Antarctic Mesoscale Prediction System (AMPS) to produce a cyclone climatology for 2001–2009 that identified 17,000 cyclone tracks. Of these cyclones, 20 percent were mesoscale in terms of their size. Mesoscale systems were common south of the Antarctic Circumpolar Trough (ACT) over the coastal oceans of the Indian Ocean sector, and in the Ross Sea. Small-scale systems occurred twice as frequently in July as in December–February. Large synoptic systems occurred most frequently in the ACT. These systems move southward from the ACT and then turn eastward along the Antarctic coast as they decay.

A notable feature of the atmospheric circulation is the Southern Annular Mode (SAM), the primary mode of atmospheric variability in high southern latitudes (Marshall 2003). Opposing pressure and geopotential height anomalies are found over Antarctica and the southern middle latitudes. The westerly circulation is strong (weak) when the index of this difference is positive (negative). The SAM was mostly negative from 1980 to 1993, positive to 2000, and then negative to 2008. There has been an increasing trend in the SAM in the austral summer since the mid-1960s, and in austral autumn since 1958, that has been linked to the intensification of the ozone hole in the stratosphere. Polyani et al. (2011) show that the impacts of ozone depletion in the second half of the twentieth century were two to three times larger than those associated with increased greenhouse gases for the southern hemisphere tropospheric summer circulation. The wester-lies have increased by 15–20 percent since the 1970s (Turner and Marshall 2011). The change in the SAM has led to a decrease in the annual and seasonal number of cyclones south of 40° S. There are now fewer but more intense cyclones in the Antarctic coastal zone between 60° and 70° S, except in the Amundsen–Bellingshausen Sea sector (Turner et al. 2014)

Another major circulation mode is the Semiannual Oscillation (SAO), which consists of a twice-yearly expansion and contraction of the circumpolar trough around Antarctica (Van Den Broeke 1998; van Loon 1967). This feature occurs in response to differences in heat storage between Antarctica and the surrounding oceans. As a result of the SAO, the surface pressure in mid- and high latitudes shows a clear half-yearly wave. The amplitude of this wave peaks at 45–50° S in each of the three oceans (with values of 2–3 hPa), has a minimum at 60° S (0.5 hPa), and peaks again over coastal Antarctica (4–5 hPa). The phase reverses at approximately 60° S from equinoctial pressure maxima (March/September) north of this latitude to solstitial maxima (January/July) south of it. Because of the meridional pressure gradient, the zonal westerlies show equinoctial maxima that are 20–30 percent stronger than those in summer and winter.

Van den Broeke (1998) has analyzed observations at twenty-seven stations that were operational during (part of) the period 1957–1979. For the annual cycle of surface pressure, the second harmonic explains 17–36 percent of the total vari-ance on the Antarctic plateau, 36–68 percent along the East Antarctic coast, and almost 80 percent on the west coast of the Antarctic Peninsula, decreasing farther to the north. A significant coupling between the half-yearly wave in surface pressure and that in surface temperature is found for coastal East Antarctica, which can be directly explained by the changes in meridional circulation brought about by the SAO.

The South Pacific sector of the Southern Ocean, comprising the Ross, Amundsen, and Bellingshausen seas between 60–70° S, is a persistent area of low pressure known as the Amundsen Sea low (ASL) (Raphael et al. 2016). It appears to result from the interaction of the southern westerlies with the elevated topography of Victoria Land, although its presence also marks the large number of low-pressure systems that move south from mid-latitudes, or form in the baroclinic zone of the circumpolar trough.

According to Fogt et al. (2012), the Amundsen–Bellingshausen Sea low accounts for more than 550 singular depressions per year coming from the Pacific Ocean. The low is centered just west of the Antarctic Peninsula in austral summer and over the Ross Sea in winter (Fogt et al. 2012; Turner et al. 2013). There is also an annual cycle in its meridional location, which is farther north in summer and farthest south in late winter.

The ASL is strongly correlated to the tropical Pacific, as is shown by the association between annual southern hemisphere mean sea-level pressure (MSLP) and the Southern Oscillation Index (SOI) (Mayewski et al. 2017). The ASL is weaker (i.e., high-pressure anomalies dominate, westerly wind speeds are reduced, less sea ice) and the Weddell Sea low is stronger (low-pressure anomalies, southerly winds, greater sea ice extent) during El Niño events when the eastern tropical Pacific is warmer; the reverse is true in La Niña events.

4.2 Oceanography and Water Balance

4.2.1 Arctic

Study of the characteristics of the Arctic Ocean began in the 1890s with the work of Nansen on the famous *Fram* drift. Rudels (2015) describes the principal contributions extending to the International Polar Year, 2007–2009. A major synthesis of data collected from the 1950s through 1970s was published by Gorshkov (1980) in the *Atlas of the Arctic Ocean.*

The Arctic Ocean has a total volume of 17×10^6 km^3, or 1.3 percent of the global ocean, and accounts for about 3 percent (9.4 million km^2) of the area. It is divided into two main basins, the Eurasian and the Canadian, which are separated by the Lomonosov Ridge; the Eurasian Basin is further separated into the Nansen and Amundsen basins by the Gakkel Ridge and the Canadian Basin into the Makarov and Canada basins by the Alpha-Mendeleyev Ridge (Carmack et al. 2016). About 36 percent of the Arctic Ocean's area is taken up by extensive continental shelves – especially off Eurasia where they may extend up to 800 km from the coast; these shallow shelves are greatly influenced by runoff from the surrounding land areas. The connection to the Pacific Ocean via Bering Strait is very shallow (less than 60 m deep) and only 85 km wide, while the four main channels linking the Arctic Ocean to the Atlantic Ocean through the Canadian Arctic archipelago vary from 85 to 900 m in depth. The Denmark Strait between Greenland and Svalbard, with a depth of 650 m, links the Arctic Ocean to the North Atlantic. There is also a very important inflow of Atlantic water into the Arctic through the Norwegian and Barents seas via the North Atlantic and West Spitsbergen currents; the latter current has salinities of 34–35 practical salinity units (psu, dimensionless). [In 2010, a new Reference Composition Salinity Scale (g kg^{-1}) was introduced, based on the thermodynamic scale of seawater 2010 (TEOS-10) (Intergovernmental Oceanographic Commission et al. 2010) (www.teos-10.org/). The new units are gradually being used. An absolute salinity of 35.2 g kg^{-1} corresponds to 35.0 psu.]

The temperature of the surface water is close to the freezing point of seawater throughout the year, −1.5 to −1.8 °C. Figure 4.10 shows the 1950–1989 mean distributions of temperature and salinity at the surface for winter and summer.

In winter, heat loss and brine rejection from growing sea ice lead to the formation of a thermocline (steep vertical temperature gradient) and pycnocline (sharp vertical density gradient) in the water column at about 25–200 m (Figure 4.11). The halocline associated with the vertical salinity gradient maintains the stable density structure. The upper 20 m of the Arctic mixed layer is isothermal and isohaline due to turbulent mixing through ice motion and convection driven by brine rejection. The temperature maximum at the top of the Atlantic water mass is likely a result of water that has experienced significant summer radiative heating and been transported from the Amundsen Gulf or Bering Sea.

While the Arctic Ocean is only 1 percent and 3 percent of the global ocean in terms of volume and surface area, respectively, it collects more than 11 percent of the global river discharge (Carmack et al. 2016). The mean freshwater budget of the entire Arctic (excluding the Canadian Arctic archipelago and Baffin Bay) has been calculated by Serreze et al. (2006). They combined terrestrial and oceanic observations with insights gained from the ERA-40 reanalysis and land surface and ice–ocean models. The major terms are annual net precipitation over the Arctic Ocean, 2,000 km^3 (24 percent); river runoff, 3,200 km^3 (38 percent); and Pacific inflow, 2,500 km^3 (30 percent) – all relative to a reference salinity of 34.8 psu. Haine et al. (2015) provided updated estimates that include the Canadian archipelago and Baffin Bay: net precipitation, 2,200 km^3; river runoff, 4,200 km^3; and Pacific inflow, 2,600 km^3.

There is about 100,000 km^3 of freshwater (FW) stored in the Arctic Ocean (Haine et al. 2015). Most freshwater exists in the Canadian Basin, and specifically in the Beaufort Gyre above it, where about 23,500 km^3 is stored. In the Eurasian

Basin, typical liquid FW thicknesses are 5–10 m, compared with 20–25 m in the western Arctic.

Anomalies of freshwater discharge into the North Atlantic via Fram Strait or Baffin Bay have occurred at irregular intervals. The Great Salinity Anomaly (GSA) originally referred to a major pulse of fresh water input to the Nordic Seas in the late 1960s and early 1970s (Dickson et al. 1988; Häkkinen 1999) (Box 4.3).

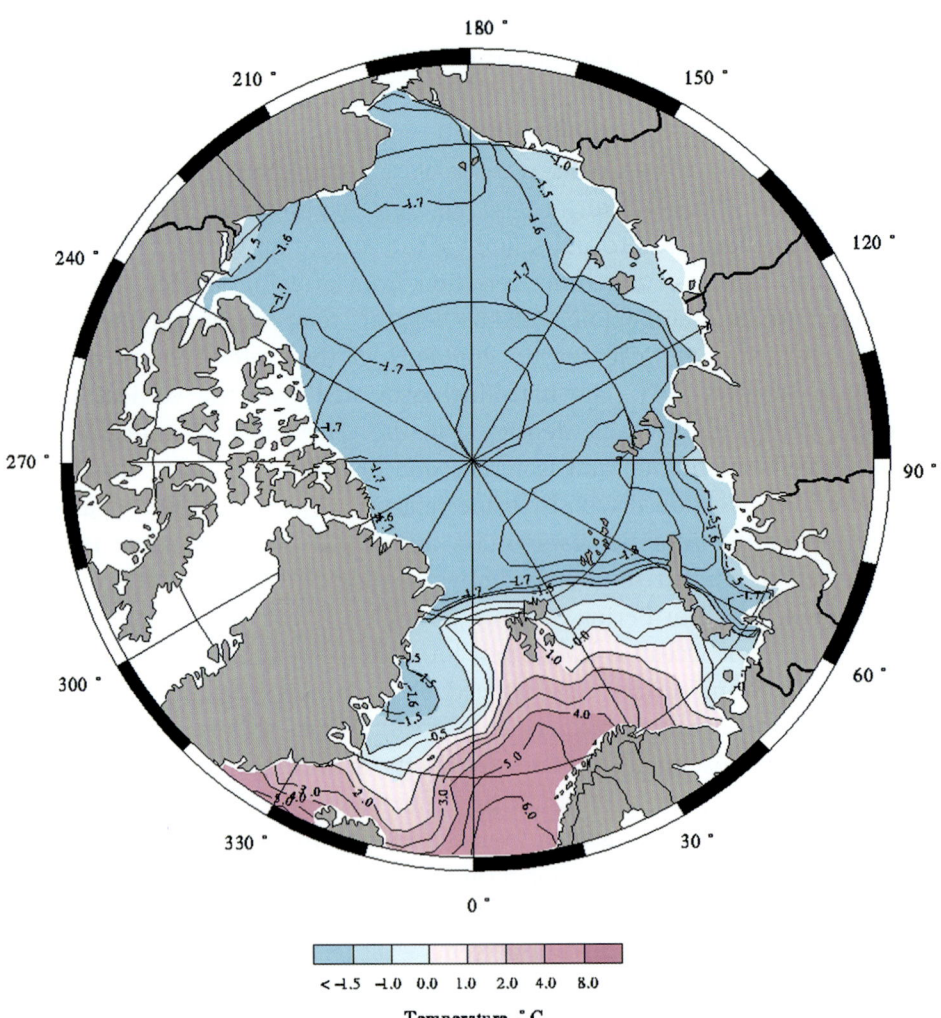

Figure 4.10 The 1950–1989 mean winter and summer temperature and salinity (psu) at the surface of the Arctic Ocean: (a) winter temperature, (b) winter salinity, (c) summer temperature, (d) summer salinity.
Source: Tanis and Timokhov 1997, 1998.

The source of much of the river discharge is in the mid-latitudes, where the headwaters of the major river systems originate. Figure 4.12 illustrates how four different definitions of the Arctic affect the area from which river runoff occurs:

(a) All Arctic Regions (AAR) (Shiklomanov 2000), which includes flow from North American rivers draining into Hudson Bay and from the Yukon River (Alaska) and Anadyrsky River (Russia), which drain into the North Pacific just south of the Bering Strait. Also included in the AAR definition is flow from all of Greenland, the western edge of Norway, and the Canadian Arctic archipelago, giving a total of 23.7×10^6 km^2.

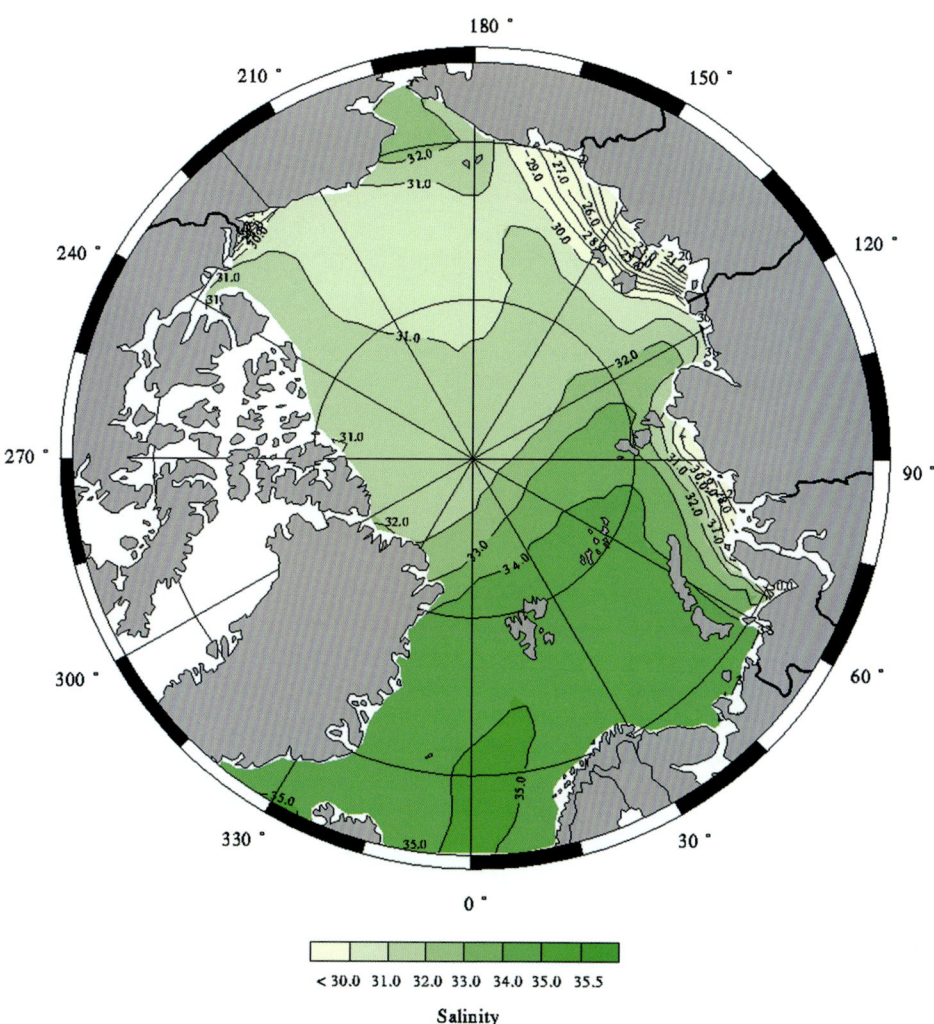

Figure 4.10 (cont.)

(b) Arctic Ocean Basin (AOB) (Shiklomanov 2000), which does not include rivers that empty into the subarctic Pacific or Hudson Bay. It has an area of 18.9×10^6 km^2.

(c) Arctic Climate System Study (ACSYS) of the World Climate Research Program (WCRP), which excludes the southern portions of Greenland and the Canadian Arctic archipelago. Its area is 17.1×10^6 km^2.

(d) Arctic Ocean River Basin (AORB) (Lewis et al. 2000), which that excludes much of the Canadian Arctic archipelago. It has an area of 15.5×10^6 km^2.

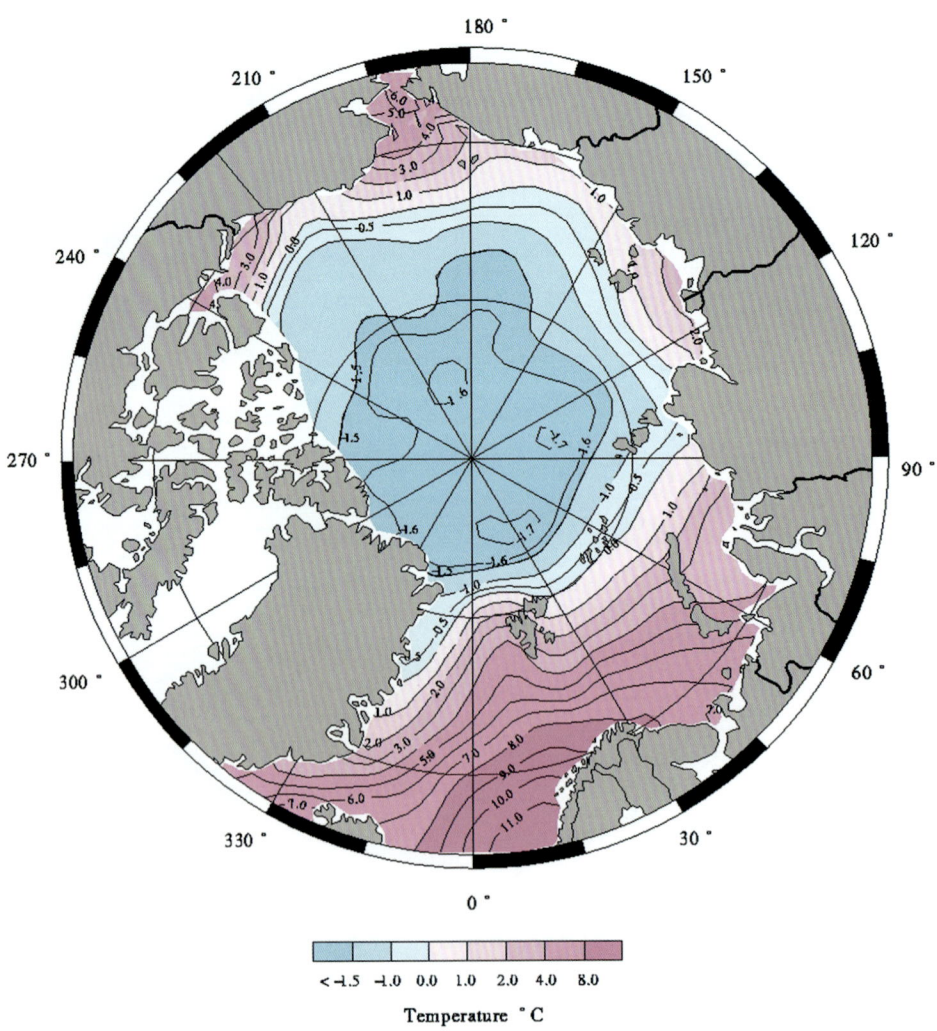

Figure 4.10 (cont.)

The runoff into the seas of the Russian Arctic is reported in Arctic Climatology Project (2000). For 1961–1990, the mean values are shown in Table 4.2. Underground water outflow to the Arctic Ocean from the continents appears to be small, on the order of 50 km^3 yr^{-1} according to Zektser and Dzhamalov (1988).

A circulation mode in the North Pacific Ocean can affect the higher latitudes. The Pacific Decadal Oscillation (PDO) features warm or cool surface waters in the North Pacific Ocean, north of 20° N. During a "warm" (positive) phase, the west

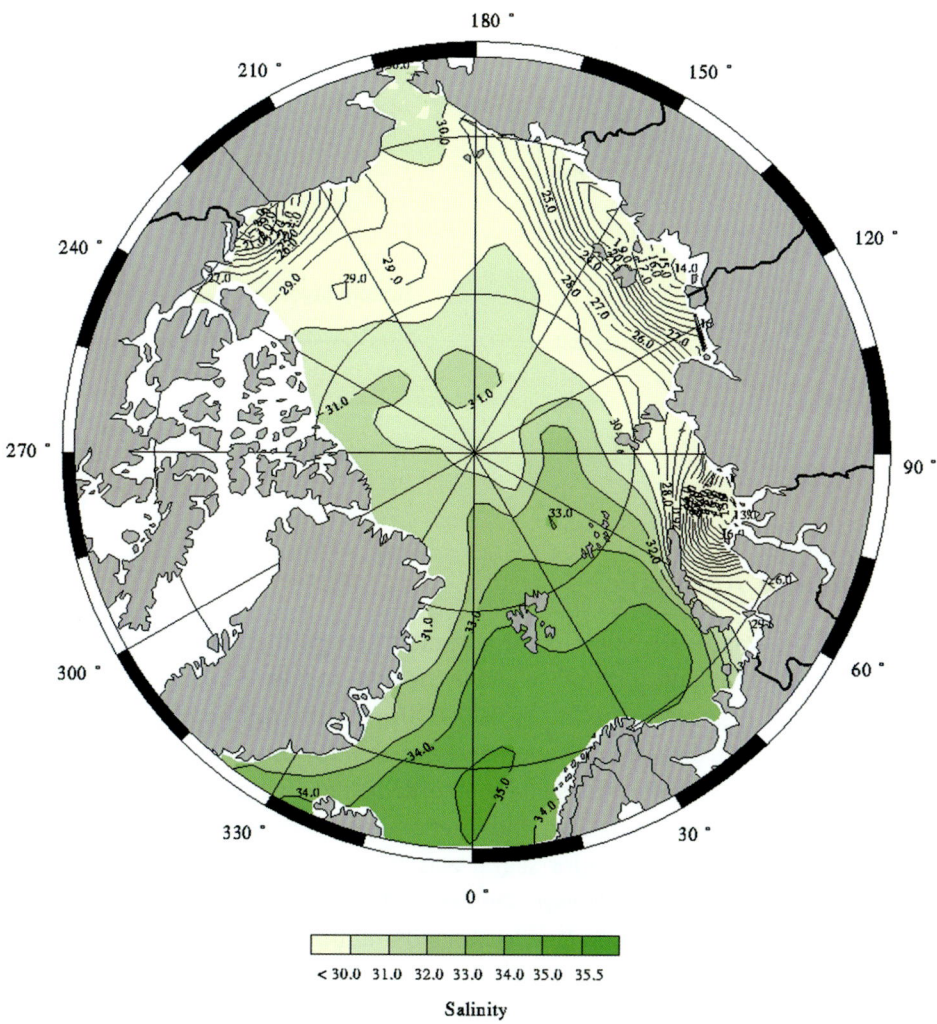

Figure 4.10 (cont.)

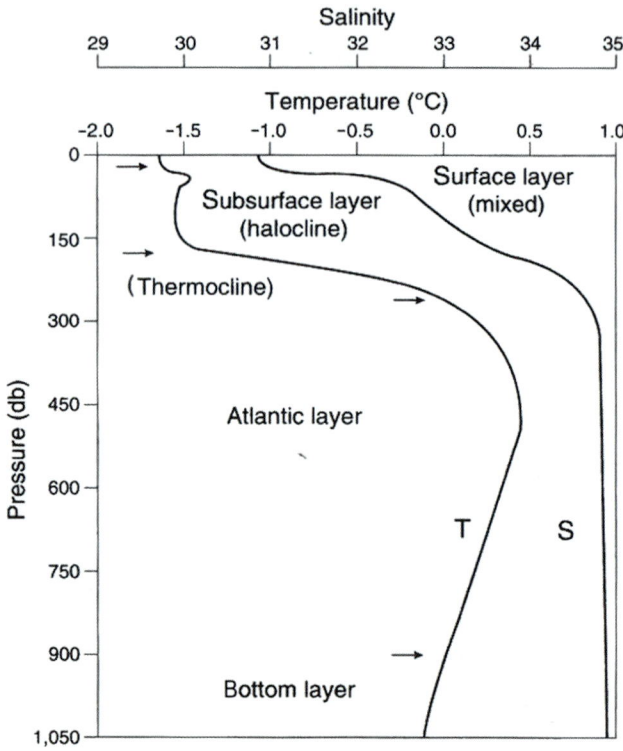

Figure 4.11
Temperature and salinity profile in the Canadian Basin of the Arctic Ocean.
Source: Melling and Lewis 1982, 968, figure 1.

Box 4.3 **The Great Salinity Anomaly**

The Great Salinity Anomaly (GSA) was an advective event, traceable around the Atlantic subpolar gyre for more than fourteen years from its origins north of Iceland in the mid- to late 1960s until its return to the Greenland Sea in 1981–1982. It affected the upper 500–800 m of the northern North Atlantic. There were successive occurrences of the same phenomenon in the 1980s and the 1990s (Belkin 2004). In 1989, a strongly positive NAO mode intensified the northwesterly winds over Baffin Bay and Labrador Sea, lasted through 1995, and enhanced the Arctic Ocean freshwater export via the Canadian archipelago to Baffin Bay and the Labrador Sea. It was transported by the Irminger Current toward Iceland and then past the Faroes to the Norwegian Sea in 1994–1995. The advection rate of the GSA of the 1970s, 1980s, and 1990s between Newfoundland and the Faroe–Shetland Channel is conservatively estimated to have been 3.5, 10, and 10 cm s^{-1}, respectively. The recovery time for reversion of such anomalies is typically on the order of several years.

Table 4.2 Mean runoff (km³), 1961–1990, into the seas of the Russian Arctic

Barents/White Sea	Kara Sea	Laptev Sea	East Siberian Sea	Chukchi Sea	Total
431	1,333	754	233	31	2,780

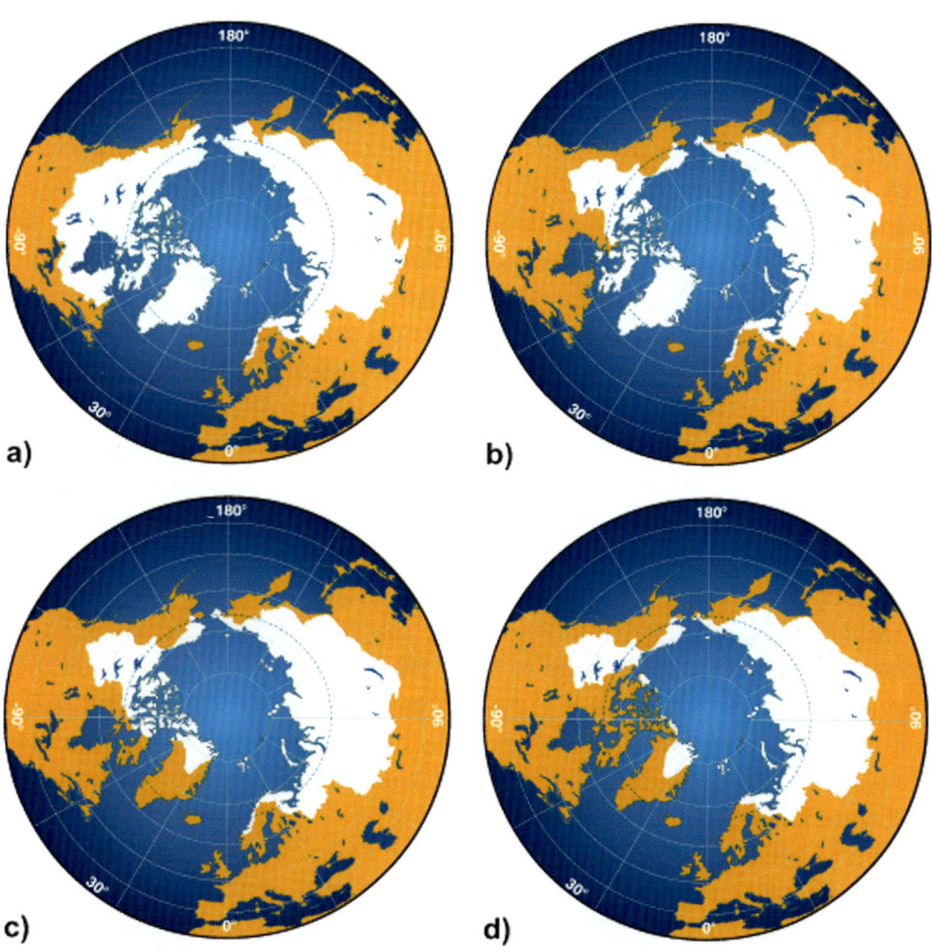

Figure 4.12 Definitions of the Arctic showing contrasting areas that contribute freshwater to the Arctic Ocean: (a) All Arctic Regions (AAR), (b) Arctic Ocean Basin (AOB), (c) Arctic Climate System Study (ACSYS), (d) Arctic Ocean River Basin (AORB).
Source: Prowse et al. 2015, 120, figure 3. Courtesy of American Geophysical Union.

Pacific cools up to 0.5 °C and part of the eastern ocean warms; during a "cool" (negative) phase, the opposite pattern occurs. The oscillation is quasi-periodic, lasting for twenty to thirty years. "Cool" PDO regimes prevailed over 1890–1924, 1947–1976 (except around 1958–1960), and since about 1998, while "warm" PDO regimes dominated in 1925–1946 and from 1977 through the late 1990s. The PDO is not a single mode of oceanographic variability, but arises through the combined influence of the ENSO in the equatorial Pacific, atmospheric variability in the extratropics, and the North Pacific Ocean gyre.

4.2.2 Antarctic

The Antarctic continent is surrounded by the vast Southern Ocean. Its northern limit is usually taken as 60° S, where there are no land interruptions. It has an area of 70 million km^2 (Kort 1964). Pronounced meridional gradients in surface properties separate the waters of the Southern Ocean from the warmer and saltier waters of the subtropical circulations. This boundary, which is termed the Subtropical Front (STF), is located at approximately 45° S.

The Antarctic Circumpolar Current (ACC) flows continuously eastward through the Southern Ocean, driven by the persistent westerly winds around the Antarctic continent, linking the Indian, Pacific, and Atlantic oceans (Wyrtki 1960) (Figure 4.13). Donahue et al. (2016) determined this flow through Drake Passage between 2007 and 2011 through continuous monitoring with a line of moored instrumentation. Annual mean near-bottom currents are remarkably stable from year to year. The mean barotropic transport, determined from the near-bottom current meter records, is 45.6 Sv. Summing this with the mean baroclinic transport, relative to zero at the sea floor of 127.7 Sv, gives a total transport through Drake Passage of 173.3 Sv, which is 30 percent greater than previous estimates. This total is 170 times that of the world's rivers. The ACC is not a broad strong flow, but rather is made up of bands of very strong jets. Nowlin et al. (1977) found four jets in the Drake Passage based on hydrography and current meter moorings.

Two well-defined fronts – the Sub-Antarctic Front (SAF) and the Polar Front (PF) – are continuous features of the ACC (Orsi et al. 1995). Temperatures at 100 m decrease by about 3 °C across both the SAF and PF. Well-established indicators of the SAF and PF can be traced in an unbroken line around Antarctica, although at some locations there are three fronts. At Drake Passage, three deep-reaching fronts account for most of the ACC transport. The third deep-reaching front observed to the south of the PF at Drake Passage also continues with similar

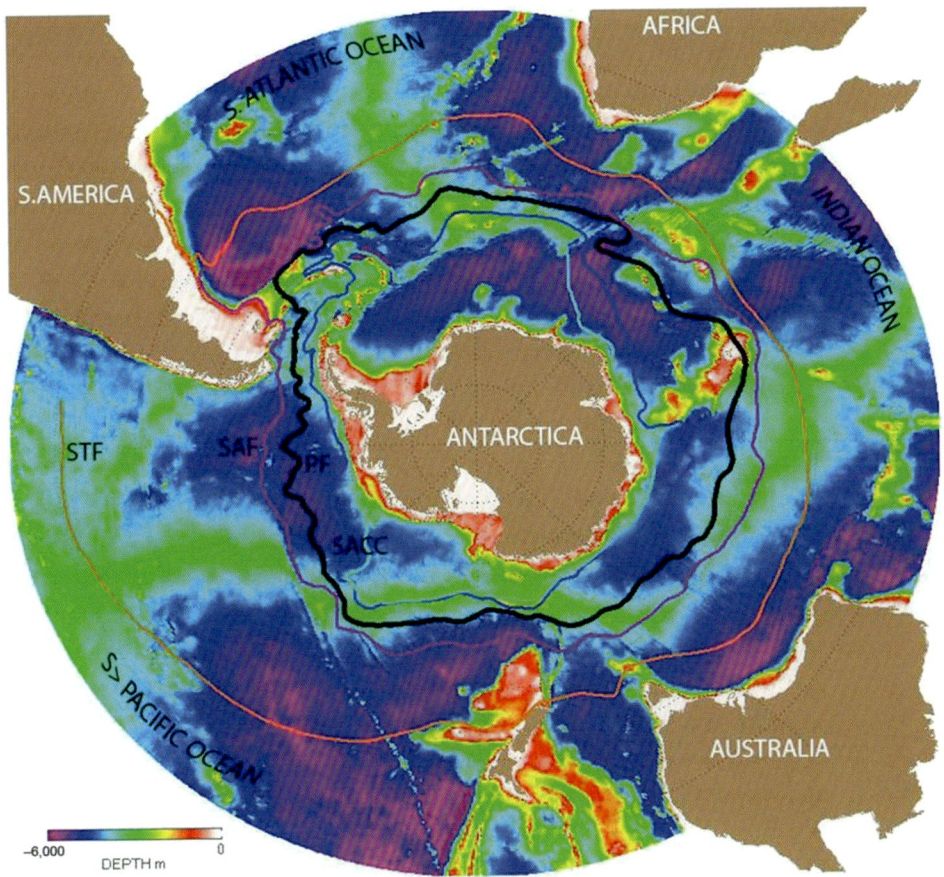

Figure 4.13 Antarctic Circumpolar Current image from GRACE Mission. Seawater density fronts after Orsi, Whitworth, and Nowlin (1995). SACC = Southern Antarctic Circumpolar Current Front, PF = Polar Front, SAF = Sub-Antarctic Front, STF = Subtropical Front.
Source: www.nasa.gov/vision/earth/lookingatearth/grace-images-20051220.html. Courtesy of NASA/ JPL-Caltech.

characteristics as a circumpolar feature; it is named by Orsi et al. as the Southern ACC (SACC) front (see Figure 4.13). This front does not separate distinct surface water masses.

From north to south in the Southern Ocean, the four water masses separating the fronts are the Sub-Antarctic Zone (SAZ), the Polar Frontal Zone (PFZ), the Antarctic Zone (AZ) and the Continental Zone (CZ).

The Polar Front meanders and forms cyclonic eddies, which are then shed and make their way across the APFZ, often to be trapped in the SAF. These eddies can be identified by the Antarctic Surface Water (AASW) flow that they carry with them. These cold features appear to widen the Antarctic Polar Frontal Zone by intensifying the SAF and moving it to the north (Sievers and Emery 1978).

AASW extends with rather uniform properties from the PF to the continental margins of Antarctica. The southern boundary of the ACC is close to the Antarctic continent between the eastern border of the Ross Sea and the Antarctic Peninsula, and around eastern Antarctica. It is farthest away from the continent at 140° W and 20° E.

The warm, saline water mass that occupies most of the deep layers of the ACC is the Circumpolar Deep Water (CDW). The poleward edge of the Upper Circumpolar Deep Water (UCDW) signal is considered by Orsi et al. (1995) to be a reasonable definition of the southern boundary of the ACC. It marks the southern extent of the only water mass found exclusively in the ACC and not in the subpolar regime.

A physical definition of the northern boundary of the ACC is provided by the location of the Antarctic Polar Frontal Zone (APFZ), which Emery (1977) identifies in six north–south XBT sections from south of Australia and east to the Drake Passage. This complex transition zone includes the cold AASW to the south of the PF. Sub-Antarctic waters are found to the north of the SAF. An earlier concept of an Antarctic Convergence is now rejected by oceanographers. The average drop in temperature from north to south is close to 3 °C – that is, from 5.6 °C to less than 2 °C. The APFZ is located between about 50° and 60° S and varies greatly in width around Antarctica.

Within the Ross and Weddell seas there are clockwise gyres, forced by the polar easterly winds, that form as a result of interactions between the circumpolar current and the continental shelf, which has an abrupt edge. The Ekman transport, which is directed outward from the centers of the gyres, leads to upwelling of cold, nutrient-rich water.

In winter, the Southern Ocean freezes outward to 65° S in the Pacific sector and 55° S in the Atlantic sector. In contrast to the Arctic, due to the high ocean waves, the sea ice cover forms mainly from accumulations of frazil ice. Also, the sea ice is almost entirely first-year ice, which means that it forms and decays through the course of an annual cycle. The physical processes involved are discussed in Section 7.3. At numerous locations along the coasts of Antarctica open water areas – polynyas – form as a result of strong offshore winds. Polynyas may also form due to the upwelling of warmer deep water induced by sea mounts (e.g., Maud Rise in the eastern Weddell Sea); their characteristics are described in Section 7.4.

The melt season is quite different between the two polar regions. In the Antarctic there are very few melt ponds on the sea ice, whereas these are common in the Arctic summer. According to Andreas and Ackley (1982), the low relative humidity (60 percent or less) owing to the relatively dry winds off the Antarctic continent and an effective radiation parameter (ø), which is smaller than that characteristic of the Arctic, are primarily responsible for the absence of melt features. The parameter ø is the ratio of the nonturbulent flux to the maximum possible value of the latent heat flux – that is, the latent heat flux that would occur if the relative humidity were zero. In essence, ø parameterizes the effective net radiation at the surface. It is inversely related to the wind speed. Antarctic wind speeds are about double those in the Arctic (10 versus 5 m s^{-1}). For example, if the wind speed is large, the rapid removal of heat by surface sublimation can preclude melting despite a large net radiation balance. In the Antarctic, surfaces can tolerate higher air temperatures without melting and temperatures must be above 0 °C for melt to begin, a situation that is rare. Sublimation supplements the small energy loss due to sensible heat flux.

4.3 Climate

4.3.1 Radiative Regimes

The radiative regimes of the polar regions exhibit the same basic annual patterns, as a result of the six-month polar day and night, but the individual components differ greatly in their specific characteristics due to the nature of their surface properties (sea ice on the Arctic Ocean versus high-elevation Antarctic ice sheet) and the cloud conditions.

In winter, the surface radiation balance is negative due to the outgoing infrared radiation from the snow and ice surfaces. The net radiation budget over the Arctic Ocean is −80 megajoules (MJ) m^{-2} in January. In July over Antarctica, it is approximately −50 MJ m^{-2}, as a result of the much lower surface temperatures that reduce the longwave emission. There is a small downward transfer of sensible heat from the relatively warmer atmosphere to the colder surface.

During spring, the Arctic receives increasing amounts of incoming solar radiation but, as a result of the ubiquitous snow cover with an albedo of 0.8–0.9, nearly all of it is reflected back (Uttal et al. 2002). Gradually, the snow cover retreats on land and begins to melt on the Arctic sea ice. Webster et al. (2015) monitored melt pond formation on first-year ice (FYI) and multiyear ice (MYI) in the Beaufort and Chukchi seas in 2011 as the Applied Physics Laboratory Ice Station (APLIS) drifted westward and in September became ice free. They used 1-m resolution panchromatic satellite imagery paired with airborne and in situ data. First-year ice is mostly undeformed with a relatively homogeneous snow cover. Surprisingly, melt ponds formed on MYI at the end of May, three weeks earlier than on FYI, which lacks the preexisting surface topography of the former. Both ice types had comparable mean snow depths (17–18 cm), but multiyear ice had 0–5 cm of snow covering 37 percent of its surveyed area (compared with 15 percent of undeformed FYI), which may have facilitated earlier melt due to its low surface albedo compared to thicker snow. Maximum pond fractions were 53 and 38 percent on FYI and MYI, respectively. Pond fraction reached its maximum of 48 percent on July 23, which was double the maximum observed during SHEBA in 1988 (Perovich et al. 2002). At SHEBA, ponds began to freeze over on August 17. The initial snow depth at APLIS averaged 18 cm, while that at SHEBA was 34 cm. For average density, 20 W m^{-2} of energy was needed to melt the snow at APLIS, or nearly half of the required 38 W m^{-2} for melting the snow cover at SHEBA.

At SHEBA, there were five distinct phases in the evolution of albedo: dry snow, melting snow, pond formation, pond evolution, and autumn freeze-up (Figure 4.14). By the end of July, the average albedo along the line was 0.4, and there was significant spatial variability, with values ranging from 0.1 for deep, dark ponds to 0.65 for bare, white ice.

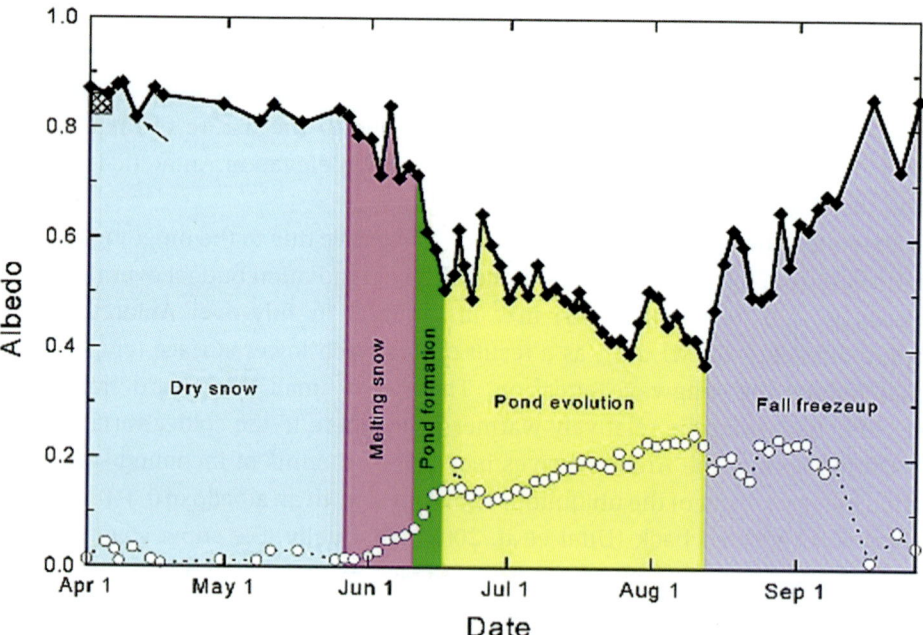

Figure 4.14 Time series of wavelength-integrated albedo from April 1, 1998 to September 27, 1998. Values are averaged over a 200-m-long albedo line. The arrow points to April 17, when the sky was clear. Also plotted is the albedo measured at the beginning of the experiment in October 1997 (solid squares). The standard deviation of albedo measured along the albedo line is plotted as open circles.
Source: Perovich et al. 2002, 20–8, figure 9. Courtesy of American Geophysical Union.

King (1996) studied L↓ at the South Pole, Mizuho, Syowa, and Neumayr stations. Relationships were derived relative to screen air temperature, T_A, and temperature at the top of the surface inversion, T_M.

$$L\downarrow = 1.09 \times 10^{-7} T_A^{3.85} \tag{4.1}$$

$$L\downarrow = 1.31 \times 10^{-13} T_M^{6.29} \tag{4.2}$$

Surface heat fluxes are small over the Antarctic ice sheets, so the fourth-power dependence on surface temperature results from a surface energy budget dominated by a near balance between upward and downward longwave radiation. The sixth-power relationship between L and T is reminiscent of Swinbank's (1963) formula for clear-sky radiation, but here monthly data were used. Two factors may explain this relationship. First, over the interior of Antarctica, clouds are relatively tenuous and radiate like gray bodies, having effective emissivities that are strongly dependent on cloud ice water content, and hence on temperature. Second, variations in the amount and type of cloud cover will lead to variations in effective emissivity.

Total cloud fraction (TCF) over the Arctic has been analyzed from observations and reanalyses by Chernokulsky and Mokhov (2012). High values of ocean-mean TCF are noted from May to October (about 0.8). In April and May, there is a rapid increase in cloud amounts as a result of both moisture advection and surface melt (Barry et al. 1987). Low TCF values are noted from November to April (0.55–0.75). The summer maximum is associated with low-level stratiform (St and Sc) cloud, while in winter cyclone-induced upper-level clouds dominate. The Arctic land-mean TCF has a prominent annual cycle, with the maximum in August–October (0.64–0.78) and the minimum in December–March (0.50–0.68). Different observational series are in better agreement in summer than in winter and over the ocean than over land for the Arctic mean TCF. Drifting station observations show that during winter, half of the cloud base heights are up to 600–700 m (Frolov et al. 2005, 47). In summer, 60–65 percent of cloud bases are below 300 m.

For the Antarctic, few cloud data are available. For the South Pole station, Town et al. (2007) presented data from five different data sources: routine visual observations (1957–2004; Cvis), surface-based spectral infrared data (2001; CPAERI), surface-based broadband IR data (1994–2003; Cpyr), the Extended Advanced Very High Resolution Radiometer (AVHRR) Polar Pathfinder (APP-x) data set (1994–1999; CAPP-x), and the International Satellite Cloud Climatology Project (ISCCP) data set (1994–2003; ISCCP). Cloud cover has been found to range from 45–50 percent during the short summer to a relatively constant 55–65 percent during the six months of winter. The surface-based broadband IR data (1994–2003; Cpyr) are recommended as the best for further use.

Black carbon (BC) aerosol is present in the Arctic atmosphere as a result industrial pollution, especially from Russia, that is transported into the Arctic in the cold season and as a result of forest fires in summer. The concentration at the Alert and Barrow ranges from about 10 ng m^{-3} in summer to 100 ng m^{-3} in winter (Sharma et al. 2006). From 1988 to 2003, however, there was a 54 percent decrease at Alert and a 27 percent decrease at Barrow for the all-year data, possibly as a result of pollution controls.

Stone et al. (2014) point out that when carbonaceous particles are deposited on snow and ice, they reduce the surface albedo, which accelerates melting, leading to a temperature–albedo feedback that amplifies Arctic warming. BC, in particular, has been implicated as a major warming agent at high latitudes. BC aerosols in the atmosphere, however, attenuate sunlight and radiatively cool the surface. A modest loading of smoke (with an aerosol optical depth of 0.1) results a net shortwave forcing of approximately -10 W m^{-2}, which radiatively cools the surface. The cumulative dimming impact of atmospheric BC aerosols over the annual cycle very likely more than offsets the darkening effects.

4.3.2 Arctic Temperature Conditions

Average January temperatures over the Arctic Ocean range from about -34 to $0\,°C$, and winter temperatures can drop below $-50\,°C$ over large parts of the Arctic. There is a large warm anomaly, however, over the Norwegian and Barents Sea region as a result of the northeastward flow of the Atlantic Current and associated penetration of North Atlantic cyclones (Figure 4.15). While the sun returns in March, absorbed radiation amounts remain low due to the high reflectivity of the persistent snow cover. Only with snow melt in May–June do temperatures begin to rise above freezing. Average July temperatures range from about 0 to $+10\,°C$, with temperatures over some land areas occasionally exceeding $30\,°C$ in summer. They remain around $0\,°C$ over the Arctic Ocean due to the melting pack ice and about $-10\,°C$ over interior Greenland. In contrast to the clear skies of spring, the Arctic Ocean has 80 percent cloud cover in summer, with low stratiform cloud due to both moisture advection from lower latitudes and evaporation from the melting pack ice and open water. The climate of Greenland is discussed in Section 6.1.

The coldest location in the northern hemisphere is not in the Arctic, but in the sub-Arctic of northeastern Russia. This is determined by the continental location, far from the moderating influence of the ocean, and by the topography, where deep valleys trap cold, dense air and create strong temperature inversions (discussed next). The record low temperature of $-67.7\,°C$ occurred in Oimyakon on February 6, 1933, and in Verkhoyansk on February 5 and 7, 1892.

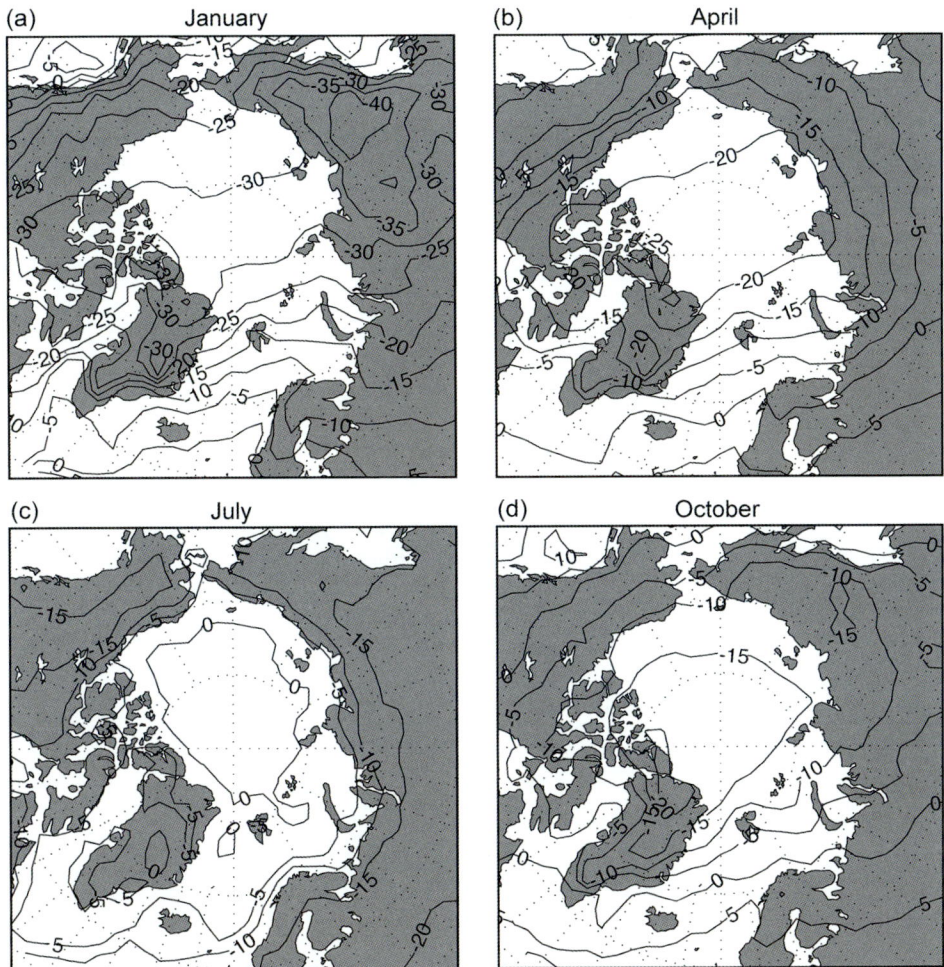

Figure 4.15 Mean air temperatures over the Arctic: (a) January, (b) April, (c) July, (d) October.
Source: Serreze and Barry 2014, 52, figure 2.24.

Areas of polar maritime climate (Ballinger et al. 2013) (Figure 4.16) are delimited by the mean temperature of the coldest month being below $-6.7\ °C$, where the poleward margins approximately coincide with the maximum winter extent of pack ice. Summers are cool (5–10 °C) and cloudy. This climatic type occurs in two regions of the northern hemisphere: (a) a southwest-to-northeast zone in the North Atlantic, from south of Greenland through the Denmark Strait to the Barents Sea north of Norway and the Kola Peninsula; and (b) the southern Bering Sea and Aleutian Islands. These areas have a high storm frequency with a winter precipitation maximum, extensive cloud cover, and strong winds.

Figure 4.16
The polar maritime climate zones in each hemisphere for 1979–2010 from ERA-Interim, CSFR, and JRA-23 reanalyses. Source: Ballinger et al. 2013, 3938, figure 2. Courtesy of American Meteorological Society. © Copyright 2013 American Meteorological Society (AMS).

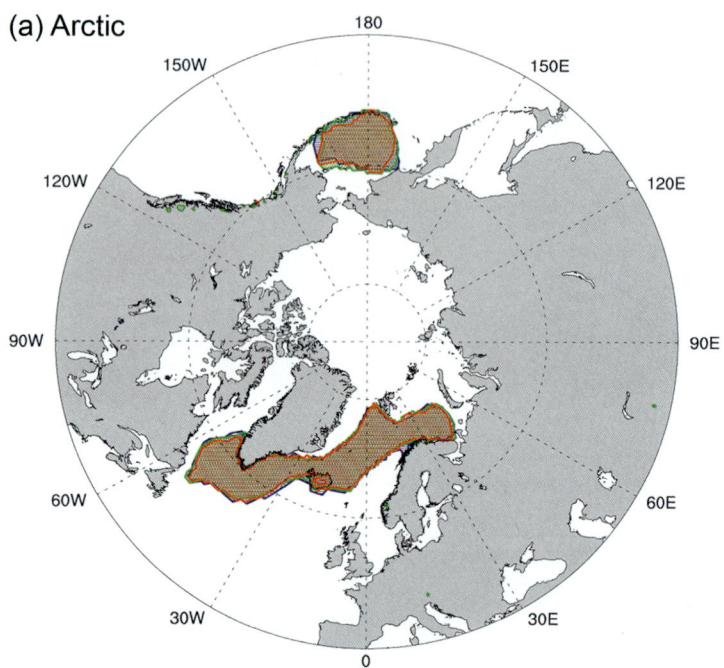

(a) Arctic

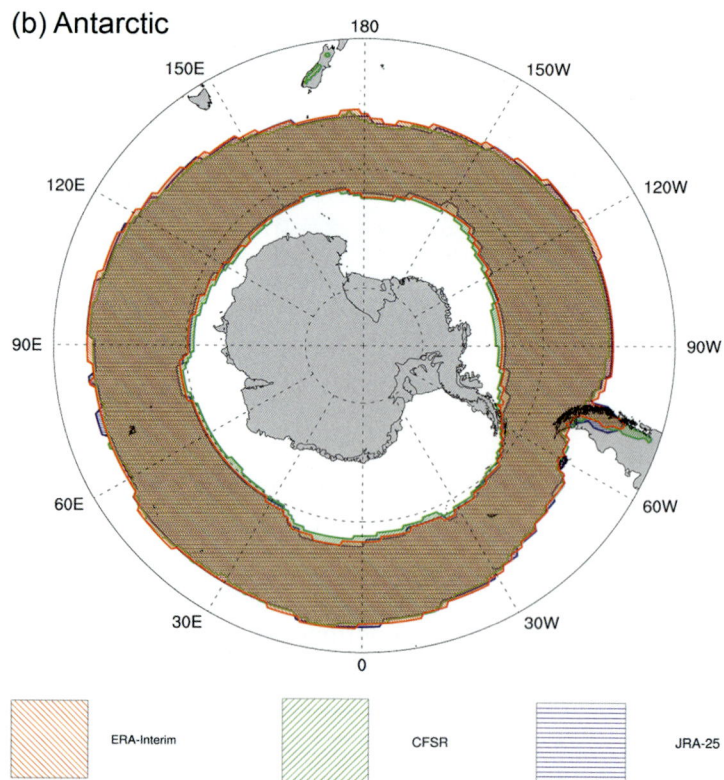

(b) Antarctic

4.3.3 Arctic Inversions

Devasthale et al. (2010) have analyzed inversions for the entire Arctic Ocean for the summer and winter seasons of 2003–2008 using the Atmospheric Infrared Sounder (AIRS) data for clear-sky conditions. Inversions are present in almost each profile retrieved over the inner Arctic region in summer, but the inversion frequency along the Arctic coastal region decreases from June to August. In winter, inversions are ubiquitous and are present in every profile analyzed over the inner Arctic region. When averaged over the entire study area (70–90° N), the inversion frequency in summer ranges from 69 to 86 percent for the ascending passes and from 72 to 86 percent for the descending passes of AQUA. For winter, the frequency values are 88–91 percent for the ascending passes and 89–92 percent for the descending passes. In summer months, the mean values of inversion strength for the entire study area range from 2.5 to 3.9 K, while in winter they range from 7.8 to 8.9 K. The mean inversion depth in winter at SHEBA was 1 km (Tjernstrom and Graversen 2009).

4.3.4 Arctic Wind

Winds at drifting stations in the Arctic Basin are typically rather light. Winds of 2–5 m s^{-1} have a 50–60 percent frequency, year round. They are most frequently 2–3 m s^{-1} in January and 4–5 m s^{-1} in July (Frolov et al. 2005, 45). Maximum speeds were 25–28 m s^{-1} in winter and 16–17 m s^{-1} in summer. At 82° N, 175° E, the mean speed was 4.5 m s^{-1} in summer and 4.7 m s^{-1} in winter. Winds are particularly weak in the central Arctic Basin, whereas in the Atlantic sector cyclonic activity leads to 30–50 percent frequency of speeds of 6–10 m s^{-1}. At Jørgen Brønlunds Fiord in Peary Land, Fristrup (1961) reported a mean wind speed of 5.2 m s^{-1} with a maximum of 40 m s^{-1}. There, the winds are channeled by the terrain and are strongest from the west. During the winter half-year, the wind is calm on 35 percent of occasions.

The topography of southern Greenland gives rise to a wide range of local wind conditions as a result of the interaction of mesoscale weather systems with the steep terrain. Typically, the maximum wind velocities of 25–40 m s^{-1} are found at an altitude of 1–2 km in tip jets and barrier winds, but there are also strong downslope katabatic winds in the fjords. Moore and Renfrew (2005) provide a wintertime climatology using QuikSAT Scatterometer data. Tip jets are caused by a combination of conservation of the Bernoulli function during orographic descent and acceleration due to flow splitting as stable air passes around Cape Farewell, while barrier winds are a geostrophic response to stable air being forced against high topography. It is proposed that reverse tip jets occur when barrier winds reach the end of the topographic barrier and move from a geostrophic to a gradient wind balance, becoming super-geostrophic as a result of their anticyclonic curvature. The winter-mean 10-m wind speed field depicts a highly localized maximum (greater than 15 m s^{-1}) just to the south and east of Cape Farewell. Other local maxima (barrier jets with the topography to the right) exist along the southeast coast of Greenland near Denmark Strait around 65° and 67° N. At Cape Farewell, the wind direction is bimodal, with winds coming most frequently from either the west and west–northwest (tip jets) or the north–northeast and northeast (reverse tip jets).

4.3.5 Arctic Moisture Budget Components

Moisture transport into the Arctic is effected by gateways in the northern North Atlantic, subpolar North Pacific, and Labrador Sea where storm tracks terminate (Dufour et al. 2016). At 70° N, transient eddies provide 90 percent of the moisture import. Vázquez et al. (2016) performed a Lagrangian analysis of moisture transport into the Arctic domain delimited by the area north of the boreal mean decadal 10 °C sea surface isotherm, the surface air 0 °C contour that encircles the North Pole, and the southern limit of terrain that drains into the High Arctic, using the ERA-Interim data for 1980–2012. Four major moisture sources were shown to be the most important: the subtropical and southern extratropical Pacific and Atlantic oceans, North America, and Siberia. Oceanic sources play an important role throughout the year, whereas continental ones take effect only in summer. The sink areas associated with each source are moderately influenced by changes in atmospheric circulation, mainly associated with the East Atlantic pattern for the Atlantic source and related to the West Pacific and Pacific/North American (PNA) teleconnection patterns for the Pacific source.

Water vapor intrusions into the Arctic via the Labrador Sea, the northwest Atlantic, the Barents–Kara Sea, and the North Pacific were investigated by Johansson et al. (2017). They found that these intrusions lead to overall average, clear-sky, Arctic surface warming of up to 5.3 °C in winter and 2.3 °C in summer. The largest impact is from the North Pacific (up to 10 K) followed by the Barents–Kara Sea. Cloudiness is increased 10–30 percent in winter and 10–20 percent in summer. Calculations of total atmospheric transport (P − E) using MERRA for 1979–2005 gave 205 mm for the polar cap (70–90° N) according to Cullather and Bosilovich (2011), in line with earlier estimates.

Precipitation is low over most of the Arctic Basin and tundra regions, where a substantial fraction of it falls as snow. Annual totals are large only where cyclonic storms penetrate northward, as in the Nordic and Barents Sea region. Serreze and Hurst (2000) blended the Legates and Willmott gridded product with measurements from Russian "North Pole" drifting stations and gauge-corrected station data for Eurasia and Canada to generate a monthly precipitation climatology. Most of the Arctic Basin received less than 200 mm yr^{-1}, with less than 150 mm yr^{-1} found over the Beaufort Sea and northern Canadian Arctic archipelago. Maximum totals greater than 1,000 mm yr^{-1} are found south of Greenland.

In a study of Russian North Pole drifting station data covering May 1955 to March 1991, Bogdanova et al. (2002) determined a mean annual precipitation value of 165 mm. Their bias-correction method was based on a model accounting for all main systematic errors of precipitation measurement by the standard Tretyakov gauge – namely, aerodynamic error; the joint effect of wetting, evaporation, and condensation at the interior of the gauge collector; trace precipitation; and the effect of "false" precipitation due to the deposition of blowing snow into the gauge. The amount was verified by comparison with SWE measurements at the same stations by Warren et al. (1999).

A study of precipitation underestimation has been made by Liljedahl et al. (2017) using snowfall observations and bias-adjusted snowfall to end-of-winter snow accumulation measurements on the ground for 1999–2014 at a tundra wetland near Utqiagvik (formerly Barrow), Alaska. Conventional snowfall gauges captured only 23–56 percent of end-of-winter snow accumulation. Bias-adjusted, long-term annual precipitation estimates were more than doubled from 123 to 274 mm.

Evaporation amounts over the Arctic are even more uncertain. The SHEBA experiment in the Beaufort Sea in 1997–1998 found that evaporation was nearly zero between October and April, and peaked in July at about 7 mm $month^{-1}$, giving an annual total of about 20 mm (Persson 2002) and a net annual precipitation of about 145 mm. An aerological calculation of P − ET from atmospheric moisture transport and storage in the air column has been made by Serreze (Serreze and Barry 2014, Figure 6.9). Averages for the central Arctic Ocean are about 200 mm, or about 50 mm greater than those implied by the estimates of Bogdanova et al. (2002). Pan-Arctic evapotranspiration (ET) averaged approximately 230 mm yr^{-1} for 1983–2005, according to Zhang et al. (2009). These researchers used satellite-visible data to estimate NDVI and applied the Penman–Monteith and Priestley–Taylor equations (see Barry and Blanken 2016) to daily reanalysis data for vegetation and water bodies, respectively. Values ranged from 159 mm yr^{-1} in the Arctic tundra to 410 mm yr^{-1} for the water bodies that represent 6 percent of the area.

4.3.6 Arctic Snow Cover

Snow cover characteristics may be described by a number of parameters, including duration, extent, depth, and snow water equivalent (SWE), that are widely measured. Other measurements that may be made in research projects include

Box 4.4 **Snow and Ice Albedo**

Albedo is a dimensionless measure of the reflectance of a surface. It ranges from 0.0 for a black body that absorbs all incident radiation to 1.0 for a white body that is a perfect reflector. Albedo is the directional integration of reflectance over all solar angles in a given time interval. It is normally measured over the majority of the solar spectrum, between 0.3 and 3 μm. A fresh snow cover is the most highly reflective natural surface, with an albedo of approximately 0.9 in the visible wavelengths. The albedo decreases as snow ages; it is about 0.4 for old melted snow and 0.2 for dirty snow. Clean ice has an albedo of 0.5–0.6. Ocean water has an albedo of approximately 0.05, forest 0.1–0.15, and grass 0.2–0.25. Snow has low reflectance in the thermal infrared wavelengths. The planetary albedo is about 0.30, mainly due to cloud cover.

A major climatic role is played by temperature–albedo feedback. When melting snow and ice reveal surfaces with lower albedo (grass, soil, trees, water), the darker surface lowers albedo, increasing local temperatures, which in turn induces more melting and thus reduces the albedo further, resulting in more heating. The reverse happens when snow and ice become more extensive, increasing albedo and leading to cooling.

albedo (Box 4.4), grain size, structural properties, liquid water content, and density. In the Arctic, there are few weather stations and snow measurement is difficult due to wind effects on the airflow around the gauge and blowing and drifting snow. It is necessary not only to install wind shields on rain gauges, but also to surround the gauge with a double snow fence (Goodison and Metcalfe 1992). Bokhorst et al. (2016) provide an overview of measurements of Arctic snowfall and its role in ecosystems.

The climatology of snow depth on the Arctic Ocean has been documented by Warren et al. (1999) based on data collected by the Soviet North Pole drifting stations. Snow accumulates rapidly in September and October, moderately in November, very slowly in December and January, then moderately again from February to May. The ice is mostly snow free during August. The deepest snow is just north of Greenland and Ellesmere Island, peaking in early June at more than 40 cm, when the snow is already melting north of Siberia and Alaska (Figure 4.17). The average snow density increases with time throughout the snow accumulation season and is 300 kg m^2, with little geographical variation.

The northern hemisphere SWE for February to March 1980–2010 has been determined from five snow data sets by Mudryk et al. (2015). Maximum values of 200 mm were observed over a large portion of north central Siberia and the northeast coast of Siberia and Kamchatka (Figure 4.18).

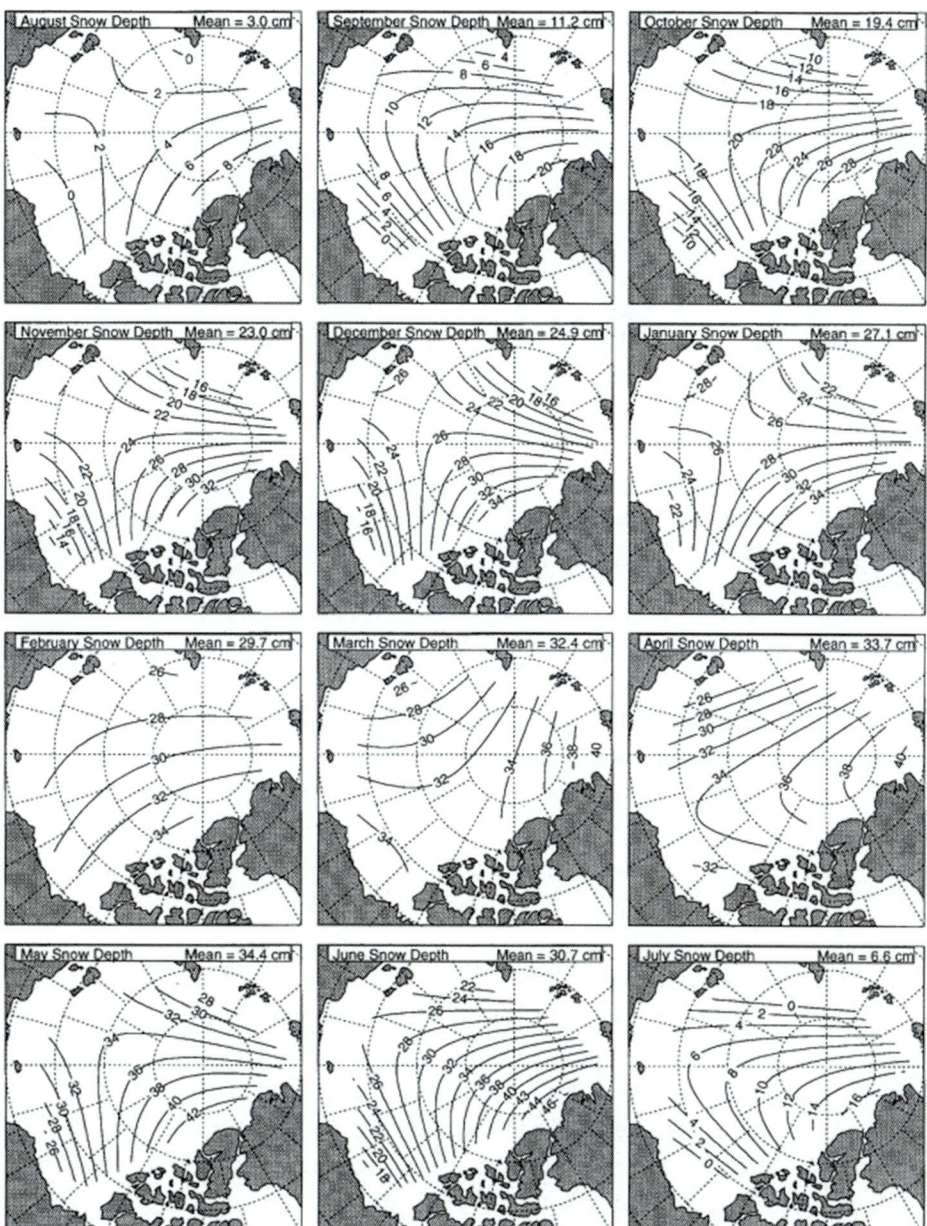

Figure 4.17 Mean snow depth for 1954–1991 on multiyear sea ice at drifting stations for each month, in centimeters of geometric depth. A two-dimensional quadratic function was fitted to all the data available for each month, irrespective of year.

Source: Warren et al. 1999, figure 9. Courtesy of American Meteorological Society.

© Copyright 1999 American Meteorological Society (AMS).

a   **Multi-Dataset Mean SWE**   b   **Mean SWE / Spread**

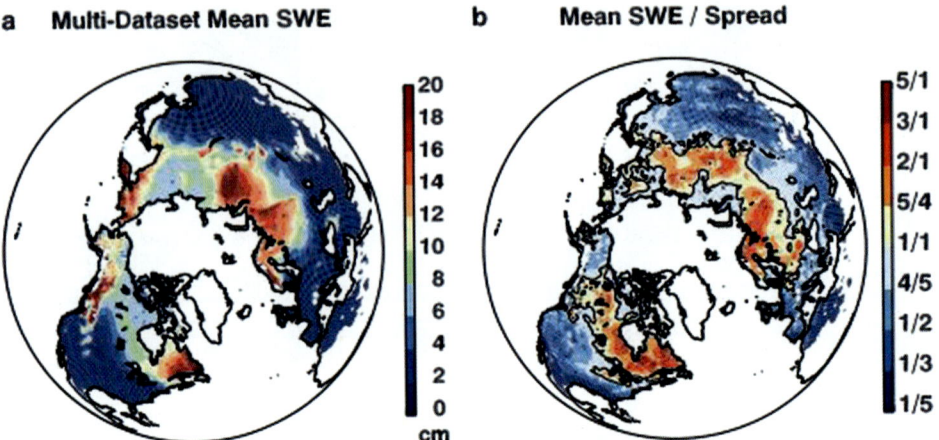

Figure 4.18 (a) Climatology of multi-data set mean SWE for February–March over the 1981–2010 period; (b) ratio of climatological SWE to spread among the five component data sets calculated for February–March over the 1981–2010 period. The black contour delineates the 1:1 ratio.
Source: Mudryk et al. 2015, 2045, figure 5. Courtesy of American Meteorological Society.
© Copyright 2016 American Meteorological Society (AMS).

The frequency of winter snow melt events has been determined for the pan-Arctic from passive microwave data by Wang et al. (2016), based on temporal variations in the brightness temperature difference between 19 and 37 GHz. For the 1988–2013 period, these researchers found that winter melt days were relatively rare, averaging fewer than seven melt days per winter over most areas, with higher numbers of melt days (around two weeks per winter) in sub-Arctic locations (central Quebec–Labrador, southern Alaska, and Scandinavia).

Various aerosol deposits on snow cover have been shown to modify its albedo. Teppei et al. (2015) modeled the snow darkening effect (SDE) of dust, black carbon, and organic carbon (OC) on boreal spring climate. The SDE was found to produce significant surface warming (more than 15 W m^{-2}) over broad areas in mid-latitudes, with dust being the most important contributor to the warming in central Asia and the western Himalayas and BC having a larger impact in Europe, the eastern Himalayas, East Asia, and North America. The contribution of OC to such warming is generally low but still significant mainly over southeastern Siberia, northeastern East Asia, and western Canada (approximately 19 percent of the total solar visible absorption by these snow impurities).

Jiao et al. (2014) observed the effect of BC in snow and sea ice in eight Arctic regions during May to August 2004–2009. For 797 samples, there was a mean BC content of 19.2 ng g^{-1}. During the sunlit season, the reduction of snow and ice albedo due to BC increases surface solar heating and can accelerate melting.

Park et al. (2013) investigated the role of declining Arctic sea ice in September on recent decreases of terrestrial Arctic snow depths during 1979–2006. Analyses of satellite measurements of sea ice extent and snow depth, simulated by a land surface model (CHANGE), suggested that an anomalously larger snow depth over northeastern Siberia during autumn and winter was significantly correlated with the declining September Arctic sea ice extent, which has resulted in both cooling and an increase in precipitation. In contrast, the sea ice reduction has amplified warming in North America, which has offset the input of precipitation to snow cover, decreasing snow depth. However, a part of the Canadian Arctic recorded an increase in snow depth driven locally by the diminishing September Arctic sea ice extent.

Wegman et al. (2015) similarly found that sea ice decline in the Barents–Kara Sea region results in increasing snow depths over western Siberia.

4.4 Antarctic Climate

4.4.1 Antarctic Temperature Conditions

Antarctica is surrounded by a circumpolar zone of maritime polar climate over the Southern Ocean and sub-Antarctic islands between about 49° and 60° S (Figure 4.16B). The poleward margin approximately coincides with the maximum winter extent of pack ice (Ballinger et al. 2013). Summers are cool (5–10 °C) and cloudy with persistent strong winds.

The temperatures in Antarctica are the lowest on earth (Figure 4.19). Their distribution largely mirrors the topography, with the lowest averages found on the East Antarctic plateau around 3,900 m elevation. Climatological data are available for all four synoptic hours (00, 06, 12, and 18) for all Antarctic stations at REference Antarctic Data for Environmental Research (READER; www.scar.org/data-products/reader).

The MET-READER database contains monthly mean surface and upper air climatological data derived from the in situ meteorological observations made at Antarctic stations with long-term records. Data are available on temperature, surface pressure, wind speed/direction, and geopotential height.

The record low air temperature and skin temperature found by Landsat 8 are described in Box 4.5. The warmest locations are on the Antarctic Peninsula, where maxima have reached 15 °C. Along the Antarctic Peninsula in winter, the temperature increases about 1 °C per degree of latitude. At the coasts, temperatures in summer rise to near freezing; they are affected by onshore winds and in the Dry Valleys they generally increase with distance inland. In winter at the coast, katabatic winds generally prevent very low temperatures as a result of

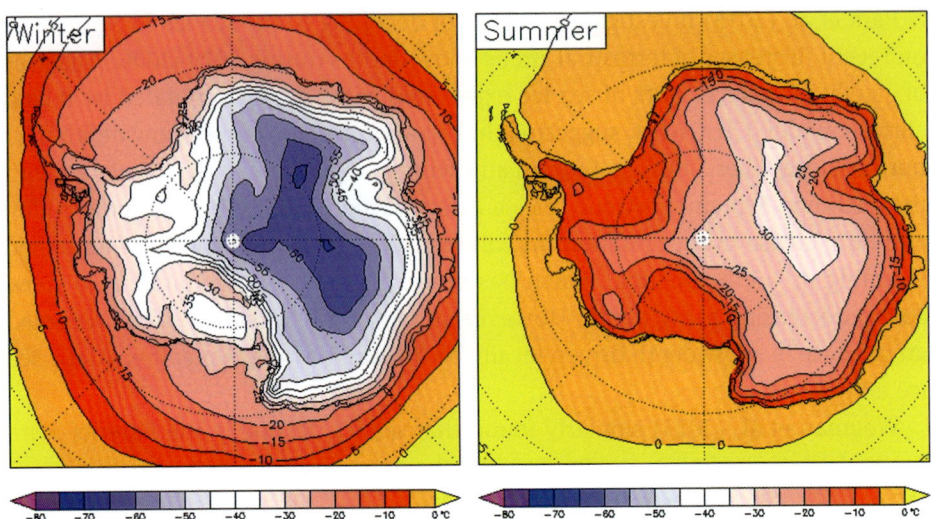

Figure 4.19 Mean winter and summer temperatures over Antarctica. Based on ECMWF data.
Source: https://commons.wikimedia.org/wiki/File:Antarctic_surface_temperature.png, William M. Comolley.

Box 4.5 Extreme Minimum Temperatures in Antarctica

The world's lowest temperatures are observed on the Antarctic plateau during winter darkness. Scambos et al. (2013) used Landsat 8 thermal emission data to determine the record low skin temperature on the plateau. They found that surface temperature reached a record low of −94.7 °C in August 2010; −93 °C was measured on July 31, 2013. The coldest locations are situated in topographic depressions between about 3,800 and 3,900 m just south of an ice ridge. The record low air temperature reading was −89.2 °C at Vostok on July 21, 1983.

turbulent mixing in the boundary layer. Monthly means at McMurdo Station on the coast range from −26 °C in August to −3 °C in January. At the South Pole, the highest temperature ever recorded was −12.3 °C.

The annual temperature regime at interior stations shows a well-known six-month (March–August) period with near-constant mean temperatures – the "coreless" (*kernlose* in German) winter (Bromwich and Parish 1998). This phenomenon is attributed to radiative cooling eventually diminishing as the temperature falls and the longwave radiation emitted by the surface decreases; it is balanced by downward longwave radiation and the downward sensible heat transfer from the warmer air above the surface.

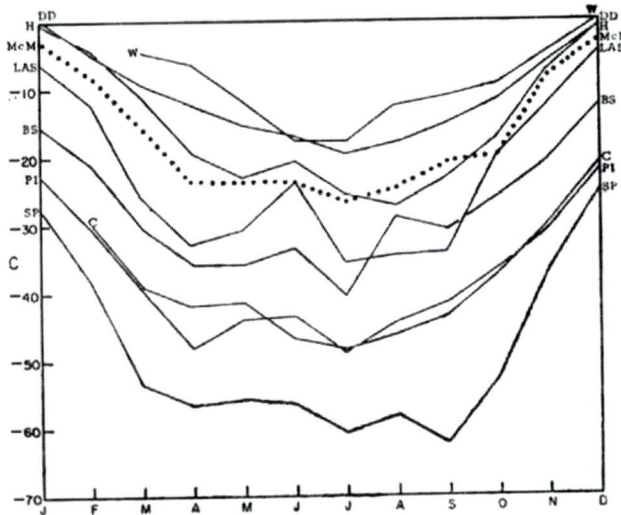

Figure 4.20

Temperature curves illustrating the "coreless" winter at coastal and interior stations in the Pacific sector of Antarctica in 1957. W = Wilkes; DD = Dumont D'Urville; H = Hallett; McM = McMurdo; LAS = Little America Station; BS = Byrd Station; C = Charcot; PI = Pioneerskaya; SP = South Pole.

Source: Wexler 1959, 201, figure 9.

The "coreless" winter feature of the annual temperature cycle is widespread at Antarctic stations (Wendler and Kodama 1993; Wexler 1959) (Figure 4.20). The reversal in the seasonal temperature trend appears to be related to more frequent destruction of the low-level surface inversion by storms, rather than an increase in the level of warm air advection into the region.

Allison et al. (1993) investigated the climate of East Antarctica, 110–140° E, based on two AWS networks. There were a total of sixteen stations in the area 55–145° E and 65–75° S, stretching from sea level to greater than 3,000 m altitude. The records of ten of these stations are sufficiently long to be adequate for a climatological study. Results include (1) an absolute lowest minimum temperature of −84.6 °C at Dome C; (2) no minimum below −40 °C at D10 near the coast; (3) a "coreless" winter regime, without seasonal temperature trends for six months, at all stations; (4) mean surface wind speeds increasing to maxima near, rather than at, the coast; (5) high directional constancy in all seasons, with directions closer to the fall line in winter and during night hours than in summer and during day hours; (6) a distinct semiannual pressure variation with a main minimum in spring (September) and a secondary minimum in autumn (March); and (7) interrelationships among surface temperature, pressure, and wind related to the ice sheet topography.

The University of Wisconsin–Madison Antarctic AWS project has been making meteorological surface observations on the Ross Ice Shelf (RIS) for approximately thirty years, mainly from the mid-1980s to 2013 (Costanza et al. 2016). This network offers the most continuous set of routine measurements of surface meteorological variables in this region. The RIS is divided into three representative regions – central, coastal, and the area along the Transantarctic Mountains – for the purpose of describing the specific characteristics of its sections. The central AWS experiences the lowest mean temperature and the lowest resultant wind speed. The AWS along the Transantarctic Mountains experiences the highest mean temperature, the highest mean sea-level pressure, and the highest mean resultant wind speed. Finally, the coastal AWS experiences the lowest mean pressure. Mean resultant wind speed varies from 1 to 5 m s^{-1} and maximum wind speed from 19 to 35 m s^{-1}. Seasonal data for the coastal AWS are winter temperatures, -31 to -40 °C; winter extreme minima, -55 to -65 °C; summer temperatures, -6 to -9 °C; and summer extreme maxima, 0.4–6.2 °C.

4.4.2 Antarctic Inversions

Radiative cooling over the ice sheet surface modifies the temperature structure of the lower troposphere, setting up strong ground-based inversions. Phillpot and Zillman (1970) investigated the characteristics of inversions over Antarctica using data from twenty-two bases. Over approximately 80 percent of the continent in winter, the average inversion strength exceeds 15 °C; it averages 20–25 °C over the high plateaus of East Antarctica. Over the interior, an inversion is almost permanent, except in December–January. Occasionally, intense cyclonic systems may briefly destroy it. Mean depths in winter are between 500 and 700 m at the high plateau stations and 300–400 m at the coasts away from ice shelves.

   Hudson and Brandt (2005) reexamined the surface-based inversion at the South Pole and Dome C (3,266 m). At the South Pole, a typical profile in winter shows a 20 °C warming in the lowest 100 m. At Dome C, in summers 2003–2004 and 2004–2005, there were strong inversions, averaging 10 °C between the surface and 30 m. The lowest temperatures and the maximum inversion occur with a wind of 3–5 m s^{-1} rather than calms, probably due to an inversion wind. Cloud cover, leading to enhanced downward longwave radiation, can cause inversions to be broken up.

   Connolley (1996) demonstrated a strong linear relationship between surface temperature (T_s) and inversion strength (I). For July, the r^2 value based on fifteen stations was 0.94 and the equation was

$$I = -0.46T_s - 4.1$$

Using output from the UK Meteorological Office (UKMO) GCM, Connolley also derived a map of inversion strength in July. Values ranged from 5 °C at the coast to 25 °C on the high plateau of East Antarctica, with 15–20 °C over the major ice shelves.

Cassano et al. (2016) have analyzed the near-surface atmospheric state for two years over the Ross Ice Shelf using data from a 30-m tower. Stable stratification dominated the surface layer, occurring 83 percent of the time. The strongest inversions occurred for wind speeds less than 4 m s^{-1}. The strongest inversions over the depth of the tower exceeded 25 °C. In summer, there were unstable stratifications half the time. The researchers grouped the profiles into thirty patterns using self-organizing maps. The strongest winds occurred for the nearly well-mixed but slightly stable patterns; the weakest winds occurred for the strongest inversion patterns. The median inversion strength between 1.85 and 0.85 m of 3 °C and 90th percentile of 12 °C was for winds less than 4 m s^{-1}.

4.4.3 Antarctic Wind

Coastal regions of Antarctica are dominated in winter by katabatic winds, as opposed to the slope flows of the interior. Weller (1969) notes that katabatic regimes dominate to about 100 km inland, and Streten (1990) estimates that off Mawson (67.6° S, 62.9° E) they extend some 10–15 km offshore.

Mawson is notorious for persistent, very strong winds. The southeasterly katabatic flow has a mean annual mean speed of 11 m s^{-1}. On average, gusts more than 41 m s^{-1} occur bimonthly, with a winter maximum. Due to the winds, drifting snow is common, occurring on ten days per month in winter.

Marshall and Turner (1997) report that strong drainage exists, with several confluence zones near the continental margins being responsible for persistently strong katabatic airflows. One such confluence zone is located in Victoria Land adjacent to Terra Nova Bay in the western Ross Sea (Parish and Bromwich 1987). Bromwich et al. (1990) showed that the two primary routes for katabatic drainage from this confluence zone are the Reeves Glacier valley and David Glacier (at the head of the Drygalski Ice Tongue). These katabatic winds from the Terra Nova Bay confluence zone (TCZ) are known to propagate up to 350 km downstream (Bromwich 1989), and are thought to be wholly or partly responsible for both the formation of the Terra Nova Bay polynya (Bromwich and Kurtz 1984) and the frequent mesoscale cyclogenesis over the southwestern Ross Sea (Carrasco and Bromwich 1994). The latter phenomenon is associated with a horizontal baroclinic zone that forms when the cold TCZ katabatic airstream moves out over the relatively warm open ocean or sea-ice surface.

Parish and Cassano (2003) question the katabatic origin of the high directional frequency of winds at Antarctic coastal stations, particularly in summer, when solar heating of the ice slope takes place. The persistent unidirectional nature of the Antarctic surface wind throughout the year implies that significant topographic influences other than those due to katabatic forcing must be involved.

Numerical simulations have shown that strong adjustments in the pressure and wind fields take place when stable air impinges on the continental orography. The Antarctic ice sheet acts to block the flow of stable air toward the continent. The subsequent adjustment process results in a horizontal pressure field that is shaped by the slope and direction of the terrain. The resulting low-level wind fields resemble katabatic winds, with their directions being tied to the underlying terrain and their speeds dependent on the slope of the ice surface.

4.4.4 Antarctic Moisture Budget

Precipitation over Antarctica is almost impossible to measure directly due to the winds and blowing snow. However, annual snow accumulation provides a good indication of the spatial pattern and amounts, because the net effects of sublimation and wind drifting are small on a continent-wide scale. Giovinetto and Bentley (1985) identified annual totals less than 50 mm over much of East Antarctica, increasing to around 200 mm in many coastal areas and up to 800 mm in narrow coastal zones of the Amundsen and Bellingshausen seas and on the west side of the Antarctic Peninsula (Figure 4.21). The snowfall characteristics of Antarctica were reviewed by Bromwich (1988). The meridional distribution is due to the orographic lifting of moist air by the ice sheet. Zonal precipitation variations are related to the quasi-stationary cyclones in the circumpolar low-pressure trough. The annual cycle of net P − E derived from ECMWF analyses shows a summer minimum and a broad March–October maximum when monthly values greater than 2,500 m elevation are around 5 mm (Bromwich and Parish 1998). Hence, most precipitation falls in winter, when the average moisture content of the air is low. The intensity of cyclonic activity is the key factor governing the amount of precipitation and its variations. Precipitation generation in coastal regions is strongly influenced by the deflection of poleward-moving, moist maritime air masses by the steep marginal ice slopes; hence the flow is

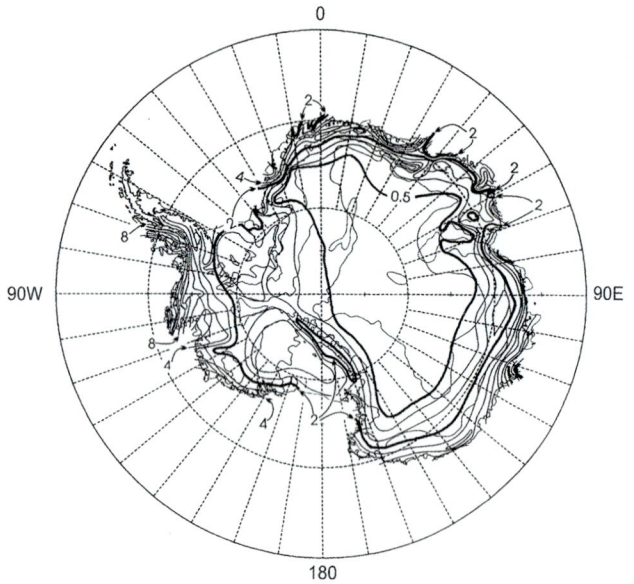

Figure 4.21 Annual accumulation over Antarctica. Isopleths are 100 mm a^{-1}: the lines for 0.5 and 2 are bold. Contours of elevation (km) are light solid lines.
Source: After Giovinetto and Bentley 1985, from Bromwich and Parish 1998, 187, figure 4.12. Courtesy of American Meteorological Society.
© Copyright 1998 American Meteorological Society (AMS).

parallel to the terrain contours. Direct orographic lifting is the dominant precipitation formation mechanism inland of the 1-km elevation contour; intrusions of moist air far into the continent are accompanied by poleward flow through a deep tropospheric layer. Above 3,000 m elevation where terrain slopes are gentle, radiative cooling is the primary mechanism by which saturation is maintained within moist air, thus leading to precipitation. Most precipitation at these elevations falls from clear skies as "diamond dust." This phenomenon does not differ from precipitation originating in clouds but is a direct result of the low moisture content of the air, which means that ice layers are optically too thin to be visible as clouds.

A number of atmospheric studies of net precipitation over the Southern Ocean and Antarctica have been performed using reanalysis products. Cullather et al. (1998) derived net precipitation (precipitation minus evaporation/sublimation) for Antarctica from the ECMWF operational analyses via the atmospheric moisture budget method. For 1985–1995, the average continental value is 151 mm yr^{-1}. Interannual variability in the Southern Ocean storm tracks is an important mechanism for enhanced net precipitation in both east and west Antarctica.

Cullather et al. (1998) used ECMWF analyses for 1985–1995 to estimate the average net precipitation (P − E) of Antarctica. They determined a continental value of 151 mm yr^{-1} water equivalent (w.e.). Earlier studies had reported a range from 135 to 197 mm yr^{-1}. Large regional differences have been identified using other data sets. Interannual variability in the Southern Ocean storm tracks has been found to be an important mechanism for enhanced precipitation minus evaporation (P − E) in both east and west Antarctica.

Tietäväinen and Vihma (2008) used the ERA-40 reanalysis for 1979–2001. The convergence of meridional water vapor transport was found to be at its largest at 64°–68° S, while the convergence of zonal transport was regionally important in areas of high cyclolysis. The ERA-40 result for the mean precipitation over the Antarctic continental ice sheet in 1979–2001 was 177 ± 8 mm yr^{-1}.

Cullather and Bosilovich (2011) determined the atmospheric moisture budget over the Antarctic continent from the Modern Era Retrospective-Analysis for Research and Applications (MERRA) for 1979–2005. The results by the aerological method indicated a value of 188 mm, within the range of previous estimates. Calculations for the Southern Ocean were considered unreliable.

Using data from sixteen radiosonde stations around the Antarctic coast, Connolley and King (1993) derived values and variability of average horizontal moisture transport and precipitation over the ice sheet. They found that the interannual variability in the moisture fluxes is mostly due to variation in the strength of the circulation. Annual total column moisture (TCM) values are similar, about 4 kg m^{-2}, at all stations around the coast of East Antarctica.

Moisture fluxes reflect the predominantly zonal easterly flow in the lower tropo-sphere, and their meridional components are generally small. As a result of interannual variations in the strength of the atmospheric circulation, and to a lesser extent in the TCM, interannual variability of the fluxes is high, suggesting that there may be large interannual variability in the precipitation over Antarctica.

The synoptic origins of precipitation over the Antarctic Peninsula during March 1992 to February 1993 were investigated by Turner et al. (1995) using meteoro-logical observations, satellite imagery, and analyses produced by the UK Meteorological Office. Precipitation at Rothera Station was found to occur at 30 percent of the synoptic reporting time, with 80 percent of precipitation reports being associated with cyclonic disturbances. Although three fourths of all pre-cipitation reports were for snow, the proximity of Rothera to the zone of max-imum cyclonic activity meant that incursions of mild air produced rain in all seasons. During the year, 95 percent of all precipitation was classified as light. Variability of precipitation on the intraseasonal time scale was highly dependent on the synoptic-scale circulation. The most common synoptic situation for pre-cipitation was a frontal cyclone over the Bellingshausen Sea, which accounted for 38 percent of all precipitation events and 62 percent of the moderate and heavy precipitation reports. Of the extratropical cyclones that gave rise to precipitation, 49 percent were found to have developed south of 60° S. None of the precipitation at Rothera was attributable to mesocyclones.

4.5 Recent Climate Changes

4.5.1 Arctic

It has been well documented that the Arctic has warmed more than any other part of the globe during the last century. According to McBean et al. (2005), Arctic land temperatures rose during the 1900–2003 period, with the exception of 1946–1965. Between 1966 and 2003, the warming trend exceeded 1–2 °C dec-ade^{-1} in northern Eurasia and northwest North America, with the largest increases noted in winter and spring. Since 1950, the warming rate in the maritime areas has been similar to that on land (Polyakov et al. 2003)

For the Svalbard (land) area, Przybylak et al. (2016) found that the period 1865–1920 was markedly colder than today (1981–2010) by about 3 °C in all seasons except summer, when air temperature was similar in both periods. This pattern shows very good correspondence with the results of similar reconstructions for other Arctic regions, as well as the entire Arctic (Pryzbylak et al. 2010). In Svalbard, it appears that both atmospheric and oceanic circulation, as well as sea-ice changes around the archipelago, whether naturally or anthropogenically driven, were the main direct causes of observed warming in the area from the nineteenth century to the present. For northern Eurasia (15° E to the Pacific Ocean and 40° N to the Arctic coasts), over the interval 1881–2014, air temperature increase by 2.1 °C in winter, 2.4 °C in spring, 0.86 °C in summer, and 1.2 °C in autumn (Groisman et al. 2016). However, study of winter air temperatures in central Eurasia shows a decrease of almost 2 °C for 1979–2012. Although this trend coincides with a loss of Barents–Kara Sea sea ice area, McCusker et al. (2016) have shown that the two are not causally related. Instead, the cooling is attributed to a circulation anomaly involving high pressure over the Barents–Kara Sea and a downstream trough.

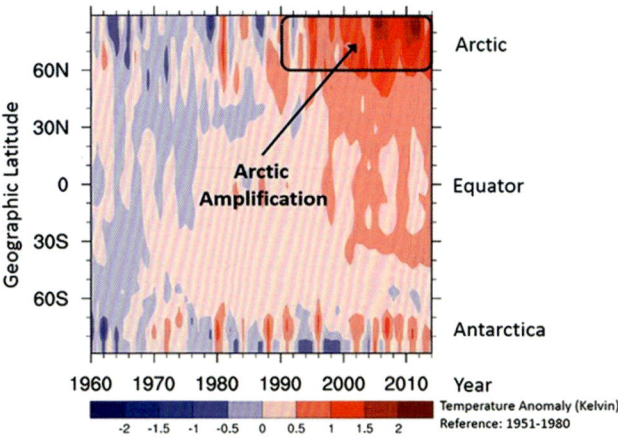

Figure 4.22 Mean temperatures, in °C (by location and year), for 1960–2012, shown as differences from 1951–1980 mean temperatures. The area inside the black box shows how warming is amplified in the Arctic. The data are from the NASA Goddard Institute for Space Studies.
Source: Wendisch et al. 2017, figure 1.
Courtesy of American Geophysical Union.

For the Arctic as a whole, the warming observed in the late twentieth century resulted from polar amplification of the general global trend (Serreze and Barry 2011). Its causes are discussed in Box 4.6. The temperature rise has been twice the global average, as shown by Wendisch et al. (2017) based on temperature departures from 1951 to 1980 (Figure 4.22). This increase has accelerated since the 1990s. January–February 2016 set new extremes, with the anomaly over the central Arctic reaching a record 5 °C in January and 4 °C in February.

4.5.2 Antarctic

Over the past fifty years, the Southern Annular Mode has become more positive in austral summer and autumn, and westerly winds have increased by 15–20 percent in strength since the 1970s (Turner et al. 2014). The summer trend in the SAM is linked to the growth of the stratospheric ozone hole and related strengthening of the polar vortex in the lower stratosphere. Cyclone frequency has decreased over the coastal zone of Antarctica (60–70° S), except for the Amundsen Sea low (ASL) region, but cyclone intensity has increased.

Box 4.6 **Polar Amplification**

Model simulations from the Coupled Model Intercomparison Project (CMIP) phase 5 archive have been used by Pithan and Mauritsen (2014) to quantify the contributions of the various feedbacks. The largest contribution to Arctic amplification has been shown to come from temperature feedbacks: As the surface warms, more energy is radiated back to space in low latitudes, compared with the Arctic. This effect can be attributed to both the different vertical structure of the warming in high and low latitudes, and a smaller increase in emitted blackbody radiation per unit warming at lower temperatures. The authors noted that the surface–albedo feedback is the second main contributor to Arctic amplification (Figure 4.23); other contributions are substantially smaller or even oppose Arctic amplification. The ice–albedo feedback mechanism warms the Arctic Ocean during the summer, and the heat gained by the ocean is released during the winter, causing the cold-season warming.

Screen and Simmonds (2010) concluded that the theory is correct by comparing trend patterns in surface air temperature (SAT), surface turbulence heat flux (HF), and net surface infrared radiation (IR). However, in this comparison, downward IR is more appropriate to use. By analyzing the same data used by Screen and Simmonds, using the surface energy budget, Lee et al. (2017) show that over most of the Arctic the skin temperature trend, which closely resembles the SAT trend, is largely accounted for by the downward IR trend, rather than the HF trend.

Figure 4.23
The spread of Arctic warming contributions in CMIP-5 models. Boxes show the median, 25th percentile, and 75th percentile; whiskers show the full ensemble spread. The Planck term refers to the radiative-temperature feedback.
Source: Pithan and Mauritsen 2014, 183, figure 3 right side.

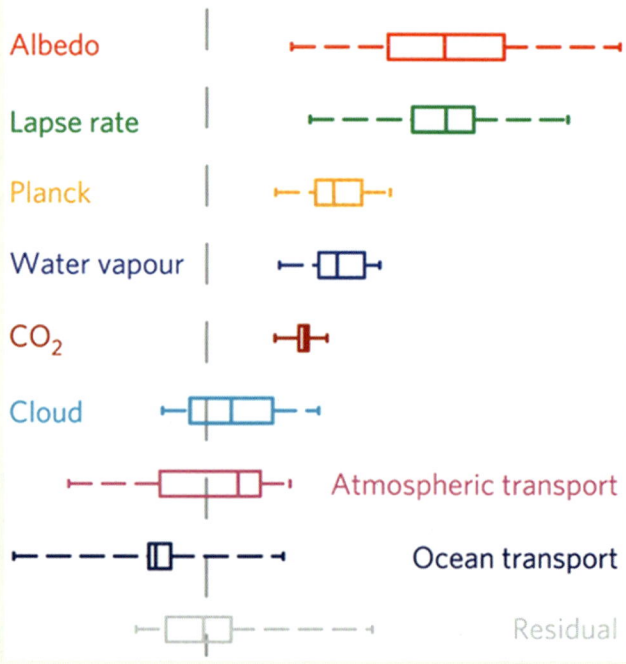

Screen and Francis (2016) demonstrate that the contribution of sea-ice loss to wintertime Arctic amplification seems to depend on the phase of the Pacific Decadal Oscillation (PDO). Their results suggest that, for the same pattern and amount of sea-ice loss, consequent Arctic warming is larger during the negative PDO phase relative to the positive phase, leading to larger reductions in the poleward gradient of tropospheric thickness and to a more pronounced reduction in the upper-level westerlies. The negative phase of the PDO occurs when the eastern Pacific cools and the western Pacific warms.

There were extensive regions of Arctic temperature extremes in January and February 2016 that continued into April. For January, the Arctic-wide averaged temperature anomaly was 2.0 °C above the previous record of 3.0 °C (Overland and Wang 2016). Mid-latitude atmospheric circulation played a major role. Extensive low geopotential heights at 700 hPa extended over the southeastern United States, across the Atlantic, and well into the Arctic. Low geopotential heights along the Aleutian Islands and a ridge along northwestern North America contributed to southerly wind flow. These two regions of low geopotential height were seen as a major split in the tropospheric polar vortex over the Arctic. Warm air advection north of central Eurasia reinforced the ridge that split the flow near the North Pole.

There have been significant changes in the snow regime, particularly during the spring season. Notably, snow cover disappeared earlier in spring at an average rate of 3.4 days decade^{-1} over the pan-Arctic terrestrial region (excluding Greenland) during 1972–2009 (Callaghan et al. 2011). This contributed to an 18 percent reduction in May–June Arctic snow cover extent over the 1966–2008 period. The changes were not uniform, however, and the rates of change in SWE and snow cover duration (SCD) have been observed to vary across the Arctic, with the largest decreases occurring in maritime regions (Alaska, northern Scandinavia, and the Pacific coast region of Russia). New analyses show stronger decreases in Arctic coastal regions than inland. A distributed 10-km resolution snow data set for 1979–2009 for the Arctic was created by Liston and Hiemstra (2011) using Modern-Era Retrospective Analysis for Research and Applications (MERRA) reanalyses as forcing, and the Micromet and Snow Model. These data clearly show a general snow decrease throughout the Arctic: Maximum winter SWE has decreased, snow cover onset is later, the snow-free date in spring is earlier, and snow cover duration has decreased. In this data set, the total number of snow days in a year averaged -2.49 days decade^{-1}, with minimum and maximum regional trends of -17.2 and -7.2 days decade^{-1}, respectively. The average trend for peak SWE in a snow season was -0.17 cm decade^{-1}, with minimum and maximum regional trends of -2.50 and -5.70 cm decade^{-1}, respectively.

Hori et al. (2017) have derived northern hemisphere SCE for 1978–2015 based on AVHRR and MODIS data and compared their JAXA Satellite Monitoring for Environmental Studies (JASMES) results with those from NOAA. According to their study, JASMES values are more accurate. The error-adjusted SCE values exhibit negative trends in all seasons as follows:

38,751 km^2 year^{-1} in winter
17,802 km^2 year^{-1} in spring
10,730 km^2 year^{-1} in summer
43,483 km^2 year^{-1} in autumn

Box 4.6 (continued)

These trends in autumn and winter are statistically significant. Western Eurasia exhibits strong negative (shorter) snow cover duration trends of up to two months over three decades. In contrast, some parts of eastern Asia and western and northern North America have positive (lengthening) trends of one month over three decades.

Surface temperatures have shown significant warming across the Antarctic Peninsula and, to a lesser extent, in the rest of West Antarctica since the early 1950s, with little change across the rest of the continent (Turner et al. 2014). Faraday/Vernadsky Station, at 65.4° S on the western side of the peninsula, has experienced the largest statistically significant trend of increased surface temperature, $+0.54$ °C decade^{-1} for the period 1951–2011, with an increase of $+1.01$ °C decade^{-1} in winter. Temperatures on the eastern side of the Antarctic Peninsula have risen most during the summer and autumn (0.39 °C decade^{-1} from 1946 to 2011 at Esperanza), linked to the strengthening of the westerlies. Bromwich et al. (2013) estimate that the Byrd Station in West Antarctica warmed by 2.4 ± 1.2 °C over the 1958–2010 period. Bore hole data from the West Antarctic (Orsi et al. 2012) show that cool conditions prevailed on the West Antarctica ice sheet (WAIS) during the Little Ice Age period, and that warming began around 1800, continuing over the past one hundred years. The rate of warming has increased sharply in the past twenty years, reaching approximately 0.7 °C decade^{-1}. Bore hole data from East Antarctic (Muto et al. 2011) covering the last twenty to fifty years on the Dronning Maud Land divide, a region without any other long-term records, show warming of 0.1–0.2 °C decade^{-1}.

Clem et al. (2016) have investigated the relationships between temperatures in the Antarctic Peninsula and the SAM and ENSO. Their work demonstrates that circulation changes associated with the SAM dominate interannual temperature variability across the entire Antarctic Peninsula during both summer and autumn, while relationships with tropical Pacific SST variability associated with ENSO are strongest and statistically significant primarily during winter and spring. Temperatures in the western Antarctic Peninsula during autumn, winter, and spring are closely tied to changes in the ASL and associated meridional wind anomalies. The interannual variability of ASL depth is most strongly correlated with the SAM index during autumn, while the ENSO relationship is strongest during winter and spring. Interannual variability of northeast Antarctic Peninsula temperatures is primarily sensitive to anomalies of zonal wind crossing the peninsula and causing leeside adiabatic warming.

SUMMARY

The two polar regions have major geographical differences – namely, in the Arctic, the sea ice is surrounded by mostly land, in the Antarctic, the sea ice and Southern Ocean surrounds the ice-sheet covered continent. The environmental forcings are very different and therefore the trends in the two regions are not expected to be the same or even similar. The geographical placement of land, ice and ocean in northern and southern hemispheres affects large scale atmospheric circulation.

The Arctic Oscillation (AO) is a major feature of circulation variability in the northern hemisphere. It is usually correlated with the North Atlantic Oscillation (NAO) between the Icelandic low and Azores high. Another source of variability is the Arctic dipole between high pressure over Arctic North America and low pressure over Arctic Eurasia.

In winter, pressure is high in a ridge over the Beaufort Sea. Only in spring is there an Arctic anticyclone. Pressure is low over the Arctic Ocean in summer. Lows are barotropic in structure.

The surface circulation in the southern hemisphere has small seasonal variability. Interannual variability in the Amundsen Sea low is linked to that in the amplitude of wave number 1, which is also influenced by El Niño. A wave number 3 pattern in 500-hPa heights in austral winter over the three southern continents is negatively correlated with heights over the Southern Ocean. Around Antarctica there is a circumpolar low-pressure trough. Lows spiral inward toward the coast, and a large number of mesocyclones arise near the coasts of the Indian Ocean sector and the Ross Sea. Circulation modes in the Southern Hemisphere include the Southern Annular Mode (SAM) and the Semi-Annual Oscillation (SAO), a twice-yearly expansion and contraction of the circumpolar trough.

The Arctic Ocean accounts for 3 percent of the world's ocean area. It is divided into the Canadian and Eurasian basins by the Lomonosov Ridge. About 36 percent of the ocean area is continental shelf, where runoff from major rivers maintains low salinities. The Denmark Strait links the Arctic to the North Atlantic, and warm, salty water enters the Arctic via the Norwegian and Barents seas, circulating around the basin at depth. The water structure in winter features a shallow thermocline and pycnocline. The vertical salinity gradient and halocline maintain the stable density structure. Net precipitation accounts for about 24 percent of the mean freshwater budget, runoff 38 percent, and Pacific inflow 30 percent. At irregular intervals, large anomalies of freshwater discharge have entered the North Atlantic via Fram Strait or Baffin Bay. The Great Salinity Anomaly of the late 1960s and 1970s circulated in the North Atlantic for more than a decade; similar events occurred in the 1980s and 1990s. The area from which river discharge enters the Arctic depends on the definition used: The defined areas range from 15 million km^2 for the Arctic River Basin to 23.7 million km^2 for "All Arctic Regions." A circulation mode in the North Pacific Ocean can affect high latitudes. The Pacific Decadal Oscillation (PDO) features a cool west Pacific and warm east Pacific in the positive ("warm") phase, and vice versa in the negative ("cool") phase. Each phase lasts twenty to thirty years. The PDO phase has been negative since about 1990.

The vast Southern Ocean surrounds Antarctica. The Subtropical Front (STF) is found at approximately 45° S. The Antarctic Circumpolar Current (ACC), which comprises bands of strong jets, flows continuously eastward in the Southern Ocean, driven by the westerlies. The deep layers of the ACC are occupied by warm, saline Circumpolar Deep Water (CDW), whose poleward edge marks the southern boundary of the ACC. The ACC's northern boundary is the Antarctic Polar Front Zone (APFZ), located between about 50° and 60 °S. The total transport through Drake Passage is 173 Sv (73 percent baroclinic).

The Sub-Antarctic Front (SAF) and the Polar Front (PF) are well-defined features. Temperatures decrease by approximately 3 °C across both. The PF meanders and forms cyclonic eddies that carry Antarctic Surface Water (AASW). AASW extends from the PF south to the continent.

In winter, the Southern Ocean freezes outward to 65° S in the Pacific sector and 55° S in the Atlantic sector. Mostly first-year ice (FYI) forms, due to frazil accumulation. Offshore winds create numerous coastal polynyas. In contrast to the Arctic, melt ponds rarely form on the sea ice surface during the melt season due to the low relative humidity and strong winds.

The radiation regimes of both polar regions have the same basic annual cycle, but the components differ owing to the surface properties (sea ice versus ice sheet) and cloud conditions. The surface net radiation is approximately −80 MJ m^{-2} in January, while but approximately −50 MJ m^{-2} in July in Antarctica, as the lower surface temperature reduces the outgoing infrared radiation. In spring in the Arctic, incoming solar radiation is mostly reflected by the high albedo of the snow cover. In the Beaufort–Chukchi seas in 2011, the melt pond fraction reached 53 percent on FYI and 39 percent on multiyear ice (MYI). At SHEBA in late July 1988, average albedo fell to 0.4. Total cloud fraction (TCF) over the Arctic Ocean increases from 0.55–0.75 during November–April to 0.8 in the summer, due to moisture advection and surface melt. Summer cloudiness consists of low-level stratiform clouds.

At the South Pole, cloud cover ranges from 45–50 percent in the short summer to 55–65 percent in the six months of winter.

Temperatures in the Arctic in January range from $-34\ °C$ to $0\ °C$, with a large positive anomaly over the Norwegian–Barents Sea region. The coldest locations are in the valleys of northeastern Russia due to persistent inversions. With snow melt in May–June, temperatures rise above freezing. July mean values range from $0\ °C$ over the ocean to $10\ °C$ in the interior. Polar maritime climates are defined by a coldest month below $-6.7\ °C$ and by cool, cloudy summers.

Inversions are ubiquitous in the Arctic winter, occurring with 90 percent frequency, a mean strength of 8–9 K, and 1 km depth. In summer, they occur with 70–86 percent frequency and mean strength of 2.5–4 K. Winds in the Arctic are generally light, especially in the central Arctic. Southern Greenland has tip jets of $25–40\ m\ s^{-1}$ at 1–2 km altitude due to the interaction of mesoscale weather systems with steep terrain. Greenland fjords have strong downslope katabatic winds.

Moisture sources for the Arctic comprise the subtropical and southern extratropical Pacific and Atlantic, year round, and North America and Siberia in summer. Total atmospheric transport $(P - E)$ for the polar cap is 205 mm. Precipitation over most of the Arctic is low, less than $200\ mm\ a^{-1}$. Evaporation rates are uncertain. Pan-Arctic evapotranspiration estimates are $230\ mm\ a^{-1}$, with $160\ mm\ a^{-1}$ over tundra. Snow depth from North Pole drifting stations reaches a maximum north of Greenland and Ellesmere Island, where it is 40 cm. Snow water equivalent (SWE) in February–March has maxima of 200 mm over a wide area of north-central and northeastern Siberia. Autumn sea ice decline is linked to higher snowfall amounts in northeast and western Siberia.

Antarctica is surrounded by a circumpolar zone of polar maritime climate over the Southern Ocean. Temperatures over the continent are the lowest on Earth and largely mirror the topography. Temperatures are near freezing along the coast in summer. Inland in the six months of winter, temperatures are near constant ("coreless" winter). Minimum skin temperatures on the East Antarctic plateau of almost $-95\ °C$ have been measured by satellite.

Inversion strength is more than $15\ °C$ over most of Antarctica and $20–25\ °C$ on the East Antarctic plateau. Coastal Antarctica is dominated by katabatic winds in winter. However, their unidirectional character, which persists year round, implies significant topographic influences. Precipitation cannot be measured directly. Annual snow accumulation is approximately 50 mm on the East Antarctic plateau and 200 mm in many coastal areas, with approximately 800 mm of precipitation occurring on the Amundsen–Bellingshausen coast. Atmospheric studies of net Antarctic precipitation $(P - E)$ indicate an annual average of about 150–190 mm. Similar estimates for the Southern Ocean are unreliable.

Temperatures in the twentieth century in the Arctic show increases except for 1946–1965. Rates of increase exceeded 1–2 °C decade^{-1}, especially in winter and spring. The temperature rise was about twice the global average as a result of polar amplification. This phenomenon occurs as a result of temperature feedback and surface albedo feedback. Snow cover disappeared three days decade^{-1} earlier in spring during 1972–2009 over the pan-Arctic terrestrial region.

Surface temperatures have warned significantly since the 1950s in the Antarctic peninsula, with little change noted over most of the continent. A West Antarctic bore hole shows warming began around 1800 and increased sharply over the last two decades.

QUESTIONS

1. Compare the planetary wave structure of the two hemispheres.
2. Describe and explain the Antarctic stratospheric ozone hole.
3. Compare the low-level circulation over the Arctic and the Antarctic.
4. Describe the major modes of atmospheric circulation in the two hemispheres.
5. Describe the ocean circulations into and out of the Arctic Ocean.
6. Which characteristics make the Arctic Ocean unique?
7. Discuss the characteristics of the Antarctic Circumpolar Current.
8. How is the sea ice regime different in the Arctic and the Antarctic?
9. What factors account for the different radiation regimes in the two polar regions.
10. Explain the main controls of temperature in the two polar regions.
11. Why are precipitation data unreliable in high latitudes?
12. Why did the Arctic and Antarctic peninsula warm at twice the global average over the twentieth century?

References

Allison, I., G. Wendler, and U. Radok. 1993. "Climatology of the East Antarctic Ice Sheet (100 °E to 140 °E) Derived from Automatic Weather Stations." *Journal of Geophysical Research* 98: 8815–23.

Andreas, E. L., and S. F. Ackley. 1982. *On the Differences in Ablation Seasons of Arctic and Antarctic Sea Ice*. CRREL Report 82–33. Hanover, NH: Cold Regions Research and Engineering Lab.

Arctic Climatology Project. 2000. *Environmental Working Group Arctic Meteorology and Climate Atlas*, edited by F. Fetterer and V. Radionov. Boulder, CO: National Snow and Ice Data Center. doi: 10.7265/N5MS3QNJ

Ballinger, T. J., T. W. Schmidlin, and D. F. Steinhoff. 2013. "The Polar Marine Climate Revisited." *Journal of Climate* 26: 3935–52.

Barry, R. G., and P. D. Blanken. 2016. *Microclimate and Local Climate*. Cambridge: Cambridge University Press.

Barry, R. G., and A. M. Carleton. 2001. *Synoptic and Dynamic Climatology*. London: Routledge.

Barry, R. G., and E. A. Hall-McKim. 2014. *Essentials of the Earth's Climate System*. Cambridge: Cambridge University Press.

Barry, R. G., et al. 1987. "Arctic Cloudiness in Spring from Satellite Imagery." *Journal of Climatology* 7: 423–51.

Belkin, I. M. 2004. "Propagation of the 'Great Salinity Anomaly' of the 1990s around the Northern North Atlantic." *Geophysical Research Letters* 31: L08306.

Bogdanova, E. G., B. M. Ilyin, and I. V. Dragomilova. 2002. "Application of an Improved Bias Correction Model to Precipitation Measured at Russian North Pole Drifting Stations." *Journal of Hydrometeorology* 3: 700–13.

Bokhorst, S., et al. 2016. "Changing Arctic Snow Cover: A Review of Recent Developments and Assessment of Future Needs for Observations, Modelling, and Impacts." AMBIO. doi: 10.1007/s13280-016-0770-0.

Bromwich, D. H. 1988. "Snowfall in High Southern Latitudes." *Reviews of Geophysics* 26: 149–68.

Bromwich, D. H. 1989. "Satellite Analyses of Antarctic Katabatic Wind Behavior." *Bulletin of the American Meteorological Society* 70: 738–49.

Bromwich, D. H., and D. D. Kurtz. 1984. "Katabatic Wind Forcing of the Terra Nova Bay Polynya." *Journal of Geophysical Research* 89(C3): 3561–72.

Bromwich, D. H., and T. R. Parish. 1998. "Meteorology of the Antarctic." In *Meteorology of the Southern Hemisphere*, edited by D. J. Karoly and D. G. Vincent, 175–200. Meteorology Monographs 27(49).

Bromwich, D. H., T. R. Parish, and A. Zormanc. 1990. "The Confluence Zone of the Intense Katabatic Winds at Terra Nova Bay, Antarctica, as Derived from Airborne Sastrugi Surveys and Mesoscale Numerical Modeling." *Journal of Geophysical Research* 95(D5): 5495–509.

Bromwich, D. H., et al. 2013. "Central West Antarctica among the Most Rapidly Warming Regions on Earth." *Nature Geoscience* 6: 139–45.

Callaghan, T., et al. 2011. "The Changing Face of Arctic Snow Cover: A Synthesis of Observed and Projected Changes." AMBIO 40: 17–31.

Carleton, A. M. 2002. "Synoptic Interactions between Antarctica and Lower Latitudes." *Australian Meteorological Magazine* 40: 129–47.

Carleton, A. M., and D. A. Carpenter. 1990. "Satellite Climatology of Polar Lows and Broad-Scale Climatic Associations for the Southern Hemisphere." *International Journal of Climatology* 10: 219–46.

Carleton, A. M., and M. Fitch. 1993. "Synoptic Aspects of Antarctic Meso-cyclones." *Journal of Geophysical Research* 98: 12997–13018.

Carmack, E. C., et al. 2016. "Freshwater and Its Role in the Arctic Marine System: Sources, Disposition, Storage, Export, and Physical and Biogeochemical Consequences in the Arctic and Global Oceans." *Journal of Geophysical Research: Biogeosciences* 121: 675–717.

Carrasco, J. F., and D. H. Bromwich. 1994. "Climatological Aspects of Mesoscale Cyclogenesis over the Ross Ice Shelf Regions of Antarctica." *Monthly Weather Review* 122: 2405–25.

Cassano, J. J., M. A. Nigro, and M. A. Lazzara. 2016. "Characteristics of the Near-Surface Atmosphere over the Ross Ice Shelf, Antarctica." *Journal of Geophysical Research: Atmosphere* 121, doi: 10.1002/2015JD024383.

Chernokulsky, A., and I. I. Mokhov. 2012. "Climatology of Total Cloudiness in the Arctic: An Intercomparison of Observations and Reanalyses." *Advances in Meteorology,* 542093. doi: 10.1155/2012/542093

Clem, K. R., et al. 2016. "The Relative Influence of ENSO and SAM on Antarctic Peninsula Climate." *Journal of Geophysical Research: Atmosphere.* doi: 10.1002/2016JD025305.

Connolley, W. M. 1996. "The Antarctic Temperature Inversion." *International Journal of Climatology* 16: 1333–42.

Connolley, W. M., and J. C. King. 1993. "Atmospheric Water Vapour Transport to Antarctica Inferred from Radiosonde Data." *Quarterly Journal of the Royal Meteorological Society* 119: 325–42.

Costanza, C. A., et al. 2016. "The Surface Climatology of the Ross Ice Shelf Antarctica." *International M.C. Climatology* 36: 4929–41.

Crawford, A., and M. C. Serreze. 2015. "A New Look at the Summer Arctic Frontal Zone." *Journal of Climate* 28: 737–54.

Crawford, A., and M. C. Serreze. 2016. "Does the Summer Arctic Frontal Zone Influence Arctic Ocean Cyclone Activity?" *Journal of Climate* 39: 4977–93.

Cullather, R. I., and M.G. Bosilovich. 2011. "The Moisture Budget of the Polar Atmosphere in MERRA." *Journal of Climate* 24: 2861–79.

Cullather, R. I., D. H. Bromwich, and M. L. Van Woert. 1998. "Spatial and Temporal Variability of Antarctic Precipitation from Atmospheric Methods." *Journal of Climate* 11: 334–67.

Devasthale, A., et al. 2010. "Quantifying the Clear-Sky Temperature Inversion Frequency and Strength over the Arctic Ocean during Summer and Winter Seasons from AIRS Profiles." *Atmospheric Chemistry and Physics* 10: 5565–72.

Dickson, R. R., et al. 1988. "The 'Great Salinity Anomaly' in the Northern North Atlantic, 1968–1982." *Progress in Oceanography* 20: 103–51.

Donahue, K. A., et al. 2016. "Mean Antarctic Circumpolar Current Transport Measured in Drake Passage." *Geophysical Research Letters* 43: 760–7.

Dufour, A., O. Zolina, and S. K. Gulev. 2016. "Atmospheric Moisture Transport to the Arctic: Assessment of Reanalyses and Analysis of Transport Components." *Journal of Climate* 29: 5061–81.

Emery, W. J. 1977. "Antarctic Polar Frontal Zone from Australia to the Drake Passage." *Journal of Physical Oceanography* 7: 811–22.

Fioletov, V. E. 2008. "Ozone Climatology, Trends, and Substances That Control Ozone." *Atmosphere and Ocean* 46: 39–67.

Fioletov, V. E., et al. 2002. "The Global Ozone and Zonal Total Ozone Variations Estimated from Ground-Based and Satellite Measurements: 1978–2000." *Journal of Geophysical Research* 107: D22.

Fitch, M., and A. M. Carleton. 1991. "Antarctic Mesocyclone Regimes from Satellite and Conventional Data." *Tellus, A* 44: 180–96.

Fogt, R. L., et al. 2012. "The Characteristic Variability and Connection to the Underlying Synoptic Activity of the Amundsen-Bellingshausen Sea Low. *Journal of Geophysical Research* 117(D7): D07111.

Fristrup, B. 1961. "Climatological Studies of Some High Arctic Stations in North Greenland." *Physical Geography of Greenland. Folia Geogr. Danica (Copenhagen)* 9: 67–78.

Frolov, I. F., et al. 2005. *The Arctic Basin: Results from the Russian Drifting Stations.* Chichester, UK: Praxis Publishing.

Giovinetto, M. B., and C. R. Bentley. 1985. "Surface Balance in Ice Drainage Systems of Antarctica." *Antarctica Journal U.S.* 20(4): 6–13.

Goodison, B. E., and J. R. Metcalfe. 1992. "The WMO Solid Precipitation Intercomparison: Canadian Assessment." In WMO Technical Conference on Instruments and Method of Observation, WMO/TD No. 462, 221–5.

Gorshkov, S. G. 1980. *Atlas of Oceans: Arctic Ocean.* [In Russian]. Moscow: Military Defense Publishing House.

Groisman, P., et al. 2016. "Northern Eurasia Future Initiative: Facing the Challenges and Pathways of Global Change in the 21st Century." [White paper]. http://neespi.org

Grytsal, A. V., et al. 2007. "Structure and Long-Term Change in the Zonal Asymmetry in Antarctic Total Ozone during Spring." *Annales Geophysicae* 25(2): 361–74.

Haine, T. W. N., et al. 2015. "Arctic Freshwater Export: Status, Mechanisms, and Prospects." *Global and Planetary Change* 125: 13–35.

Häkkinen, S. 1999. "A Simulation of Thermohaline Effects of a Great Salinity Anomaly." *Journal of Climate* 12: 1781–95.

Hartmann, D. L., A. M. G. Klein Tank, and M. Rusticucci. 2013. "Observations, Atmosphere and Surface." In *Climate Change 2013: The Physical Science Basis. Contribution of Working Group I to the Fifth Assessment Report of the Intergovernmental Panel on Climate Change*, edited by T. F. Stocker et al., 159–254. Cambridge: Cambridge University Press.

Hori, M., et al. 2017. "A 38-Year (1978–2015) Northern Hemisphere Daily Snow Cover Extent Product Derived Using Consistent Objective Criteria from Satellite-Borne Optical Sensors." *Remote Sensing of the Environment* 191: 402–18.

Hudson, S. R., and R. E. Brandt. 2005. "A Look at the Surface-Based Temperature Inversion on the Antarctic Plateau." *Journal of Climate* 18(11): 1673–96.

Intergovernmental Oceanographic Commission (IOC), Scientific Committee on Oceanic Research (SCOR), and International Association for the Physical Sciences of the Oceans (IAPSO). 2010. *The International Thermodynamic Equation of Seawater – 2010: Calculation and Use of Thermodynamic Properties.* IOC, UNESCO.

James, I. N. 1989. "The Antarctic Drainage Flow: Implications for Hemispheric Flow on the Southern Hemisphere." *Antarctic Science* 1: 279–90.

Jiao, C., et al. 2014. "An AeroCom Assessment of Black Carbon in Arctic Snow and Sea Ice." *Atmospheric Chemistry and Physics* 14: 2399–417.

Johansson, E., et al. 2017. "Response of the Lower Troposphere to Moisture Intrusions into the Arctic." *Geophysical Research Letters* 44: 2527–36.

Karoly, D. A., and D. G. Vincent. 1998. *Meteorology of the Southern Hemisphere*. Boston: American Meteorological Society.

King, J. C. 1996. "Longwave Atmospheric Radiation over Antarctica." *Antarctic Science* 8: 105–9.

King, J. C., and J. Turner. 1997. *Meteorology and Climatology of the Antarctic*. Cambridge: Cambridge University Press.

Kondratyev, K. Ya. 2003. "Arctic Atmosphere." In *Arctic Environment Variability in the Context of Global Change*, edited by L. P. Bobylev et al., 17–88. Chichester, UK: Springer-Praxis.

Kort, V. G. 1964. "Antarctic Oceanography." In *Research in Geophysics*. Vol. 2, edited by H. Odishaw, 309–33. Cambridge, MA: MIT Press.

Krebs, S. J., and R. G. Barry. 1970. "The Arctic Front and the Tundra–Taiga Boundary in Eurasia." *Geography Review* 60: 548–54.

Lee, S., et al. 2017. "Revisiting the Cause of the 1989–2009 Arctic Surface Warming Using the Surface Energy Budget: Downward Infrared Radiation Dominates the Surface Fluxes." *Geophysical Research Letters* 44. doi: 10.1002/2017GL075375.

Lewis, E. L., et al. 2000. *The Freshwater Budget of the Arctic Ocean*. Dordrecht: Kluwer Academic.

Liljedahl, A. K., et al. 2017. "Tundra Water Budget and Implications of Precipitation Underestimation." *Water Resources Research* 53. doi: 10.1002/2016WR020001.

Liston, G. E., and C. A. Hiemstra. 2011. "The Changing Cryosphere: Pan-Arctic Snow Trends (1979–2009)." *Journal of Climate* 24: 5691–712.

Lorenz, E. N. 1951. "Seasonal and Irregular Variations of the Northern Hemisphere Sea-Level Pressure Profile." *Journal of Meteorology* 8(1): 52–9.

Marshall, G. J. 2003. "Trends in the Southern Annular Mode from Observations and Reanalyses." *Journal of Climate* 16: 4134–43.

Marshall, G. J., and J. Turner. 1997. "Katabatic Wind Propagation over the Western Ross Sea Observed Using ERS-1 Scatterometer Data." *Antarctic Science* 9: 221–6.

Mayewski, P. A., et al. 2017. "Ice Core and Climate Reanalysis Analogs to Predict Antarctic and Southern Hemisphere Climate Changes." *Quaternary Science Reviews* 155: 50–66.

McBean, G., et al. 2005. "Arctic Climate: Past and Present." In *Arctic Climate Impact Assessment*, 19–60. Cambridge: Cambridge University Press.

McCusker, K. E., J. C. Fyfe, and M. Sigmond. 2016. "Twenty-Five Winters of Unexpected Eurasian Cooling Unlikely Due to Arctic Sea-Ice Loss." *Nature Geoscience* 9: 838–42.

Melling, H., and E. L. Lewis. 1982. "Shelf Drainage Flows in the Beaufort Sea and Their Effect on the Arctic Ocean Pycnocline." *Deep-Sea Research* 29: 967–85.

Moore, G. W. K., and I. A. Renfrew. 2005. "Tip Jets and Barrier Winds: A QuikSCAT Climatology of High Wind Speed Events around Greenland." *Journal of Climate* 18: 3713–25.

Mudryk, L. R., et al. 2015. "Characterization of Northern Hemisphere Snow Water Equivalent Datasets, 1981–2010." *Journal of Climate* 28: 8037–51.

Muto, A., et al. 2011. "Recent Surface Temperature Trends in the Interior of East Antarctica from Borehole Firn Temperature Measurements and Geophysical Inverse Methods." *Geophysical Research Letters* 38: L15502.

Nowlin, W. D. Jr., T. Whitworth III, and R. D. Pillsburry. 1977. "Structure and Transport of the Antarctic Circumpolar Current at Drake Passage from Short-Term Measurements." *Journal of Physical Oceanography* 7: 778–802.

Orsi, A. H., T. Whitworth III, and W. D. Nowlin. 1995. "On the Meridional Extent and Fronts of the Antarctic Circumpolar Current." *Deep-Sea Research* 42: 641–73.

Orsi, A. J., B. D. Cornuelle, and J. P. Severinghaus. 2012. "Little Ice Age Cold Interval in West Antarctica: Evidence from Borehole Temperature at the West Antarctic Ice Sheet (WAIS) Divide." *Geophysical Research Letters* 39: L0971.

Overland, J. E. 2009. "Meteorology of the Beaufort Sea." *Journal of Geophysical Research* 114(C1): C00A07.

Overland, J. E., and M.-Y. Wang. 2016. "Recent Extreme Arctic Temperatures Are due to a Split Polar Vortex." *Journal of Climate* 29: 5609–16.

Parish, T. R., and D. H. Bromwich. 1987. "The Surface Wind Field over the Antarctic Ice Sheets." *Nature* 328: 51–4.

Parish, T. R., and J. J. Cassano. 2003. "The Role of Katabatic Winds on the Antarctic Surface Wind Regime." *Monthly Weather Reviews* 131: 317–33.

Park, H., et al. 2013. "The Role of Declining Arctic Sea Ice in Recent Decreasing Terrestrial Arctic Snow Depths." *Polar Science* 7: 174–82.

Perovich, D. K., et al. 2002. "Seasonal Evolution of the Albedo of Multiyear Arctic Sea Ice." *Journal of Geophysical Research* 107(C10): 8044.

Persson, P. O. G. 2002. "Measurements near the Atmospheric Surface Flux Group Tower at SHEBA: Near Surface Conditions and Surface Energy Budget." *Journal of Geophysical Research* 107(C10): C000705.

Phillpot, H. R., and J. W. Zillman. 1970. "The Surface Temperature Inversions over the Antarctic Continent." *Journal of Geophysical Research* 75(21): 4161–9.

Pithan, F., and T. Mauritsen. 2014. "Arctic Amplification Dominated by Temperature Feedbacks in Contemporary Climate Models." *Nature Geoscience* 7: 181–4.

Polyakov, I. V., et al. 2003. "Variability and Trends of Air Temperature and Pressure in the Maritime Arctic." *Journal of Climate* 16: 2078–85.

Polyani, L. M., et al. 2011. "Stratospheric Ozone Depletion: The Main Driver of Twentieth-Century Atmospheric Circulation Changes in the Southern Hemisphere." *Journal of Climate* 24: 795–812.

Prowse, T. D., et al. 2015. "Arctic Freshwater Synthesis: Introduction." *Journal of Geophysical Research: Biogeoscience* 120(11): 2121–31.

Pryzbylak, R., et al. 2010. "Air Temperature Changes in the Arctic from 1801 to 1920." *International Journal of Climatology* 30: 791–812.

Pryzbylak, R., et al. 2016. "Air Temperature Changes in Svalbard and the Surrounding Seas from 1865 to 1920." *International Journal of Climatology* 36: 2899–916.

Qian, W.-H., K.-J. Wu, and H.-Y. Liang. 2015. "Arctic and Antarctic Cells in the Troposphere." *Theoretical and Applied Climatology*. doi: 10.1007/s00704-015-1485-z.

Raphael, M. N., et al. 2016. "The Amundsen Sea Low: Variability, Change, and Impact on Antarctic Climate." *Bulletin of the American Meteorological Society* 97(1): 111–21.

Rasmussen, E. A., and J. Turner, eds. 2003. *Polar Lows: Mesoscale Weather Systems in the Polar Regions*. Cambridge: Cambridge University Press.

Rudels, B. 2015. "Arctic Ocean Circulation, Processes and Water Masses: A Description of Observations and Ideas with Focus on the Period Prior to the International Polar Year 2007–2009." *Progress in Oceanography* 132: 22–67.

Scambos, T., et al. 2013. "The Coldest Place on Earth." https://fallmeeting.agu.org/2013/files/2013/12/ColdestPlaceOnEarth.pdf

Screen, J. A., and J. A. Francis. 2016. "Contribution of Sea-Ice Loss to Arctic Amplification Is Regulated by Pacific Ocean Decadal Variability." *Nature Climate Change* 6: 856–60.

Screen, J., and I. Simmonds. 2010. "The Central Role of Diminishing Sea Ice in Recent Arctic Temperature Amplification." *Nature* 464: 1334–7.

Serreze, M. C., and A. P. Barrett. 2008. "The Summer Cyclone Maximum over the Central Arctic Ocean." *Journal of Climate* 21: 1048–65.

Serreze, M. C., and A. P. Barrett. 2011. "Characteristics of the Beaufort Sea High." *Journal of Climate* 24: 159–82.

Serreze, M. C., and R. G. Barry. 2011. "Processes and Impacts of Arctic Amplification: A Research Synthesis." *Global and Planetary Change* 77: 85–96.

Serreze, M. C., and R. G. Barry. 2014. *The Arctic Climate System*. 2nd ed. Cambridge: Cambridge University Press.

Serreze, M. C., and C. M. Hurst. 2000. "Representation of Mean Arctic Precipitation from NCEP–NCAR and ERA Reanalyses." *Journal of Climate* 13: 182–201.

Serreze, M. C., A. H. Lynch, and M. P. Clarke. 2001. "The Arctic Frontal Zone as Seen in the NCEP–NCAR Reanalysis." *Journal of Climate* 14: 1550–67.

Serreze, M. C., et al. 2006. "The Large-Scale Freshwater Cycle of the Arctic." *Journal of Geophysical Research* 111: C11010.

Sharma, E., et al. 2006. "Variations and Sources of the Equivalent Black Carbon in the High Arctic Revealed by Long-Term Observations at Alert and Barrow: 1989–2003." *Journal of Geophysical Research* 111: D14208.

Shiklomanov, I. A. 2000. "The Dynamics of River Water Inflow to the Arctic Ocean." In *The Freshwater Budget of the Arctic Ocean*, edited by E. L. Lewis et al., 281–96. Dordrecht: Kluwer Academic Publishers.

Sievers, H. A., and W. J. Emery. 1978. "Variability of the Antarctic Polar Frontal Zone in the Drake Passage – Summer 1976–77. *Journal of Geophysical Research* 83: 3010–22.

Simmonds, I. 1998. "The Climate of the Antarctic Region." In *Climates of the Southern Continents: Past, Present and Future*, edited by J. E. Hobbs, J. A. Lindesay, and H. A. Bridgman, 137–60. Chichester, UK: Wiley and Sons.

Stolarski, R. S., and S. Frith. 2006. "Search for Evidence of Trend Slowdown in the Long-Term TOMS/SBUV Total Ozone Data Record: The Importance of Instrument Drift Uncertainty." *Atmospheric Chemistry and Physics* 6: 4057–65.

Stone, R. S., et al. 2014. "A Characterization of Arctic Aerosols on the Basis of Aerosol Optical Depth and Black Carbon Measurements." *Elementa: Science of the Anthropocene* 2: 27.

Streten, N. A. 1990. "A Review of the Climate of Mawson: A Representative Strong Wind Site in East Antarctica." *Antarctic Science* 2: 79–89.

Swinbank, W. C. 1963. "Long-Wave Radiation from Clear Skies." *Quarterly Journal of the Royal Meteorological Society* 89: 339–48.

Tanaka, H. L., et al. 2012. "The Structure and Behavior of the Arctic Cyclone in Summer Analyzed by the JRA-25/JCDAS Data." *Polar Science* 6: 55–69.

Tanis, F., and Timokhov, I., eds. 1997. *Joint U.S.–Russian Atlas for the Arctic Ocean: Oceanographic Atlas for the Winter Period.* [Digital media.] Boulder, CO: Environmental Working Group, University of Colorado, National Snow and Ice Data Center.

Tanis, F., and Timokhov, I., eds. 1998. *Joint U.S.–Russian Atlas for the Arctic Ocean: Oceanographic Atlas for the Summer Period.* [Digital media.] Boulder, CO: Environmental Working Group, University of Colorado, National Snow and Ice Data Center.

Teppei, J., et al. 2015. "Impact of Snow Darkening via Dust, Black Carbon, and Organic Carbon on Boreal Spring Climate in the Earth System." *Journal of Geophysical Research: Atmosphere* 120: 5485–503.

Thompson, D. W. J., and J. M. Wallace. 1998. "The Arctic Oscillation Signature in the Wintertime Geopotential Height and Temperature Fields." *Geophysical Research Letters* 25(9): 1297–300.

Tietäväinen, H., and T. Vihma. 2008. "Atmospheric Moisture Budget over Antarctica and the Southern Ocean Based on the ERA-40 Reanalysis." *International Journal of Climatology* 28: 1977–95.

Tjernstrom, M., and R. G. Graversen. 2009. "The Vertical Structure of the Lower Arctic Troposphere Analysed from Observations and ERA-40 Reanalysis." *Quarterly Journal of the Royal Meteorological Society* 135(639): 431–43.

Town, M. S., V. P. Walden, and S. G. Warren. 2007. "Cloud Cover over the South Pole from Visual Observations, Satellite Retrievals, and Surface-Based Infrared Radiation Measurement." *Journal of Climate* 20: 544–59.

Turner, J., and G. J. Marshall. 2011. *Climate Change in the Polar Regions.* Cambridge: Cambridge University Press.

Turner, J., et al. 1995. "The Synoptic Origins of Precipitation over the Antarctic Peninsula." *Antarctic Science* 7: 327–37.

Turner, J., et al. 2013. "The Amundsen Sea Low." *International Journal of Climatology* 33: 1818–29.

Turner, J., et al. 2014. "Antarctic Climate Change and the Environment: An Update." *Polar Record* 50: 237–59.

Turner, J., et al. 2016. "Variability and Trends in the Southern Hemisphere High Latitude, Quasi-Stationary Planetary Waves." *International Journal of Climatology*. doi: 10.1002/joc.4848.

Uotila, P., et al. 2011. "Relationships between Antarctic Cyclones and Surface Conditions as Derived from High-Resolution Numerical Weather Prediction Data." *Journal of Geophysical Research* 116: D07109.

Uttal, T., et al. 2002. "Surface Heat Budget of the Arctic Ocean." *Bulletin of the American Meteorological Society* 83: 255–75.

Van Den Broeke, M. R. 1998. "The Semi-Annual Oscillation and Antarctic Climate, Part I: Influence on Near-Surface Temperatures." *Antarctic Science* 10: 175–83.

van Loon, H. 1967. "The Half-Yearly Oscillation in the Middle and High Southern Latitudes and the Coreless Winter." *Journal of Atmospheric Science* 24: 472–86.

Vázquez, M., et al. 2016. "Moisture Transport into the Arctic: Source–Receptor Relationships and the Roles of Atmospheric Circulation and Evaporation." *Journal of Geophysical Research: Atmosphere.* doi: 10.1002/2016JD025400.

Verezemskaya, P., et al. 2017. "Southern Ocean Mesocyclones and Polar Lows from Manually Tracked Satellite Mosaics." *Geophysical Research Letters* 44. doi: 10.1002/2017GL074053.

Wang, L., et al. 2016. "Frequency and Distribution of Winter Melt Events from Passive Microwave Satellite Data in the Pan-Arctic, 1988–2013." *Cryosphere Discussion.* doi: 10.5194/tc-2016-126.

Warren, S. G., et al. 1999. "Snow Depth on the Arctic Sea Ice." *Journal of Climate* 12: 1814–29.

Weber, M., et al. 2007. "Improved SCIAMACHY WFDOAS Total Ozone Retrieval: Steps towards Homogenising Long-Term Total Ozone Datasets from GOME, SCIAMACHY, and GOME2." In *Proceedings of the Envisat Symposium*. ESA SP-636. Montreux, Switzerland: European Space Agency.

Weber, M., et al. 2012. "Stratospheric ozone." In State of the Climate in 2011. *Bulletin of the American Meteorological Society* 93(7 suppl): S46–9.

Webster, M. A., et al. 2015. "Seasonal Evolution of Melt Ponds on Arctic Sea Ice." *Journal of Geophysical Research: Oceans* 120: 5968–82.

Wegman, M., et al. 2015. "Arctic Moisture Source for Eurasian Snow Cover Variations in Autumn." *Environmental Research Letters* 10: 054015.

Weller, G. 1969. "A Meridional Surface Wind Speed Profile in MacRobertson Land Antarctica." *Pure and Applied Geophysics* 77: 193–200.

Wendisch, M., et al. 2017. "Understanding Causes and Effects of Rapid Warming in the Arctic." *Eos* 98. doi: 10.1029/2017EO064803.

Wendler, G., and Y. Kodama. 1993. "The Kernlose Winter in Adelie Coast." In *Antarctic Meteorology and Climatology: Studies Based on Automatic Weather Stations*, edited by D. H. Bromwich and C. R. Stearns, 139–47. Washington, DC: American Geophysical Union.

Wexler, H. 1959. "Seasonal and Other Temperature Changes in the Antarctic Atmosphere." *Quarterly Journal of the Royal Meteorological Society* 85: 196–205.

Wu, B., J. Wang, and J. E. Walsh. 2006. "Dipole Anomaly in the Winter Arctic Atmosphere and Its Association with Sea Ice Motion." *Journal of Climate* 19: 210–25.

Wyrtki, K. 1960. "The Antarctic Circumpolar Current and the Antarctic Polar Front." *Dtsch. Hydr. Z.* 13(4): 153–74.

Zektser, I. S., and R. G. Dzhamalov. 1988. *Role of Ground Water in the Hydrological Cycle and in Continental Water Balance*. Paris: UNESCO International Hydrological Programme IHP-III.

Zhang, J.-K., et al. 2016. "Persistent Shift of the Arctic Polar Vortex towards the Eurasian Continent in Recent Decades." *Nature Climate Change* 6: 1094–9.

Zhang, K., et al. 2009. "Satellite Based Analysis of Northern ET Trends and Associated Changes in the Regional Water Balance from1983 to 2005." *Journal of Hydrology* 379(1): 92–110.

Terrestrial Environments and Surface Types of the Polar Regions

5

There are seven major surface types in polar regions: polar desert; tundra; permafrost; lakes and rivers; glaciers; ice sheets; and ice shelves. The last two types are treated in Chapter 6. The first four are harsh environments in which plants and animals have evolved with some shared characteristic adaptations. The ice and snow surfaces of glaciers have only microorganisms. The environmental characteristics of these surface types are discussed in turn in this chapter.

There are three broad types of landforms in the northern polar region: (1) flat plains and plateaus, which are largely covered by alluvial, glacial, and marine deposits; (2) rugged uplands, which have ice-scoured rock surfaces and are cut by deep fiords; and (3) glaciated mountains, which include alpine terrain and rounded slopes. Similar landforms occur in the Antarctic Peninsula and sub-Antarctic islands, but there are also volcanic peaks in coastal Antarctica and nunataks (unglaciated peaks) that protrude through the ice sheet.

5.1 Polar Desert

Polar deserts are areas with annual precipitation less than 250 mm and a mean temperature during the warmest month of less than 5 °C. Some of the precipitation falls as rain in the warm season. Polar deserts experience five to six months of polar night and a corresponding polar summer with months of 24-hour daylight. They cover nearly 5 million km^2 of the Earth. Figure 5.1 shows their distribution in the northern hemisphere based on an analysis by Charlier (1969). He defines polar desert soils as mature and well drained, occurring in the "high arctic," sparsely covered by higher plants, and saline or alkaline in character; see also Tedrow (1966) for more details. Desert pavement is usually present. Charlier also accepts a July mean temperature of 4.5 °C as the equatorward limit of polar deserts, though this may be a little too restrictive. For northern Ellesmere Island, King (1981) lists summer temperatures at Eureka, Tanquary Fiord, and Lake

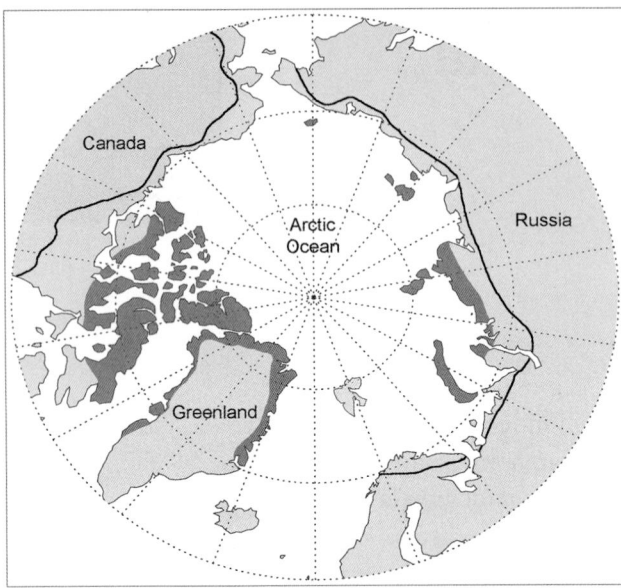

Figure 5.1 Polar desert regions (shaded) in the northern hemisphere (redrawn after Charlier 1969). The line demarcates the southern temperature boundary.
Source: Charlier 1969, 1991, figure 1. Courtesy of Geological Society of America.

Hazen. Mean July values for Eureka were 5.3 °C for 1962–1978 and 4.4 °C for 1963–1967, compared with 6.0 °C for Tanquary Fiord (1963–1967) and 5.6 °C for Lake Hazen (1963–1967). The latter two stations are essentially inland, while Eureka is on the coast. Absolute maxima at the three stations for 1963–1967 were 14.4, 17.1, and 13.5 °C, respectively.

Bovis and Barry (1973) used a discriminant analysis of seven climate variables to group forty-eight Arctic stations as desert or desert–tundra transition. Seven stations (Nord, Isachsen, Alert, Mould Bay, Resolute, Eureka, and Craig Harbour) were grouped as polar desert, and seventeen as transitional. Using Charlier's July temperature criterion, twenty-one stations were placed in each group and many Eurasian Arctic stations were designated as polar desert. Bovis and Barry also found that estimates of evaporation, based on the work of Budyko (1956) and Turc (1955), applied as aridity indices, did not adequately delimit polar desert areas. The Budyko values of water surplus were too large and the Turc values too low.

Much of the polar desert surface is covered with fell fields (felsenmeer), comprising angular blocks derived from freeze–thaw splitting of the bedrock, and gravel plains. Figures 5.2 and 5.3 illustrate typical polar desert surfaces.

Stones and fine sediments are often organized as patterned ground. This may consist of stone circles and ice-wedge polygons up to 10–20 m across or, alternatively, stone stripes on sloping surfaces (French 2007). These unusual surface forms originate as a result of freeze–thaw processes affecting the surface materials. These processes tend to displace rocks and soil into geometric patterns. The ground is typically underlain by thick permafrost (see Section 5.3) and the summer thawed layer is only 15–30 cm deep.

In the Arctic, 1–5 percent of the polar desert surface has vegetation. There are fewer than 100 vascular plant species, bryophytes are abundant, and lichens are common, as well as algae and mosses. In the Eurasian Arctic, polar desert is found in the Arctic islands and in the northernmost Taimyr Peninsula. The summer climate of Franz Josef Land is very cold (mean July temperature = 1 °C), foggy, and wet (mean total annual precipitation = 300 mm) due to the influence of the Barents Sea. There are 57 species of flowering plants, 115 lichen species, and 102 mosses (Alexandrova 1988). On the northern islands of Severnaya Zemlya, the number of flowering plant species is only 17. The growth forms of plants are typically graminoid, cushion, and rosette. Sparse shrubs have heights ranging from 5 to 100 cm, while forbs range from 2 to 10 cm.

In Antarctica, polar desert occupies a small fraction (0.03 percent) of the continent, outside the ice sheet, although Korotkevich (1972) included the ice sheet in his definition. Here we consider only the 4,800 km² of the Dry Valleys near McMurdo (Box 5.1), where numerous hypersaline lakes are found (see Section 5.4). These areas are essentially snow free and have very low humidity due to subsiding air in katabatic airflow. Cold desert soils, which occur in Antarctica, are well drained and have no organic matter (Tedrow and Ugolini 1966).

Figure 5.2 "Polar desert" tundra of maritime areas of Arctic Bioclimate Subzone A (CAVM Team 2003), Hayes Island, Franz Jozef Land, Russia. The vegetation is composed mostly of cryptobiotic soil crusts, and cushion forms of forbs, lichens, and mosses covering wet sandy soils. Dominant vascular plant species include *Papaver dahlianum* spp. *polar* (yellow flowers), *Stellaria edwardsii, S. crassipes, Draba micropetala, Saxifraga cespitosa,* and *Phippsia algida*. August 10, 2010. Source: Courtesy of D. A. "Skip" Walker, Alaska Geobotany Center, Institute of Arctic Biology and Department of Biology and Wildlife, University of Alaska, Fairbanks.

A number of meteorological stations have been built, and some continue to operate in polar desert regions. A meteorological station in Peary Land, north Greenland, was operated at Jørgen Brønlunds Fiord (82.2° N, 30.5° W) during August 1948–August 1950 (Fristrup 1952, 1961). The air temperature ranged from −31 °C in January to 6.2 °C in July. Corresponding values for Nord (81.6° N, 16.7° W) on the northeast coast were −30 and 4.3 °C, respectively; those for Eureka (80.1° N, 86.4° W) on northwest Ellesmere Island were −37.6 and 5.7 °C, respectively. There were sixty-two frost free days at Jørgen Brønlunds Fiord, compared with seventy-seven at Eureka and only thirty-three at Alert on the north coast of Ellesmere Island. During the International Geophysical Year (IGY), 1957–1958, a synoptic weather station was operated at Lake Hazen, Ellesmere Island, and demonstrated the contrast in temperature conditions between this inland station and the coastal stations of Alert and Eureka (Jackson 1959). Low temperatures during the long winter were much more common inland than at the coastal stations. The temperature reached −46 °C on only 1 day at Alert during the winter of 1957–1958, but on 73 days at Lake Hazen, whereas temperatures

Figure 5.3 Lichen-rich polar desert landscape on gentle south-facing hillslope, Hayes Island, Franz Jozef Land, Russia. The brown lichens are mainly *Cetrariella delis* and *Cetraria islandica*. The white lichens are *Stereocaulon alpine* and *Thamnolia subuliformis*. The yellowish lichens are mainly *Flavocetraria cucullata*. The rich lichen cover is due to the lack of reindeer and other herbivores on the island.
Source: Courtesy of D. A. "Skip" Walker, Alaska Geobotany Center, Institute of Arctic Biology and Department of Biology and Wildlife, University of Alaska, Fairbanks.

of −40 °C or below were recorded at the inland station on 121 days compared with 83 days at Eureka and 29 days at Alert. Also, it is worth noting that January 1958 was anomalously mild. The low temperatures are attributed to the very light winds and frequent calms at Lake Hazen, as a result of its sheltered location.

Also during the IGY, a climate station was operated on the nearby Gilman Glacier. Calms were reported more than 20 percent of the time in the summers of 1957 and 1958 (Lotz and Sagar 1962). The mean temperature for July 1958 was 6.5 °C at Lake Hazen.

5.2 Tundra

Tundra is a Finnish–Sami word referring to a treeless plain. The biome is controlled by a short growing season and low temperatures. It occurs in both polar regions and at high altitudes in lower latitudes (alpine tundra) and accounts for about 8 percent of the global land surface. The vegetation in the Arctic is composed of grasses, heath, moss and lichen cushions, sedges, and dwarf willow shrubs. A recent account of circumpolar descriptions of Arctic tundra and plant biomass is provided by Walker et al. (2016). Vegetation is often discontinuous, with fellfield being a common surface type. In the Antarctic, tundra is found only on the Antarctic Peninsula and sub-Antarctic islands. Its biome consists mainly of lichens, mosses, and algal species. The tundra surface is underlain by perennially frozen ground, or permafrost (see Section 5.3), but the uppermost 20–60 cm or so thaws in summer, forming an active layer. Tundra climates are described by Barry et al. (1981) based on data for the International Biological Programme (IBP) Tundra Biome sites. Winters last six months, with mid-winter temperatures averaging −25 to −30 °C and extremes of −50 °C. In summer, with 24-hour daylight, temperatures may reach 12–15 °C, but no month has an average temperature above 10 °C. Precipitation falls mainly as light rain in summer and as snow during the rest of the year. Annual precipitation totals are typically only 150–250 mm. Blowing snow is common across the open tundra surfaces, and snow remains on the ground for eight or nine months. In summer,

Box 5.1 **The McMurdo Dry Valleys**

The Victoria, Wright, and Taylor Dry Valleys form an ice-free area of 4,800 km^2 near McMurdo Sound. Two large valley glaciers (Mackay and Ferrar) serve as outlets to the East Antarctic ice sheet, flowing through the Dry Valleys region to the coast. There are numerous alpine glaciers, some of which contribute to two piedmont glaciers (Bowers and Wilson). There are a wide variety of lakes: Some freeze to the bottom in winter, while others retain basal liquid water. Except for a few hypersaline water bodies, all have permanent ice cover.

The McMurdo Dry Valleys in Antarctica (77.0° S, 162.9° E) became a National Science Foundation (NSF) Long-Term Ecological Research (LTER) area in 1992. This area has been the site of a wide range of biogeochemical, hydrological, glaciological, ecological, limnological, and microbiological research (https://lternet.edu/sites/mcm/research-topics).

A network of automatic weather stations (AWS) has operational stations at Explorer's Cove; on the shores of Lakes Fryxell, Hoare, Bonney, Brownworth, and Vanda; and on the Commonwealth, Howard, and Taylor glaciers. Mass balance measurements are available from a stake network, GPS measurements, and ice radar on the Canada, Commonwealth, Howard, and Taylor glaciers. Surface energy balance calculations have been performed on the Canada Glacier. Stream flow and annual water budgets for lakes Bonney (east and west lobes), Hoare, and Fryxell, are monitored in Taylor Valley.

A publication list can be found at www.mcmlter.org. Some representative papers in the present context are Fountain et al. (1998), Doran et al. (2002), and Green and Lyons (2009).

the thawing active layer makes the ground soggy and snow melt is often unable to drain away, such that marshes, bogs, and lakes are prominent features.

A circumpolar vegetation map has been prepared by Walker et al. (2005) at a scale of 1:7.5 million (Figure 5.4). Within the Arctic (total area of 7.1×10^6 km^2), about 5.05×10^6 km^2 is vegetated; the remainder is ice covered. About 26 percent of the vegetated area is erect shrubland, 18 percent peaty graminoid tundra, 13 percent mountain complexes, 12 percent barrens, 11 percent mineral graminoid tundra, 11 percent prostrate-shrub tundra, and 7 percent wetlands. Canada has by far the most terrain in the High Arctic domain; this area is mostly associated with abundant barren types and prostrate dwarf-shrub tundra (Figure 5.5). *Dryas integrifolia* and *Salix arctica* are typical of Arctic bioclimate subzone C. Russia has the largest area in the Low Arctic domain, which predominantly consists of low-shrub tundra.

Intensive study of tundra environments was conducted through the IBP Tundra Biome project, 1970–1974, which involved eleven nations (Bliss et al. 1981). Site descriptions have been given for the Truelove Lowlands, Devon Island, Point Barrow, Prudhoe Bay, and Eagle Summit, Alaska; Agapa, western

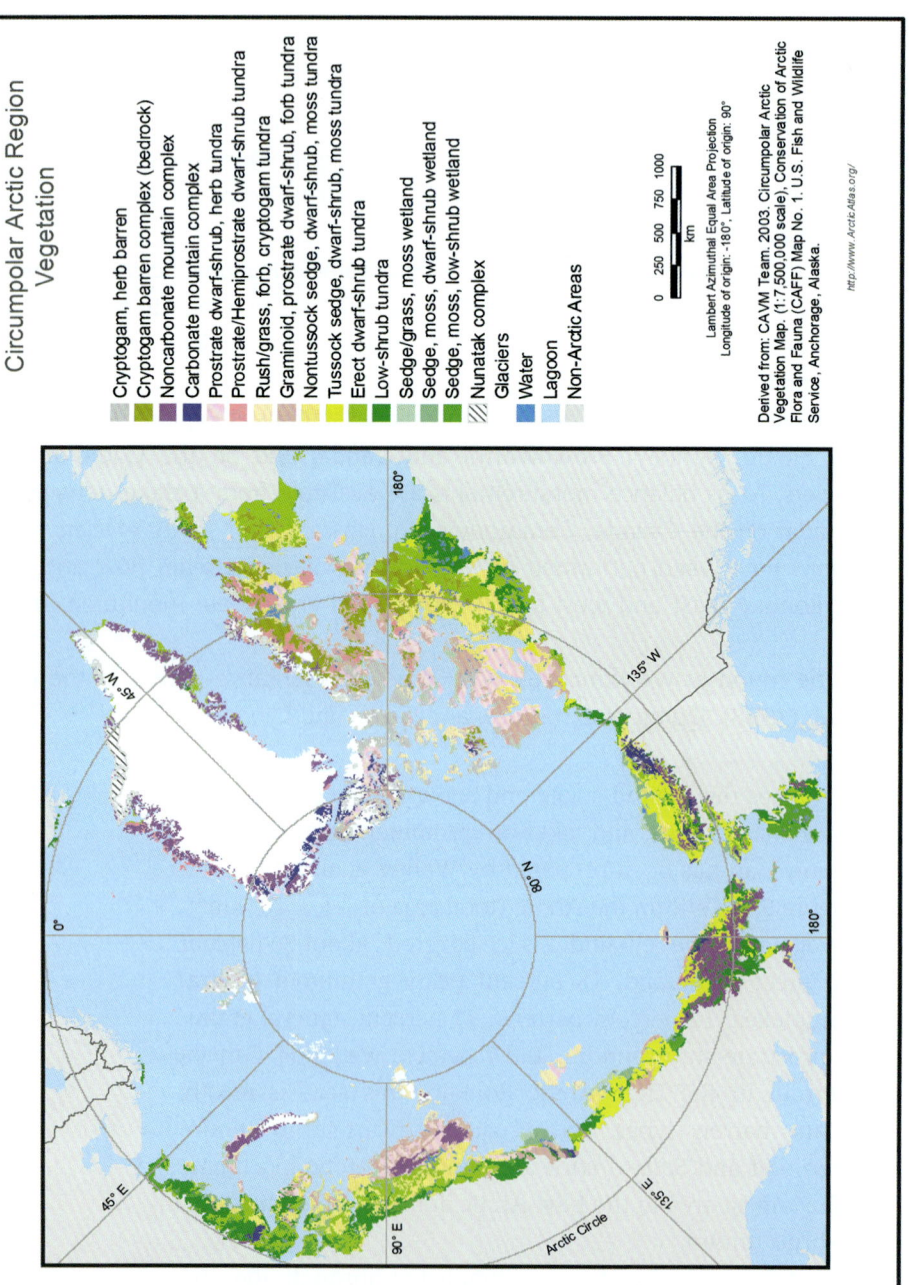

Circumpolar Arctic Region Vegetation

- Cryptogam, herb barren
- Cryptogam barren complex (bedrock)
- Noncarbonate mountain complex
- Carbonate mountain complex
- Prostrate dwarf-shrub, herb tundra
- Prostrate/Hemiprostrate dwarf-shrub tundra
- Rush/grass, forb, cryptogam tundra
- Graminoid, prostrate dwarf-shrub, forb tundra
- Nontussock sedge, dwarf-shrub, moss tundra
- Tussock sedge, dwarf-shrub, moss tundra
- Erect dwarf-shrub tundra
- Low-shrub tundra
- Sedge/grass, moss wetland
- Sedge, moss, dwarf-shrub wetland
- Sedge, moss, low-shrub wetland
- Nunatak complex
- Glaciers
- Water
- Lagoon
- Non-Arctic Areas

0 250 500 750 1000
km

Lambert Azimuthal Equal Area Projection
Longitude of origin: -180°, Latitude of origin: 90°

Derived from: CAVM Team. 2003. Circumpolar Arctic
Vegetation Map. (1:7,500,000 scale), Conservation of Arctic
Flora and Fauna (CAFF) Map No. 1. U.S. Fish and Wildlife
Service. Anchorage, Alaska.

http://www.Arctic.Atlas.org/

Figure 5.4 Circumpolar Arctic vegetation.

Source: CAVM Team. 2003. *Circumpolar Arctic Vegetation Map. Conservation of Arctic Flora and Fauna (CAFF) Map No. 1.* Anchorage, AK: US Fish and Wildlife Service.

Taimyr, Nizhne Kolymskiy, Yakutia, and Disko Island, West Greenland; Kevo, Finland; and Stordalen (Abisko), Sweden; as well as for Signy Island, South Georgia, and Macquarie Island in the southern hemisphere (Table 5.1). The last two sites are in the sub-Antarctic and are dominated by tussock grassland.

The coastal tundra at Barrow has been described in detail by Brown et al. (1980) and Brown (1980). The coastal plain is made up of marine, fluvial, alluvial, and eolian deposits. Monthly temperatures range between 4 °C in July and −26 °C in January. Adjusted annual precipitation (for the effect of gauge undercatch) is 170 mm, of which 63 percent falls as snow (Dingman et al. 1980). The tundra at Barrow, where the pH is acidic, is mainly grasses and sedges, with nineteen lichen taxa. On the calcareous substrate at Prudhoe Bay, *Carex* species are noticeable (Figure 5.6). The vas-

Figure 5.5 Melting ice wedges forming a polygon pattern just off Dalton Highway, Alaska, north of Happy Valley Airport. The road blocked the normal flow of surface water, triggering the melting of the ice wedges, and resulting in deep trenches between the raised polygons. Vegetation on the raised polygon centers is dry tundra dominated by prostrate dwarf shrubs.
Source: Courtesy of Dr. Kevin Schaefer, National Snow and Ice Data Center, University of Colorado, Boulder, CO, August 2014.

cular plant species number only 124 at Barrow and 168 at Prudhoe Bay, but increase to 308 species on the inland coastal plain and 516 species in the foothills north of the Brooks Range (Brown et al. 1980). The corresponding numbers for mosses are 137, 209, 245, and 357 species, respectively. As much as 40 percent of the coastal plain is occupied by lakes. Figure 5.7 illustrates the forb (herb)-rich tundra of the foothills zone.

In the Russian Arctic, three latitudinal zones of tundra are commonly distinguished: Arctic tundra, typical tundra, and southern tundra (Shahgedanova and Kuznetsov 2002). Arctic tundra occupies the southern island of Novaya Zemlya, Vaygach, Novosibirski and Wrangel islands, and the Yamal–Gydan and Taimyr peninsulas, and occurs locally on the banks of the Anabar and Kolyma rivers. The predominant plant cover is a moss layer in which flowering plants and dwarf willow take root. The moss holds moisture and insulates the ground below, allowing ice lenses to persist. Two subtypes are polygonal tundra and spotted tundra; in both of these subtypes, bare ground occupies as much as 50 percent of the surface. The bare spots have diameters of 40–50 cm. There are differences longitudinally with, for example, more *Carex* species to the west of the Indigirka River. In Novaya Zemlya, there are *Deschampsia* tussock associations. On the Yamal-Gydan and Taimyr

Table 5.1 Characteristics of IBP Tundra Biome sites (based on Bliss et al. 1981, appendix)

Place	Location	Plant species		
Truelove lowland	75.3° N, 84.7° W	41% sedge-moss meadow	20% cushion plant lichens and moss on raised beaches	12% granite outcrops dwarf shrub heath and grasses	22% lakes and ponds
Point Barrow	71.3° N, 156.7° W	75% grasses and sedges	Salix spp. on drier sites		
Prudhoe Bay	70.25° N, 148.6° W	Sedges and mosses on low-center ice-wedge polygons.			
Eagle Summit	65.5° N, 145.4° W	Subsite1: sedge-moss Dryas wet meadow	Subsite 2: Dryas-moss mesic meadow	Subsite 3: sparse forb-lichen fellfield	
Agapa, western Taimyr	71.4° N, 88.9° E	On hills: dwarf shrubs, moss, and lichen tundra	On terraces and floodplains: Betual exils shrub tundra, willow stands, and Spagnum bog	Polygonal ground	
Nizhne-Kolymskiy, Yakutia	69.1° N, 170.0° E	Plot 1: above-ground live material – 71% woody plants, 25% monocotyledons	Plot 2: above-ground live material – 65% woody plants, 18% bryophytes and lichens, and 17% monocotyledons		

Location	Coordinates	Site 1	Site 2	Site 3	Site 4
Disko Island	69.25° N, 53.5° W	Site 1: herb slope; closed canopy of bryophytes, *Salix* stands	Site 2: fellfield; scattered mat and cushion species, prostrate *Salix*	Site 3: snow bed; mat-forming plants with ericaceous species	Site 4: crypto-gamic heath; moss and lichens with few vascular plants
Kevo	69.75° N, 27.0° E	Dwarf *Betula* shrub and lichen heath	Ericaceous dwarf shrub; lichens in dry sites, mosses in mesic sites		
Abisko	68.3° N, 18.7° E	Mire	87% bryophytes	*Spagnum* on the hummock slopes	Dwarf shrubs 33%, monocotyledons 20%, forbs 10% cover
Signy Island	60° S, 45° W	200 lichen species and 75 species of mosses; 2 vascular plants in wet sites			
South Georgia	54.9° S, 34.7° W	*Poa* tussock grassland in the coastal zone; *Festuca* tussocks inland; bryophytes and dwarf shrubs			
Macquarie Island	54.5° S, 158.9° E	*Poa* tussock grassland on the coastal slope	Closed herbfield community, on beach terrace and flat upland	Fen (*Juncus*) and bog (bryophytes) on flat areas	Fellfield on most of the plateau, vegetation stripes perpendicular to the wind

Figure 5.6 Hummocky tundra of bioclimate subzone C, near Eureka, Ellesmere Island, Nunavut, Canada. The dominant vascular plants include *Carex rupestris, Ceerastium beeringianum, Dryas integrifolia, Kobresia myosuroides, Salix arctica,* and *Saxifraga tricuspidata.*
Source: Courtesy of D. A. "Skip" Walker, Alaska Geobotany Center, Institute of Arctic Biology and Department of Biology and Wildlife, University of Alaska, Fairbanks.

peninsulas, *Dryas* tundra and *Eriophorum* are found. Wrangel Island is unique in having 400 species of vascular plants due to it not being totally glaciated and owing to its connection to Beringia during glacial periods of lower sea level.

Typical tundra is prominent on the Taimyr peninsula and between the Yana and Kolyma rivers. It is dominated by moss sward that is 5–12 cm high, along with lichens. The herbaceous tier is mainly *Carex* sedge. Other plants are *Eriophorum, Saxifraga,* dwarf *Salix, Dryas,* and *Vaccinium.* In eastern Taimyr, dwarf *Betula* and *Salix* species are less than 20 cm tall (Figure 5.8).

Southern tundra is characterized by bushes of *Betula* (birch), *Salix* (willow), and *Alnus* (alder) up to 50 cm high (Figure 5.9). *Larix* (larch) trees occur along rivers. Bryophytes form a continuous cover with meadow associations in river valleys. Arid northeast Asia is mainly steppe tundra (Box 5.2). West Siberia has widespread *Betula nana* with mosses and *Carex.* In the Yamal–Gydan–Taimy region, *Alnus* is admixed with dwarf birch and willow. In the East Siberian zone, dwarf shrub and herb dwarf shrub tundras dominate, with lichens being more prevalent than in the west. Tussock tundra is dominant in the Chukchi peninsula, with *Eriophorum* and *Carex* being the predominant species.

Mires (acidic peat bogs and neutral fens) are common elements of tundra landscapes. They occupy 560,000 km^2 in the typical and southern tundras (Bliss and Matyeeva 1992), occurring on coastal plains, in lowlands, and along rivers. In the West Siberian lowlands, wetlands nay occupy as much as 50 percent of the surface (Koronkevich 2002). There are three types: homogeneous mires, flat-topped hummocky mires with wet depressions, and polygonal mires. In hummocky mires, peat hillocks form from the moss cover. Polygonal mires have polygons, 5–10 m across, that are flat or concave in the center with slightly elevated rims. Mires and lakes are numerous on the Yamal–Gydan and western Taimyr peninsulas. Bring et al. (2016) note that as a result of underlying permafrost, the water storage in tundra soils is limited to the seasonally thawed active layer. Peak discharge is in spring and summer.

Shrub expansion has occurred throughout the Arctic tundra during the past fifty years as a result of higher soil and air temperatures (Tape et al. 2006). This has given rise to Arctic-wide "greening' that is apparent in the more than thirty years of satellite records of the Normalized Difference Vegetation Index (NDVI) (Xu et al. 2013). Zhu et al. (2016) determined that the positive trends in the Leaf Area Index (LAI) in the northern high latitudes and the Tibetan Plateau are attributable to rising temperatures, which enhances photosynthesis and lengthens the growing season. This pattern stands in contrast to most of the global trend, which is driven primarily by CO_2 fertilization. However, Phoenix and Bjerke (2016) note that greening can be viewed as a steady and gradual process, whereas recent browning of some tundra areas may arise from a large number of drivers, occurring as either changes in the prevailing trend or often as biotic or weather events, reducing greenness temporarily in different parts of the landscape, at different times. Events that can promote browning include extreme winter warming, which leads to mid-winter bud burst and loss of freeze tolerance. Other drivers include warm autumns, which reduce winter hardening; rain-on-snow, which results in plant ice encasement; lack of snow cover combined with high irradiance, leading to frost; drought; and snow falling on unfrozen ground, which enhances snow mold growth and respiratory losses from subnivean plants. In addition, outbreaks of defoliating insects and rust fungi may cause browning.

Figure 5.7 Forb-rich tundra at the southern edge of Arctic bioclimate subzone D, on nonacidic loess soils, near Sagwon, Arctic Foothills, northern Alaska. Mean July temperature is approximately 12 °C. Common vascular-plant species include *Anemone parviflora, Arctagrostis latifolia, Arctostaphylos rubra, Astragalus umbellatus, Cardamine hyperborea, Carex bigelowii, C. membranacea, C. scirpoidea, C. vaginata, Cassiope tetragona, Castilleja caudata, Dryas integrifolia, Eriophorum angustifolium* spp. *triste, Equisetum arvense, E. variegatum, E. scirpoidea, Kobresia myosuroides, Lupinus arcticus, Minuartia arctica, Oxytropis maydelliana, Papaver macounii, Parrya nudicaulis, Pedicularis capitata, P. kanei, P. langsdorfii, Polygonum bistorta* spp. *plumosum, P. viviparum, Pyrola grandiflora, Rhododendron lapponicum, Salix glauca, S. lanata* spp. *richardsonii, S. arctica, S. reticulata, Saussurea angustifolia, Saxifraga hieracifolia, Senecio atropurpureus, S. resedifolius, Stellaria longipes,* and *Tofieldia pusilla.*
Source: Courtesy of D. A. "Skip" Walker, Alaska Geobotany Center, Institute of Arctic Biology and Department of Biology and Wildlife, University of Alaska, Fairbanks.

5.3 Permafrost

Permanently frozen ground, or permafrost, is ground that remains frozen for at least two summers and the intervening winter. It is defined solely by the ground

Figure 5.8 Tundra at the northern edge of Arctic bioclimate subzone E, Sagwon vicinity, Arctic Foothills, northern Alaska. Mean July temperature is approximately 12 °C. This tundra is typical of moist acidic soils. The dominant vascular-plant species in the foreground are *Betula nana, Salix pulchra, Carex bigelowii, Arctous alpina, and Rhododendron decumbens.* The vegetation in the lighter-toned areas in the background is growing on nonacidic soils.
Source: Courtesy of D. A. "Skip" Walker, Alaska Geobotany Center, Institute of Arctic Biology and Department of Biology and Wildlife, University of Alaska, Fairbanks.

temperature. It occupies about 24 percent (25.5 million km^2) of the northern hemisphere, although the actual area underlain by permafrost is approximately 12–18 percent of the exposed land area: In the permafrost regions, the frozen ground becomes discontinuous equatorward. Permafrost is divided into four classes based on the percentage of the ground that it underlays: continuous, representing more than 90 percent of the area; discontinuous, 50–90 percent; sporadic, 10–50 percent; and isolated, less than 10 percent (Figure 5.10). Areas accounted for by the first three categories are 9.6–10.7 million km^2, 2.2–4.0 million km^2, and 0.4–2.0 million km^2, respectively (Heginbottom et al. 2013; Zhang et al. 2000).

In continental regions of the northern hemisphere, the southern limit of continuous (discontinuous) permafrost is approximated by the position of the mean annual air temperature (MAAT) isotherm of −8 °C (−1 °C). In the continuous zone, thicknesses may range up to 700 m in northern Alaska and 1,000 m in north-central Siberia. In much of northern Canada, however, thick permafrost is lacking due to the past cover of the Laurentide ice sheet. Permafrost is also distinguished as "warm" or "cold" according to whether the temperature is greater than or less than −1 °C.

Permafrost temperatures (at 10–20 m depth) in Yakutia have been mapped by Beer et al. (2013) based on a map of landscapes and permafrost conditions. Figure 5.11a shows the large 11 °C north–south range of permafrost temperatures in Yakutia.

Mountain permafrost is widespread above about 4,500 m at 20° N, with a linear decrease in altitude to 1,000 m at 60° N (Cheng and Dramis 1992). Much of the northern part of the Tibetan plateau is underlain by continuous permafrost. The lower altitudinal limit is about 4,200 m, increasing to 4,800 m in the south (French 2007). The majority of the permafrost on the Qinghai–Tibet plateau is less than 100 m thick, with substantial areas less than 50 m thick. Near the southernmost extent of continuous permafrost in North America, the thickness averages about 15–30 m.

In the southern hemisphere, permafrost is present at high elevations in the southern Andes, on the sub-Antarctic islands, and in Antarctica. Bockheim (1995) published a map of permafrost in Antarctica indicating that permafrost exists throughout the ice-free areas of the Antarctic Peninsula and the Dry Valleys – an estimated 49,000 km^2 – with subglacial permafrost occurring only in areas where the ice sheet is thin enough to maintain frozen subglacial conditions (Figure 5.12). The ice-free area of Antarctica contains continuous permafrost, although permafrost soils account for only 0.35 percent of the Antarctic continent (Campbell and Claridge 2009).

Subsea permafrost can be ice-bonded (cemented by ice) or ice-bearing (containing some ice). Freezing-point depression by saline pore solution can result in unfrozen sediment. Subsea permafrost is widespread on the continental shelves of the Arctic Ocean as a result of their exposure due to the lower sea levels of the Pleistocene glaciations. Recent seismic reflection studies in the Beaufort Sea indicate that the seaward extent of continuous ice-bearing permafrost is within 37 km of the modern shoreline at water depths less than 25 m (Brothers et al. 2016). For the Eurasian Arctic, Romanovskii et al. (1998) have mapped subsea

Figure 5.9 Shrub tundra of southern part of Arctic bioclimate subzone E near treeline, southwest-facing slope, near Council, Seward Peninsula, Alaska. Dominant shrubs are *Betula glandulosa*, *Salix glauca*, *Pentaphylloides floribunda*, *Vaccinium uligiosum*, and *Salix reticulata*, mixed with the grass *Festuca altaica* and the moss *Hylocomium splendens*. Source: Courtesy of D. A. "Skip" Walker, Alaska Geobotany Center, Institute of Arctic Biology and Department of Biology and Wildlife, University of Alaska, Fairbanks.

Box 5.2 **Steppe Tundra**

The concept of steppe tundra originated in the USSR in the 1930s (Yurtsev 1982). This term refers to an assemblage that includes both steppe (cold dry grassland) and Arcto-alpine tundra plants. Xerophytes such as *Artemesia* are also abundant. Areas of relict steppe plants occur in Yakutia as well as steppe tundra. Yurtsev (1982, 159–70) illustrates these vegetation types and their distribution. Mean annual temperatures in this region are −12 to −15 °C, July means are 12–15 °C, and annual precipitation is 150–250 mm.

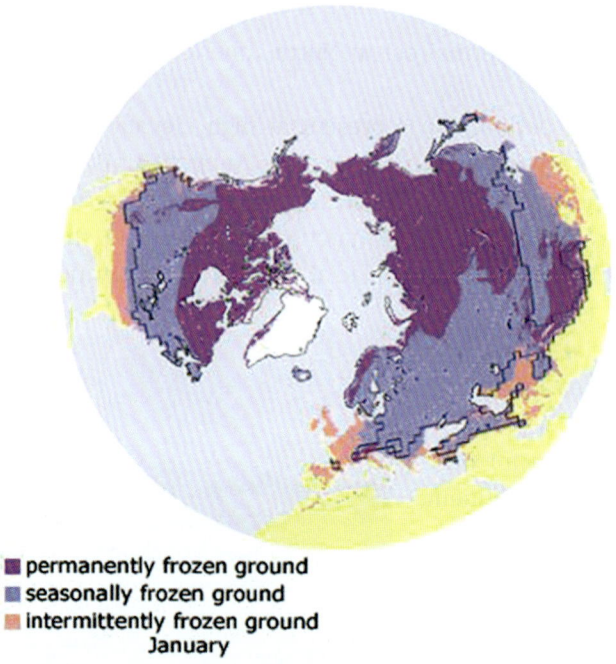

■ permanently frozen ground
■ seasonally frozen ground
■ intermittently frozen ground
January

Legend for EASE-Grid Permafrost and Ground Ice Map

Permafrost Extent (percent of area)	Ground Ice Content (visible ice in the upper 10-20 m of the ground; percent by volume				
	Lowlands, highlands, and intra-and intermontane depressions characterized by thick overburden cover (>5-10 m)			Mountains, highlands, ridges, and plateaus characterized by thin overburden cover (<5-10 m) and exposed bedrock)	
	High (>20%)	Medium (10–20%)	Low (0–10%)	High to medium (>10%)	Low (0–10%)
Continuous (90–100%)	Ch	Cm	Cl	Ch	Cl
Discontinuous (50–90%)	Dh	Dm	Dl	Dh	Dl
Sporadic (10–50%)	Sh	Sm	Sl	Sh	Sl
Isolated Patches (0–10%)	Ih	Im	Il	Ih	Il

Variations in the extent of permafrost are shown by the different colors; variations in the amount of ground ice are shown by the different intensities of color. Letter codes assist in determining to which basic permafrost and ground ice class any particular unit belongs. Letter codes are defined in the documentation that accompanies the data files.

Ice caps and glaciers

Figure 5.10 (a) The circumpolar distribution of permafrost and ground ice conditions in the northern hemisphere, (b) legend. Based on Brown et al. (1998, updated 2001).
Source: Courtesy of National Snow and Ice Data Center, Boulder, CO.

permafrost on the Laptev Sea shelf to the following zones:

1. Almost continuous ice-bonded relic permafrost to the −60 m and −70 m isobaths
2. Widespread discontinuous permafrost to the −100 m isobath
3. Open linear taliks (unfrozen ground) above active faults with high geothermal heat flux

From the modern seashore to the −15 m and −20 m isobaths, both deep supra-permafrost sub-sea taliks and bodies of ice-rich permafrost exist. The maximum thickness of relic permafrost (200–600 m) is a characteristic of the lithospheric blocks with low heat flux that are predicted to occur within the area from the north coast of Kotel'nyi Island to the −45 m isobath.

In the western part of the East Siberian Sea, the greater part of the shelf – out to the 50–60 m isobath – is occupied by continuous ice-bearing and ice-bonded relic off-shore permafrost with thickness ranging

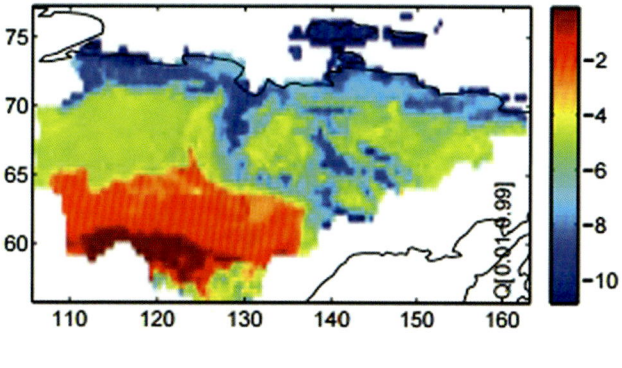

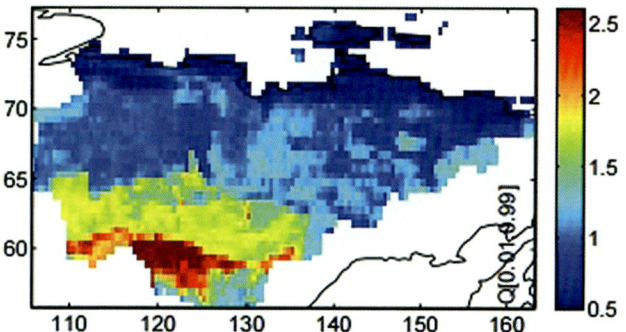

Figure 5.11 Distribution of (a) permafrost temperatures and (b) active layer thickness in Yakutia, eastern Siberia.
Source: Beer et al. 2013, 307, figures 1a and 3a. Courtesy of Copernicus Publication.

from 300 to 600 m (Romanovskii et al. 2005). The thickest values are around the Novosibirskie Islands. In the peripheral part of the shelf, degradation of the massifs and islands of ice-bearing relic offshore permafrost is taking place.

The upper layer of the ground in permafrost terrain is subject to seasonal thaw, forming what is termed an "active layer." Active layer thickness (ALT) may range from approximately 10 cm to 4 m depending on the energy regime and environmental conditions (i.e., snow depth, surface water, vegetation cover, and albedo) as well as the soil thermal properties. Figure 5.11b shows the range of ALT in Yakutia, which varies from 0.5 m in the Arctic to 2.5 m in the sub-Arctic. As shown by Streletsky et al. (2008), in the Alaska coastal plain the ALT ranges from 40 cm on moist nonacidic tundra to 63 cm on wet graminoid tundra. The presence of peat is especially important due to its low thermal conduction relative to mineral soil and its moisture content (represented by ice content). This property necessitates more energy to supply latent heat of fusion. As a consequence of these conditions, ALT is negatively correlated with peat thickness

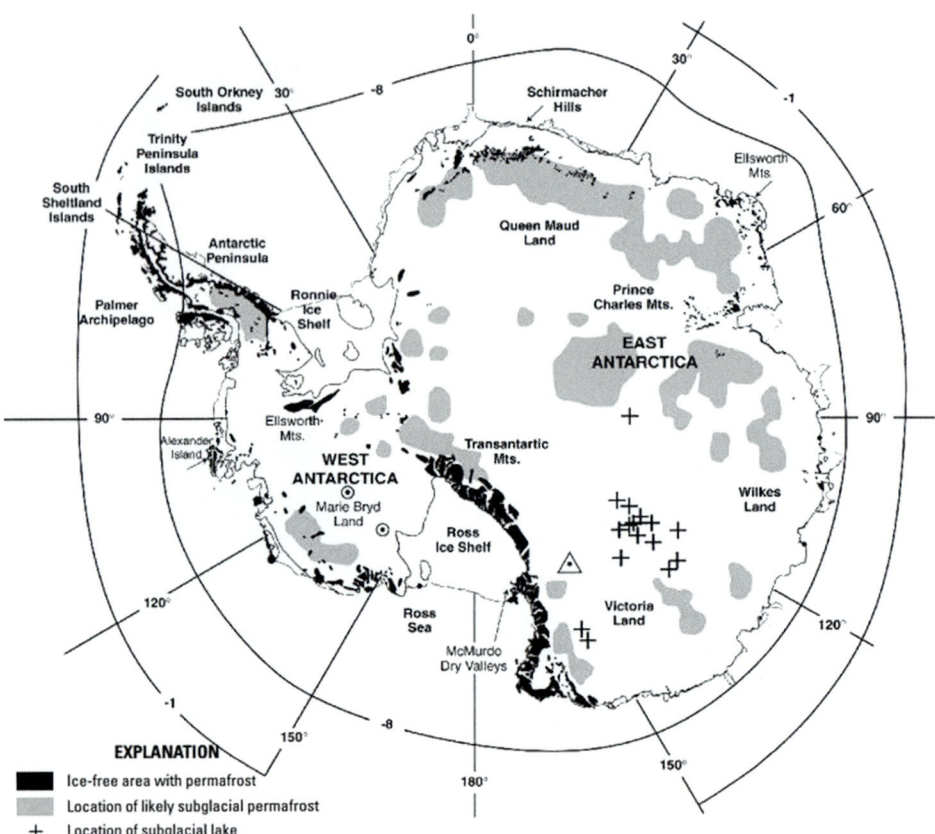

Figure 5.12 Permafrost distribution in the Antarctic (modified from Bockheim 1995). Black areas are ice-free areas with permafrost, shaded areas are locations of likely subglacial permafrost, and (+) indicates locations of subglacial lakes.
Source: Heginbottom et al. 2013, figure 4A.

(Atchley et al. 2016). Snow depth affects winter ground temperature, due to its insulating properties, thereby raising winter soil temperature, which leads in turn to warmer soils in summer and increased ALT (Zhang 2005).

5.3.1 Ground Ice

There are four major types of ground ice: pore ice, segregated ice, intrusive ice, and vein (or wedge) ice (Mackay 1972). Pore ice provides the bonding that holds soil grains together. Segregated ice forms layers or lenses in the soil that vary in thickness from a few millimeters to tens of meters; it is created by the migration and subsequent freezing of pore water. Intrusive ice forms by the intrusion of water, usually under pressure, into the frozen zone. It occurs as sill ice and dome-shaped pingos (discussed later). Vein ice forms when water drains into fissures formed by thermal contraction of the ground surface and freezes. Repeated events can lead to the growth of ice wedges 4 m or more deep. These veins are thickest and deepest in central Siberia, where there are high-ice-content sediments known as "yedoma" (Vasil'chuk and Vasil'chuk, 1997). Figure 5.13 maps the ground ice content in Siberia. Areas in central Siberia have values that are more than 60 percent by volume.

Samoylov Island, which is centrally located within the Lena River delta at 72° N, 126° E, offers a helpful case study of ground ice. Boike et al. (2013) describe in detail the characteristics of the island's meteorological parameters (temperature, radiation, and snow cover), soil temperature, soil moisture, vegetation, and permafrost. The annual air temperature ranges between −33 °C in February and 10 °C in July. The land surface characteristics have been described using high-resolution aerial images in combination with data from ground-based observations. The polygonal tundra on Samoylov Island lies within ice-rich permafrost terrain that is characterized by ubiquitous water bodies. The mean annual permafrost temperature is −8.6 °C at 10.7 m depth. The deeper permafrost temperatures have increased between 0.3 and 1.3 °C over the last five years. However, no clear increases in air and active layer temperatures have been detected since 1998, although winter air temperatures during recent years have not been as low as in earlier years.

5.3.2 Periglacial Regions

Periglacial regions where freeze–thaw processes are active can be delimited by the +3 °C isotherm of mean annual air temperature; this line distinguishes areas of frost-induced solifluction (the collective name for gradual mass-wasting slope processes) from areas of patterned ground (French 2007). Within the periglacial zone, French uses the −2 °C isotherm to create a subdivision between environments in which frost action does or does not dominate. Periglacial processes give rise to the movement of soil particles and unconsolidated sediments that result in solifluction lobes, stone polygons, soil stripes, and turf hummocks. The earliest substantial research on these features was carried out by the German geographer, Carl Troll; his 1944 monograph was later translated into English (Troll 1958).

Permafrost terrain exhibits a wide range of mesoscale to microscale surface landforms. Mesoscale forms include thermal contraction-crack polygonal ground (patterned ground) that ranges from 15 to 40 m across in unconsolidated

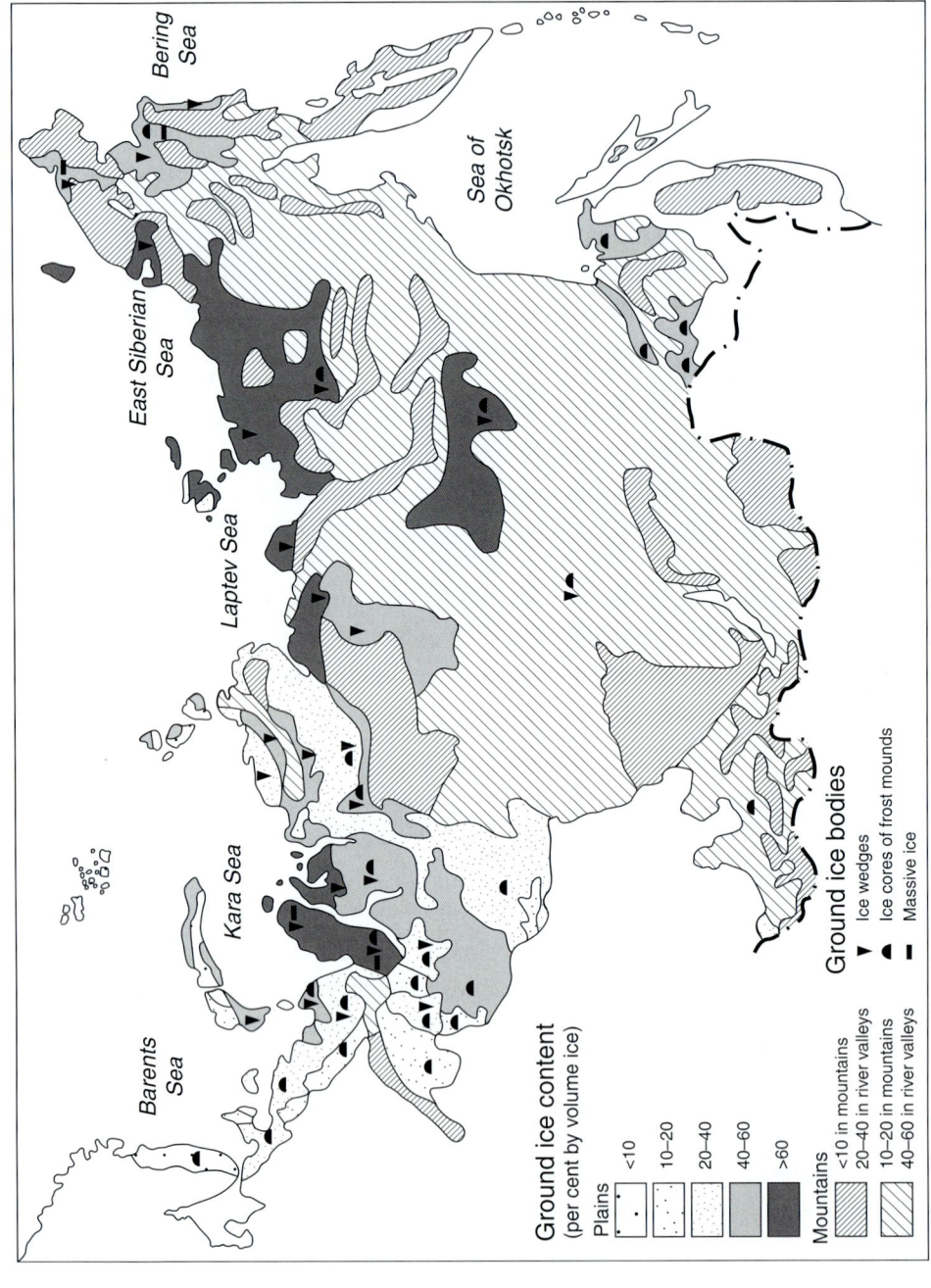

Ground ice content
(per cent by volume ice)

Plains

- <10
- 10–20
- 20–40
- 40–60
- >60

Mountains

- <10 in mountains
- 20–40 in river valleys
- 10–20 in mountains
- 40–60 in river valleys

Ground ice bodies

- ◄ Ice wedges
- ◖ Ice cores of frost mounds
- ▬ Massive ice

Barents Sea

Kara Sea

Bering Sea

East Siberian Sea

Laptev Sea

Sea of Okhotsk

Figure 5.13 The distribution of ground ice content in Siberia.
Source: Tumel 2002, 156, figure 6.2.

sediments and from 5 to 15 m in bedrock. The cracks may fill with mineral soil or water that freezes into ice wedges, which in turn gradually force the crack to widen.

Several types of perennial frost mound occur in areas of continuous and discontinuous permafrost (Mackay 1986). Palsa mounds form in peaty organic soils in discontinuous and sporadic permafrost. They are typically 0.5–2 m high and generally 5–25 m in diameter, often forming in groups. Sometimes they form elongated ridges (Figure 5.14). Winter freezing locally penetrates deeper than in the adjacent ground, forming an ice lens if the ground is saturated. The freezing is assisted by thinner snow cover on the mounds. Palsas comprise layers of segregated ice, representing annual increments, and peat or mineral soil. Overlying soil layers are subject to frost heaving. Palsas depend on a water source, usually groundwater, to form the ice layers; hence they are common in bogs.

Figure 5.14 Palsa mounds in tundra of the coastal plain, Alaska, 200 m off the Dalton Highway just south of Pump Station 3. The wet conditions cause the ice blisters to heave up the turf, which eventually causes the turf to crack, triggering the palsas to melt.
Source: Courtesy of Dr. Kevin Schaefer, National Snow and Ice Data Center, University of Colorado, Boulder, CO, August 2014.

Earth hummocks (*thufur*, an Icelandic name) are a dome-shaped, non-sorted net or circle feature widely distributed in the Arctic and sub-Arctic. As described by Tarnocai and Zoltai (1978), they have an average diameter of 80–160 m and an average height of 40–60 cm. They form from materials having 60–99 percent clay and silt content and a high ice concentration in the near-surface permafrost. Cryogenic in origin, their development is controlled by the texture, temperature, and moisture of the soil.

Pingos are typically isolated features in the continuous permafrost zone. These mounds of earth-covered ice are typically about 5m in height, but can reach nearly 50 m, with a diameter of up to 500 m (Figure 5.15).

Grosse and Jones (2011) identified 6,000 pingos in northern Asia and estimated that there may be 11,000 globally. Tuktoyaktuk in the Mackenzie Delta has 1,350 pingos, more than 60 percent of which occur in continuous permafrost with high ice content and thick sediments. Most pingos are located in regions with mean annual ground temperatures between -3 and $-11\ ^\circ$C.

Müller (1959) studied pingos in East Greenland and in the Mackenzie Delta. Mackay (1979) showed that most pingos in the Mackenzie Delta have grown in residual ponds left behind by rapid lake drainage through erosion of ice-wedge

Figure 5.15 Ybyuk pingo, Tuktoyaktuk Peninsula, Northwest Territories, Canada. This is one of the largest pingos in the world; it is 49 m in height and 300 m in diameter. The vegetation on the pingo is more similar to that found in alpine areas than on the wet coastal plain and the drained thaw lake where the pingo occurs. The willows (*Salix pulchra*) in the foreground and on the pingo are about 1 m tall.
Source: Courtesy of D. A. "Skip" Walker, Alaska Geobotany Center, Institute of Arctic Biology and Department of Biology and Wildlife, University of Alaska, Fairbanks.

polygon systems. Permafrost aggradation in saturated lake-bottom sediments creates the high pore water pressures necessary for these hydrostatic (closed system) pingos' growth. As permafrost rises to the drained body's former floor, pore water is expelled in front of the rising permafrost; the resulting pressure causes the frozen ground to rise and an ice core to form.

Hydraulic (open system) pingos are relatively common in East Greenland and Svalbard, and also occur in central Alaska and the central Yukon (Heginbottom et al. 2013). These pingos result from groundwater flowing from an outside source (i.e., subpermafrost or intrapermafrost aquifers). Hydrostatic pressure, associated with the surrounding mountains, initiates the formation of the ice core as water is pushed up, forming a hydrolaccolith, which subsequently freezes (Müller 1959). Many of those in East Greenland are conical in shape with steep sides. They occur especially in the gravel beds of wide glacial valleys, but also on slopes of up to 8°. The ice bodies are covered by a 3- to 20-m-thick layer of mineral matter composed of bedrock or frozen gravel, sand, silt and clay. In some pingos, salt-rich water of local meteoric origin and gas rise up through the ice body.

Soil movement due to freeze–thaw cycles gives rise to a phenomenon known as cryoturbation. It involves frost heave during autumnal freeze-up of the active layer and thaw settlement in spring. The vertical displacement is typically 2–5 cm. Fine and coarse particles may be sorted by uplift due to frost heave, by the preferential migration of fine particles ahead of a freezing plane, and by mechanical sorting when larger particles migrate downward under gravity. Patterned ground may develop as circles or polygons on level ground (Figure 5.16), and as elongated stone stripes on slopes. These features have a typical dimension of 1–2 m, with a relief of 10–30 cm (Figure 5.17).

Another type of permafrost-related landscape is thermokarst. It develops when the melting of ground ice leads to the ground subsiding or slumping on slopes. Subsidence leads to the formation of circular or oval hollows, known as alases in

Siberia, with lakes in their bottom. These hollows may be 3–40 m deep, with a diameter of 100 m.

Another large-scale flow feature is the rock glacier – a tongue-like or lobate body of angular boulders, generally occurring in mountainous terrain; it usually has ridges and furrows on its surface, and has a steep front at the angle of repose (Barsch 1996; Martin and Whalley 1987; Washburn 1979; Whallet and Azizi 2003). The rock glacier may be either periglacial, talus-derived, occurring below talus slopes, or glacier-derived, found below mountain glaciers or below areas that alpine glaciers formerly occupied. The periglacial type requires a mean annual air temperature of $-1.5\,°C$ or less. The ice component, which typically accounts for 60 percent of the rock glacier's volume, allows for deformation of the ice-rich matrix, providing a mechanism for the rock glacier's movement downslope. These features are typically 200–800 m long and 20–100 m thick, with flow rates of approximately $0.1\ \mathrm{m\ a^{-1}}$.

A related feature is the protalus lobe of nonglacial origin. In this feature, the ice may originate as snowbanks and then become buried by debris from cliffs above that slides down the bank. Protalus ramparts are generally attributed to debris accumulating at the front of snowbanks or small "glacierettes." They are 100–300 m long and 2–8 m high. Active ones are found on Southampton Island (Bird 1967, 230).

Figure 5.16 Stone circles.

Figure 5.17 Patterned ground near Eureka, Ellesmere Island, June 1963.
Source: R. G. Barry.

5.3.3 Recent Permafrost Changes

Permafrost environments are undergoing rapid changes as a result of global warming. Ground temperatures are rising in many Arctic regions in response to changes in air temperature and changes in snow cover depth and duration. At

West Dock, Alaska, measurements show a significant long-term warming trend in mean annual temperatures at all levels: air, ground surface, and permafrost at 20 m depth. The linear trend in air and ground surface temperatures was 3 °C between 1987 and 2007, and the increase in permafrost temperature at 20 m depth was 2 °C (Romanovsky et al. 2008). Ground temperature measurements between the 1950s and 1980s show that permafrost in Alaska warmed between 0.3 and 0.6 °C over this period (Romanovsky et al. 2010; Smith et al. 2010). Temperatures in cold permafrost show the greatest response.

In the central Mackenzie Valley, near Norman Wells, permafrost is up to 50 m thick and has a temperature of about −1 °C. Smith (2011) reported warming of 0.33 °C decade^{-1} during 1984–2008 at a depth of 10 m. "Warm" permafrost is especially vulnerable to disturbance where it is currently protected by the eco-system but could not form under the prevailing climatic conditions. Studies in the Kenai Peninsula lowlands of south-central Alaska using aerial photographs from 1950 indicate that residual permafrost plateaus covered 920 ha across portions of four wetland complexes when the photos were taken. However, between 1950 and circa 2010, the permafrost plateau extent decreased by 60 percent to 370 ha, with degradation of lateral features along the plateau margins accounting for nearly all of the area reduction (Jones et al. 2016). The mean annual air temperature in the area was about 1.5 °C and the permafrost was close to the point of thawing, having temperatures in the range of −0.04 to −0.08 °C. The depth to the permafrost table averaged 1.48 m, but was as shallow as 0.5 m. Measured permafrost thickness ranged from 0.33 to more than 6.90 m, with seasonal thawing to 0.45 m. Apart from the effects of increasing air temperatures (+0.4 °C decade^{-1}) on permafrost degradation, wildfires that remove the pro-tective forest cover are an important factor in the fate of the permafrost in this area.

Climate warming leads to permafrost degradation and the formation of thermo-karst landforms. For example, north of the Brooks Range, Alaska, volumetric ice content ranges from 40 to 90 percent. Thermokarst lakes are widely distributed in areas of Arctic tundra (see Section 5.4).

Another change is occurring at Arctic coasts where ice-rich sediments are undergoing rapid wave erosion, leading to coastline retreat (Box 5.3). This phenomenon is apparent in much of northern and western Alaska and northern Siberia. Observations on Vize Island in the Kara Sea indicated a shoreline retreat of 1–1.5 m a^{-1} in the 1950s. This accelerated to approximately 10 m a^{-1} in the 2010s (Gertcyk 2016). Gorokobich and Leiserowiz (2012) used 112 aerial photo-graphs from 1950 to 2003 to monitor shoreline erosion in the vicinity of Kotzebue Sound, Alaska. They found mean erosion rates of 0.12–0.08 m a^{-1} in the region from 1950 to 2003.

Box 5.3 **Coastal Recession in Alaska**

Shishmaref, Alaska, has become a poster child for the impacts of global warming. This city of about 560 Inupiat residents is located on a barrier island 8 km off the west coast of Alaska, just north of Bering Strait. Rising air and sea temperatures are decreasing the sea ice that formerly protected the coast from storm waves and tides. This factor, together with thawing of the permafrost in the cliff sediments, has resulted in shoreline retreat rates of 3 m annually. Houses have toppled into the sea, and the water system and infrastructure are at risk. In a 2016 poll, a majority of residents voted to relocate inland, but a suitable site and necessary funding have not yet been identified. A similar situation is believed to be affecting some thirty indigenous settlements in Alaska.

Lantuit et al. (2012) describe a geomorphological classification scheme for many coasts in the Arctic (101,447 km of coastline in 1,315 segments). The cryolithological characteristics were obtained from the northern hemisphere permafrost map developed by Brown et al. (1998). The total length of the coastline affected by the presence of permafrost in the northern hemisphere is 407,680 km. Coastal retreat rates were derived from literature reports based on remote sensing analyses from the second half of the twentieth century. The average rate of erosion for arctic coasts is $0.57 \mathrm{~m~a}^{-1}$, but reaches $0.87 \mathrm{~m~a}^{-1}$ in the East Siberian Sea and $1.1 \mathrm{~m~a}^{-1}$ in the Beaufort Sea. Rates greater than $3 \mathrm{~m~a}^{-1}$ were identified mostly in the Laptev Sea, and in the East Siberian and Beaufort seas.

5.4 Lakes

Lakes play a significant role in the surface energy and water balances. They absorb light energy that penetrates into the water, converting it into heat energy and thereby raising the water temperature. The light penetration is a result of the low albedo of the water, which typically reflects only 0.06–0.10 of the incident solar radiation. Lakes in the Arctic/sub-Arctic usually freeze for a large part of the year. Lake ice data were originally collected by ground observations, with some records beginning in the mid-nineteenth century (Magnuson et al. 2000). More recently satellite remote sensing has been employed, partly because the cost of ground observations led to the cessation of those programs during the 1980s. (Duguay et al. 2015).

Both optical and passive microwave data have been employed to detect lake freeze-up and ice-out dates (lake ice phenology). Using QuikSCAT data for 2000–2006, Howell et al. (2009) determined freeze-up and breakup dates for

Great Slave Lake (61.7° N) and Great Bear Lake (66° N). Freeze onset was on JD 330 and 331, respectively; melt onset was on JD 123 and 139, respectively; and the water was clear of ice on JD 164 and 191, respectively.

Recently, Kang et al. (2014) showed that monthly ice thickness could be derived from AMSR-E 18.7-GHz vertical polarization data for Great Bear Lake and Great Slave Lake from January to April. Ice thickness in the High Arctic typically approaches a seasonal maximum of 2 m. In western Siberia, a steep increase is noted, from 100 cm at 65° N to 200 cm at 75° N (Gronskaya 2000). By comparison, in the Russian Far East and eastern Siberia, ice thickness increases from 100 cm at 45° N to 180 cm at 65° N, and in European Russia, it increases from 40 cm at 40° N to 60 cm at 65° N.

Verpoorter et al. (2014) used Landsat-7 Enhanced Thematic Mapper Plus (ETM+) data to estimate that there are approximately 117 million lakes worldwide, with a combined surface area of 5 million km^2, which is 3.7 percent of the Earth's nonglaciated land area. In the Mackenzie River basin, lakes are estimated to cover 11 percent of the surface (Rouse et al. 2008b). In the North American Arctic and sub-Arctic, in addition to Great Bear Lake, Great Slave Lake, and Lake Hazen, there are countless small shallow lakes on the Arctic coastal plain, where the areal coverage can reach 40–50 percent. The average depths of Great Slave Lake and Great Bear Lake are 88 and 76 m, respectively, and their surface areas are 28,500 and 31,100 km^2, respectively (Rouse et al. 2008b). Lake Hazen has the largest volume among lakes north of the Arctic Circle and has the third largest surface area, 538 km^2 (after Lake Taymyr, Russia, and Lake Inari, Finland) (Duguay et al. 2015; Jeffries et al. 2012). Lake Taymyr has a surface area of 5,500 km^2 and a mean depth of only 10 m.

Fresh water is unusual in having its maximum density at a temperature of 3.98 °C. Hence, as it is heated in summer, less dense water accumulates at the surface; conversely, as it cools in winter, a shallow, lighter layer forms at the surface until freezing occurs. Thus, lakes are typically stratified. In summer, thermally stratified lakes are warmer at the top and cooler at the bottom. The warm surface layer is called the epilimnion, and the cooler bottom layer the hypolimnion. The zone of rapid temperature decrease in the water column is called the thermocline. The stratification breaks down in the autumn and a uniform temperature profile develops.

An assessment of the role of lakes in the energy and water balance of the central Mackenzie River valley was provided by Rouse et al. (2005, 2008a, 2008b). The evaporation season for upland and small, medium, and large lakes lasts for nineteen, twenty-two, twenty-four, and thirty weeks, respectively. Net radiation is substantially greater over lake-dominated surfaces. Table 5.2 shows the energy balance components for the entire region, and for the uplands that occupy 55 percent of the region, in early and late season. On Great Slave Lake,

Table 5.2 Energy balance components (MJ m^{-2}) for uplands and the entire region in the two "seasons" (from Rouse et al. 2005, 303, table 7)

Early season	Rn	G	LE	H
Uplands	933	57	509	367
Entire region	973	286	433	254
Late season				
Uplands	161	−57	51	167
Entire region	205	−278	305	178

Blanken et al. (2000) determined the amount of evapotranspiration based on eddy flux measurements. By extrapolation, they determined that evaporation for the entire ice-free periods totaled 386 ± 127 mm and 485 ± 144 mm in 1997 and 1998–1999, respectively.

Thermokarst lakes are widespread in the ice-rich continuous permafrost regions of northern Alaska and northeast Siberia and Sakha–Yakutia. Farquharson et al. (2016) studied a 300 km^2 area of Alaska and reported that 63 percent of the area is occupied by thermokarst landforms, though the distribution of thermokarst land-forms and overall landscape complexity vary markedly with surficial geology. Areas underlain by ice-rich marine silt are the most affected by thermokarst (97 percent of the total area), whereas areas underlain by glacial drift are the least affected (14 percent). Drained thermokarst-lake basins are the most widespread thermokarst landforms, covering 33 percent of the entire study region, with greater prevalence in areas of marine silt (48 percent coverage), marine sand (47 percent), and eolian silt (34 percent). Thermokarst-lakes are the second most common thermokarst landform, covering 16 percent of the study region, with highest coverage in the areas underlain by marine silt (39 percent coverage). Alas valleys are widespread in areas of eolian silt (14 percent), which are located in gradually sloping uplands. Areas of marine silt have been particularly vulnerable to thaw in the past because of their ice-rich nature and low-gradient topography, which facilitates the repeated development of thermokarst-lakes. Ice-rich eolian, upland terrain (yedoma) will be particularly susceptible to thawing in the future because it still contains massive concentrations of ground ice in the form of syngenetic ice wedges that have remained largely intact since the Pleistocene.

5.4.1 Proglacial Lakes

Proglacial lakes are numerous in the Himalaya mountains of Nepal and Bhutan. They are forming and expanding currently as a result of rapid glacier retreat –

Figure 5.18 Dudh Pokhari glacial lake in the Hinku Valley, Makalu Barun National Park and Buffer Zone, 2012.
Source: Byers et al. 2013, 120, figure 2.

changes that are discussed later in this chapter. Such lakes are dammed by loose boulders and soil that often contains ice (Byers et al. 2013). According to Bajra-charya et al. (2007), twenty-four new glacial lakes have formed and thirty-four major lakes have grown substantially during the past several decades in the Mt. Everest and Makalu Barun National Parks (Figure 5.18). Twelve lakes in the Hongu valley between 4,400 and 5,500 m were identified as potentially dangerous. However, field work by Byers et al. (2013) lowered the risk level for ten of the twelve lakes, but identified one that had been missed in remote sensing analysis. Glacial lake outburst floods (GLOFs) in Nepal were studied by Ives (1986; Ives et al. 2010) and for glacial Lake Imja by Watanabe et al. (2009). Lake Imja shows several growth stages: Stage 1 (1956–1975), slowest, coalescence of several small supraglacial ponds; Stage 2 (1975–1978), a short period of most rapid expansion; Stage 3 (1978–1997), slow, gradual expansion of a single lake; and Stage 4 (1997–2007), renewed acceleration, mainly eastward expansion into the glacier surface. The lake's water level fell from 5,041 to 5,004 m during 1964–2006. The results show that there is no immediate danger of a catastrophic outburst.

5.4.2 Perennially Frozen Lakes

In both polar regions, a small number of lakes are perennially frozen. Doran et al. (2004) list five such lakes in the Arctic and eleven general locations in the Antarctic. The Arctic lakes are in Greenland (77.0° N, 66.1° W), central and northern Ellesmere Island (78° N and 82° N), Ward Hunt Island (83° N, 74° W), and Ziegler Island, Franz Josef Land (81.0° N, 56.0° E). Ice thicknesses range from 2 to 5.5 m. Nine of the perennially frozen lakes in the Antarctic are in the eastern hemisphere between 66° and 71° S. Exception include those in the

McMurdo Dry Valleys at 77–78° S, 160–164° E, where there are about twenty such lakes. The climate of the Dry Valleys is extreme polar desert; mean annual temperatures range between −15 and −30 °C and annual precipitation (snowfall) is less than 50 mm and highly variable.

Ice thickness on Lake House in the Pearse Valley is 11 m, and that on Lake Vida in Victoria Valley is 19 m (Doran et al. 2004). Lake Vida is a "sealed" lake, in which the ice covers a highly saline water layer (Doran et al. 2002).

Perennial ice reduces light penetration and gas exchange between the water and atmosphere, and alters sediment pathways (Doran et al. 2004). Wind-blown sediments are trapped in the ice and gradually sink into it due to solar heating, with the depth of sinking being related to grain size. A sediment layer forms at the dynamic equilibrium between downward-moving particles and the upward motion of the ice cover due to ablation and growth. Associated with the sediment layer are inclusions of liquid water at about 2-m depth.

Lake levels in the McMurdo Dry Valleys have risen substantially and steadily since the first observations were made by Scott's *Discovery Expedition* in 1903. Detailed records of level change in the lakes have been made for nearly 50 years (Castendyck et al. 2016). Lake Vanda in the Wright Valley is the deepest and largest of the Dry Valley lakes, occupying 7.45×10^6 m^2, and has the most complete direct record of level and inflow volume. The modern vertical profile of conductivity shows a basal brine, the relic of a late Holocene (1,000–2,000 years ago) low-stand hypersaline pool, that is overlain by two layers of fresher water from recent refilling derived from the Onyx River. The lake ice cover has been about 3 m since 1980. The most recent filling trend began abruptly eighty years ago, in the early 1930s, and has averaged 22 cm yr^{-1}. There is an average (1969–2014) annual discharge of 4.07×10^6 m^3 yr^{-1}, equivalent to approximately 2 percent of the lake's current estimated volume. Water loss occurs only through ice ablation, at a rate of about 0.32 m yr^{-1}. The cause of the rising water level is attributed by Castendyck et al. to an increased frequency of föhn winds between the polar plateau and the Ross Sea, associated with the deepening of the Amundsen Sea low pressure center.

Hodgson et al. (2004) note that many Antarctic lakes were formed as a result of isostatic rebound of the continent after deglaciation, which cut off an arm of the ocean and led to the formation of closed lake basins. In closed (*endorheic*) lakes (with no inflow or outflow), the long-term excess of evaporation over precipitation has given rise to high salinity in the lake water. These are common in the Dry Valleys and in the Vestfold Hills of Princess Elizabeth Land (68.5° S, 78.25° E). The latter area contains thirty-seven permanently stratified (meromictic) water bodies that range in salinity from 4 to 235 g l^{-1}, in temperature from −14 to 24 °C, in depth from 5 to 110 m, and in area from 3.6 to 146 ha. Open lakes that have outflow and inflow are fresh water (Gibson 1999). Both types of lake are common in Antarctic "oases" like the Bunger Hills of Wilkes Land (66.3° S, 100.8° E, 950 km^2), Larsemann Hills in Prydz Bay (69° S, 75° E), and Vestfold Hills, and on the Antarctic Peninsula and islands.

Another type of Antarctic lake is the epishelf lake (Hodgson et al. 2004), which that forms in fjords and marine embayments that have become dammed by glaciers or ice shelves. Over time the seawater is replaced by fresh glacier melt water, but often a thick freshwater layer overlies seawater. The thickness of this layer is determined by the ice shelf thickness, and the layer undergoes tidal mixing (Gibson and Anderson 2002). An example is Beaver Lake, Antarctica (70.8° S, 68.3° E). Epishelf lakes are found in the Bunger Hills, Shimmacher oasis, and Alexander Island off the west coast of the Antarctic Peninsula. Five epishelf lakes in the Bunger Hills investigated by Gibson and Anderson (2002) were found to exhibit a wide range of conditions. In contrast to other Antarctic lakes, temperatures in these epishelf lakes were mostly within a degree of freezing. In January 2000, lake ice thickness was between 3.5 and 3.9 m. Isolated saline water evident in three of the lakes is thought to have entered the basins through the connection to marine waters during drier periods of reduced fresh-water input. Saline water enters the lake to compensate for the difference between glacier melt inflow and losses by ablation from the ice cove.

Biologically, the closed basin lakes contain the highest biomass (mainly in cyanobacteria and diatoms) and species diversity of Antarctic "terrestrial" habitats (Hodgson et al. 2004). Aquatic mosses are among the highest forms of plant life. Hodgson et al. cite Willmotte's surveys of benthic microbial mats in the Larsemann Hills, Vestfold Hills, and McMurdo Dry Valleys; they were found to contain 1,500 strains of bacteria (in 9 lakes), 60 strains of cyanobacteria (in 24 lakes), 230 strains of fungi (in 17 lake), 91 strains of algae (in 3 lakes), and 50 protozoans (in 6 lakes). Doran et al. (2004) report that the microbial mat growing on lake bottoms becomes buoyant as a result of high levels of dissolved gases, and "lifts off" by rising through the water column. This biomass may become frozen at the base of the lake ice and gradually migrate up to the ice surface through ablation at the top and freezing at the bottom of the ice cover.

5.4.3 Subglacial Lakes

Lakes exist below ice sheets because geothermal heat balances heat loss at the ice surface. Geothermal heating on the order of 50 mW m^{-2} in East Antarctica, and 70 mW m^{-2} in West Antarctica, is sufficient to melt the ice sheet base in its thicker regions. At least 379 such lakes are found beneath the Antarctic ice sheet

Box 5.4 Lake Vostok

Lake Vostok is at least 240 km long and 50 km wide, and lies between 3,750 m (in the south) and 4,150 m (in the north) in a deep subglacial trough beneath the central East Antarctic ice sheet (Siegert et al. 2001; Zotikov 2006). It has an area of 14,000 km^2 and reaches a maximum depth of 1,000 m. All subglacial lakes are subject to high pressure (approximately 350 bars), low temperatures (about -3 °C), and permanent darkness. Subglacial lakes can be identified by three distinct characteristics on radio echo-sounding records: (1) an especially strong reflection from the ice-sheet base; (2) echoes of constant strength along the track indicating that the surface is very smooth; and (3) a very flat and horizontal character (mirror-like), with slopes less than 1 percent.

(Wright and Siegert 2012). They are commonly located close to the ice divide, where the surface slopes and therefore hydrological potential gradients are low. They may also be associated with the stoss face of major subglacial obstacles, or beneath the heads of ice stream tributaries. The weight of the ice generates pressure that lowers the melting point. The ceiling of the lake is determined by the intersection of the melting point and the vertical temperature gradient. The lake has a hydraulic seal along its entire perimeter, which is created when the ice around the lake is higher so that the equipotential surface dips into the ground below.

The possibility of fresh water beneath Antarctica was first theorized by Russian scientist Peter Kropotkin in the 1890s. The largest and deepest subglacial lake is Lake Vostok (Box 5.4)

5.4.4 Recent Changes in Lakes

Lake surfaces around the globe are warming in summer at a fast rate, according to O'Reilly et al. (2015). The global mean value was 0.34 °C decade^{-1} between 1985 and 2009. The changes, which depend on climatic and local characteristics, range from seasonally ice-covered lakes in areas where temperature and solar radiation are increasing with diminishing cloud cover (0.72 °C decade^{-1}), to ice-free lakes that are experiencing increases in air temperature and solar radiation (0.53 °C decade^{-1}).

In the predominantly continuous permafrost zone of eastern Siberia, Boike et al. (2016) noted recent increases in lake area. Their study area covered the central part of the Lena River catchment, extending from east of Yakutsk to the central Siberian plateau, and from the southern Lena River to north of the Vilyui

River. Remote sensing products were used to analyze changes in water bodies, land surface temperature (LST), and leaf area index (LAI), as well as the occurrence and extent of forest fires, and the area and duration of snow cover. The remote sensing analyses (for LST, snow cover, LAI, and fire) were based on Moderate Resolution Imaging Spectroradiometer (MODIS)–derived National Aeronautics and Space (NASA) products (250–1,000 m) for 2000–2011. Changes in water bodies were calculated from two mosaics of Landsat (30 m) satellite images from 2002 and 2009. Within the 315,000 km^2 study area, the total lake coverage increased by almost 18 percent between 2002 and 2009, but this increase varied in different parts of the study area, ranging between 11 and 42 percent. The land surface temperatures showed a consistent warming trend, with an average increase of about 0.12 $°C\ yr^{-1}$. The average rate of warming during the April–May transition period was 0.17 and 0.19 $°C\ yr^{-1}$ but ranged up to 0.49 $°C\ yr^{-1}$ in the September–October period. Regional differences in the rates of LST change suggest that the spring warming trend is very likely due to changes in snow-covered area. The warming trend observed in autumn does not, however, appear to be directly related to any changes snow cover, to the atmospheric conditions, or to the proportion of the land surface that is covered by water (i.e., due to wetting and drying).

Using Landsat data for 1999–2014, Nitze et al. (2017) determined lake area changes in the permafrost zone of Alaska North Slope, western Alaska, central Yakutia, and the Kolyma lowland. There was a 2.8 percent decrease in western Alaska, as well as small changes (less than 0.7 percent) on the North Slope and in Kolyma, mainly due to lake drainage. However, in central Yakutia there was a 48 percent increase in area, a trend attributed to warmer and wetter climatic conditions.

The southern slope of the Himalaya is a prime location for lakes dammed by glacier moraines. In a recent study, Nie et al. (2017) mapped the current distribution of glacial lakes across the entire Himalaya and monitored the spatially explicit evolution of glacial lakes over five time periods from 1990 to 2015 using a total of 348 Landsat 30-m resolution images. They showed that 4,950 glacial lakes in 2015 covered a total area of 455.3 km^2, mainly located between 4,000 and 5,700 m above sea level (a.s.l.). These glacial lakes expanded approximately 14 percent from 1990 to 2015. The changing patterns of supraglacial lakes and proglacial lakes are rather complex, involving both lake disappearance and emergence. Many emergent glacial lakes are found at higher elevations, especially the new proglacial lakes that have formed as a result of glacier retreat. The most significant expansion occurred along the southern slopes of the central Himalaya. Increasing glacier meltwater induced by atmospheric warming is a primary cause for the observed lake expansion.

By integrating optical imagery, satellite altimetry, and a digital elevation model (DEM), Song et al. (2016) conducted a regional-scale investigation of glacial lake dynamics across two river basins of the southeast Tibet plateau during 1988–2013. In total, 1,278 and 1,396 glacial lakes were inventoried in 1988 and 2013, respectively. Approximately 92 percent of the lakes in 2013 were not in contact with modern glaciers, and the remaining 7.6 percent included twenty-seven debris-contact lakes (in contact with debris-covered ice) and eighty cirque lakes. Debris-contact proglacial lakes experienced much more rapid expansions (approximately 75 percent) than cirque lakes and non-glacier-contact lakes. To explore the cause of this rapid expansion, Song et al. investigated the mass balance of the parent glaciers and elevation changes in lake surfaces and debris-covered glacier tongues using time-series Landsat images, ICESat altimetry, and DEM. Their results showed that the upstream expansion of debris-contact proglacial lakes was not directly associated with rising water levels, but rather with a geomorphological alteration of upstream lake basins caused by melting-induced debris subsidence at glacier termini. This suggests that the hydrogeomorphic process of glacier thinning and retreat, rather than direct glacial meltwater alone, may have played a dominant role in the recent glacial lake expansion observed.

Lake freeze-up and breakup are highly correlated with air temperature. It has been shown that a 1 °C warming corresponds to about a five-day delay (advance) in freeze-up (break-up) for lakes in Finland (Palecki and Barry 1986). Magnuson et al. (2000) used this relationship for lakes and rivers in the northern hemisphere to show that from 1846 to 1995, changes in freeze dates averaged 5.8 days per 100 years later, and changes in breakup dates averaged 6.5 days per 100 years earlier.

Ice-out dates on eleven Arctic lakes in the Canadian Arctic archipelago have been monitored using radar data for 1997–2011 by Surdu et al. (2014). Ten of the lakes experienced earlier summer-ice minimum and water-clear-of-ice (WCI) dates, with greater changes being observed for lakes located in polar oases: nine to twenty-four days earlier WCI dates for these lakes, versus two to twenty days earlier WCI dates for polar-desert lakes.

5.5 Rivers

Arctic rivers are unusual in that they flow mainly northward toward the Arctic Ocean. The three largest are in Eurasia – the Ob, Yenisei. and Lena – followed by the Mackenzie in Canada. The Mackenzie transports sediment averaging 420×10^5 tons yr^{-1} to the Arctic Ocean, while the Lena transports 210×10^5 tons yr^{-1}, although both are frozen for up to eight months of the year. The Olenek and Kolyma rivers flow for nearly 2,000 km over continuous permafrost, and the Lena flows similarly over 1,500 km.

A major landscape feature of Arctic rivers is the presence of large deltas along the rivers that drain into the arctic coastline of Eurasia and western North America. The largest, the Lena delta, covers 32,000 km², while the Mackenzie is second with an area of 13,000 km² (Walker 1998). Mudflats and sandbars are common in arctic deltas. Because of the extreme seasonality of river discharge, most flats and bars are submerged for only a short period of time. However, because of the climatic conditions, they are usually covered with snow for much of the year. After snow melt, flats and bars serve as source regions for silt and sand transport up onto the adjacent tundra surface. In deltas with ice-wedge polygons, ponds are present in the low center of polygons and the troughs between adjacent polygons. These ponds are typically shallow and freeze to the bottom in winter. Ice breakup in the deltas usually occurs in late May or early June, and the maximum discharge is in June. In the Colville, Khatanga, Kolyma, and Olenek, 40 percent of the annual discharge occurs in June (Walker 1998). The floodwaters flow out over landfast ice and near-shore sea ice until they eventually find a crack or hole, where they then form a freshwater wedge.

The Mackenzie delta, according to Lewis (1991), can be divided into a southern part, which is tree covered and dominated by riverine processes, and a northern part, which is treeless and influenced by tides and storm-surge flooding. The northern part has numerous small ponds, while large lakes dominate in the upper delta. Within the delta, lakes occupy 25 percent of the area. Overall, the delta is characterized by an elongate network of lakes and anastomosing channels. The area has more than 1,400 pingos that vary in size, ranging up to 600 m in diameter and up to 45 m in height (Mackay 1963). They are present in the distal portion of the modern delta and in the Pleistocene deposits to the east, where the largest number is found. There they occur in sands and silts and generally in drained lakes.

The delta of the Colville River in northern Alaska covers only 600 km², although its drainage basin covers 29 percent of the North Slope and flows entirely over continuous permafrost. There are nine main channels, seventy-one secondary channels, and forty-two subsidiary ones (Walker 1998). Surface forms feature ice-wedge polygons, thermokarst lakes, and thermo-erosional niches. The main channel is up to 12 m deep and so remains open during winter, which allows sea water to penetrate up to 60 km upstream. These deep channels do not freeze to the bottom; thus, during pre-breakup flooding, which lasts about three weeks, river ice floats on top of the flood.

The Lena River has its headwaters in the Baikal Mountains and is 4,250 km in length. It has an annual discharge of 16,650 m^3 s^{-1} into the Laptev Sea and a sediment load of 17.6×10^6 t yr^{-1}. Over one third of the annual discharge occurs in June. There are 10 major channels and about 800 in all. During most of the year, when the water level is low, numerous sandbars are exposed.

The deltas of the Ob and Yenisei rivers are located at the head of long, narrow estuaries that extend inland from the Kara Sea – for 800 km, in the case of the Ob. They have areas of 3,200 and 4,500 km^2, respectively. The discharge of the Yenisei at 19,600 m^3 s^{-1} is the largest in the Arctic; that of the Ob, 13,500 m^3 s^{-1}, is the third largest. There is a large annual range. For the Yenisei, the maximum is 154,000 m^3 s^{-1}, which contrasts with its minimum of only 2,100 m^3 s^{-1}. Many channels freeze to the bottom in winter.

The Indigirka and Kolyma rivers flow into the East Siberian Sea. Their deltas cover 5,000 and 3,200 km^2, respectively. That of the Indigirka enters a broad, shallow bay and is fan shaped with a breadth of 150 km at the coast, whereas that of the Kolyma is mainly confined to a long, narrow bay like the Ob. In the Indigirka, more than 80 percent of the discharge and 92 percent of the sediment transport takes place in the three summer months. During November to April, only 1.5 percent of the annual discharge occurs. Both deltas have two main channels, as well as oxbow lakes, thermokarst lakes, ice-wedge polygons, and pingos in drained lakes. In the upper parts of the delta, the vegetation includes dwarf birch and willow; in the middle Kolyma, there are larch and birch trees.

Peterson et al. (2002) have suggested that the net discharge from the six largest Eurasian rivers flowing into the Arctic Ocean (Severnaya Dvina, Pechora, Ob, Yenisei, Lena, and Kolyma) increased by approximately 128 km^3 a^{-1} (7 percent) from 1936 to 1999. The greatest relative increase in discharge was observed in the winter (December–February). However, McDonald et al. (2007) demonstrated, based on dendro-hydrological estimates of past flow for the period 1800–1990, that the increased annual discharges of the mid- to late twentieth century were not significantly greater than the discharges experienced at other times of higher flow over the preceding 200 years and, therefore, are still within the range of long-term natural variability.

Trends in discharge of Canadian rivers flowing into the Arctic Ocean during 1964–2013 have been analyzed recently by Déry et al. (2016). They found that the overall trend is strongly positive into the western Arctic Ocean (Mackenzie, Coppermine, Back), slightly positive into the Bering Sea (Yukon and Porcupine), and strongly negative into the eastern Arctic Ocean (Koksoak). There was no significant trend in the 1964–2013 annual discharge into the Bering Strait or western Arctic Ocean. The relatively constant discharge into the eastern Arctic

Ocean in the 2000s arose from large data gaps in this region and the infilling strategy used. Note that the Canadian Arctic archipelago is ungauged.

Trends in water temperature at the mouths of Arctic rivers have been analyzed by Park et al. (2017) by combining a land process model with models of river discharge, ice cover, and water temperature dynamics. For 1979–2013, the annual trends at the main river mouths were 0.020 °C for the Severnaya Dvina (Ust Pinega), 0,012 °C for the Pechora (Oksino), 0.012 °C for the Ob (Salekhard), 0.010 °C for the Yenisei (Igarka), 0.014 °C for the Lena (Kusar), 0.015 °C for the Indigirka (Vorontsovo), and 0.015 °C for the Kolyma (Kolymskoye). The trend for the Mackenzie was −0.008 °C, and the pan-Arctic average was 0.16 °C decade^{-1}.

Arctic rivers cool in the autumn primarily by convective heat loss to the air. A small degree of supercooling (–0.01 °C is needed for turbulent flow to form frazil ice crystals that aggregate into slush balls and eventually form small pancake floes). Shore-fast ice forms in the slower-moving water by the banks and grows laterally. Pancakes freeze together into a sheet, and ice "bridges" develop at bends and constrictions (Barry and Gan 2011, 202–11). Breakup involves several stages; onset is determined by the first movement. The transport of ice blocks is termed "drive," and "wash" is the final clearance (Beltaos 2008). There may also be a breakup ice jam. On the Mackenzie River, breakup lasts an average of four days in the south and twelve days in the delta (De Rham et al. 2008).

Freeze-up dates on the ten largest Russian rivers with records spanning 160–286 years show delays of two to eight days per century and corresponding advances for breakup (Ginsburg and Soldatova 1997). Vuglainsky (2006) compared ice conditions for 1980–2000 with those for 1950–1979 for Russian rivers. In northwestern European Russia, freeze-up (breakup) occurred five to seven days later (earlier). For Siberian rivers, freeze-up was two to three days later and breakup thirty-three to thirty-five days earlier.

5.5.1 Icings

A special feature of many Arctic river basins is the presence of perennial icings. Water seeps onto the land or existing ice at subfreezing temperatures, depositing layers of ice. In Russia, two types are distinguished: river icings (naleds) and spring-fed ground icings (naryn). The initiation of icing growth in a stream is due to the weight of snow on the initial ice cover, which increases the potential water level in the ice cover (Kane 1981). Icings can build up thicknesses of several meters. River icings are most common in flat braided stream channels where there is a change of gradient. Ground icings usually occur where there is a break of slope; water appears on the ground surface from a seepage and freezes when the ground temperature drops below 0 °C (Carey 1973). They tend to be small, whereas spring icings can be extensive. An exceptional case in the Moma River valley (a tributary of the Indigirka River) in eastern Siberia grows to 25 km long, 5–8 km wide, and up to 4 m thick in places (Carey 1973; Chekotillo et al. 1960). It is estimated that the total area of icings in northern Russia is 238,000 km^2 (Kotlyakov 1997). River icings occupy 101,100 km^2 and ground icings 26,700 km^2.

For Alaska and northern Canada, Pavelsky and Zarnetske (2017) used MODIS images for 2000–2015 to identify trends in icing characteristics. They found that 70 of 122 icings were disappearing earlier by 1.6 d a^{-1}, while 14 of 25 that persisted through the summer had a smaller area (decreasing by 2.6 percent a^{-1}).

5.6 Glaciers and Ice Caps

A glacier is formed by snow accumulation that is converted into ice by densification, due to its weight, and by freeze–thaw processes near the surface. It flows slowly downslope due to gravity. The upper accumulation zone is where snow builds up and the lower ablation zone is where there is summer melt and runoff. The equilibrium line separates these two zones, but the equilibrium-line altitude (ELA) varies from year to year. A detailed account of glaciers and glacial processes can be found in Benn and Evans (1998).

Monitoring of the length of glaciers began in 1895 and gradually became adopted worldwide. Measurement of glacier mass balance began in the 1940s and in 2015–2016 included data from 130 glaciers and 24 countries. Glacier fluctuations and mass data are archived and disseminated by the World Glacier Monitoring Service in Zurich, Switzerland (Zemp 2012).

Worldwide, there are an estimated 198,000 glaciers and ice caps with an area of 727,000 km^2, according to the Randolph Inventory (Pfeffer et al. 2014).The latest Global Land Ice Measurements from Space (GLIMS) database at the National Snow and Ice Data Center (NSIDC) lists approximately 200,000 glaciers with a total area of approximately 750,000 km^2 (http://glims.colorado.edu/glacierdata) (Box 5.5). Glaciers and ice caps are prominent features of polar landscapes. They occupy about 425,000 km^2 in the Arctic, occurring mainly in the Canadian Arctic archipelago (approximately 150,000 km^2); around the Greenland ice sheet (approximately 89,000 km^2); in Svalbard, Novaya Zemlya, and Franz Josef Land (approximately 85,000 km^2); and in Alaska (approximately 87,000 km^2) (Table 5.3). The Randolph Glacier Inventory developed by Pfeffer et al. (2014; RGI Consortium 2017), shown in the right column of Table 5.3, is undoubtedly

Table 5.3 Areal extent (km^2) and distribution of glaciers and ice caps in the polar regions (Berthier et al. 2010; Dyurgerov and Meier 2005; Pfeffer et al. 2014*; Radik and Hock 2010; Williams and Ferrino 2012)

Canadian Arctic archipelago	151,433	145,767*
Eurasian Arctic islands	92,386	84,592*
Greenland outside the Greenland ice sheet	70,000	89,721*
Alaska	79,260	86,715*
Antarctica outside the Antarctic ice sheet	169,000	132,867* (including sub-Antarctic)
Sub-Antarctic islands	15,419	

Box 5.5 The Global Land Ice Measurements from Space (GLIMS) Database

The GLIMS database at the National Snow and Ice Data Center (NSIDC) in Boulder, Colorado, had its origin in an Advanced Spaceborne Thermal Emission and Reflection Radiometer (ASTER) Science Team project in 1995. Hugh Kieffer, a member of the ASTER Science Team at the US Geological Survey in Flagstaff, Arizona, planned ASTER data acquisition over ice bodies with Jeff Kargel and Bruce Raup. A network of glaciological collaborators, who were experts in a particular geographic region, assumed responsibility for the production of digital outlines and metadata for the glaciers in their region. In 1999, Raup moved to NSIDC to develop a database of glacier data in open access software with NASA funding. Regional Centers, which now number sixty, began submitting data in 2005. The database now includes numerous glacier attributed and multitemporal outlines starting in 1850.

In 2012, the Randolph Glacier Inventory (RGI) was developed for the Fifth IPCC Assessment Report as a globally complete collection of digital outlines of glaciers. It is being merged with the GLIMS database. There have been six versions as of 2017.

the most consistent across the different regions. The distribution of Arctic glaciers is shown in Figure 5.19 (Sharp 1956).

In the southern hemisphere, glaciers and ice caps are extensive in the Antarctic Peninsula and around the Antarctic ice sheet, occupying some 169,000 km^2. Bliss et al. (2013) provide an inventory of the Antarctic Peninsula based on the Antarctic Digital Database (ADD). They identified 1,133 ice caps covering 91,000 km^2 and 1,619 mountain glaciers covering 41,800 km^2. More than 99 percent of these glaciers terminate either in the ocean or at an ice shelf.

In the sub-Antarctic, defined as lying between the southern hemisphere limit of trees and the Antarctic continental shelf (Mercer 1967, 243), there are some fifteen groups of volcanic islands with glaciers and ice caps. Williams (2013) tabulates their areas and glacier cover. The total glacier area is estimated to be 15,419 km^2; the largest glacier areas are the following:

Adelaide Island	67° 15′ S, 68° 30′ W	3,888 km^2
South Shetlands islands	62° 00′ S, 58° 00′ W	2,950 km^2
Anvers Island	64° 33′ S, 63° 25′ W	2,500 km^2
South Georgia	53° 50′ S, 35° 50′ W	2,225 km^2
Brabant Island	64° 15′ S, 62° 20′ W	1,369 km^2

Glaciers can be classified according to their thermal state or their morphology. There are three types of thermal regimes: (1) polar, where the ice is always below freezing point from the surface to its base; (2) temperate, where the ice is at the pressure melting point from the surface to its base for most of the year; and (3) polythermal, where there are both temperate and polar ice, depending on depth beneath the surface and position along the glacier's length. Glacier morphology distinguishes cirque glaciers, which are restricted to a glacier basin or cirque in alpine terrain (Figure 5.20); valley glaciers, which flow along the length of a valley; and ice caps, which cover a mountain range or plateau and do not exceed an area of 50,000 km^2. Other descriptors include hanging glaciers, which descend abruptly from a mountain range; tributary glaciers, which feed into a major valley glacier; tidewater glaciers, which terminate in the sea; and surging glaciers, which periodically undergo rapid acceleration followed by quiescent intervals. Worldwide, about 4 percent of glaciers are known to surge.

5.6.1 Glacier Changes in the Arctic

Glaciers and ice caps in the Arctic have undergone extensive retreat since the end of the Little Ice Age, between about 1850 and 1900. Based on former glacial trimlines, indicated by poorly vegetated terrain, Wolken et al. (2008) showed that between the end of the Little Ice Age and 1960, the ice extent on the Queen Elizabeth Islands (QEI) shrank by 37 percent overall. While most of the reduction occurred in the eastern QEI, in central and western islands the low relief led to the almost complete removal of ice and snow by 1960. Meighen Island in the west was an exception, with a 40 percent reduction.

For the Canadian Arctic, a regional signal from long-term surface balance (SMB) records from the Devon Ice Cap, Meighen Ice Cap, Melville Island South Ice Cap, and White Glacier (Axel Heiberg Island) shows that the SMB became increasingly negative after 1987 (Sharp et al. 2014, 2015). Cumulative mass

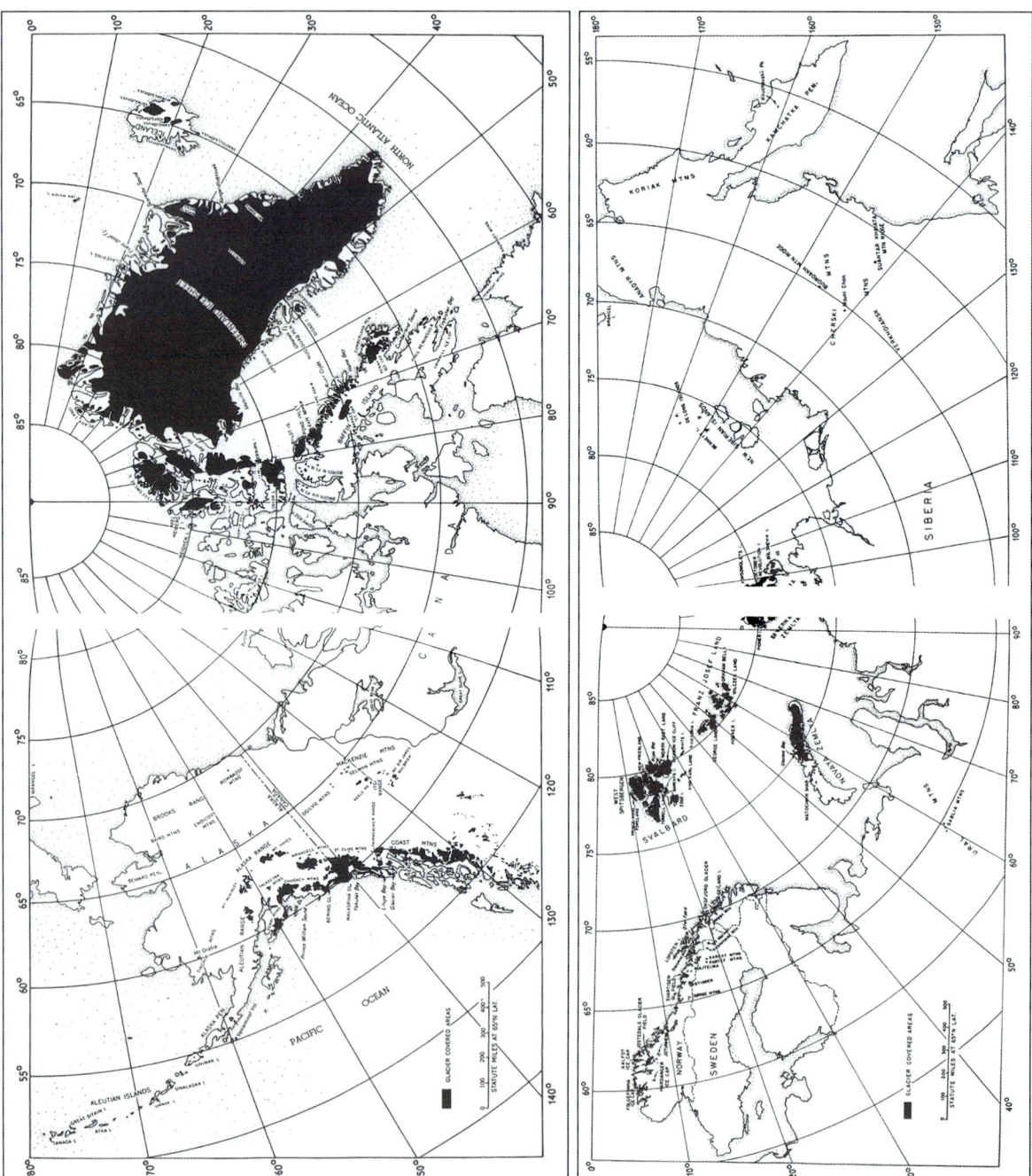

Figure 5.19 The distribution of glaciers in (a) the western Arctic and (b) the eastern Arctic.
Source: After Sharp 1956, 81, figure 1 and 82, figure 2. Courtesy of Arctic Institute of North America.

balances over the periods of measurement are −5.20 m w.e. (water equivalent) for Devon Ice Cap (1961–2009) and −7.62 m w.e. for White Glacier (1960–2008). Changes in the summer climate of the Queen Elizabeth Islands (for 1948–2008) have been based on 700-hPa air temperatures derived from National Centers for Environmental Prediction (NCEP)/National Center for Atmospheric Research (NCAR) reanalyses, which are strongly correlated with conditions at summit elevations. They show strong positive anomalies in the 1950s and the 2000s, but negative anomalies in other decades – particularly the 1960s and 1970s. The warmest decade was the 1950s over Axel Heiberg Island and northern Ellesmere Islands (0.7 °C), and the 2000s for Baffin Island, Devon Island, and the rest of Ellesmere Island (0.7–0.9 °C). The latter period has seen strong warm advection from the northwest Atlantic.

Figure 5.20 Cirque glacier Parlang Zangbo, Tibet. 29.38° N, 96.98° E.

Source: H. S. Li. 2011. *Glacier Photograph Collection*. [Digital media]. Boulder, CO: National Snow and Ice Data Center.

In northeastern Ellesmere Island, Serreze et al. (2017) found that the small St. Patrick Bay ice caps and the Murray and Simmons ice caps are rapidly shrinking. Vertical aerial photographs of these Little Ice Age relics taken during August 1959 show that the larger of the St. Patrick Bay ice caps had an area of 7.48 km², and the smaller one an area of 2.93 km². The Murray and Simmons ice caps covered 4.37 and 7.45 km², respectively. Outlines determined from ASTER satellite data for July 2016 showed that, compared to 1959, the larger and the smaller of the St. Patrick Bay ice caps had both been reduced to only 5 percent of their former area, while the Murray and Simmons ice caps fared better at 39 and 25 percent, respectively, likely reflecting their higher elevation (950–1,100 versus 750–900 m for the St. Patrick Bay caps).

On Baffin Island and Bylot Island, Gardner et al. (2012) reported an increase in mass loss from 11.1 Gt yr⁻¹ for the period 1963–2006 to 23.8 Gt yr⁻¹ for the period 2003–2011. Summer temperatures accounted for almost all the loss, with the Barnes ice cap being especially sensitive as a result of its hypsometry. Gilbert et al. (2016) have shown that the Barnes reached a quasi-equilibrium state approximately 2,000 years ago and has remained similar in size since then, with a small increase during the Little Ice Age. It will disappear under current climate conditions within the next millennium. In a model analysis, Gilbert et al. (2017) project that the ice cap will disappear within 300 years.

Gardner et al. (2011) calculate that the Canadian Arctic archipelago has recently lost 61 ± 7 Gt yr^{-1} of ice, contributing 0.17 ± 0.02 mm yr^{-1} to sea-level rise. Separate determinations using surface mass balance, ICESat data, and GRACE data show strong agreement. Between 2004–2006 and 2007–2009, the rate of mass loss sharply increased from 31 ± 8 to 92 ± 12 Gt yr^{-1} in direct response to warmer summers (64 ± 14 Gt yr^{-1} per 1 K increase).

Local ice masses around the Greenland ice sheet occupy an area of approximately 90,000 km^2 (Rastner et al. 2012). Their distribution is shown in Figure 5.21. About 50,000 km^2 is located in central East Greenland, 67–72° N (Stearns and Jiskoot 2014). Geikie plateau is the largest local ice cap. About 90 percent of the total glacierized area drains through 120 tidewater glaciers. A comparison of ice margin locations from the 2004–2005 ASTER data with the 1986 maps showed a net calving area loss of 70 km^2, which translates into a mean overall annual retreat rate between 3 and 7 km. Surge-type glaciers have shown a tidewater margin retreat of 62 km^2 since 1995, and non-surge-type glaciers a retreat of only 8 km^2 since 1985. Thus, it appears that some of the calving margin retreat could be related to surge dynamics rather than just to changes in mass balance or ocean water temperature.

For Alaska, Arendt et al. (2002) used airborne laser altimetry to estimate volume changes of sixty-seven glaciers from the mid-1950s to the mid-1990s. Extrapolation to all glaciers in Alaska yielded an estimated total volume change of -52 ± 15 km^3 a^{-1} (w.e.), equivalent to a sea level rise (SLE) of 0.14 ± 0.04 mm a^{-1}. Repeat measurements on twenty-eight glaciers from the mid-1990s to 2000–2001 indicated an increased average rate of thinning, leading to an extrapolated annual volume loss from Alaskan glaciers of -96 ± 35 km^3 a^{-1}, or 0.27 ± 0.10 mm a^{-1} SLE over that time interval. Berthier et al. (2010) subsequently revised the estimates of Arendt et al. downward by 34 percent. Combining a comprehensive glacier inventory with elevation changes derived from sequential digital elevation models, they found that between 1962 and 2006 Alaskan glaciers lost 41.9 ± 8.6 km^3 a^{-1} (w.e.), and contributed 0.12 ± 0.02 mm a^{-1} SLE. The lower values are attributed to the higher spatial resolution of the glacier inventory and to the reduction of ice thinning beneath debris cover.

The sub-Arctic Lemon Creek Glacier and the Taku Glacier, both part of the Juneau Icefield in southwest Alaska, have the longest record of mass balance observations in North America. Pelto et al. (2013) reported that the annual surface mass balance (SMB) measured on Taku Glacier averaged 0.40 m a^{-1} for 1946-1985, and -0.08 m a^{-1} for 1986-2011. The recent annual mass balance decline has resulted in the cessation of the long-term thickening of the glacier. Mean SMB on Lemon Creek Glacier has declined from -0.30 m a^{-1} for the

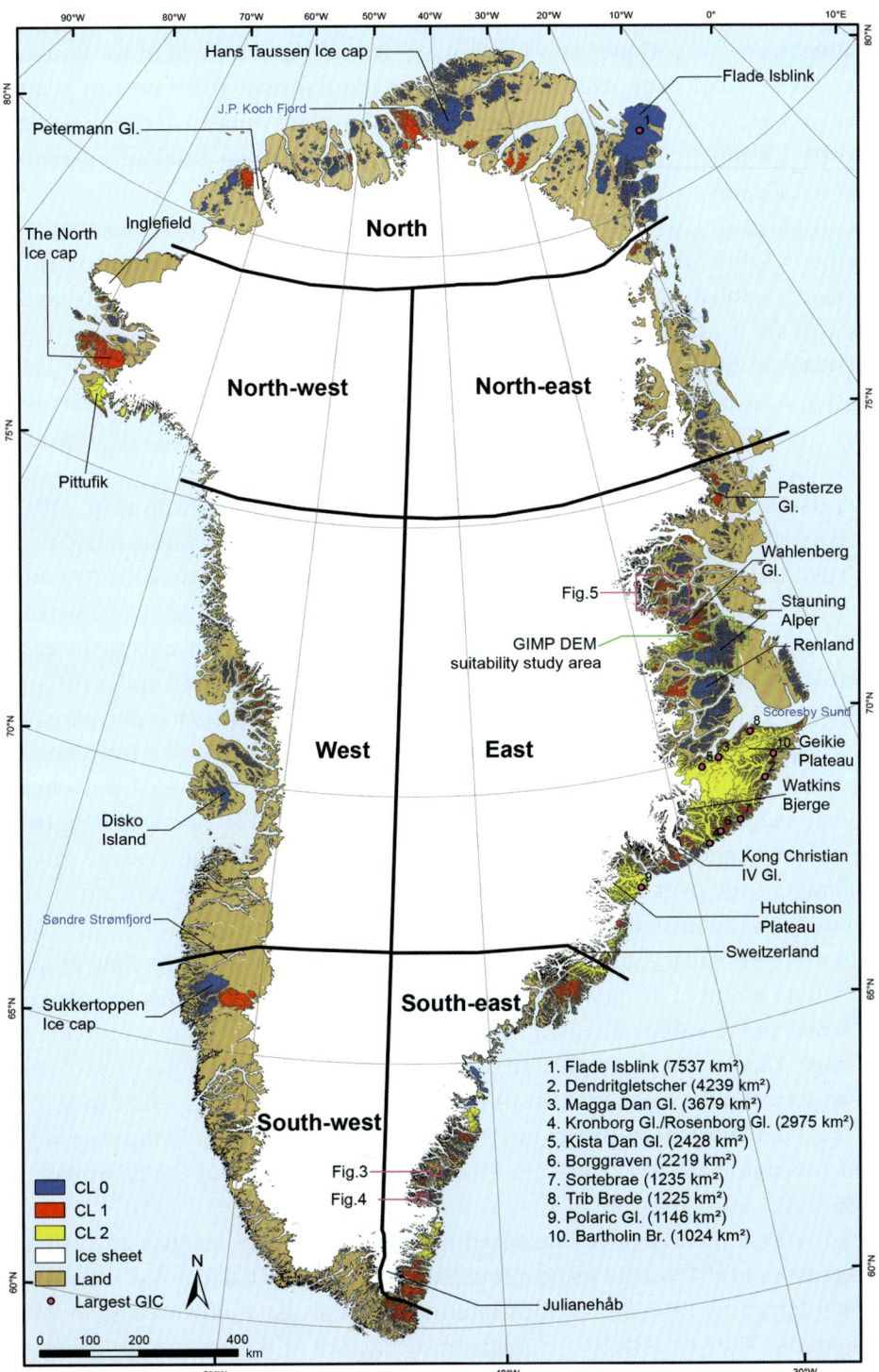

Figure 5.21 Map of Greenland showing all local glaciers and ice caps.
Source: Rastner et al. 2012, 1485, figure 1, philipp.rastner@geo.uzh.ch.

1953–1985 period to -0.60 m a^{-1} during the 1986–2011 period. The cumulative change in annual surface mass balance is -26.6 m w.e., a 29 m of ice thinning over the 55 years. The relationship between the transient snow line on Lemon Creek Glacier and Taku Glacier to that on the other Juneau Icefield glaciers (Norris, Mendenhall, Herbert, and Eagle) is strong, with correlations exceeding 0.82 in all cases.

Barrand et al. (2016) have reported on the glaciers in the Torngat Mountains of northern Labrador. Field surveys and remote-sensing analyses were used to measure regional glacier shrinkage of 27 percent from 1950 to 2005, substantial rates of ice surface thinning (up to 6 m yr^{-1}), and negative geodetic mass balances at Abraham, Hidden and Minaret Glaciers between 2005 and 2011. Glacier mass balances appear to be controlled by variations in winter precipitation and, increasingly, by strong summer and autumn atmospheric warming since the early 1990s.

A glacier inventory for Svalbard performed in the 2000s by Nuth et al. (2013) used Satellite Pour l'Observation de la Terre (SPOT) 5 high-resolution images and ASTER data. It identified 1,668 glaciers, with a glacier area of 33,775 km^2 covering 57 percent of the total land area of the archipelago. About 60 percent of the glacierized area is on Spitsbergen and 40 percent on the islands to the east–northeast. Currently, 68 percent of the glacierized area in Svalbard drains through tidewater glaciers. The glacierized area of the entire archipelago has decreased by an average of 80 km^2 yr^{-1} over the past 30 years, representing a reduction of 7 percent. Nuth et al. (2013) compared ICESat data for 2003–2007 with topographic maps and digital elevation models for 1965–1990 to calculate long-term elevation changes of glaciers on the Svalbard archipelago. The average rate of volume change over the past 40 years for Svalbard (excluding Austfonna and Kvitøya) was estimated to be -9.71 km^3 yr^{-1}.

In northern Spitsbergen, Sobota et al. (2016) found that from the time of their maximum extent in the late nineteenth and early twentieth centuries to 2015, the total area of the Kaffiøyra region valley glaciers decreased by about 43 percent on average. Glaciers in the region retreated at an average rate of 12 m yr^{-1}. The mean annual mass balance of the three analyzed glaciers was (1) -0.72 m w.e. in the case of the Waldemarbreen in 1996–2015, (2) -0.84 m w.e. in the case of the Irenebreen in 2002–2015, and (3) -0.65 m w.e. in the case of the Elisebreen in 2006–2013.

Østby et al. (2017) have diagnosed the decline in mass balance of Svalbard glaciers from 1957 to 2014 using a coupled surface energy balance and snow pack model forced by ERA-40 and ERA-Interim reanalysis data, downscaled to 1-km resolution. They calculated a climatic mass balance of 8.2 cm w. e. yr^{-1}, which has a linear trend of -1.4 cm w. e. yr^{-1} with a shift from a positive to a negative regime around 1980. For 2004–2013, they found a climatic mass balance of -21 cm w. e. yr^{-1}. Accounting for frontal ablation, this gives a total Svalbard mass balance of -39 cm w. e. yr^{-1}.

Moholdt et al. (2012) analyzed glacier changes in the Russian High Arctic for 2003–2009 using ICESat laser altimetry and GRACE gravimetry. The 51,500 km^2 glacier area on the archipelagos lost ice at a rate of -9.1 Gt yr^{-1}, about 80 percent of which was from Novaya Zemlya and the remainder from Franz Josef Land and Severnaya Zemlya.

Grant (2010) analyzed the ice cover on Novaya Zemlya using Landsat and ASTER imagery. For an area representing approximately 70 percent of the glacierized area, she found that the ice had receded from the Little Ice Age maximum by 7 percent in 2001. Glaciers on the northwest Barents Sea coast had retreated more than those on the Kara Sea side, and steep valley glaciers terminating at low elevations had the greatest retreat rates.

Using Landsat8, Sentinel 2, and Sentinel 1 A/B SAR data, together with the new ArcticDEM, Rastner et al. (2017) have compiled a new glacier inventory for Novaya Zemlya. They used Landsat-8 optical imagery acquired between 2013 and 2016 to automatically map glaciers with the band ratio method, and Landsat-5 Thematic Mapper scenes from 1999 to identify and remove seasonal snow. Drainage divides were calculated from the new ArcticDEM. The inventory gives a total glacierized area of 22,379 km^2, with 1,474 glacier entities being larger than 0.05 km^2. Within the entire sample, forty-one are marine-terminating, covering an area of 16,064 km^2 (or 72 percent). The largest ice cover is found between an elevation of 600 and 700 m a.s.l. A large number of glaciers (909 or 62 percent) were smaller than 0.5 km^2, covering an area of only 156 km^2, while 49 glaciers covered an area of 18,724 km^2 (84 percent). For the South Island of Novaya Zemlya, Rastner et al. calculated a total area loss of approximately 76.6 km^2 since 2001, with a mean retreat for land-terminating glaciers of approximately 650 m (or approximately 50 m yr^{-1}).

Helm (2007) studied ice cover in western Franz Josef Land using visible and near-infrared ASTER data at 15-m resolution during 2000–2004. He compared DEMs derived from ASTER stereo-image pairs with 1:100,000 scale maps made by the Institute of Geography, Russian Academy of Sciences in 1952–1953. The two largest ice caps in 1952 were George Land, 2,190 km^2, and Hooker Island, 449 km^2. In 2004, these ice caps had shrunk to 2,089 km^2 and 401 km^2, respectively, representing decreases of 4.6 and 10.7 percent. The total ice area of western Franz Josef Land decreased from 3,485 km^2 in 1952 to 3,267 km^2 in 2004, a 6.3 percent reduction.

Changes in the area of thirty small glaciers (mostly smaller than 1 km^2) in the northern Polar Urals (67.5–68.25° N) between 1953 and 2000 were assessed using historic aerial photography from 1953 and 1960, ASTER and panchromatic Landsat ETM+ imagery from 2000, and data from 1981 and 2008 terrestrial surveys (Shahgedanova et al. 2012). In total, glacier area declined by 22.3 percent in the 1953–1960 to 2000 period. Over 45 years, the geodetic mass balances of the Obruchev and IGAN glaciers were −20.66 and −13.54 m w.e., respectively. Glacier shrinkage in the Polar Urals is related to a summer warming of 1 °C between 1953–1981 and 1981–2008; these rates are consistent with other regions of northern Asia. While glacier shrinkage intensified in the 1981–2000 period relative to 1953–1981, increasing winter precipitation and shading effects slowed glacier wastage in 2000–2008.

Gardner et al. (2013) summarize the mass change for 2003–2009 for nineteen regions of the globe. The global glacier total was −215 Gt yr^{-1} For the Arctic areas, the changes (in Gt yr^{-1}) were Alaska, −50; Arctic Canada north, −33: Arctic Canada south, −27: Greenland periphery, −38; Svalbard, −5; Russian Arctic, −11; and North Asia, −2.

5.6.2 Glacier Changes in the Antarctic

In the Antarctic Peninsula, 87 percent of 244 glaciers considered by Cook et al. (2005) retreated, on average, from the mid-twentieth century (circa 1953) through 2004. Advances were more common through 1964 and retreats more common after 1964. In 2000–2004, 75 percent of the glaciers were in retreat. The Antarctic Peninsula is the region where warming has been most pronounced since the mid-twentieth century. However, recent work by Cook et al. (2016) indicates that ocean warming has been the primary factor forcing glacier retreat along the western peninsula. These authors identified a strong correspondence between ocean temperatures around 150-m depth and glacier-front changes along the approximately 1,000-km western coastline. Ninety percent of 574 marine-terminating glaciers are found to have retreated. In the south, glaciers that terminate in warm Circumpolar Deep Water have undergone considerable retreat, whereas those in the far northwest along Bransfield Strait, which terminate in cooler waters, have not. Furthermore, a mid-ocean warming since the 1990s in the south is coincident with widespread acceleration of glacier retreat.

Recently, Fieber et al. (2018) analyzed detailed elevation and volume data for sixteen glaciers in the Antarctic Peninsula from 61° to 64° S, 54° to 63° W, using WorldView-2 satellite stereo imagery with archival stereo aerial photography since the 1950s. The mean annual mass loss for all glaciers was 0.24 ± 0.08 m.w. e. yr^{-1}. The combined mass balance of all sixteen glaciers was -1.86 Gt, which corresponds to -0.005 mm sea level equivalent (SLE) over the fifty-seven-year observation period.

In the Amundsen Sea sector of West Antarctica, Rignot et al. (2014) determined an extensive grounding line retreat between 1992 and 2011 based on InSAR analysis. The Smith/Kohler Glaciers retreated 35 km along the ice plain, Pine Island Glacier retreated 31 km at its center, Thwaites Glacier retreated 14 km along its fast-flow core, and Haynes Glacier retreated 10 km along its flanks. The rapid retreat is attributed to the presence of a retrograde submarine bed that promotes marine ice sheet instability. There was also an increase in ocean heat flux in this sector in the early 2000s.

5.7 Glacial Landscapes

During the Last Glacial Maximum (LGM) it is estimated that nearly one third of Earth's landmass was covered with ice from glaciers and continental sized ice sheets. Glacial landscapes can be thought of as including two different settings: areas that are still covered with ice (glacierized) and those that have been previously covered by glaciers (glaciated). The Greenland and Antarctic ice sheets have the most complete remaining ice cover with smaller percentage ice coverage found at Svalbard and many Arctic archipelagos, areas of Alaska-Yukon region, the Patagonian icefield and the Tibetan plateau. Glaciated areas, where glaciers and ice sheets once existed, but have now retreated, include much of Canada, Scandinavia, the United Kingdom, and northern United States, along with mountainous areas such as Glacier National Park, Montana, and Yosemite, California.

Glaciers have had a profound effect on humanity. Glaciers provide water for irrigation, hydroelectric power, and water supplies such as streams and aquifers. Glacial deposits are commonly found to be highly mineral-rich, hence glacial erosion and deposition have had profound effects on the development of agriculture at the end of the last ice age (Hambrey and Alean 2017, xxxviii). Glacial landscapes are some of the most beautiful in the world and display a wide range of glaciological phenomena largely produced from the erosional and depositional processes of glaciers and ice sheets (see Hambrey and Alean 2017; Shroder 2013). The understanding of glacier dynamics, that is obtained from evidence revealed in the study of glacial landscapes, provides some of the most important sources of information on environmental change.

Figure 5.22 Vast landscapes of (a) areal scouring, (b) carved and scoured bedrock, along the Mara River, Canadian Shield, Northwest Territories.
Source: Photo courtesy of Eileen Hall-McKim.

5.7.1 Erosional Landforms

The main types of glacial erosion are (1) areal flow, where the ice flows across the landscape in a relatively unconfined way; (2) linear flow, where a major stream of ice flows in a rock channel; and (3) an interaction of glacial and periglacial activity. Some landforms created by areal flow include whalebacks, drumlins, roches moutonnées, and grooves [see the discussion later in this chapter and also Munroe-Stasiuk et al. (2013) and Stroeven et al. (2013)]. Landscapes of areal scouring occur where an ice sheet has removed the preexisting cover of loose material and eroded into the underlying bedrock over large areas. Many spectacular examples of areally scoured landscapes occur on the Canadian Shield, in west Greenland, and in northwest Scotland (Gordon 1981; Rea and Evans 1996; Sugden 1974) (Figure 5.22).

Processes at the ice–bed interface play a major role in the movement and behavior of glaciers, and in the formation of glacial erosion. Essentially all erosion created by the movement of a glacier occurs along its bed and the walls and sides of the channel that is confining it or a base of bedrock. Direct observation of this process is very difficult. Some direct observations have been made by descending into crevasses or entering marginal cracks so as to gain access to the affected areas. In other studies, bore holes and various remote geophysical sensing techniques have proved helpful, but are limited in scope. Much of what is known about glacial erosion has been obtained from studies of exposures of freshly glaciated bedrock features and surfaces and glacial till (Benn and Evans 2010; Bennet and Glasser 2009; Sugden and John 1976) (Harbor, 2013).

The three fundamental methods of glacial erosion are abrasion, plucking (quarrying), and physical and chemical erosion by subglacial meltwater. Abrasion is generally described as the process whereby bedrock is scoured by debris carried in the lower layers of a glacier. Close inspection of recently glaciated bedrock will reveal the work of abrasion by the smoothing and rounding of protuberances and the scratching and polishing of rock surfaces (Sugden and John 1976) Striations are scratches or gouges etched into rock surfaces by grinding of sand and rock particles carried in the base of glaciers. Abrasion can be found in the form of smoothed or polished bedrock surfaces and appears as a shine, gloss or lustre on scoured rock outcrops. Benn and Evan (2010) note that polishing of surfaces occurs once the asperities on abraded rocks have been worn down. Another line of evidence of abrasion is rock flour, consisting of fine-grained, silt-sized particles of rock, generated by mechanical grinding of bedrock by glacial action. These fine-grained particles contribute to the characteristic blue-green color of glacial meltwater streams. Because it consists of fresh mineral fragments, rock flour is difficult to explain except as a product of grinding (Vivian 1970).

Plucking or quarrying is the process by which glaciers loosen, pick up, and transport rocks ranging in size from fine-grained particles and fragments to boulders and large blocks of rock. The small amount of water that forms by pressure melting around a rock on a glacier's bed is a very effective agent of erosion (Hallet 1996). Along with the freeze–thaw process, if the bedrock is at all permeable, some of this water penetrates the rock's cracks and upon freezing shatters the rock (Gudmundsson 2011). The broken fragments are then picked up and carried away by the glacier. Another form of plucking involves removal of large, joint-defined blocks. This occurs largely in valley glaciers as the glacier moves down the valley. Where there are lines of weakness, freeze-thaw processes form cracks, allowing meltwater to infiltrate, causing them to widen. Eventually, even large joint blocks loosen and become entrained in the glacier as it descends a steep face.

Another important method of glacial erosion is the breakdown of rock by meltwater. According to Sugden (1974) glacial meltwater is a very effective agent for chemical and mechanical erosion due to high pressure and high velocity flow as it moves through large subglacial channels. All rocks are traversed by cracks on a number of scales from large to microscopic. Boulton (1974) and Iverson (1991) show that repeated cycles of freezing and thawing breaks rocks into finer particles, wherefore they loosen, and begin to move. Additionally, in some instances fluctuations in pressure on the bed of a glacier can be great enough to crush, shatter, or fracture brittle bedrock, producing fragments that the glacier removes.

Factors influencing glacial erosion are complex and depend largely on conditions at the ice-rock interface. Some of these factors include: the temperature of the base of the glacier, the nature of debris in the ice, the amount of water contained in the ice, the weight of the overlying ice mass, and the lithology and structure of the bedrock materials (Harbor 2013; Sugden 1974). For example, a cold polar glacier, solidly frozen to its bed, slips only occasionally, if at all, in contrast, a polythermal glacier at the pressure melting point, is continuously slipping over the bed. Water at the base of a glacier can enhance or inhibit the abrasion process depending on its temperature. Flowing water, however, can be an asset to plucking; indeed, basal glacial streams are major locations of fluvioglacial erosion (Iverson 1991; Boulton 1974). Understanding that the factors controlling glacial erosion are complex, following is a brief review of some of the major landscape features found in glaciated landscapes.

5.7.2 Small-Scale Features of Glaciated Bedrock

The smaller-scale features of glaciated bedrock include glacial striae, grooves, and chattermarks. Striations are linear scratches etched into many rock surfaces by glacial ice and indicate grinding by sand and rock particles moving under considerable ice pressure. These consist of small-scale scratches, mostly a fraction of a millimeter deep to a few to several tens of centimeters long. Striae have long been recognized as evidence for scoring by particles embedded in glacier ice (e.g., Agassiz 1838a; Chamberlin 1888; Gilbert 1903, 1906). Striae are generally clustered in groups. In some cases, the direction of flow of basal ice over such a surface can be useful in reconstruction of paleo ice flows of ice sheets and glaciers. On even surfaces of rock, these striae or striations are mostly straight and parallel. On uneven surfaces, many striae are curved in response to microtopographic relief, indicating that temperate glacier ice under stress and highly confining pressure at the base of a glacier is remarkably mobile (Iverson 1991).

At times, striae of different ages may intersect or crosscut. Crossing striae of multiple striae of different ages may indicate a glacial history created by ice advances separated by thousands of years (Bennet and Glasser 2009). In these instances, the ice advance creating the younger striae was modest, leaving the older set still discernable. Most commonly, but not always, striae widen gradually or end abruptly at their terminus in the down-glacier direction, and thus are good indicators of flow direction (Benn and Evans 1998; Iverson 1991). Glacial striations can be found worldwide on formerly glaciated landscapes. Striae examples are found in Central Park, New York City, Glacier National Park, Montana; and along the former path of the Moiry Glacier, Switzerland.

Associated with striae, but less abundant, are larger, longer, deeper, linear, U-shaped grooves with smooth bottoms and walls and rounded edges. A typical groove is a few centimeters to 10–20 cm deep, twice as wide, and up to several tens of meters long, but can be as much as 1–2 m deep and 50–100 m in length. Smith (1948) notes that some of the largest grooves are observed on bedrock in the Mackenzie Valley, northwestern Canada, which have depths to 30 m, widths of 100 m, and lengths to 12 km. Another remarkable example of grooves is found in the United States at Glacial Groove State Memorial, Kelleys Island, Ohio (Figure 5.23).

Chattermarks are a series of closely spaced, crescent-shaped fractures that are concave in the downside direction, and are considered to be a direct result of quarrying, due to stresses exerted by overlying ice (Hallet 1996; Laverdiere and Bernard 1970). According to Munroe-Stasiuk et al. (2013), chattermarks are oriented perpendicular to ice movement and thus, may provide information on the former ice flow direction of the region (5.24).

Figure 5.23 Glacial grooves at Glacial Groove State Park, Kelleys Island, Ohio.
Source: Virginia Hill, Glacier Photograph Collection, National Snow and Ice Data Center, Boulder, CO.

5.7.3 Large-Scale Features of Glaciated Bedrock

Roches moutonnées can be described as asymmetrical hills of exposed bedrock with a gently sloping upstream side, where all sides and edges have been polished smooth and eroded in the direction the glacier was moving, and a steep downstream side, where rock was plucked by the glaciers. They are often marked with glacial striations or grooves. Roches moutonnées range in size from less than 1 m to several hundred meters across. Examples of large roches moutonnées are found at the Great Slave Lake, Canada; the Cairngorms of northern Scotland; and the asymmetrical granite hills of New England. Very large asymmetrical hills, known as *flyggbergs*, occur in parts of Sweden and are up to 3 km long and 350 m high (Benn and Evans 2010).

Whalebacks are smoothed, elongated bedrock features that lack the quarried lee faces characteristic of roches moutonnées (Evans 1996; Sugden and John 1976). These features are bedrock knolls, typically several meters high and rounded and

Figure 5.24 Chattermarks, Greenland, 15 cm across.
Source: Mike Hughes, Glacier Photograph Collection, National Snow and Ice Data Center, Boulder, CO.

smoothed on all sides by a glacier. Whalebacks tend to have low height-to-length ratios. They are relatively high (1–2 m) in comparison to their length (1.5–3 m). Striations and other small-scale features of glacial abrasion may be superimposed on any surface of a whaleback, and are often continuous along the entire length of the whaleback. The absence of quarried lee faces on whalebacks is thought to imply that there was continuous ice-bed contact without presence of low pressure cavities during their formation (Evans 1996; Glasser and Warren 1990).

A trimline is another glacial erosion feature that forms when an ice stream moves down-valley and then recedes, leaving a sharp line between barren ground and vegetation cover above it. The upper limits of glacier occupancy in a glaciated valley can be preserved by these erosional features (Flint 1971). In time, vegetation cover will regrow in the glaciated area, but the trimline remains well defined. In subpolar areas, the line may be marked by trees of different ages. The oldest trees growing inside the trimline provide a minimal date for glacier advance.

5.7.4 Glaciated Mountains

As glaciers move through mountain ranges, they leave behind often spectacular eroded landscape (Figure 5.25). Most valleys in mountainous terrains have been carved by stream erosion. But in mountains that have been glaciated, some have been dramatically modified by glacial erosion, but few, if any, are created by glacial erosion alone. Valley glaciers take the path of least resistance as they move from the highlands to lower ground, and those routes are usually preexisting canyons cut by stream erosion. Stream-cut valleys with steep walls and floors form a V-shaped valley. Glacier ice flow will modify a river canyon by making it wider and straighter, eventually forming a U-shaped valley. A valley glacier that flows out of a mountainous area and spreads out over broad or sloping surrounding terrain is a piedmont glacier. A valley glacier may flow all the way to a coastline, creating a deep glacial trough submerged with seawater, known as a fjord.

Cirques are concave, semicircular basins carved by the base of a glacier as it erodes the landscape. *Cirque* is a French word meaning "circle," and this landform is an amphitheater-shaped basin with steep walls, formed at the head of a glacial valley (Figure 5.26). Synonymous terms are *corrie* (in Scotland) and *cwm* (in Wales). Cirque lakes form when cirques are overdeepened by glacial excavation and can be as deep as 100 m, contained behind a bedrock sill. The divide between two opposed cirques is narrowed, and may be lowered by hundreds of meters, forming a saddle-shaped gap, known as a col. Cirques occur preferentially on the lee side of mountain ranges with respect to predominant wind directions (Vilborg 1977). In some places, two or more adjacent basins have merged into a compound cirque, with scalloped walls and outline. According to Stroeven et al. (2013), as cirques become enlarged and intersect, they can remove and reshape the original topography and create sharp crested topography such as arêtes and horns.

Arête is a French word meaning "a narrow crest of rock." An arête forms when two adjacent glaciers have worn a steep ridge into the rock between them. Sharp (1988) notes that when the cirque wall is being eroded, if the bedrock is homogeneous, the skyline profile of the arête may be smooth; in contrast, if the rock is heterogeneous and irregular in structure, particularly if it is jointed, the arête's profile will be rough and jagged or a saw-toothed ridge. A horn results when glaciers erode three or more arêtes, usually forming a sharp-edge peak. In time, the peak erodes into a steep-sided pyramidal spire rising to a sharp point. Examples include the Matterhorn in the Pennine Alps on the Swiss-Italian border, and Pilot Peak in the northwestern corner of Wyoming in the United States.

Hanging valleys can form when a glacial valley erodes at a rate more rapid than its tributaries, lowering the floor of the valley and leaving the tributary abruptly cut off from it junction and hanging above the valley floor below. The stream in the hanging tributary then cascades down to

Figure 5.25 Glaciated and glacierized mountain landscape south of Inugsuin Fiord, Baffin Island. Total relief is about 1,700 m (Jack Ives).
Source: J. D. Ives and R. G. Barry, eds. 1974. Frontispiece. In *Arctic and Alpine Environments*. London: Methuen.

Figure 5.26 Deep cirques, Disko Island Greenland.
Source: Ole Mikkelsen, Glacier Photograph Collection, National Snow and Ice Data Center, Boulder, CO.

Figure 5.27 Geirangerfjord in Norway.
Source: David Wilkinson, Glacier Photograph Collection, National Snow and Ice Data Center, Boulder, CO.

the valley floor. One of the most famous hanging valleys in the United States is Bridal Veil Falls in Yosemite National Park.

The largest and most spectacular manifestations of glacial erosion are troughs and fjords carved by ice flow through major rock channels. Troughs are valleys eroded by glaciers, with smoothed, steep valley sides and a rounded, U-shaped cross-section. Fine examples are found at the Cairngorms of Scotland, Snowdonia, Wales and Glacier National Park, US. Fjords are glacial troughs that reach out into the ocean; typically, they are the largest troughs. The Lambert and Thiel troughs, the largest troughs on Earth, are presently occupied by outlet glaciers of the Antarctic ice sheet and are approximately 1,000 km long and more than 50 km wide (Drewry 1983). The Lambert trough system reaches a depth of 2300 m below sea level (BAS/BEDMAP2 2002) (See Appendix A) Fiords are coastal troughs, and most are deeper than the adjacent sea (Figure 5.27). Sognefjord in Norway reaches a depth of 1,300 m below sea level. Deep open fjords occur along the coasts of Greenland, Norway and the Canadian Arctic Islands of Ellesmere and Axel Heiberg. Scoresby Sound, a large fiord system on the east coast of Greenland has a length of nearly 350 km and depth of 1,450 m. Chatham Strait and Lynn Canal in south-east Alaska, form the longest and deepest fiords in North America of at least 370 km long.

5.7.5 Depositional Landforms

Depositional landforms arise from the transport and deposit of glacial sediments or till, and processes involving ice motion or flowing water and ice acting together (Boulton 1996). The latter are termed fluvioglacial processes.

Moraines are ridges and mounds of debris (silt, sand, gravel, and boulders) deposited both on valley glaciers and along the margins of glaciers and ice sheets. Valley glaciers form lateral moraines comprising debris of basal and supraglacial origin and rockfalls along its edges. End moraines are deposited when a glacier is stationary for a long period; if an end moraine lies at the point of farthest advance, it is called a terminal moraine. The most prominent moraines are often from the Little Ice Age advances (~ 1300–1850). Glaciers deposit debris as they advance and recede. These deposits, which are known as ground moraine, are usually only a few meters thick but can cover vast areas. Extensive ground moraine is commonly termed till or till sheets (Flint 1971).

Till is sediment that has been deposited by the ice of valley glaciers and ice sheets and has not undergone subsequent reworking. It can consist of mud, sand, gravel, and boulders in varying proportions. Till occupies vast areas that were formerly covered by continental ice sheets (Clark 1987).

A striking product of glacial deposition is groups of elongated, streamlined hillocks called drumlins. A drumlin is deposited till that has been streamlined in the direction of continental ice movement, with a blunt end upstream and tapered end downstream.

Figure 5.28 Large perched erratic block on Pyramid Peak, New Jersey.
Source: Wally Gobetz, Glacier Photograph Collection, National Snow and Ice Data Center, Boulder, CO.

Multiple drumlins, known as swarms, can contain as many as 10,000 individual features; they occur in fields in Central Canada, England, Ireland, and the northern United States, among other areas. Drumlins are widely distributed in glaciated areas (Flint 1971) and occur in a very great variety of shapes and sizes (Chorley 1959; Gravenor 1953). They are believed to be a product of an ice sheet's motion. It is speculated that drumlins form near the edge of an ice sheet under conditions of unusually heavy debris load in the basal ice. Many theories of drumlin formation have been proposed; some of these are summarized by Boulton (1987), Ebers (1926), and Embleton and King (1975).

Erratic boulders, also known as erratics, are boulders or blocks of various sizes that have been transported by ice far from their source and often deposited in very unusual and out-of-place ways (Figure 5.28). They are readily identified when they are resting on rock of different lithology. Indicator erratics are those for which a definite source is known. The occurrence of boulders in areas far removed from their bedrock source has long been recognized as firm evidence for glacial transportation (Agassiz 1838b). Erratics can be carried for hundreds of miles and can range in size from microgranite fragments to very large boulders. Some of the largest glacially transported blocks are the quartzites that lie on the Cretaceous bedrock prairie of southern Alberta, Canada, more than 375 km from their source in Jasper National Park. The largest block is estimated to weigh approximately 16,000 tons (Trenhaile 1990). Erratics and erratics trains can be used to construct a history of ice flow and ice sheet dynamics, although debris transport histories may involve several glacial cycles (Benn and Evans 2010).

Nonetheless, the distribution of moraines, drumlins, erratics, and fine-grained components of till, provides a powerful tool for reconstructing paleo-environments and the history of ice dispersal (Boulton 1996a).

5.7.6 Fluvioglacial Landforms

The main types of landform resulting from meltwater deposition are eskers, kames, and kame terraces. An esker is a narrow ridge of well-sorted sand and gravel that was deposited by a subglacial stream beneath a valley glacier or an ice sheet. Occasionally, an esker may have originated englacially. Eskers form parallel to ice flow and generally show a close correspondence with the most recent direction regional of ice flow (Bennett and Glasser 2009). Those formed by ice sheets may be up to 50 m high and several hundred kilometers long. Such features are widespread in northern Canada (Figure 5.29). Those formed beneath valley glaciers are much smaller.

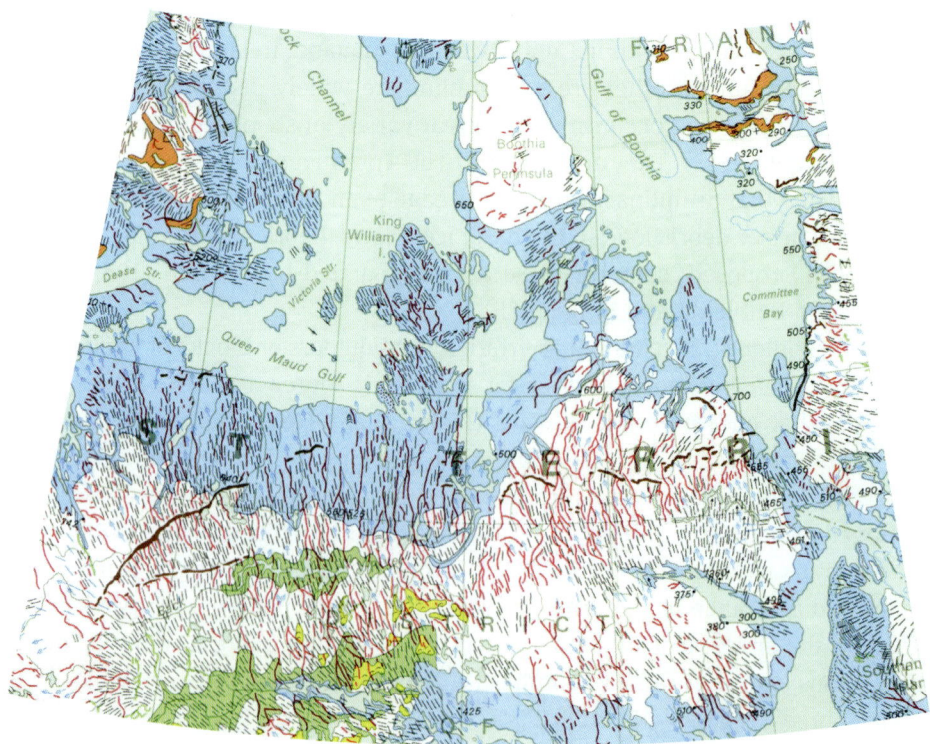

Figure 5.29 (a) A section of the Glacial Map of Canada, 64–72° N, 84–104° E, (b) legend. Source: Prest et al. 1968, map 1253A.

Kames (from Scottish *Kaim* or hill) are steep-sided, variously shaped mounds or ridges composed of stratified sand and gravel that has been deposited by glacial meltwater. Generally, kames develop in low-lying terrain and often occur in association with water filled hollow (kettles) in *kame and kettle topography* (Holmes 1947). Kame terraces often form along side a valley glacier or accumulate in ponds or lakes. Sugden and John (1976) observed that terraces may be "paired" on opposite valley walls, and a series of terraces on a slope may indicate successive positions of the downwasting ice margin. Outwash fans occur where meltwater from a glacier or ice sheet has deposited sand and gravel debris into a fan-shaped body of sediment. Several outwash fans may merge, coalescing into vast outwash plains.

LEGEND

Western and northern limit of Wisconsin (last or classical) Laurentide ice sheet; ticks on glacier side of line (approximate, assumed)

Eastern limit of Wisconsin (last or classical) Cordilleran ice sheet: ticks on glacier side of line (approximate, assumed)

Limit of Cochrane ice-advance

Glacial striation (sense of direction known or inferred, unknown)

Glacial lineation parallel to ice-flow direction; includes glacial fluting, drumlinoid ridge, drumlin, crag and tail hill, roches moutonnées (arrowhead indicates sense of direction)

Glacial lineation mainly transverse to ice-flow direction; includes morainal features variously termed minor, annual, washboard, DeGeer, cyclical, ice-crack, cross-valley, recessional, ribbed, ribbled till, ice-push, ice-thrust ridges

    Irregular to well developed rib-like ridges (mainly on Island of Newfoundland and in Shield regions)

    Corrugated and arcuate ridges; includes ice-thrust ridges (commonly outlines ice lobations on the Interior Plains)

    DeGeer moraines — successions of minor, straight to arcuate ridges, restricted to areas where an active ice-front terminated in moderately deep lake or sea water (mainly Baffin Island, and east and south of Hudson and James Bays)

    Undifferentiated transverse ridges

Glacial lineation mainly oblique to ice-flow directions; believed to be crevasse-fillings (shown only in western Saskatchewan)

Trace of end moraine (compilers' air photo interpretation); in part mantled by younger sediments

Meltwater channel, spillway; includes parts of pre-Wisconsin drainage channels (direction indicated only where not obvious)

Direction of transport of erratics in Arctic Islands:

    Last glaciation . . . . N, W, NW

    Far-travelled Laurentide erratics of an earlier glaciation . . . . NW

    Limit of Greenland erratics . . . . W W W W

Ice-flow direction inferred from various data . . . .

Approximate elevation of marine limit in feet 400

    with locality . . . . •425

Prominent strand lines (marine and lacustrine)

Bathymetric contours (interval 500 metres)

Compiled by V. K. Prest, D. R. Grant and V. N. Rampton, 1964-1966

For sources of information see reverse side of this map

Cartography by the Geological Survey of Canada, 1967

Existing glacier; includes ice cap, montane, piedmont and valley glacier, local areas of permanent snow

Unglaciated area

Area of Wisconsin (last or classical) glaciation; mainly ground moraine (other features shown as geomorphic subdivisions)

Area of pre-Wisconsin glaciation beyond the limit of last glaciation [1]; mainly ground moraine (other features shown as geomorphic subdivisions)

Area in part unglaciated, in part covered by ice of one or more glaciations

GEOMORPHIC SUB-DIVISIONS

Area of maximum marine overlap [2]

Area of maximum glacial lake coverage (confined largely to areas of mapped lake deposits on the Interior Plains but locally omitted in hummocky terrain)

Outwash area — commonly dune-covered; includes outwash plain, valley train, delta

Esker, kame, kame-complex

Ribbed moraine[3] — areas with irregular to arcuate ribbed pattern more or less transverse to ice-flow direction

Hummocky terrain[4] -hummocky, dead-ice and disintegration moraine, includes prairie mounds and some transversely lineated ground moraine; local pitted lacustrine deposits

End moraine[5,6]; includes interlobate, lateral, marginal and kame moraines

[1] Includes local mountain areas that were not completely covered by glaciers

[2] On northeastern Baffin Island the surface marine deposits marked with an asterisk (•) pre-date the last glaciation

[3] Mainly from compilers' air photo interpretations; includes some areas mapped by others as either end moraine complex or ground moraine

[4] Mainly from compilers' air photo interpretations; includes areas mapped by others as either hummocky or ground moraine or as end moraine complex

[5] Mainly as mapped by others but locally extended by compilers' air photo interpretations

[6] In lower Fraser Valley 'end moraine' denotes a significant ice-frontal position marked by a local thickening of glaciomarine stony clay, ice-contact deposits and outwash

Figure 5.29 (cont.)

SUMMARY

Polar desert and tundra are the main land surfaces in the Arctic. Both are underlain by permafrost. Polar deserts have a warmest month mean temperature of less than 5 °C and mean annual precipitation of less than 250 mm, mostly falling as warm-season rain. Winter temperatures at inland Lake Hazen, Ellesmere Island, in 1957–1968, were substantially lower than at the coastal stations due to its sheltered location. Much of the polar desert surface is fell field and gravel plains. Cushion plants, lichens, and mosses cover only 1–5 percent of the surface and patterned ground is common. Some 20–100 vascular plant species are found in polar deserts.

Tundra occupies approximately 8 percent of the global land surface. It ranges from barrens, graminoids, prostate and erect shrubs, to wetlands. Mid-winter temperatures average −25 to −30 °C, and no month has a mean temperature above 10 °C. Annual precipitation is 150–250 mm yr^{-1}, falling mainly as summer rain. Snow covers the ground for 8–9 months and melt is often unable to drain away from the (20–60 cm) thawed active layer.

Barrow, Alaska, was the location of an intensive Tundra Biome project site. At this site, the number of vascular plants was found to be two to three times greater on the inland coastal plain than at the coast, where there are mainly grasses and sedges with dwarf shrubs. In the Russian Arctic, there are three latitudinal zones: Arctic tundra, typical tundra, and southern tundra. The last of these has birch, willow, and alder shrubs. There are longitudinal differences in the zones, and the arid northeast consists of steppe tundra. Mires are common in the typical and southern tundra. Rising air and soil temperatures over the last five decades have led to widespread shrub expansion ("greening") across the Arctic tundra. Local browning seems to have multiple causes.

Permafrost zones occupy approximately 24 percent of the northern hemisphere; permafrost underlies 12–18 percent of the exposed land surface in these zones because it is not all continuous. The southern boundary of continuous permafrost is approximated by a mean annual air temperature of −8 °C. Thicknesses are up to 1,000 m in north-central Siberia, but thick permafrost is absent from northern Canada due to the Laurentide ice sheet of the last glacial cycle. Permafrost is continuous on the northern Tibet plateau, with a lower limit of 4,200 m in this area that rises to 4,800 m in the south. Permafrost occurs in ice-free areas of the Antarctic Peninsula and McMurdo Dry Valleys. Subsea permafrost (ice-bonded and ice-bearing sediment) is widespread on Arctic continental shelves, descending to the −20-m isobaths in the Beaufort Sea and to the −100-m isobaths in the

Laptev Sea. The active layer thickness can range from 10 cm to 4 m, depending on energy regime, snow and vegetation cover, and soil thermal properties. Peat is an especially important element of permafrost due to its low thermal conductivity.

There are four types of ground ice: pore, segregated, intrusive, and vein. Intrusive ice leads to palsa mounds and pingos, while vein ice can form ice wedges. Periglacial freeze–thaw processes give rise to movement of soil particles and gravel that form solifluction lobes, stone polygons and stripes, and earth hummocks. Earth-covered ice mounds, or pingos, are typically 5 m high, but can reach more than 45 m. There are 1,350 pingos in the Mackenzie delta and about 6,000 pingos in northern Asia. Closed system pingos grow by permafrost aggradation in drained lake bed sediments; open system pingos (found in East Greenland) arise when sub- or intra-permafrost aquifers create an ice core as water is pushed up in a hydolaccolith.

Larger-scale features include thermokarst, created by melting ground ice, that forms hollows (alases in Siberia), 3–40 m deep and 100 m in diameter, and rock glaciers. Rock glaciers may be talus-derived periglacial forms or glacier-related features, occurring below current or past alpine glaciers. They form a tongue or lobe of angular blocks, 200–800 m long and 20–100 m thick, with high ice content.

Permafrost ground temperatures are rising due to global warming and changes in snow cover. Warming is, in turn, leading to thermokarst formation, especially on ice-rich marine silts. Arctic coasts with ice-rich sediments are experiencing rapid wave erosion and retreat. The average erosion rate for Arctic coasts is 0.6 m yr^{-1} but rates greater than 3 m yr^{-1} have been recorded locally.

Lakes cover 11 percent of the Mackenzie River basin. They absorb more than 90 percent of incoming solar radiation, and evapotranspiration rates are high when they are unfrozen. Arctic and sub-Arctic lakes are frozen for a large part of the year. In the Northwest Territories, Great Bear Lake has ice for 225 days of the year. Thermokarst lakes were found to occupy 16 percent of a 300 km^2 study area in Alaska. Of the almost 1,400 glacial lakes in southeast Tibet identified in 2013, 92 percent were not in contact with a glacier. The few that were showed rapid expansion attributed to melt-induced debris subsidence at glacier termini, rather than as a result of direct glacier melt. There are about five permanently frozen lakes in the Arctic and eleven such locations in Antarctica, with twenty perennially frozen lakes in the Dry Valleys. Ice thickness ranges from 2 to 5.5 m in the Arctic, and up to 19 m on Lake Vida. Lake Vanda, the deepest and largest perennially frozen lake, has 3-m-thick ice and experiences water loss by ice ablation.

Closed lakes have high salinity due to their high evaporation rates. Open lakes with inflow and outflow have freshwater. Epishelf lakes form when glaciers or ice shelves dam a fjord or embayment. A freshwater layer, from glacier melt, may overlie the seawater.

Subglacial lakes exist because of geothermal heating (50–70 mW m^{-2}) at the base of an ice sheet. Some 379 subglacial lakes have been identified in Antarctica. The largest, Lake Vostok, has an area of 14,000 km^2 and a maximum depth of 1,000 m.

Global lake surfaces have warmed since 1985. For seasonally ice-covered lakes, the rate of warming is 0.7 °C decade^{-1}. In the central Lena River basin, lake area increased overall by 18 percent during 2002–2009. Land surface temperatures increased in spring, suggesting snow cover changes were involved in this trend.

Glacier-dammed lakes across the entire Himalaya number 4,950, covering 455 km^2. They expanded by 14 percent from 1990 to 2015, especially on the southern slopes of the Central Himalaya due to glacier retreat. Some pose a high risk of glacier lake outburst floods.

Northern hemisphere lake and river freeze-up occurred 5.6 days per century later over 1846–1995, and breakup 6.5 days per century earlier. Lakes in polar oases in the Canadian Arctic had water-clear-of-ice dates that were nine to twenty-four days earlier than historical records during 1997–2011.

Arctic rivers flow mainly northward to the Arctic Ocean, where they have large deltas. Ice breakup occurs in late May or early June and peak discharge in June. The Mackenzie delta has a northern treeless portion and a southern tree-covered part. Lakes occupy 25 percent of this area, mainly in the southern part.

The Colville River flows entirely over continuous permafrost. The 12-m-deep main channel, one of nine, remains open in winter. The 4,250-km-long Lena River has ten main channels and about 800 in total. Over one third of the annual discharge occurs in June into the Laptev Sea. The deltas of the Ob and Yenisei are at the head of long estuaries. The Yenisei has the largest discharge into the Arctic Ocean but a large interannual variation. Suggested increases in discharge of the six largest Eurasian rivers appear not to be significant. For Canadian rivers during 1964–2013, discharge trends were strongly positive into the western Arctic and strongly negative into the Eastern Arctic Ocean.

Icings due to water seeps that freeze are common in Arctic river valleys. They may be river fed or spring fed. In northern Russia, they occupy 238,000 km^2 and are several meters thick. In Alaska and northern Canada, many icings are disappearing earlier, and ones that persist in summer are smaller.

Arctic glaciers and ice caps occupy approximately 425,000 km^2 and occur in the Canadian Arctic archipelago; around the Greenland ice sheet; in Svalbard, Novaya Zemlya, and Franz Josef Land; and in Alaska. In the southern hemisphere, they are extensive in the Antarctic Peninsula, around the ice sheet, and in the sub-Antarctic islands. Glaciers can be classified according to their thermal state or morphology.

Ice in the Canadian Arctic has undergone massive retreat since the end of the Little Ice Age. There was almost complete removal of ice on the low-lying central and western Canadian Arctic archipelago by 1960. Surface mass balance (SMB) on the Canadian ice caps became increasingly negative after 1987. A revised analysis for Alaskan glaciers indicated that they contributed 0.12 mm yr^{-1} of sea level equivalent (SLE) for 1962–2006. In the Torngat Mountains, northern Labrador, glaciers shrank by 27 percent between 1950 and 2005. Glaciers in northern Spitsbergen decreased by 43 percent between their maxima around 1900 and 2015. Mass changes for 2003–2009, in Gt yr^{-1}, totaled −50 for Alaska, −38 for the Greenland periphery, −33 for Arctic Canada north, −27 for Arctic Canada south, and −11 for the Russian Arctic, with a global glacier total mass change of −215 Gt yr^{-1}. Most glaciers in the Antarctic Peninsula retreated from the 1950s to 2004.

Glacial landscapes are either currently glacierized or were glaciated in the past. Glacial landforms are either erosional or depositional. Glacial erosion is accomplished by abrasion, in which debris transported by the glacier scours the bedrock; by plucking, in which the glacier loosens, picks up, and transports rocks and fragments; and by freeze–thaw processes, in which freezing of basal water in cracks can shatter the bedrock. Basal streams are also locations of fluvioglacial erosion.

Small-scale features of glaciated bedrock are grooves and striations. Mesoscale features are roches moutonnées and whalebacks. Large-scale features of erosion are found in mountainous terrain, where there are glacial cirques, U-shaped valleys, glacial steps, and coastal fjords.

Depositional glacial landscapes include moraines, basal till sheets, drumlins, and eskers. Fluvioglacial forms include eskers, kames, and kame terraces.

QUESTIONS

1. Describe the environmental conditions in polar desert areas.
2. How variable is Arctic tundra?
3. How is northern hemisphere permafrost changing, and what are the causes of these changes?
4. In what ways are lakes important features of the polar regions?
5. How are Arctic rivers distinctive?
6. Account for the general distribution of glaciers and ice caps across the Arctic.
7. Suggest reasons for the large observed differences in glacier retreat rates across the Arctic.
8. Compare the glacial landforms of Keewatin and Baffin Island.

References

Agassiz, L. 1838a. "On the Polished and Striated Surfaces of the Rocks Which Form the Beds of Glaciers in the Alps." *Proceedings of the Geological Society of London* 3: 321–2.

Agassiz, L. 1838b. "On Glaciers and the Evidence of Their Having One Existed in Scotland, Ireland and England. *Proceedings of the Geological Society of London* 3: 327–32.

Alexandrova, V. D. 1988. *Vegetation of the Soviet Polar Deserts*. Cambridge: Cambridge University Press.

Arendt, A., et al. 2002. "Rapid Wastage of Alaska Glaciers and Their Contribution to Rising Sea Level." *Science* 297(5580): 382–6.

Atchley, A. L., et al. 2016. "Influences and Interactions of Inundation, Peat, and Snow on Active Layer Thickness." *Geophysical Research Letters* 43(10): 5116–23.

Bajracharya, S. R., P. K. Mool, and B. R. Shrestha. 2007. *Impact of Climate Change on Himalayan Glaciers and Glacial Lakes: Case Studies on GLOF and Associated Hazards in Nepal and Bhutan*. Kathmandu, Nepal: International Centre for Integrated Mountain Development.

Barrand, N. E., et al. 2016. "Recent Changes in Area and Thickness of Torngat Mountain Glaciers (Northern Labrador, Canada)." *Cryosphere Discussion*. doi: 10.5194/tc-2016-171/.

Barry, R. G., G. M. Courtin, and C. Labine. 1981. "Tundra Climate." In *Tundra Ecosystems: A Comparative Analysis*, edited by L. C. Bliss and O. W. Heal, 81–114. Cambridge: Cambridge University Press.

Barry, R. G., and T. Y. Gan. 2011. *The Global Cryosphere: Past, Present and Future*. Cambridge: Cambridge University Press.

Barsch, D. 1996. *Rock Glaciers*. Berlin: Springer.

Beer, C., A. N. Fedorov, and Y. Torgovkin. 2013. "Permafrost Temperature and Active-Layer Thickness of Yakutia with 0.5-Degree Spatial Resolution for Model Evaluation." *Earth System Science Data* 5: 305–10.

Beltaos, S., ed. 2008. *River Ice Breakup*. Highlands Ranch, CO: Water Resources Publications.

Benn, D. I., and D. J. A. Evans. 1998. *Glaciers and Glaciation*. London: Arnold.

Benn, D. I., and D. J. A. Evans. 2010. *Glaciers and Glaciation*. 2nd ed. London: Routledge.

Bennett, M. R., and N. F. Glasser. 2009. *Glacial Geology: Ice Sheets and Landforms*, 2nd ed. Chichester, UK: Wiley Blackwell.

Berthier, E., et al. 2010. "Contribution of Alaska Glaciers to Sea-Level Rise Derived from Satellite Imagery." *Nature Geoscience* 3: 92–5.

Bird, J. B. 1967. *The Physiography of Arctic Canada*. Baltimore, MD: John Hopkins Press.

Blanken, P. D., et al. 2000. "Eddy Covariance Measurements of Evaporation from Great Slave Lake, Northwest Territories, Canada." *Water Resources Research* 36: 1069–77.

Bliss, A., R. Hock, and J. G. Cogley. 2013. "A New Inventory of Mountain Glaciers and Ice Caps for the Antarctic Periphery." *Annals of Glaciology* 54(63): 191–9.

Bliss, L. C., O. W. Heal, and J. J. Moore, eds. 1981. *Tundra Ecosystems: A Comparative Analysis*. Cambridge: Cambridge University Press.

Bliss, L. C., and N. V. Matyeeva. 1992. "Circumpolar Arctic Vegetation." In *Arctic Ecosystem in a Changing Climate: An Ecophysiological Perspective*, edited by F. S. Chapin, 59–90. San Diego, CA: Academic Press.

Bockheim, J. G. 1995. "Permafrost Distribution in the Southern Circumpolar Region and Its Relation to the Environment: A Review and Recommendations for Further Research." *Permafrost and Periglacial Processes* 6: 27–45.

Boike, J., et al. 2013. "Baseline Characteristics of Climate, Permafrost, and Land Cover from Samoylov Island, Lena River Delta, Siberia." *Journal of Geophysical Research: Biogeoscience* 10: 21054–128.

Boike, J., et al. 2016. "Satellite-Derived Changes in the Permafrost Landscape of Central Yakutia, 2000–2011: Wetting, Drying, and Fires." *Global and Planetary Change* 139: 116–27.

Boulton, G. S. 1974. "Processes and Patterns of Subglacial Erosion." In *Glacial Geomorphology*, edited by D. R. Coats, 41–87. Binghampton, NY: University of New York.

Boulton, G. S. 1987. "A Theory of Drumlin Formation by Subglacial Sediment Deformation. In *Drumlin Symposium*, edited by J. Rose, 25–80. Potsdam: Balkema.

Boulton, G. S. 1996. "Theory of Glacial Erosion, Transport and Deposition as a Consequence of Subglacial Sediment Deformation." *Journal of Glaciology* 42: 43–62.

Bovis, M. J., and R. G. Barry. 1973. "A Climatological Analysis of North Polar Desert Areas." In *Polar Deserts and Modern Man*, edited by T. L. Smiley and J. H. Zumberge, 23–31. Tucson, AZ: University of Arizona Press.

Bring, A., et al. 2016. "Arctic Terrestrial Hydrology: A Synthesis of Processes, Regional Effects and Research Challenges." *Journal of Geophysical Research: Biogeoscience*. doi: 10.1002/2015JG003131.

Brothers, L. L., et al. 2016. "Subsea Ice-Bearing Permafrost on the U.S. Beaufort Margin: 1. Minimum Seaward Extent Defined from Multichannel Seismic Reflection Data." *Geochemistry, Geophysics, Geosystems*. doi: 10.1002/2016GC006584.

Brown, J., ed. 1980. *An Arctic Ecosystem: The Coastal Tundra at Barrow, Alaska*. Stroudsburg, PA: Dowd, Hutchinson and Ross.

Brown, J., et al. 1980. "The Coastal Tundra at Barrow. In *An Arctic Ecosystem: The Coastal Tundra at Barrow, Alaska*, edited by J. Brown, 1–29. Stroudsburg, PA: Dowd, Hutchinson and Ross.

Brown, J., et al. 1998, revised February 2001. *Circum-Arctic Map of Permafrost and Ground Ice Conditions*. [Digital media]. Boulder, CO: National Snow and Ice Data Center.

Budyko, M. I. 1956. *The Heat Balance of the Earth's Surface*. Leningrad: Gidromet. Izdat. Translated by N. A. Stepanova, US Weather Bureau, Washington, DC, 1958.

Byers, A. C., et al. 2013. "Glacial Lakes of the Hinku and Hongu Valleys, Makalu Barun National Park and Buffer Zone, Nepal." *Natural Hazards* 69: 115–39.

Campbell, I. B., and G. G. C. Claridge. 2009. "Antarctic Permafrost Soils." In *Permafrost Soils*, edited by R. Margesin, 17–31. Berlin: Springer-Verlag.

Carey, K. 1973. *Icings Developed from Surface and Groundwater*. CRREL Monograph, Vol. III-D3. Hanover, NH: US Army Cold Regions Research and Engineering Laboratory.

Castendyck, D. N., et al. 2016. "Lake Vanda: A Sentinelf for Climate Change in the McMurdo Sound Region of Antarctica." *Global and Planetary Change* 144: 213–27.

CAVM Team. 2003. *Circumpolar Arctic Vegetation Map. Conservation of Arctic Flora and Fauna Map (CAFF) Map No. 1*. Anchorage, AK: US Fish and Wildlife Service.

Chamberlin, T. C. 1888. "The Rock Scourings of the Great Ice Invasions." *US Geologic Survey Annual Report* 7: 155–248.

Charlier, R. H. 1969. "The Geographic Distribution of Polar Desert Soils in the Northern Hemisphere." *Geological Society of America Bulletin* 80: 1985–96.

Chekotillo, A. M., A. A. Tsvid, and V. N. Makarov. 1960. *Naledy na territorii SSSR i bor'ba s nim. (Icings in the USSR and Their Control)*. Blaovashchensk: Amur Knizhn Izdat. Translated for CRREL, US Army, Hanover, NH, 1965.

Cheng, G.-D., and F. Dramis. 1992. "Distribution of Mountain Permafrost and Climate." *Permafrost and Periglacial Processes* 3: 83–9.

Chorley, R. J. 1959. "The Shape of Drumlins." *Journal of Glaciology* 3: 339–44.

Clark, P. U. 1987. "Subglacial Sediment Dispersal and Till Composition." *Journal of Geology* 95: 527–41.

Cook, A. J., et al. 2005. "Retreating Glacier Fronts on the Antarctic Peninsula over the Past Half-Century." *Science* 308: 541–44.

Cook, A. J., et al. 2016. "Ocean Forcing of Glacier Retreat in the Western Antarctic Peninsula." *Science* 353: 283–6.

De Rham, L. P., T. D. Prowse, and B. P. Bonsal. 2008. "Temporal Variations in River Ice Break-Up over the Mackenzie River Basin, Canada. *Journal of Hydrology* 349: 441–54.

Déry, S. J., et al. 2016. "Recent Trends and Variability in River Discharge across Northern Canada." *Hydrology and Earth Systems Sciences* 20: 4801–18.

Dingman, S. L., et al. 1980. "Climate, Snow Cover, Microclimate and Hydrology." In *An Arctic Ecosystem: The Coastal Tundra at Barrow, Alaska*, edited by J. Brown, 30–65. Stroudsburg, PA: Dowd, Hutchinson and Ross.

Doran, P. T., et al. 2002. "Valley Floor Climate Observations from the McMurdo Dry Valleys, Antarctica, 1986–2000." *Journal of Geophysical Research* 107: 4772–84.

Doran, P. T., et al. 2004. "Paleolimnology of Extreme Cold Terrestrial and Extraterrestrial Environments." In *Long-Term Change in Arctic and Antarctic Lakes*, edited by R. Pienitz et al., 475–507. Dordrecht: Kluwer.

Drewry, D. J. 1983. *Antarctica, Glaciological and Geophysical Folio*. Cambridge: University of Cambridge Press.

Duguay, C., et al. 2015. "Remote Sensing of Lake and River Ice." In *Remote Sensing of the Cryosphere*, edited by M. Tedesco, 273–306. New York: John Wiley and Sons.

Dyurgerov, M., and M. Meier. 2005. *Glaciers and the Changing Earth System: A 2004 Snapshot*. Occasional Paper No. 58. Boulder, CO: Institute of Arctic and Alpine Research, University of Colorado.

Ebers, E. 1926. "Dir bisherigen Ergebrisse der Drumlinforschung. Eine Monographie der Drumlins. Neues Jahrbuch Min." *Geology and Palaeontology* 53(A): 153–270.

Embleton, C. 1964. "Subglacial Drainage and Supposed Ice-Dammed Lakes in Northeast Wales." *Proceedings of the Geologists' Association* 75(1): 31–8.

Embleton, C., and C. A. M. King. 1975. *Glacial Geomorphology*. London: Edward Arnold, London.

Ericson, K. 2004. "Geomorphological Surfaces of Different Age and Origin in Granite Landscapes: An Evaluation of the Schmidt Hammer Test." *Earth Surface Processes and Landforms* 29: 495–509.

Evans, I. S. 1996. "Abraded Rock Landforms (Whalebacks) Developed under Ice Streams in Mountain Areas." *Annals of Glaciology* 22: 9–16.

Farquharson, L. M., et al. 2016. "Spatial Distribution of Thermokarst Terrain in Arctic Alaska." *Geomorphology*. doi: 10.1016/j.geomorph.2016.08.007.

Fieber, K. D., et al. 2018. "Rigorous 3D Change Determination in Antarctic Peninsula Glaciers from Stereo WorldView-2 and Archival Aerial Imagery." *Remote Sensing of the Environment* 205: 18–31.

Flint, R. F. 1971. *Glacial and Quaternary Geology*. New York: Wiley.

Fountain, A. G., et al. 1998. "Glaciers of the McMurdo Dry Valleys, Southern Victoria Land, Antarctica." In *Ecosystem Processes in a Polar Desert: The McMurdo Dry Valleys, Antarctica*. Antarctic Research Series No. 72, edited by J. C. Priscu, 65–75. Washington, DC: American Geophysical Union.

French, H. M. 2007. *The Periglacial Environment*. Chichester, UK: John Wiley and Sons.

Fristrup, B. 1952. "Physical Geography of Peart Land. I. Meteorological Observations for Jørgen Brønlunds Fiord." *Meddelelser om Grønland* 127(4): 1–143.

Fristrup, B. 1961. "Climatological Studies of Some High Arctic Stations in North Greenland." *Physical Geography of Greenland. Folia Geogr. Danica (Copenhagen)* 9: 67–78.

Fu, P., and J. Harbor. 2011. "Glaciological Variables Controlling Glacial Erosion." In *Encyclopedia of Snow, Ice and Glaciers*, edited by V. P. Singh et al. Amsterdam: Springer.

Gardner, A. S., et al. 2011. "Sharply Increased Mass Loss from Glaciers and Ice Caps in the Canadian Arctic Archipelago." *Nature* 473: 357–60.

Gardner, A. S., et al. 2012. "Accelerated Contributions of Canada's Baffin and Bylot Island Glaciers to Sea Level Rise over the Past Half Century." *Cryosphere* 6: 1103–25.

Gardner, A. S., et al. 2013. "Reconciled Estimate of Glacier Contributions to Sea Level Rise: 2003 to 2009." *Science* 340: 852–7.

Gertcyk, O. August 10, 2016. "Arctic's Climate on a Cliff-Edge." *The Siberian Times*. http://siberiantimes.com/ecology/casestudy/news/n0704-arctics-climate-on-a-cliff-edge/

Gibson, J. A. E. 1999. "The Meromictic Lakes and Stratified Marine Basins of the Vestfold Hills, East Antarctica." *Antarctic Science* 11: 175–92.

Gibson, J. A. E., and D. T. Anderson. 2002. "Physical Structure of Epishelf Lakes of the Southern Bunger Hills, East Antarctic." *Antarctic Science* 14: 253–61.

Gilbert, A., et al. 2016. "Sensitivity of Barnes Ice Cap, Baffin Island, Canada, to Climate State and Internal Dynamics." *Journal of Geophysical Research: Earth Surface*. doi: 10.1002/2016JF003839.

Gilbert, A., et al. 2017. "The Projected Demise of Barnes Ice Cap: Evidence of an Unusually Warm 21st Century Arctic." *Geophysical Research Letters* 44: 2810–16.

Gilbert, G. K. 1903. *Glaciers and Glaciation of Alaska*. New York: Doubleday, Page & Co.

Gilbert, G. K. 1906. "Crescentic Gouges on Glaciated Surfaces." *Geological Society of America Bulletin* 17: 303–16.

Ginsburg, B. M., and I. I. Soldatova. 1997. "Long-Term Variability of Ice Phenomena Dates on Rivers as an Indicator of Climate Variations in Transitional Seasons." *Soviet Meteorology and Hydrology* 11: 71–8.

Glasser, N. F., and C. R. Warren. 1990. "Medium Scale Landforms of Glacial Erosion in South Greenland: Process and Form." *Geografiska Annaler* 72A: 211–15.

Glen, J. W., and W. V. Lewis. 1961. "Measurements of Side-Slip at Austerdalsbreen, 1959." *Journal of Glaciology* 3(30): 1109–22.

Gordon, J. E. 1981. "Ice-Scoured Topography and Its Relationship to Bedrock Structure and Ice Movements in Parts of Northern Scotland and West Greenland." *Geografiska Annaler* 63A: 55–65.

Gorokhovich, Y., and A. Leiserowiz. 2012. ."Historical and Future Coastal Changes in Northwest Alaska." *Journal of Coastal Research* 28: 174–86.

Grant, L. 2010. *Changes in Glacier Extent since the Little Ice Age and Links to 20th/21st Century Climatic Variability on Novaya Zemlya, Russian Arctic*. PhD dissertation. Reading, UK: University of Reading, Department of Geography.

Gravenor, C. P. 1953. "The Origin of Drumlins." *American Journal of Science* 251: 674–81.

Green, W. J., and W. B. Lyons. 2009. "The Saline Lakes of the McMurdo Dry Valleys, Antarctica." *Aquatic Geochemistry* 15: 321–48.

Gronskaya, T. P. 2000. "Ice Thickness in Relation to Climate Forcing in Russia." *Verhandlungen des Internationalen Verein Limnologie* 27L: 2800–2.

Grosse, G., and B. M. Jones. 2011. "Spatial Distribution of Pingos in Northern Asia." *Cryosphere* 5: 13–33.

Gudmundsson, A. 2011. *Rock Fractures in Geological Processes*. Cambridge: Cambridge University Press.

Hallet, B. 1996. "Glacial Quarrying: A Simple Theoretical Model." *Annals of Glaciology* 22: 1–8.

Hambrey, M. G., and J. G. Alean. 2017. *Colour Atlas of Glacial Phenomena*. Boca Raton, FL: CRC Press.

Harbor, J. 2013. "Glacial Erosion Processes and Rates." In *Treatise on Geomorphology. Vol. 8: Glacial and Periglacial Geomorphology*, edited by J. Shroder, R. Giardino, and J. Harbor, 74–82. Cambridge, MA: Academic Press.

Harbor, J., and J. Warburton. 1993. "Relative Rates of Glacial and Nonglacial Erosion in Alpine Environments." *Arctic and Alpine Research* 25: 1–7.

Heginbottom, J. A., et al. 2013. "Permafrost and Periglacial Environments." Professional Paper 1386-A. In *State of the Earth's Cryosphere at the Beginning of the 21st Century: Glaciers, Global Snow Cover, Floating Ice, and Permafrost and Periglacial Environment*, edited by R. S. Williams, Jr., and J. Ferrino, A425–96. Reston, VA: US Geological Survey.

Helm, C. W. 2007. *Glacier Changes in Franz Josef Lan, 1952–2004*. [MA thesis]. Boulder, CO: University of Colorado, Department of Geography.

Hodgson, D. A., P. T. Doran, and D. Roberts. 2004. "Paleolimnological Studies from the Antarctic and Subantarctic Islands." In *Long-Term Change in Arctic and Antarctic Lakes*, edited by R. Pienitz et al., 419–74. Dordrecht: Kluwer.

Holmes, C. D. 1947. "Kames." *American Journal of Science* 245: 240–9.

Howell, S. E., et al. 2009. "Variability in Ice Phenology on Great Bear Lake and Great Slave Lake, Northwest Territories, Canada from SeaWinds/QuikSCAT; 2000–2006." *Remote Sensing of the Environment* 113: 816–34.

Iverson, N. R. 1991. "Potential Effects of Subglacial Water-Pressure Fluctuations on Quarrying." *Journal of Glaciology* 37: 27–36.

Ives, J. D. 1986. *Jokulhlaup Disasters in the Himalaya and Their Identification*. ICIMOD Occasional Paper. Kathmandu, Nepal: International Centre for Integrated Mountain Development.

Ives, J. D., R. B. Shrestha, and P. K. Mool. 2010. *Formation of Glacial Lakes in the Hindu Kush-Himalayas and GLOF Risk Assessment*. Kathmandu, Nepal: International Centre for Integrated Mountain Development.

Jackson, C. I. 1959. "Coastal and Inland Weather Contrasts in the Canadian Arctic." *Journal of Geophysical Research* 64: 1451–5.

Jahns, R. H. 1943. "Sheet Structure in Granites, It Origin and Use as a Measure of Glacial Erosion in New England." *Journal of Geology* 51: 71–98.

Jones, B. M., et al. 2016. "Presence of Rapidly Degrading Permafrost Plateaus in South Central Alaska." *Cryosphere*. doi: 10.5194/tc-2016-91.

Kane, D. L. 1981. "Physical Mechanics of Aufeis Growth." *Canadian Journal of Civil Engineering* 8: 188–95.

Kang, K.-K., et al. 2014. "Estimation of Ice Thickness on Large Northern Lakes from AMSR-E Brightness Temperature Measurements." *Remote Sensing of the Environment* 150: 1–19.

King, L. 1981. "The Summer of 1978 in Northern Ellesmere Island: A Comparison of Climatic Records of Weather Stations." In *Ergebnisse der Heidelberg–Ellesmere Island: Expedition*, edited by D. Barsch and L. King, 69, 77–107. Heidelberg: Geogr. Arbeit.

Koronkevich, N. 2002. "Rivers, Lakes, Inland Seas and Wetlands." In *The Physical Geography of Northern Eurasia*, edited by M. Shahgedanova, 122–48. Oxford: Oxford University Press.

Korotkevich, E. S. 1972. *Polyarnie Pustini (Polar Deserts)*. [In Russian.] Leningrad: Gidrometeoizdat.

Kotlyakov, V. M., ed. 1997. *Atlas of Snow and Ice Resources*. Moscow: Institute of Geography, Russian Academy of Sciences.

Lantuit, H., et al. 2012. "The Arctic Coastal Dynamics Database: A New Classification Scheme and Statistics on Arctic Permafrost Coastlines." *Estuaries and Coasts* 35: 383–400.

Laverdiere, C., and C. Bernard. 1970. "Bibliographie Annotee sur les Broutures Glaciaires." *La Revue de geographie de Montreal* 24: 79–89.

Lewis, P. 1991. "Sedimentation in the Mackenzie Delta." In *Mackenzie Delta: Environmental Interactions and Implications of Development*, edited by P. Marsh and C. Ommaney. Saskatoon: Environment Canada.

Lotz, J. R., and R. B. Sagar. 1962. "Northern Ellesmere Island: An Arctic Desert." *Geography Annals* 44: 366–77.

Mackay, J. R. 1963. *The Mackenzie Delta Area, N.W.T.* Memoir 8. Ottawa: Geographical Branch, Mines and Technical Surveys.

Mackay, J. R. 1972. "The World of Underground Ice." *Annals of the Association of American Geographers* 62: 1–22.

Mackay, J. R. 1979. "Pingos of the Tuktoyaktuk Peninsula Area, Northwest Territories." *Geographie Physique et Quaternaire* 33: 3–6.

Mackay, J. R. 1986. "Frost Mounds." In *Focus: Permafrost Geomorphology*, edited by H. M. French. *Canadian Geographer* 30: 363–4.

Magnuson, J. D., et al. 2000. "Historical Trends in Lake and River Ice Cover in the Northern Hemisphere." *Science* 289(5485): 1743–4.

Martin, H. E., and W. B. Whalley. 1987. "Rock Glaciers, Part 1: Rock Glacier Morphology: Classification and Distribution." *Progress in Physical Geography* 11: 260–83.

McDonald, G. M., et al. 2007. "Recent Eurasian River Discharge to the Arctic Ocean in the Context of Longer-Term Dendrohydrological Records." *Journal of Geophysical Research* 112: G04S50.

Mercer, J. H. 1967. *Southern Hemisphere Glacier Atlas*. Technical Report 67–76-ES, Series ES-33. Natick, MA: US Army Natick Laboratories, Earth Sciences Laboratory, prepared by the American Geographical Society, New York.

Moholdt, G., B. Wouters, and A. S. Gardner. 2012. "Recent Mass Changes of Glaciers in the Russian High Arctic." *Geophysical Research Letters* 39: L10502.

Moran, S. R. 1971. "Glaciotectonic Structures in Drift." In *Till: A Symposium*, edited by R. P. Goldthwait, 127–48. Columbus: Ohio State University Press.

Morgan, V. I., and W. R. Budd. 1975. "Radio Echo Sounding of the Lambert Glacier Basin." *Journal of Glaciology* 15: 103–11.

Müller, F. 1959. "Beobachtung uber Pingos: Detailuntersuchungen in Ost Grønland und in der Kanadisschen Arktis. (Observations on Pingos: Detailed Investigations in East Greenland and in the Canadian Arctic)." [In German, English summary]. *Meddelelser om Grønland* 153.

Munroe-Stasiuk, M. J., J. Hayman, and J. Harbor. 2013. "Erosional Features." In *Treatise on Geomorphology. Vol. 8: Glacial and Periglacial Geomorphology*, edited by J. R. Giardino and J. M. Harbor, 83–99. London: Academic Press.

Nie, Y., et al. 2017. "A Regional-Scale Assessment of Himalayan Glacial Lake Changes Using Satellite Observations from 1990 to 2015." *Remote Sensing of the Environment* 189: 1–13.

Nitze, I., et al. 2017. "Landsat-Based Trend Analysis of Lake Dynamics across Northern Permafrost Regions." *Remote Sensing* 9: 640–67.

Nuth, C., et al. 2013. "Decadal Changes from a Multi-Temporal Glacier Inventory of Svalbard." *Cryosphere* 7: 1603–21.

O'Reilly, C. M., et al. 2015. "Rapid and Highly Variable Warming of Lake Surface Waters around the Globe." *Geophysical Research Letters* 42(24): 10773–81.

Østby, T. I., et al. 2017. "Diagnosing the Decline in Climatic Mass Balance of Glaciers in Svalbard over 1957–2014." *Cryosphere* 11: 191–215.

Palecki, M. A., and R. G. Barry. 1986. "Freeze-Up and Break-Up of Lakes as an Index of Temperature Changes during the Transition Seasons: A Case Study for Finland." *Journal of Applied Meteorology and Climatology* 25: 893–902.

Park, H., et al. 2017. "Warming Water in Arctic Terrestrial Rivers." *Journal of Hydrometeorology* 18: 1983–95.

Pavelsky, T. M., and J. P. Zarenetske. 2017. "Rapid Decline in River Icings Detected in Arctic Alaska: Implications for a Changing Hydrologic Cycle and River Ecosystems." *Geophysical Research Letters* 44: 3328–35.

Pelto, M., J. Kavanaugh, and C. McNeil. 2013. "Juneau Icefield Mass Balance Program 1946–2011." *Earth System Science Data* 5: 319–30.

Peterson, B. J., et al. 2002. "Increasing River Discharge to the Arctic Ocean." *Science* 298: 2171–3.

Pfeffer, W. T., et al. 2014. "The Randolph Glacier Inventory: A Globally Complete Inventory of Glaciers." *Journal of Glaciology* 60: 537–52.

Phoenix, D. R., and J. W. Bjerke. 2016. "Arctic Browning: Extreme Events and Trends Reversing Arctic Greening." *Global Change Biology* 22: 2960–2.

Prest, V. K., D. R. Grant, and V. N. Rampton. 1968. *Glacial Map of Canada: Map 1253A*. Ottawa: Geological Survey of Canada.

Radik, V., and R. Hock. 2010. "Regional and Global Volumes of Glaciers Derived from Statistical Upscaling of Glacier Inventory Data." *Journal of Geophysical Research* 115. doi: 10.1029/2009JF001373.

Rastner, P., et al. 2012. "The First Complete Inventory of the Local Glaciers and Ice Caps on Greenland." *Cryosphere* 6: 1483–95.

Rastner, P., T. Strozzi, and F. Paul. 2017. *New Opportunities for Creating Glacier Inventories in the Russian Arctic Exemplified for Novaya Zemlya*. Abstract 76A2561. Polar Ice, Polar Climate Polar Change, International Glaciology Society Symposium, Boulder, CO.

Rea, B. R., and D. J. A. Evans. 1996. "Landscapes of Aerial Scouring in NW Scotland." *Scottish Geographical Magazine* 112: 47–50.

RGI Consortium. 2017. *Randolph Glacier Inventory: A Dataset of Global Glacier Outlines: Version 6.0: Technical Report*. [Digital media]. Colorado: Global Land Ice Measurements from Space. doi: 10.7265/N5-RGI-60

Rignot, E., et al. 2014. "Widespread, Rapid Grounding Line Retreat of Pine Island, Thwaites, Smith, and Kohler Glaciers, West Antarctica, from 1992 to 2011." *Geophysical Research Letters* 41: 3502–9.

Romanovskii, N. N., et al. 1998. "Map of Predicted Offshore Permafrost Distribution on the Laptev Sea Shelf." In *Permafrost: 7th International Conference*, Yellowknife, NWT, Canada. Collection Nordicna No. 55, 967–72.

Romanovskii, N. N., et al. 2005. "Offshore Permafrost and Gas Hydrate Stability Zone on the Shelf of East Siberian Seas." *Geo-Marine Letters* 25: 167–82.

Romanovsky, V. E., et al. 2008. "Soil Climate and Frost Heave Along the Permafrost/ Ecological North American Arctic Transect." In *9th International Permafrost Conference, Fairbanks, Alaska*, edited by D. L. Kane and K. M. Hinkel, 1519–24. Fairbanks, AK: Institute of Northern Engineering.

Romanovsky, V. E., S. L. Smith, and H. H. Christiansen. 2010. "Permafrost Thermal State in the Polar Northern Hemisphere during the International Polar Year 2007–2009: A Synthesis." *Permafrost and Periglacial Processes* 21: 106–16.

Rouse, W. R., et al. 2005. "Role of Northern Lakes in a Regional Energy Balance." *Journal of Hydrometeorology* 6: 291–305.

Rouse, W. R., et al. 2008a. "The Influence of Lakes on the Regional Heat and Water Balance of the Central Mackenzie River Basin." In *Cold Region Atmospheric and Hydrologic Studies: The Mackenzie GEWEX Experience, Volume 1: Atmospheric Dynamics*, 309–25. Berlin: Springer-Verlag.

Rouse, W. R., et al. 2008b. "Climate–Lake Interactions." In *Cold Region Atmospheric and Hydrologic Studies: The Mackenzie GEWEX Experience, Volume 2: Hydrologic Processes*, 139–60. Berlin: Springer-Verlag.

Serreze, M. C., et al. 2017. "Rapid Wastage of the Hazen Plateau Ice Caps, Northeastern Ellesmere Island, Nunavut, Canada." *Cryosphere* 11: 169–77.

Shahgedanova, M., and M. Kuznetsov. 2002. "The Arctic Environment." In *The Physical Geography of Northern Eurasia*, edited by M. Shahgedanova, 191–215. Oxford: Oxford University Press.

Shahgedanova M., G. Nosenko, I. Bushueva, and M. Ivanov. 2012. Changes in Area and Geodetic Mass Balance of Small Glaciers, Polar Urals, Russia, 1950–2008." *Journal of Glaciology* 58(211): 953–64.

Sharp, M., et al. 2014. "Recent Glacier Changes in the Canadian Arctic." In *Global Land Ice Measurements from Space: Satellite Multispectral Imaging of Glaciers*, edited by J. S. Kargel et al., 205–28. New York: Springer-Praxis.

Sharp, M., et al. 2015. "Trends in Permafrost Conditions and Ecology in Northern Canada [Arctic] Glaciers and Ice Caps (Outside Greenland)." *State of the Climate in 2014: Bulletin of the American Meteorological Society* 96(7): 135–7.

Sharp, R. P. 1956. "Glaciers in the Arctic." *Arctic* 9: 78–117.

Sharp, R. P. 1988. *Living Ice: Understanding Glaciers and Glaciation*. Cambridge: Cambridge University Press.

Shroder, J. F., ed.-in-chief. 2013. *Treatise on Geomorphology. Vol. 8: Glacial and Periglacial Geomorphology*, edited by J. R. Giardino and J. M. Harbor. London: Academic Press, Elsevier.

Siegert, M. J., et al. 2001. "Physical, Chemical and Biological Processes in Lake Vostok and Other Antarctic Subglacial Lakes." *Nature* 414: 603–9.

Sissons, J. B. 1967. *The Evolution of Scotland's Scenery*. Edinburgh: Oliver and Boyd.

Smith, H. T. U. 1948. "Giant Glacial Grooves in Northwest Canada." *American Journal of Science* 246: 503–14.

Smith, S. 2011. *Trends in Permafrost Conditions and Ecology in Northern Canada. Canadian Biodiversity: Ecosystem Status and Trends 2010*. Technical Thematic Report No. 9. Ottawa, Ontario: Canadian Councils of Resource Ministers.

Smith, S. L., et al. 2010. "Thermal State of Permafrost in North America: A Contribution to the International Polar Year." *Permafrost and Periglacial Processes* 21: 117–35.

Sobota, I., M. Novak, and P. Weckwerth. 2016. "Long-Term Changes of Glaciers in North-Western Spitsbergen." *Global and Planetary Change* 144: 182–97.

Song, Ch.-Q., et al. 2016. "Glacial Lake Evolution in the Southeastern Tibetan Plateau and the Cause of Rapid Expansion of Proglacial Lakes Linked to Glacial–Hydrogeomorphic Processes." *Journal of Hydrology* 540: 504–14.

Stearns, L. A., and Jiskoot, H. 2014. "Glacier Fluctuations and Dynamics around the Margin of the Greenland Ice Sheet." In *Global Land Ice Measurements from Space*, edited by J. S. Kargel et al., 183–204. Berlin: Springer-Verlag.

Streletsky, D. A., et al. 2008. "13 Years of Observations at Alaskan CALM Sites: Long-Term Active Layer and Ground Surface Temperature Trends." In *Ninth International Conference on Permafrost: Fairbanks: Institute of Northern Engineering*, edited by D. L. Kane and K. M. Hinkel, 1727–32. Fairbanks, AK: University of Alaska, Fairbanks.

Stroeven, A. P., J. Hayman, and J. Harbor. 2013. "Erosional Landscapes." In *Treatise on Geomorphology. Vol. 8: Glacial and Periglacial Geomorphology*, edited by J. R. Giardino and J. M. Harbor, 100–12. London: Academic Press, Elsevier.

Sugden, D. E. 1974. "Landscapes of Glacial Erosion in Greenland and Their Relationship to Ice, Topographic and Bedrock Conditions." In *Progress in Geomorphology* (Special Publication No. 7), edited by R. S. Waters and E. H. Brown, 177–95. London: Institute of British Geographers.

Sugden, D. E. 1978. "Glacial Erosion by the Laurentide Ice Sheet." *Journal of Glaciology* 20: 367–91.

Sugden, D. E., and B. S. John. 1976. *Glaciers and Landscape: A Geomorphological Approach.* London: Edward Arnold.

Sugden, D. E., et al. 1992. "Evolution of Large Roches Moutonnées." *Geografiska Annaler* 74A: 253–64.

Surdu, C. M., et al. 2014. "Response of Ice Cover on Shallow Lakes of the North Slope of Alaska to Contemporary Climate Conditions (1950–2011): Radar Remote-Sensing and Numerical Modeling Data Analysis." *Cryosphere* 8: 167–80.

Tape, K., M. Sturm, and C. Racine. 2006. "The Evidence for Shrub Expansion in Northern Alaska and the Pan-Arctic." *Global Change Biology* 12(4): 686–702.

Tarnocai, C., and S. C. Zoltai. 1978. "Earth Hummocks of the Canadian Arctic and Subarctic." *Arctic and Alpine Research* 10: 581–94.

Tedrow, J. C. F. 1966. "Polar Desert Soils." *Soil Science Society of America, Proceedings* 30: 381–7.

Tedrow, J. C. F., and F. C. Ugolini. 1966 "Antarctic Soils." In *Antarctic Soils and Soil-Forming Processes.* Antarctic Research Series No. 8, 161–77. Washington, DC: American Geophysical Union.

Trenhaile, A. S. 1990. *The Geomorphology of Canada: An Introduction.* Toronto: Oxford University Press.

Troll, C. 1958. *Structure Soils, Solifluction and Frost Climates of the Earth.* Translation No. 43. Wilmette, IL: US Army Snow, Ice and Permafrost Research Establishment, Corps of Engineers.

Tumel, N. 2002. "Permafrost." In *The Physical Geography of Northern Eurasia*, edited by M. Shagedanova, 149–68. Oxford: Oxford University Press.

Turc, L. 1955. *Le bilan d'eau des sols: Relation entre les precipitations et l'ecoulement.* Versailles, France: Lab. des Sols.

Vasil'chuk, V. K., and A. C. Vasil'chuk. 1997. "Radiocarbon Dating and Oxygen Isotope Dating in Late Pleistocene Syngenetic Ice Wedges in Northern Siberia." *Permafrost and Periglacial Processes* 8: 335–45.

Verpoorter, C., et al. 2014. "A Global Inventory of Lakes Based on High-Resolution Satellite Imagery." *Geophysical Research Letters* 41: 6396–402.

Vilborg, L. 1977. "The Cirque Forms of Swedish Lapland." *Geografiska Annaler* 59A: 89–150.

Vivian, R. 1970. "Hydrologie et erosion sous-glaciaires." *Revue. Geogr. Alp* 58(2), 241–64.

Vuglinsky, V. S. 2006. "Ice Regime in the Rivers of Russia, Its Dynamics during Last Decade and Possible Future Changes." In *Proceedings of the 18th International Symposium on Ice,* Vol. 1, edited by H. Saeki, 93–8. Sapporo, Japan: Nakanishi Publishing.

Walker, D. A., et al. 2005. "The Circumpolar Arctic Vegetation Map." *Journal of Vegetation Science* 16: 267–82.

Walker, D. A., et al. 2016. "Circumpolar Arctic Vegetation: A Hierarchic Review and Roadmap toward an Internationally Consistent Approach to Survey, Archive and Classify Tundra Plot Data." *Environmental Research Letters,* 11: 055005.

Walker, H. J. 1998. "Arctic Deltas." *Journal of Coastal Research* 14: 718–38.

Washburn, A. L. 1979. *Geocryology: A Survey of Periglacial Processes and Environments.* 2nd ed. London: Edward Arnold.

Watanabe, T., D. Lamsal, and J. D. Ives. 2009. "Evaluating the Growth Characteristics of a Glacial Lake and Its Degree of Danger of Outburst Flooding: Imja Glacier, Khumbu Himal, Nepal." *Norsk Geografisk Tidsskrift* 63: 255–67.

Whallet, W. B., and F. Azizi. 2003. "Rock Glaciers and Protalus Landforms: Analogous Forms and Ice Sources on Earth and Mars." *Journal of Geophysical Research* 108(E4): 8032.

Williams, R. S., Jr. 2013. "Glaciers of the Subantarctic Islands." In *State of the Earth's Cryosphere at the Beginning of the 21st Century: Glaciers, Global Snow Cover, Floating Ice, and Permafrost and Periglacial Environment. Satellite Image Atlas of Glaciers of the World.* Professional Paper 1386-A, edited by R. S. Williams, Jr., and J. Ferrino, 105–9. US Geological Survey.

Williams, R. S. Jr., and J. Ferrino, eds. 2012. *State of the Earth's Cryosphere at the Beginning of the 21st Century: Glaciers, Global Snow Cover, Floating Ice, and Permafrost and Periglacial Environment. Satellite Image Atlas of Glaciers of the World.* Professional Paper 1386-A. US Geological Survey.

Wolken, G. J., J. H. England, and A. S. Dyke. 2008. "Changes in Late-Neoglacial Perennial Snow/Ice Extent and Equilibrium-Line Altitudes in the Queen Elizabeth Islands, Arctic Canada." *Holocene* 18: 615–27.

Wright, A., and M. Siegert. 2012. "A Fourth Inventory of Antarctic Subglacial Lakes." *Antarctic Science* 24: 650–4.

Xu, L., et al. 2013. "Temperature and Vegetation Seasonality Diminishment over Northern Lands." *Nature Climate Change* 3: 581–6.

Yurtsev, B. A. 1982. "Relics of the Xerophyte Vegetation of Beringia in Northeastern Asia." In *Paleoecology of Beringia*, edited by D. M. Hopkins et al., 1157–77. New York: Academic Press.

Zemp, M. 2012. "The Monitoring of Glacier at Local, Mountain and Global Scales." In *Schriftung Physische Geographie*, 65. Zurich: Geography Institute, Universität Zürich.

Zhang, T.-J. 2005. "Influence of the Seasonal Snow Cover on the Ground Thermal Regime: An Overview." *Reviews of Geophysics* 43: RG 4002.

Zhang, T.-J., et al. 2000. "Further Statistics on the Distribution of Permafrost and Ground Ice in the Northern Hemisphere." *Polar Geography* 24: 126–31.

Zhu, Z.-Ch., et al. 2016. "Greening of the Earth and Its Drivers." *Nature Climate Change* 6: 791–5.

Zotikov, I. A. 2006. *The Antarctic Subglacial Lake Vostok.* New York: Springer.

Ice Sheets and Ice Shelves 6

This chapter presents the physical characteristics of the two global ice sheets in Greenland and Antarctica and their mass balance regimes. It then discusses ice shelves, which are now mainly found in the Antarctic, and the processes involved in their maintenance and breakup.

6.1 The Greenland Ice Sheet

Greenland is mainly covered by the Greenland ice sheet (GrIS), which itself is surrounded by numerous small ice caps and glaciers. The only extensive ice-free areas are along the coast in the southwest, in the northeast, and in the far north. The GrIS lies north of the main belt of westerly winds but is affected by cyclonic pressure systems moving northeastward from the east coast of North America. Some systems move in to Baffin Bay, where they may stall before eventually crossing the ice sheet at levels above 700 hPa. Hamilton (1958a, 1958b) reported fifteen disturbances per month crossing north Greenland in winter, but few gave precipitation. Other cyclonic systems move toward Cape Farewell, where they may split around Greenland, with a trough forming over Baffin Bay and the main low moving northeastward along the east coast of Greenland and then into the East Greenland–Norwegian Sea. Occasionally, blocking ridges of high pressure become established over the island. A Greenland Blocking Index (GBI) of the mean 500-hPa geopotential height for the 60–80° N, 20–80° W region has been developed by Hanna et al. (2016) for 1851–2015. It shows a significant decrease in this mean in autumn since 1851 and a significant increase in all seasons since 1981, especially in July and August. This trend appears to be linked to the summer sea ice decline and has implications for ice sheet climate and mass balance.

The climate of Greenland is primarily Arctic, but is strongly determined by latitude and altitude. Northern Greenland, north of the ice sheet, is a polar desert (see Section 5.1). On the west coast, temperatures are moderated by the presence of Baffin Bay and the north-flowing West Greenland Current. In contrast, the east coast is chilled by the East Greenland Current, which transports pack ice south to Cape Farewell and even northward along the coast of southwest Greenland.

Since 1995, extensive climatic data have been collected by automatic weather stations (AWS) on the ice sheet near the equilibrium-line altitude (ELA) through the Program on Arctic Regional Climate Assessment (PARCA) (Steffen and Box 2001). Since about 2007, Program for Monitoring the Greenland Ice Sheet (PRO-MICE) stations have also been in operation. As of 2016, there were twenty-six such stations around the ice sheet (www.promice.org/WeatherStations.html). Monthly mean slope lapse rates vary between -0.4 °C per 100 m in June and -1.0 °C per 100 m in November, with an annual mean value of -0.71 °C per 100 m.

Box (2002) used records from twenty-four coastal stations and three ice sheet locations to reconstruct temperatures for 1873–2001. The trends at Ilulissat/ Jakobshavn indicate statistically significant warming in all seasons, up to 5 °C in winter. In contrast, trends over 1901–2000 in southern Greenland indicate statistically significant spring and summer cooling of up to 1 °C.

Shuman et al. (2001) summarized twelve years of combined AWS and Special Sensor Microwave Imager (SSM/I) brightness temperature data from sites close to Summit (approximately 3,200 m), providing a picture of the annual cycle and its interannual variability. Monthly mean temperatures ranged between -48 and -10 °C and exhibited unexpectedly strong daily variability during October– March related to synoptic disturbances. There was alternation between down-slope conditions and synoptic systems that bring cloud cover and, depending on the system's track and the local orography, may bring upslope winds. The daily data showed an absolute range between about -59 and -3 °C.

Hall et al. (2012) used Moderate-Resolution Imaging Spectroradiometer (MODIS) data for 2000–2010 to develop a clear-sky surface temperature record for Greenland. These researchers developed monthly maps for 2010. Average u cover was approximately 30 percent annually, with the lowest values (approxi-mately 20 percent) occurring in spring.

Estimates of precipitation on Greenland have been based on annual accumula-tion measurements combined with coastal measurements and several modeling approaches (see Eterna et al. 2009; Ohmura et al. 1999; Serreze and Barry 2014, 183). Chen et al. (1997) used the omega equation to calculate vertical motion over Greenland and thereby take account of orographic influences. Their map of mean annual precipitation is shown in Figure 6.1. Bromwich et al. (2001) updated that analysis using improved topographic data, giving slight improve-ments in the results.

A new map for 1958–2007 was produced by Burgess et al. (2010) using solid precipitation output from the Fifth Generation Mesoscale Model modified for polar climates (Polar MM5) calibrated by 125 firn cores and 28 coastal meteorological stations (Figure 6.2). The mean annual accumulation in this map was 337 mm water equivalent (w.e.) (591 Gt yr^{-1}).

Maximum annual precipitation totals exceed 2,000 mm near the coast in southeast Greenland. This region accounts for 31 percent of the Greenland total precipitation and dominates the interannual variability. On the west and northwest slopes of the ice sheet, there is a maximum of 600 mm of precipitation. The lowest totals, less than 100 mm, are found in the northern interior around 77–78° N. Koenig et al. (2016) reported on the use of airborne snow radar data from NASA's IceBridge missions, 2009–2012, to map precipitation accumulation, finding good agreement with contemporaneous ice cores. Precipitation over the north coastal and central interior regions primarily occurs in summer. If a cyclone exists in the Labrador Sea, heavy precipitation will fall over Greenland during that period. By contrast,

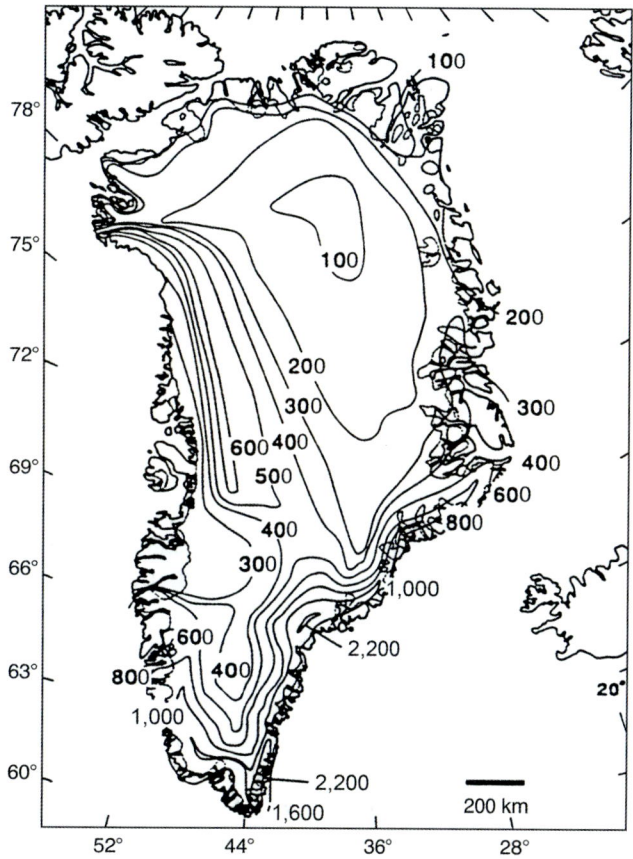

Figure 6.1 Mean annual precipitation (mm water equivalent [w.e.]) over Greenland.
Source: Modified after Bromwich et al. 2001, 33897, figure 2b. Courtesy of American Geophysical Union.

if a cyclonic pattern exists near Iceland, precipitation over Greenland is reduced. Losses by sublimation (direct ice to water vapor transition) both from the snow surface and from blowing snow over the ice sheet have been calculated to be between 12 and 23 percent of annual precipitation using different formulations (Box and Steffen 2001).

The Greenland ice sheet extends from approximately 60° to 80° N latitude, rising to 3,290 m in its main central dome (72° N) and to nearly 3,000 m in its southern dome (63–65° N). The GrIS covers an area of 1.7 million km^2 and has a volume of almost 3.0 km^3, equivalent to 7.5 m of sea level.

MacGregor et al. (2016) assessed the basal thermal state of the GrIS. Both thermo-mechanical modeling and remote inferences indicate that the northeast

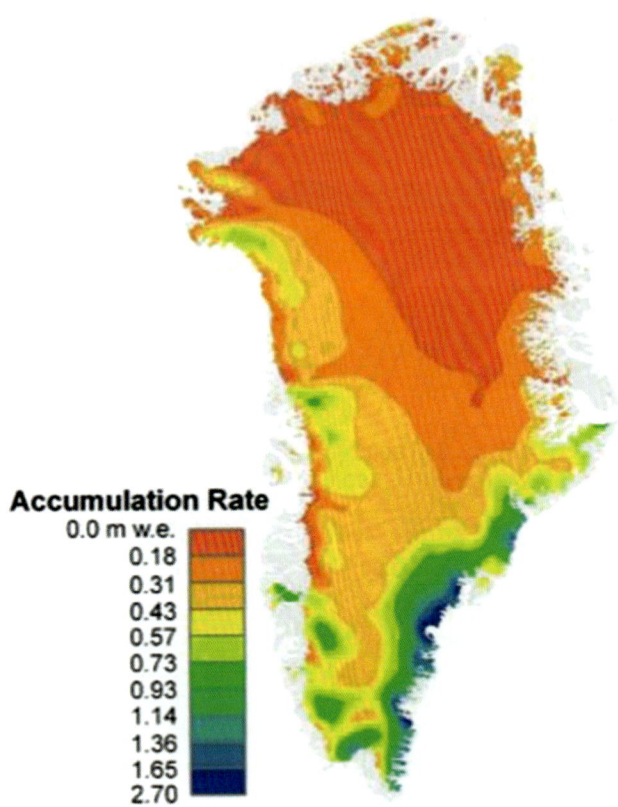

Accumulation Rate

0.0 m w.e.
0.18
0.31
0.43
0.57
0.73
0.93
1.14
1.36
1.65
2.70

Figure 6.2 Calibrated Polar MM5 average annual accumulation map (m water equivalent [w.e.]) for Greenland, 1958–2007.
Source: Burgess et al. 2010, 9, figure 9. Courtesy of American Geophysical Union.

Greenland ice stream and large portions of the southwestern ice-drainage systems are thawed at the bed (43 percent), whereas the bed beneath the central ice divides, particularly their west-facing slopes, is frozen (24 percent). The status of the remaining 33 percent is uncertain. Using airborne ground penetrating radar and ice sheet surface elevation data, Bamber et al. (2013) identified a mega-canyon extending from central Greenland northward to the fjord of Petermann Glacier. It is 750 km in length, up to 800 m deep, and 10 km wide. The authors proposed that it is a major conduit for the basal meltwater to reach the ocean and suggested that this drainage accounts for the virtual absence of subglacial lakes in Greenland. Two small (8–10 km² area) subglacial lakes have been detected by radar in northwest Greenland (Palmer et al. 2013). Refreezing of both surface and basal meltwater beneath a wide area of northern Greenland has been identified by Bell et al. (2017) using radar data. These authors determined that basal ice units in the interior are 200–1,100 m thick.

Noël et al. (2016) have described the characteristics of four small regions of the GrIS. Central east Greenland about 70° N has a large body of interconnected valley glaciers that mostly terminate in narrow glacial fjords. Central west Greenland about 70° N has a wide, gently sloping western ablation zone (where losses occur by melting, evapo-sublimation, blowing snow, and other means) where most glaciers are land-terminating. The northwestern part includes several marine-terminating glaciers that calve icebergs. Southern Greenland, south of about 62° N, is a rugged region, characterized by multiple topographically forced precipitation maxima and narrow marginal ablation zones. In north-central Greenland, the climate is dry, and most glaciers are marine-terminating. The ice sheet surface is relatively smooth and homogeneous, with a wide ablation zone.

Fresh snow compacts under its own weight until its initial near-surface density is about 300 kg m^{-3}. As it further compacts and ages, snow becomes more granular. Snow density in the top 1 m of the Greenland surface layer averages 338 ± 39 kg m^{-3} according to Koenig et al. (2016).

Grain size radii are typically in the range 0.05–1.0 mm (50–1,000 μm). Nolin and Dozier (2000) developed an inversion technique for estimating the grain size in a snowpack's surface layer from imaging spectrometer data. Using a radiative transfer model, their method relates an ice absorption feature, centered at 1.03 μm, to the optically equivalent snow grain size. This is the diameter of optically equivalent ice spheres that have the same optical properties (surface-to-volume ratio, or specific surface area [SSA]) as the original particles. For Greenland, Nolin and Stroeve (1997) measured grain radii at ETH-CU camp (70° N, 49.5° W) in May 1994 and May–June 1995. Mean values in the upper 5 cm were in the range 100–300 μm until a melt event, when they increased to 1,000 μm. This coincided with an albedo decrease at 1.03 μm wavelength from 0.75 to 0.62–0.65. The SSA of snow ranges from 2 m^2 kg^{-1} for refrozen snow layers to approximately 150 m^2 kg^{-1} for fresh, dendritic snow (Domine et al. 2007).

The snow and ice facies (observable attributes) of the surface layer of the Greenland ice sheet were first described by Carl Benson (1961). He identified the following facies:

1. Ablation facies from the glacier terminus to the annual firn line
2. Soaked facies from the firn line to the upper limit of complete wetting at the saturation limit
3. Percolation facies that is subject to localized percolation of surface melt water down to 10 m depth
4. Dry snow facies above the dry snow line that rarely undergoes any melt

Percolation causes densification and refreezes as superimposed ice. Figure 6.3 shows a slightly different classification scheme. The dry snow zone covers about 40 percent of the ice sheet surface, but its extent varied from 25 to 52 percent during 2000–2008, based on QuikSCAT data (Moon 2012).

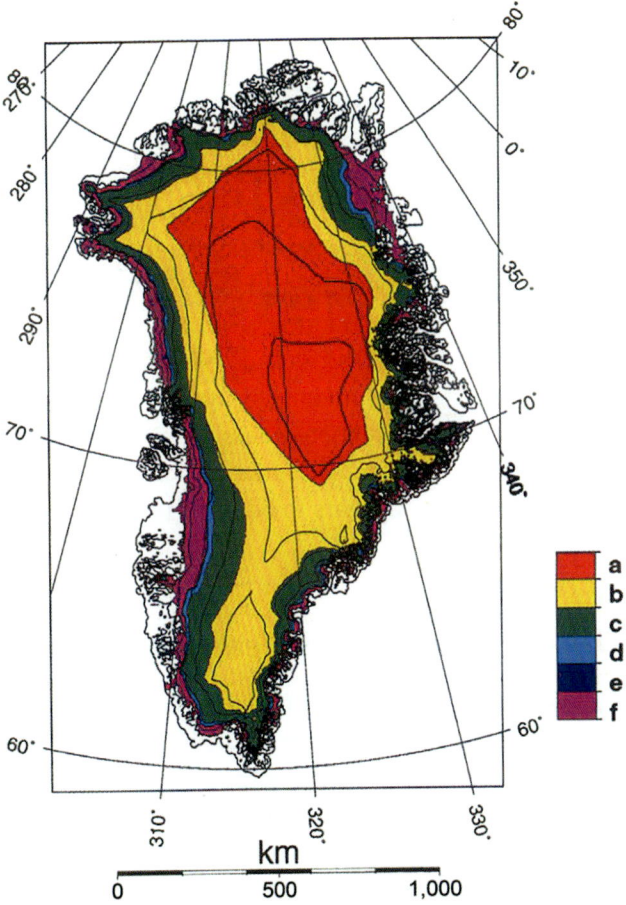

Figure 6.3 Glaciological zones on the Greenland ice sheet: (a) dry snow zone, (b) percolation zone, (c) wet snow zone, (d) slush zone, (e) superimposed ice zone, (f) ablation zone. Source: Janssens and Huybrechts 2000, 137, figure 4.

In western Greenland. the average equilibrium-line altitude is at 1,553 m above sea level (a.s.l.) (Van de Wal et al. 2012). At the end of the ablation season, superimposed ice is exposed at the surface at 1,520 m and extends to about 1,750 m (the "firn line") (Van den Broeke et al. 2008). At higher elevations, up to about 2,500 m a.s.l., is the percolation zone. Finally, above about 2,500 m is the dry snow zone (Charalampadis 2016).

Benson (1961) carried out detailed stratigraphic studies of the ice sheet. Pit studies and ramsonde measurements were performed across western Greenland from 1952 to 1955. Benson found mean annual accumulation for 1946–1955 of 0.355 m w.e. yr^{-1} at Station Centrale (3,020 m elevation). Joint European expeditions across central Greenland (EGIG) were conducted in 1958–1959 (EGIG1) and 1967–1968 (EGIG2). Mean annual accumulation for 1959–1969 was 0.329 m w.e. yr^{-1} at Station Centrale and 0.243 m w.e. a^{-1} at Crête (Merlivat et al. 1973). Based on EGIG line ground measurements and Airborne SAR/Interferometric Radar Altimeter System (ASIRAS) data for 1985–2004, Overly et al. (2016) reported mean annual accumulation at Station Centrale of 0.385 m w.e. Below 3,000 m elevation, mean accumulation increased by 20 percent over the period 1995–2004 (0.465 m w.e.) compared to the 1985–1994 period (0.387 m w.e.). Above this elevation, mean annual accumulation increased by 13 percent over the period 1995–2004 (0.335 m w.e.) compared to the 1985–1994 period (0.296 m w.e.).

Morris and Wingham (2011) observed a transition in surface conditions approximately 10 km uphill from Station Centrale. Elevations upslope from there experience less persistent winds, leaving a smooth surface with undisturbed summer surface hoar. Downslope from Station Centrale, the upper snow layer appears to be wind-packed, with sastrugi marking the surface (Box 6.1).

Box 6.1 Snow Surface Features

Several microscale features of snow and ice surfaces merit brief discussion.

There are many descriptive terms for snow. A fresh snowfall is loose and powdery, and granular when it has undergone metamorphism by freeze–thaw processes. Surface hoar is feathery ice crystals deposited on snow or the ground surface directly from water vapor in the lower air layer during calm, clear weather. Depth hoar is very large crystals that grow by vapor transport in the snowpack due to a strong vertical temperature gradient from the warm ground to the cold snow surface.

Sastrugi are snow surface features sculpted by wind into ridges and grooves typically 20–60 cm high.

Cryoconite is powdery windblown dust and soot that accelerates local melting of snow and ice due to its low albedo, forming small cryoconite holes.

Mosley-Thompson et al. (2001) reported average annual accumulation for 1979–1984 at forty-nine PARCA sites. Values ranged from 0.665 m w.e. at South Dome (63.1° N, 44.8° W, 2,850 m) to 0.31 at NASA-U (73.8° N, 49.5° W, 2,369 m), to 0.077 (for 1963–1996) at Tunu W25 (78.0° N, 35.1° W, 2,460 m). Hawley et al. (2014) used ground-penetrating radar and pit density profiles to determine accumulation along the route of the Greenland Inland Traverse team from Thule to Summit during April–May 2011. Accumulation rates varied from approximately 0.1 m w.e. yr^{-1} in the interior to 0.7 m w.e. yr^{-1} near the coast. Comparison with Benson's data from 1952 to 1955 indicated a 10 percent increase over this time span. Koenig et al. (2016) used 2–6.5 GHz airborne snow radar data collected during 2009–2012 as part of the National Aeronautics and Space Administration's (NASA) Operation Ice-Bridge mission to estimate accumulation rate over the GrIS, with an average uncertainty of 14 percent. Values over most of the interior averaged around 0.3 m w.e. and approached 1 m w.e. near the northwest and southeast coasts.

Drifting snow frequencies over the GrIS range from less than 20 percent at high elevations, where calm conditions dominate, to more than 50 percent in windy coastal areas (Lenearts et al. 2012). The highest frequencies are observed along the northeastern, southeastern, and western margins, where the surface snow density is low. Its horizontal transport ranges from less than 0.5×10^6 kg m^{-1} per year on the ice sheet plateau to more than 2×10^6 kg m^{-1} per year in the regions where drifting snow is most active.

Average annual sublimation over the GrIS was calculated by Box and Steffen (2001) using hourly meteorological data for 1995–2004 from the Greenland Climate Network. Annual net water vapor flux is as great as -87 mm at 960 m elevation and -74 mm at the ELA in western Greenland. The flux is positive at high elevations: $+32$ mm at the North Greenland Ice Core Project (NGRIP) and $+6$ mm at Summit. The ice sheet total sublimation is -0.62×10^{14} kg yr^{-1} (62 Gt), representing 12 percent of the ice sheet's precipitation. Lenearts et al. (2012) analyzed drifting snow with a regional climate model (RACMO2) and calculated surface sublimation of 16 Gt yr^{-1} and drifting snow sublimation of 24 Gt yr^{-1}, two thirds of the total determined by Box and Steffen.

Surface melt on the ice sheet has been monitored using passive microwave data by Abdalati and Steffen (1997). They used a cross-polarized gradient ratio (XPGR), which is a normalized difference between the 19-GHz horizontally polarized brightness temperature and the 37-GHz vertically-polarized brightness temperature. Mote and Anderson (1995) derived a similar time series from 37-GHz data. The MEaSUREs (Making Earth System data records for Use in Research in Earth sciences) Greenland Surface Melt Daily 25 km EASE-Grid 2.0 data set is available from the National Snow and Ice Data Center (NSIDC) (Mote 2014) for 1979–2012. The maximum melt area has been expanding irregularly since the 1980s. July 2012 witnessed the largest melt extent over the GrIS of the satellite era (Nghiem et al. 2012; Tedesco et al. 2013, 2015) based on the Mote and Anderson (1995) algorithm. Melt briefly affected approximately 97 percent of the ice sheet and lasted up to about two months longer than the 1979–2011 mean (Figure 6.4). The mean summer temperature anomalies over the ice sheet in 2012 reached $+4$–5 °C. The linear increase in melt extent during 1979–2012 was 20,000–22,000 km^2 yr^{-1}. In 2016, melt began in early April (11–12) and affected 12 percent of the ice sheet; this development occurred in a response to an omega blocking pattern in the atmosphere (Mottram 2016).

Liu et al. (2016) have shown that an important factor contributing to surface melt is the decrease in Arctic sea ice. Reduced summer sea ice favors stronger and more frequent occurrences of blocking-high pressure events over Greenland. These blocking highs enhance the transport of warm, moist air over Greenland, which increases downwelling infrared radiation, contributes to increased extreme heat events, and accounts for the majority of the observed warming trends. These findings are supported by analyses of observations and reanalysis data, as well as by independent atmospheric model simulations using a state-of-the-art atmospheric model that is forced by varying only the sea ice conditions. Reduced sea ice conditions in the model favor more extensive Greenland surface melting. There is a positive feedback between the variability in the extent of summer Arctic sea ice and the melt area of the summer Greenland ice sheet, which affects the ice sheet mass balance.

Recently, Mouginot et al. (2017) used Landsat-8 (optical) data for 2013–2016, Sentinel-1 data for 2014–2016, and RADARSAT-2 data for 2013–2014 (SAR missions) to map the ice velocity of the GrIS. Figure 6.5 shows the composite map, where the speed is on a logarithmic scale. NSIDC has used Landsat 8 panchromatic images to generate ice velocity products at sixteen-day intervals between 82° N and 82° S from May 1, 2013 to the present. The Global Land Ice Velocity Extraction from Landsat (GoLIVE) project data are accessible at https://nsidc .org/data/golive. Ice velocity vectors and gridded components at 300 m spacing are available for both ice sheets and all glaciers larger than 5 km^2.

6.1.1 Mass Balance

The mass balance of the GrIS has received considerable attention over the last two decades, with its calculation being made possible by the availability of Gravity Recovery and Climate Experiment (GRACE) gravimetric data since 2002 and more recently CryoSat-2 radar altimetry. McMillan et al. (2016) showed a cumulative mass loss of 3,000 Gt between 2003 and 2015. During 2011–2014, Greenland mass loss averaged 269 ± 51 Gt yr^{-1}, with an annual contribution of

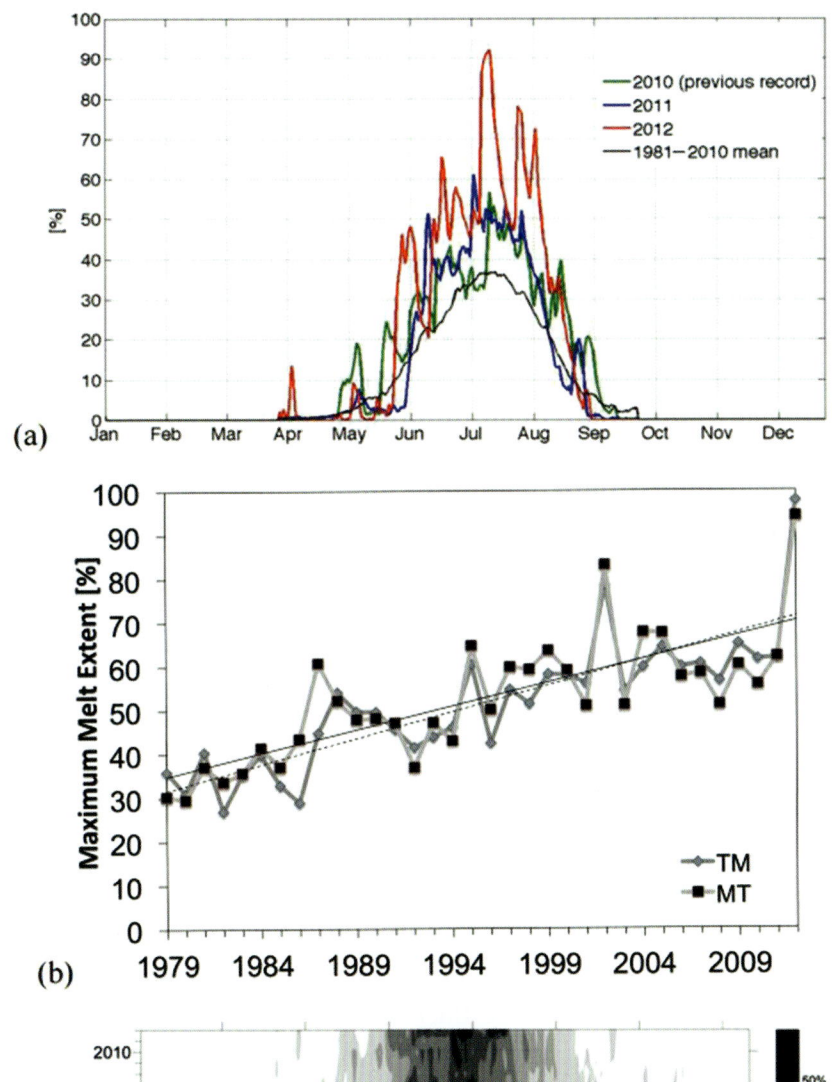

Figure 6.4
(a) Melt extent (as a percentage of the Greenland ice sheet) derived from passive microwave data using the algorithm in Tedesco (2013) for 2012 (red), 2011 (blue), and 2010 (green), and for the 1981–2010 mean (black). (b) Maximum melt extent for the period 1979–2012 using the algorithm in Mote and Anderson (1995), denoted as TM, and in Tedesco (2009), denoted as MT. (c) Daily simulated GrIS melt extent (as a percentage of the ice sheet area) from January–December, 1979–2012, from passive microwave data using the algorithm of Mote and Anderson (1995).
Source: Tedesco et al. 2013, 619, figure 2.

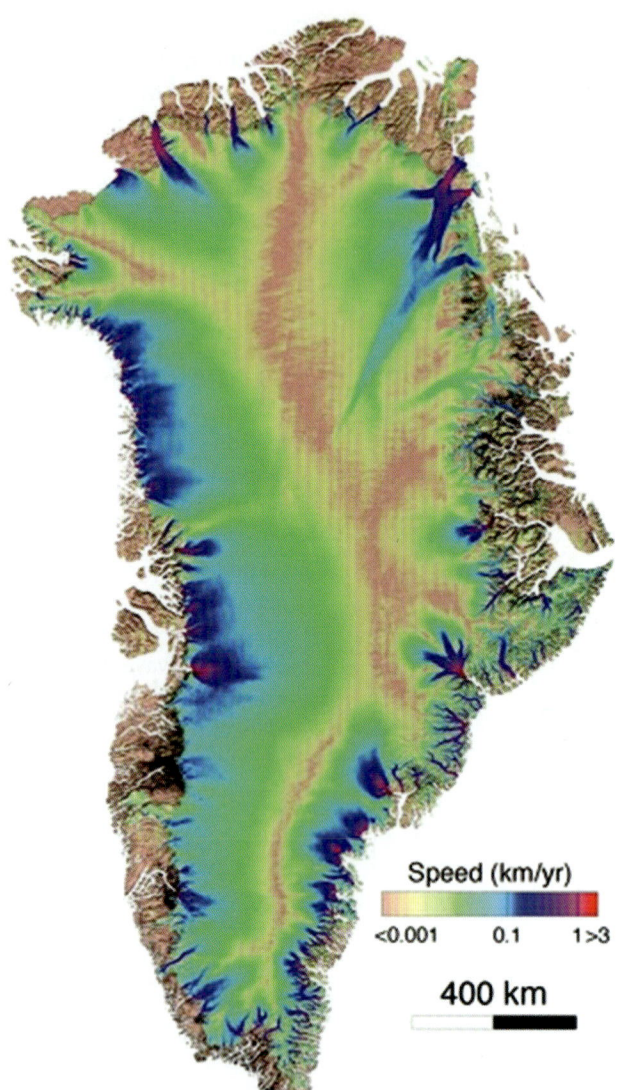

Speed (km/yr)

<0.001   0.1   1 >3

400 km

Figure 6.5 Composite mosaic of GrIS surface
velocity generated from ERS-1 and -2/ESA,
RADARSAT-1 and -2/CSA, ENVISAT/ESA,
ALOS/JAXA, TerraSAR-X/DLR, Sentinel-1/
ESA, and Landsat-8/USGS data
acquired between 1992 and 2016 in
Greenland. The ice speed is coded on
a logarithmic scale overlaid on a
MOG background mosaic (Haran et al.
2013).
Source: Mouginot et al. 2017, 17, figure 10.

0.74 ± 0.14 mm yr^{-1} to global mean sea
level. Mass balance was highly variable
during these four years, with a record loss
of 439 ± 62 Gt observed in the warm year of
2012, followed by only a moderate loss of
116 ± 65 Gt in 2013. Van den Broeke et al.
(2016) reported that the estimated
1995 value of grounding line discharge (D)
and the 1958–1995 average value of surface
mass balance (SMB) were similar at 411 and
418 Gt yr^{-1}, respectively, suggesting that ice
flow in the mid-1990s was well adjusted to
the average annual mass input. Starting in
the early to mid-1990s, SMB decreased while
D increased, leading to a quasi-persistent
negative mass balance. About 60 percent of
the associated mass loss since 1991 has been
caused by changes in SMB and the remainder by D. The decrease in SMB is fully driven
by an increase in surface melt and subsequent meltwater runoff. With a 1991–2015
average annual mass loss of approximately 0.47 ± 0.23 mm sea level equivalent
(SLE) and a peak contribution of 1.2 mm SLE
in 2012, the GrIS has recently become a
major source of global mean sea level rise
(see Section 7.3). Enderlin et al. (2014) used
ice thickness data collected by Operation
IceBridge (Box 6.2) to determine ice discharge by 178 marine-terminating glaciers.
They showed that only four glaciers contributed half, and fifteen glaciers 77 percent, of
the 739 Gt of ice lost since 2000. The relative
contribution of ice discharge to total loss
decreased from 58 percent before 2005 to
32 percent for 2009–2012. Additionally,
84 percent of the increase in mass loss after
2009 was due to increased surface runoff.

Using the regional climate Modèle Atmo-
sphérique Régional (MAR, version 3.5.2),

Box 6.2 Operation IceBridge

Operation IceBridge is a NASA mission using aircraft to map polar snow and ice that has been under way since the end of CryoSat1 in 2009. Flights are conducted over Greenland in March–May and over Antarctica in October–November. Operations are carried out over coastal Greenland, coastal Antarctica, the Antarctic Peninsula, interior Antarctica, southeast Alaskan glaciers, and Antarctic and Arctic sea ice. Data are available at https://nsidc.org/data/icebridge.

fully coupled with a snow energy balance model, Fettweis et al. (2016) have forced the MAR with eight different reanalyses over Greenland at a resolution of 20 km for the period 1900–2015. They estimated SMB before 1950 with two new reanalyses covering the whole twentieth century: 20CRv2c over 1851–2014, 2-degree resolution, and ERA-20C over 1900–2010, 1.25-degree resolution. Results from all forcing simulations indicate that the period 1961–1990, which is commonly chosen as a stable reference period for Greenland SMB, is actually a period when the SMB was anomalously positive (+10 percent) compared to the last 120 years. SMB decreased significantly after this reference period due to increasing and unprecedented melt, which has reached the highest rates in the 120-year record. Moreover, the ERA-20C forced simulation suggests that SMB during the 1920–1930 warm period over Greenland was comparable to the SMB of the 2000s due to both higher melt and lower precipitation than normal.

6.1.2 Energy Balance

From May 2001 to July 2002, scientists from the Federal Technical Institute (ETH), Zurich, carried out energy balance measurements at Summit, Greenland ($72.6°$ N, $34.5°$ W, 3,230 m a.s.l.). Figure 6.6 shows tautachrones of the monthly mean temperatures to 15 m depth.

During winter months, the surface is cooled (-24 W m^{-2}) due to a negative longwave radiation balance. This cooling effect is mainly balanced by a positive sensible heat flux (20 W m^{-2}) and the subsurface heat flux (5 W m^{-2}). In summer, monthly mean net radiation is positive (17 W m^{-2}) throughout the larger part of the day. The monthly mean sensible heat flux is close to zero. Positive net radiation is balanced in equal parts by the cooling due to sublimation (loss of latent heat, -10 W m^{-2}) and by a negative subsurface heat flux (-9 W m^{-2}). Warming of the surface by a positive latent heat flux (condensation) can be observed during clear-sky summer nights.

Miller et al. (2017) have reported the surface energy budget (SEB) at Summit using measurements for July 2013–June 2014. The maximum surface temperature

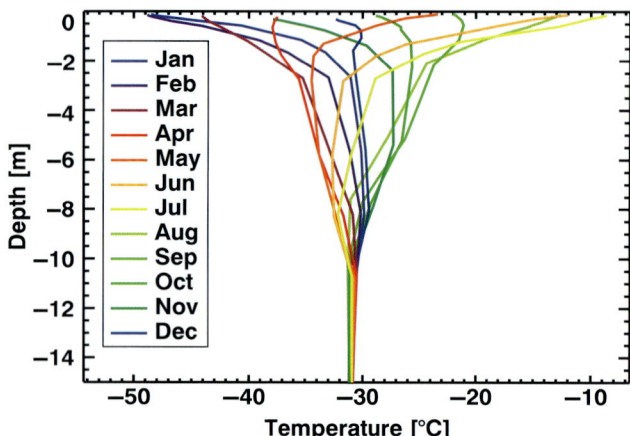

Figure 6.6 Tautachrones for Summit, Greenland, during 2001–2002.

Source: www.inscc.utah.edu/~hoch/eth_summit_experiment.html.

was −3.1 °C on July 10, 2013; the minimum was −68 °C on March 23, 2014. In spring, autumn, and winter, surface-based temperature inversions are prevalent in this area. The annual cycle and seasonal diurnal cycles of all SEB components indicate that the non-radiative terms are anticorrelated to changes in the total radiative flux; hence, they are responding to cloud radiative forcing (CRF). The annual value of CRF is 32.9 W m^{-2}. Clouds increased the surface temperature by 6.9 °C annually during January 2011–October 2013. November and August case studies illustrate that surface radiative forcing is driven by synoptically forced cloud characteristics, especially by low-level, liquid-bearing clouds. The nonradiative SEB terms and the upwelling longwave radiation component compensate for changes in downwelling radiation. Surface warming from low-level clouds typically leads to a change from a very stable to a weakly stable near-surface profile with no solar radiation, or from a weakly stable to neutral/unstable profile with solar radiation. Relationships between forcing terms and responding surface fluxes show that the upwelling longwave radiation produces 55–75 percent (40–50 percent) of the total response in winter (summer), while the nonradiative terms compensate for the remaining change in the combined downwelling longwave and net shortwave radiation. Sun angle, and the associated change to the net radiative flux, is a main driver of energy fluxes at the surface. The annual response in the outgoing infrared (IR) term is the largest (0.70) of all the response terms, as its magnitude is directly proportional to the surface temperature to the fourth power.

A study of cloud cover over Greenland using Advanced Very High Resolution Radiometer (AVHRR) and MODIS data revealed that from 1995 to 2009, summer cloud cover decreased by 0.9 percent per year (Hofer et al. 2017). AVHRR data for 1982–2009 showed decreased cloud cover over 84 percent of Greenland after 1996, while MODIS data for 2002–2015 showed a decrease over 77 percent of Greenland. Hofer et al. related this decrease to more frequent anticyclonic conditions associated with strongly negative NAO circulation patterns. Output from the MAR model indicates that the GrIS summer melt increases by 27 Gt per 1 percent reduction in summer cloud cover, principally because of the impact of increased shortwave radiation over the ablation zone, which has a low albedo.

Over the entire ice sheet, the albedo (shortwave reflection) of fresh snow reaches 0.90, but this value is reduced as a result of aging, increased snow grain size due to melt, and the deposition of dust particles. Dust layers may have albedo values of about 0.3, and dust clusters can form water-filled cryoconite holes with albedo values of approximately 0.1 (Chu 2014). At 80.3° N, in northeast Greenland, Bøggild et al. (2010) measured clear-sky broadband ice albedos ranging from 0.2 (for ice with heavy loading of uniform debris) to 0.6 (for ice hummocks with cryoconite holes). Surface concentrations of impurities were about 16 g m^{-2} on surfaces with low impurity loading, while heavily loaded surfaces had concentrations as high as 1.4 kg m^{-2}.

Summer albedo over the GrIS decreased by 0.02 per decade from 1981 to 2012, according to Tedesco et al. (2016). This trend was confined to regions with summer melt and was attributed to both surface warming that increased snow grain size and exposure of dirty ice bands (especially the "dark band" region of southwest Greenland), rather than to aerosol deposition. Moustafa et al. (2015) analyzed albedos and ablation rates using spectroradiometer data along a 1.25-km transect in June 2013 and MODIS data for summers 2012 and 2013. They found that seasonal changes in ablation area albedos are controlled by changes in the fractional coverage of snow, bare ice, and impurity-rich surface types.

Causes of mesoscale (10^2–10^3 m) albedo variability have been assessed by Ryan et al. (2016) using decimeter-scale digital imagery and broadband albedo data acquired by an unmanned aerial system. These authors characterized the reflectance properties and terrain roughness associated with six distinct surface types identified from a 25-km longitudinal transect across the ablating dark region of the Kangerlussuaq sector. Analysis of the fractional area of each surface type versus coincident MODIS albedo data revealed the relative importance of each surface type. The highest correlation with mesoscale albedo was found in the fractional area of distributed impurities. Although not the darkest surface type, the distributed impurities' extensive coverage meant that they could explain 65 percent of the albedo variability across the survey transect, including the presence of the dark region. In contrast, the 2 percent mean surface water coverage (supraglacial lakes and streams) across the survey transect explained only 12 percent of albedo variation and crevasses, only 17 percent. Localized cryoconite patches had the lowest albedo signature but accounted for less than 1 percent of the survey area; they did not appear to reduce mesoscale albedo. Coverage by clean ice averaged 51 percent in the lower western half of the transect. In the eastern half, clean ice covered 27 percent, coincident with the so-called dark region. Uniformly distributed impurities varied inversely with clean ice, with a higher fractional area found in the eastern half (70 percent) than in the western half of the transect (56 percent).

Figure 6.7 Supraglacial stream disappearing into a moulin in West Greenland.
Source: R. G. Barry and T. Y. Gan. 2011. *The Global Cryosphere*, 115, figure 3.8. Cambridge: Cambridge University Press. K. Steffen, Zurich.

New studies led by M. Tranter, of the University of Bristol in the United Kingdom, are investigating the distribution and albedo effects of algal growth on the ice sheet. Results from these investigations are not yet available.

Supraglacial streams flow briefly across the surface in the ablation zone, terminating in meltwater ponds and lakes, which periodically drain into crevasses and shafts, termed moulins, that penetrate deep into the ice sheet, occasionally to bedrock (Figure 6.7). Chu (2014) notes that moulins provide rapid drainage of large upstream areas into englacial and subglacial systems, while crevasses provide a slower, more spatially distributed drainage. In contrast to the intermittent meltwater supply from lake drainages into moulins, supraglacial streams provide a steady supply of large volumes of meltwater into moulins during the melt season. The supraglacial river flow vastly exceeds the potential lake storage, indicating the efficient removal of meltwater from the surface.

The area occupied by meltwater ponds covered about 1.4 percent of a study area in southwest Greenland (66.7–65.6° N, 49° W) of 5,328 km^2 in mid-July 2012 (Smith et al. 2015). The mean water depth was 2.0 m. During July 18–23, 2012, the area had 523 densely spaced, coalescent supraglacial stream networks, all of which terminated in moulins that were mostly located outside drained lake basins and topographic depressions.

A high-resolution estimate of supraglacial lake depth and water volume has been performed by Moussavi et al. (2016) using World View (WV)-2 multispectral measurements over an area of 1,250 km^2 in West Greenland with depth-retrieval models for June 12, 2011. The maximum lake depth, area, and volume were 7 m, 3.48 km^2, and 0.08 × 10^{-2} km^3, respectively. To estimate the total amount of meltwater in lakes, streams, and rivers captured by the WV-2 images, the researchers applied a calibrated depth-retrieval model to areas of images where water pixels were present. The analysis indicated that the total volume of meltwater stored in supraglacial lakes, streams, and rivers was $0.76 \pm 0.01 \times 10^{-2}$ km^3, approximately 65 percent of which was contained in twenty-two lakes.

While lakes are prevalent in western Greenland, in southern Greenland meltwater is stored or routed through englacial aquifers not visible from the surface. Forster et al. (2013) used observations from ground and airborne radar, as well as ice cores, to detect liquid water within the firn in southeastern Greenland. Based on a regional climate model, they estimated the area of the aquifer as approximately 70,000 km^2 and the depth to the top of the water table as averaging 23 m (5–50 m). Koenig et al. (2014) have estimated that the water volume is 140 Gt, representing approximately 0.4 mm of sea level rise. The cause of the aquifer seems to be associated with high accumulation and high melt rates.

When Harig and Simons (2016) analyzed GRACE data for Greenland, the Canadian archipelago, and Gulf of Alaska since 2003, they found that prior to 2013, interannual ice mass variability in the Gulf of Alaska and in regions around Greenland remained within the average estimated over the whole data span. Beginning in summer 2013, ice mass in regions around Greenland departed positively from its long-term trend. Over Greenland, this anomaly reached almost 500 Gt through the end of 2014. Overall, long-term ice mass loss from Greenland and the Canadian archipelago continues to accelerate, while losses around the Gulf of Alaska region continue but remain steady with no significant acceleration.

6.2 The Antarctic Ice Sheet

The Antarctic ice sheet (AIS), with an area of 14 million km^2, covers 97.6 percent of the continent of Antarctica, with a total grounded area of 10.7 million km^2. It has a volume of 22.1 million km^3 and a sea level equivalent of 56.8 m. It contains about 80 percent of the global freshwater supply. The East Antarctic ice sheet (EAIS), which rests mainly on bedrock, is far larger and higher than the marine-based West Antarctic ice sheet (WAIS), which is mostly grounded well below sea level. The mean surface elevation of the ice sheet is 2,200 m and 25 percent of the surface is above 3,000 m. The East Antarctic Plateau rises to 4070 m. The two ice sheets are demarcated by the Transantarctic Mountains (Box 6.3), which run 3,500 km from the western Ross Sea to the eastern Weddell Sea.

A recent study by Van Wyk de Vries et al. (2017) identified 138 subglacial volcanoes beneath the WAIS in Marie Byrd Land, extending from about 160° to 90° W. They have an average height of about 900 m and a diameter of approximately 20 km. While they do not appear to influence the ice sheet stability currently, they do have a potential to enhance basal melt rates in the future.

Box 6.3 Transantarctic Mountains

The Transantarctic Mountains (TAM) are a series of ranges that cross the continent from Cape Adare in northern Victoria Land to Coats Land on the Weddell Sea, separating the EAIS and the WAIS. They are 100–300 km wide, with peaks rising above 4,500 m. Ice from the EAIS flows through the Transantarctic Mountains via a series of outlet glaciers into the Ross Sea, Ross Ice Shelf, and the WAIS. Beardmore Glacier is one of the largest valley glaciers in the world, being 200 km long and 40 km wide. It descends about 2,200 m from the Antarctic Plateau to the Ross Ice Shelf. Byrd Glacier is another major glacier, about 136 km long and 24 km wide, draining an extensive area of the polar plateau through the TAM.

The climate of Antarctica is extreme in its degree of cold and aridity (Bromwich and Parish 1998). South Pole station, at 2,800 m elevation, experiences mean temperatures of −29 °C in January and −59 °C in July. Corresponding values at Vostok station (78.46° S, 106.86° E, 3,488 m), on the higher, more extensive East Antarctic plateau, are −33 and −67 °C, respectively (Simmonds 1998). Vostok recorded the world's official record low air temperature of −89 °C on July 21, 1983. However, lower extremes have been determined from Landsat 8 surface radiation data (see Box 4.5). Coastal stations are less extreme but show a high degree of daily variability from May through September with fluctuations of 20 °C being common.

Nearly all of the Antarctic ice sheet lacks any summer melting, so it mainly belongs to the dry snow zone. Surface snow melt in Antarctica has been mapped by Tedesco and Monaghan (2009) using passive microwave brightness temperature data from Scanning Multifrequency Microwave Radiometer (SMMR) and SSM/I. About 9–12 percent of the Antarctic surface experiences melt annually. The average snow melt extent for the period 1980–2009 was approximately 1,294,000 km^2.

Mouginot et al. (2017) mapped the surface velocity for the ice sheet using Landsat 8, Sentinel 1, and RADARSAT-2 data, as described for Greenland (Section 6.1). Figure 6.8 shows the composite mosaic where the speed is plotted on a logarithmic scale.

The surface mass balance of Antarctica as depicted in climate models and observations has recently been examined by Wang et al. (2016). Snow accumulation reaches a maximum around 1,200 m elevation, with a secondary maximum occurring around 1,750 m. Precipitation peaks in autumn and is low in both summer and winter. Annual precipitation averaged over the entire continent is about 166 mm (approximately 2,500 Gt ice equivalent mass, or 7 mm of sea level equivalent), according to Vaughan et al. (1999). There is little evidence of any trends but large interannual variability is apparent (Turner et al. 2014). Favier et al. (2011) provided a mass balance data set for Antarctica. Table 6.1 contains a selection of surface measurements.

The snow facies of Antarctica as mapped by Tran et al. (2008) differ from those found in Greenland. The wet zone is spatially insignificant and largely confined to the Antarctic Peninsula. Tran et al.'s classification partitions the ice sheet surface into seven categories in terms of snow accumulation and/or snow drift redistribution and snow layering set up by the topographically influenced drainage winds over the ice sheet. A different perspective has been adopted by Frezzotti et al. (2002) and Scambos et al. (2012). The surface microrelief comprises three categories: (1) depositional features formed from friable wind-transported snow (dune fields); (2) redistribution features formed as a result of the erosion of depositional features (sastrugi, pits); and (3) erosional features formed from the long-term exposure to katabatic winds (glazed surfaces). In low-accumulation areas, megadunes up to 2–4 m in height and spaced 1–3 km apart

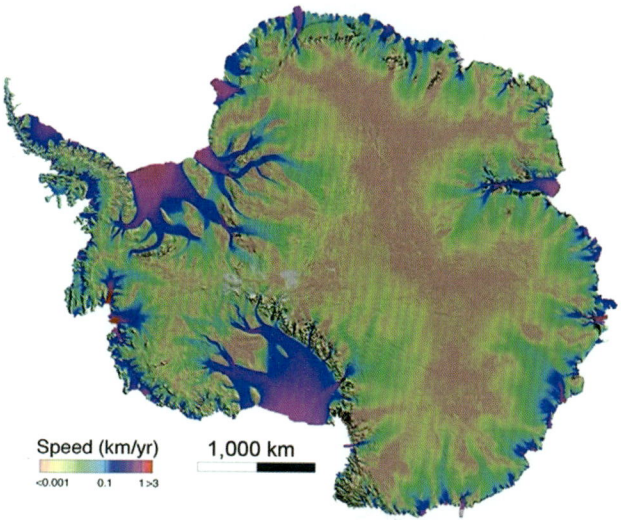

Figure 6.8 Composite mosaic of surface ice speed generated from ERS-1 and -2/ESA, RADARSAT-1 and -2/CSA, ENVISAT/ESA, ALOS/JAXA, TerraSAR-X/DLR, Sentinel-1/ESA, and Landsat-8/USGS data in Antarctica acquired between 1992 and 2016. The ice speed is coded on a logarithmic scale overlaid on MOA background mosaic (Scambos et al. 2007).
Source: Mouginot et al. 2017, 16, figure 9.

cover more than 500,000 km^2 of the East Antarctic plateau (Fahnestock et al. 2000). They have height-to-width ratios of approximately 1:200 (Scambos et al. 2017b), are perpendicular to the prevailing katabatic wind direction, and migrate slowly upwind. According to Scambos et al. (2017b), megadunes occur almost exclusively in mid- to upper elevations of the East Antarctic ice sheet, in areas of near-uniform regional slope and highly persistent katabatic airflow. In higher wind regimes, and on slightly steeper slopes, glaze and megadune areas become more chaotic. At the other end of the wind and slope regime, near ice divides, surface roughness varies according to wind speed. There are rougher surfaces when slopes are steeper, and smoother layering on ice-divide crests with low wind speeds. Glazed areas in East Antarctica cover up to 30 percent of the surface at any given instant according to Frezzotti et al.; permanent areas occupy approximately 6 percent of the surface according to Scambos et al. (2012, 2017b). Glazed surfaces, which consist of a single snow-grain thickness layer cemented by thin (0.1–2 mm) films of re-gelated ice, represent surfaces that have negligible accumulation and are strongly windswept.

Table 6.1 Description of selected data sets in low-elevation areas (from Favier et al. 2011, 592)

Location	Number of observations	Number of cells*	Time coverage	Mean elevation (m a.s.l.)	Mean SMB (mm w.e. yr^{-1})
Byrd	143	15	1955–1994	700	100
Zhongshan, Dome A	249	40	1994–2008	2,216	120
Dumont d'Urville, Dome C	27	18	1955–2009	1,815	298
Dronning Maud Land	22	21	1948–1999	138	200
Glacioclim, SAMBA	90	11	2004–2010	990	357
Law Dome	29	9	1973–1986	1,207	704
Mirny, Vostok	9	8	1955–1998	2,215	215
Mawson, Lambert West	249	40	1990–1995	2,531	100
Antarctic Peninsula	26	22	1953–1986	1,212	546
Syowa, Dome Fuji	245	37	1955–2010	2,068	106

* Number of 15×15 km^2 grid cells containing field measurements.

Palm et al. (2017) utilized eleven years of CALIOP lidar on the CALIPSO satellite to calculate amounts of blowing snow (Box 6.4) and sublimation over Antarctica. They estimated that the average integrated blowing snow sublimation is about 393 Gt yr^{-1}, considerably more than previous model-derived estimates. The maximum blowing snow transport is 5 megatons km^{-1} per year over parts of East Antarctica, and the average snow transport from continent to ocean is estimated to be about 3.68 Gt yr^{-1}.

6.2.1 Blue Ice Areas

Blue ice areas (BIAs) were defined by Bintjana (1999) as areas where surface mass balance is negative, sublimation serves as the main ablation process, and surface albedo is relatively low. Blue ice areas were first reported in the 1950s. Crary and Wilson (1961) described the surface characteristics of BIAs and discussed their formation by horizontal compressive forces, with katabatic winds removing snow

Box 6.4 **Blowing Snow**

Snow transport by the wind occurs by saltation and suspension. In the former process, particles are dislodged, carried a short distance in the lowest 50 cm or so, and then redeposited, where they dislodge other particles. Suspension of snow particles requires wind speeds of 10–15 m s^{-1}, or more, and raises the snow to heights of tens of meters. Blowing snow is subject to strong sublimation. Together, these processes account for the removal of up to half of the precipitation in the coastal and slope convergence zones around Terra Nova Bay, Antarctica (Scarchili et al. 2010).

Palm et al. (2011) used satellite lidar data from Cloud–Aerosol LIdar with Orthogonal Polarization (CALIOP) observations to determine the spatial and temporal frequency, layer height, and optical depth of blowing snow events over Antarctica for 2007–2009. The frequency was up to 70 percent in winter. Layer thickness had an average depth of 120 m but could reach 1,000 m. The distribution was mainly determined by the katabatic wind pattern, except in the megadune region in East Antarctica, where the most persistent and largest area of blowing snow occurs.

accumulation. The mass balance of BIAs is related to (1) low rates of snow accumulation as a result of mountain wind shadow effects, which cause snow drift divergence and (2) high ablation rates due to sublimation caused by katabatic winds (Bintjana and Van den Broeke 1995). The initial interest in BIAs was due to the discovery of meteorites on their surface. Subsequently, the related studies of ice flow led to paleoclimatological research. BIAs vary in size from a few hectares in the coastal mountains of East Antarctica to thousands of square kilometers for the South Yamato BIA (72° S, 35° E). They cover about 1 percent of the Antarctic surface (Bintjana 1999) and tend to form where mountains or nunataks disturb the ice flow.

Winther et al. (2001) used AVHRR scenes to map blue ice areas over the continent. They calculated a total area of approximately 120,000 km^2. In a later study, Hui et al. (2014) used Landsat-7 ETM+ images with 15-m spatial resolution from the 1999–2003 austral summers and covering the area north of 82.58° S, and a snow grain-size image from the MODIS-based Mosaic of Antarctica (MOA) data set with 125-m grid spacing acquired during the 2003–2004 austral summer from 82.58° S to the South Pole. They estimated that the total area of BIAs in Antarctica during the data acquisition period was 234,549 km^2, or 1.67 percent of the area of the continent.

Bintanja and Van den Broeke (1995) point out that meteorological conditions over the BIA differ from those over the snow–covered surroundings due to two factors. First, there are differences in surface characteristics, such as albedo (0.56 for bare ice versus 0.80 for snow), extinction characteristics for solar radiation, and surface aerodynamic roughness, between blue ice and snow. Second, there may be differences in topographic setting or nearby orography.

Dust or tephra bands are present on the surface of many BIAs. Dust sources are subglacial bedrock debris, cosmic particles, and marine and continental aerosols. Cryoconite holes are present on the surface of low-altitude BIAs. They form due to absorption of solar radiation by dark particles or stones, causing their temperature to rise above the melting point. BIA surfaces typically have shallow (2–10 cm) ripples with wavelengths between 5 and 24 cm; these ripples form perpendicular to the strongest winds. The basic mechanism for blue-ice ripple formation seems to be sublimation, occurring whenever there is wind forcing.

BIAs were first classified by Takahashi et al. (1992) into four types based on geographical setting and ice flow characteristics:

Type I BIAs – the most common type – are associated with mountains protruding through the ice. They are situated in the lee of an obstacle, which acts as a barrier for snowdrift. Their length can be estimated as 50–100 times the height of the obstacle relative to the ice surface.

Type II BIAs are located on a valley glacier. Descending katabatic winds cause net erosion of the surface and a local divergence of snowdrift, eventually leaving bare blue ice.

Type III BIAs are located on relatively steep slopes where downslope katabatic winds cause a divergence of snowdrift transport. South Yamato BIA is of this type.

Type IV BIAs are situated at the lowest part of a glacier basin. Accelerating katabatic winds in the basin remove snow from the surface.

Grinsted et al. (2003) categorized BIAs as open or closed. Flow in an open BIA is not dammed by mountains, nunataks, or bedrock topography, and ice flows through the BIA so the oldest ice does not reach the surface. Closed BIAs are located at mountain ranges where flow is dammed. In these areas, the oldest layers of ice will be found in the surface closest to the mountains, as if the layers had climbed up the mountain slope.

Sinisalo and Moore (2010) note that BIAs from different regions have different histories of formation and preservation. Moreover, blue ice development is intimately linked to the response of the surrounding ice sheet to climate variability on glacial–interglacial time scales.

The surface energy balance of a BIA was studied by Bintjana and Van den Broeke (1995) near the Swedish station Svea (74.6° S, 11.2° W) from December 28, 1992 to February 10, 1993. The energy components for the measurement period over blue ice and a snow site 6 km to the northwest are given in Table 6.2; the diurnal cycles of the energy budget components are shown in Figure 6.9.

The radiative differences give rise to latent heat losses that are twice as large over blue ice as those occurring over snow. However, the sublimation rate was found to be only 30 percent larger over the ice than over a snow site outside the valley 30 km away. Hence, the researchers concluded that the accumulation gradient, which is controlled by the wind regime, is more important than the ablation gradient in determining the extent of BIAs.

Table 6.2 Energy balance components (W m^{-2}) over blue ice and snow near Svea station, Antarctica in austral summer (after Bintjana and Van den Broeke 1995, 916, table 4)

	Blue ice site	Snow site
Net shortwave	128	64
Net longwave	−86	−56
Net radiation	42	8
Sensible heat	−7	9
Latent heat	−29	−15
Conductive heat flux	19	5
Total subsurface energy flux	−7	−2
Surface temperature (°C)	−7	−13

6.2.2 Ice Sheet Changes

Investigations and monitoring of mass balance changes in Antarctica are of increasing importance to the study of rising global sea level and have been revolutionized by new satellite technologies. The most significant changes are currently occurring in West Antarctica.

Mass loss of the WAIS has been pronounced in the Amundsen–Bellinghausen Sea sectors. Between 1992 and 2006, estimated contributions from the Amundsen Sea sector were approximately 64 Gt yr^{-1} (46 percent of total loss), while those from the Bellingshausen Sea region were approximately 49 Gt yr^{-1}, accounting for 35 percent of total mass loss from the WAIS (Rignot et al. 2008). Rignot et al. (2014) measured the grounding line retreat of the glaciers that drain the Amundsen Sea embayment using ERS-1/2 satellite radar interferometry from 1992 to 2011. Pine Island Glacier retreated 31 km at its center, with the greatest retreat occurring in 2005–2009, when the glacier ungrounded from its ice plain. Thwaites Glacier retreated 14 km along its fast-flow core and 1–9 km along its sides. Haynes Glacier retreated 10 km along its flanks. The Smith–Kohler glaciers retreated the most, 35 km along their ice plain, and the ice shelf pinning points are vanishing. These rapid retreats all proceeded along regions of retrograde bed elevation. Scambos et al. (2017a) highlight the Thwaites Glacier system (Box 6.5) as meriting special attention in the immediate future. They note that this glacier's retreat is currently contributing 0.25 mm yr^{-1} (89 Gt yr^{-1}) to global sea level rise.

There has been recent rapid retreat of the grounding line in the Bellingshausen Sea sector. Christie et al. (2016) used Landsat imagery from 1975, circa 1985, circa 1990, 2000, 2005, 2010, and 2015 to map the break-in-slope, otherwise known as

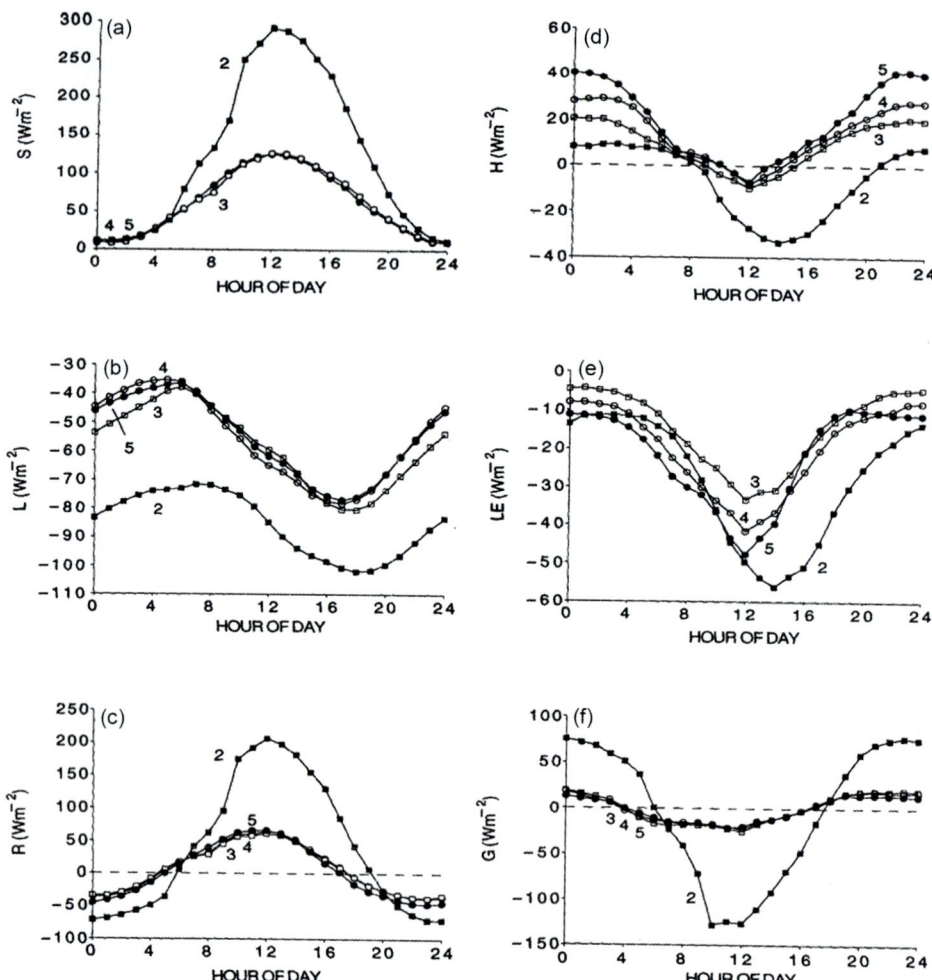

Figure 6.9 Diurnal cycle of (a) net shortwave radiation, (b) net longwave, (c) net radiation, (d) sensible heat flux, (e) latent heat flux, and (f) total subsurface heat flux (W m^{-2}), January 1–February 10, 1993, at site 2 blue ice and sites 3, 4, and 5 snow, located 5, 9, and 16 km, respectively, from site 2.
Source: Bintjana and Van den Broeke 1995, 914, figure 11.
© Copyright 1995 American Meteorological Society (AMS).

the "inflection point," which is defined as the most seaward continuous slope break detectable via satellite imagery. They also used INSar data from ERS-1 and -2 between 1992 and 2011 to map the location of tidal flexure acting on the grounding zone. The most significant changes to the grounding line position between 1990 and 2015 are located at Ferrigno and Fox Ice Streams (2.77 $\pm$ 0.50 km and 1.79 $\pm$ 0.14 km, respectively). For the WAIS as a whole, Shepherd et al. (2012) found that over 1992–2011 the mass of the ice sheet changed by -65 ± 26 Gt yr^{-1}.

Box 6.5 Glaciers Flowing into the Amundsen Sea

Two major fast-flowing glaciers – the Pine Island and Thwaites glaciers – account for around one fourth of the WAIS area and terminate in the Amundsen Sea. Their physical characteristics are somewhat different. Thwaites Glacier, which covers an area of 180,000 km^2 and moves at 10 m day^{-1}, now accounts for approximately 8 percent of global sea level rise. It has accelerated since the 1990s and its bed slope is retrograde (sloping down inland), meaning it is dynamically unstable. Currently, it is stabilized on a relatively high bedrock sill, but retreat off this sill will trigger rapid and irreversible retreat on a centuries time scale.

Pine Island Glacier drains an area of 162,000 km^2 and is 2 km thick. It has a small ice shelf, has less access to the deepest submarine basins of the WAIS, and is characterized by greater bedrock control (Alley et al. 2015). Both the Pine Island and Thwaites glaciers are subject to basal melt from upwelling Circumpolar Deep Water.

Glaciers and ice caps on Antarctic and sub-Antarctic islands, as well as on the Antarctic Peninsula, are showing retreat. On South Georgia, for example, 97 percent of 103 coastal glaciers surveyed from the 1950s to the present have retreated, and the average retreat rate has increased from 8 m yr^{-1} in the 1950s to 35 m yr^{-1} today (Cook and Vaughan 2010). Since 1953, 87 percent of marine-terminating glaciers that drain the northern sector of the Antarctic Peninsula ice cap have shown overall retreat (Turner et al. 2014). Shepherd et al. (2012) found that over 1992–2011, the ice mass of the Peninsula dropped by 20 ± 14 Gt yr^{-1}. However, retreat rates seem to be declining. The total glaciated area in the northern Antarctic Peninsula decreased by 11.1 percent during 1988–2001 but by only 3.3 percent during 2001–2009. Glacier recession on the southern part of the west coast of the peninsula appears to be responding to rising ocean temperatures, which is not the case farther north.

East Antarctica had shown only small changes in ice sheet extent until recently. Over 1992–2011, the mass of the East Antarctic ice sheet changed by $+14 \pm 43$ Gt yr^{-1}, according to Shepherd et al. (2012). Nevertheless, recent investigations suggest that the Totten Glacier, inland of Casey station, in East Antarctica is potentially unstable (Aitken et al. 2016), with water volume up to 2 m of sea level available in the retreated position. Possible collapse of the Totten Glacier into interior basins during past warm periods, most notably the Pliocene Epoch, has been identified. Totten Glacier has the largest ice discharge in East Antarctica, and 21 percent of its basin is grounded more than a kilometer below sea level.

Li et al. (2016) analyzed the ice velocity of Totten Glacier from 1989 to 2015 using Landsat and InSAR data, combined with ice thickness from Operation IceBridge, and surface mass balance from the Regional Atmospheric Climate Model. According to their results, the glacier speed exceeded its balance speed in 1989–1996, slowed down by 11 ± 12 percent in 2000 to bring its ice flux in balance with accumulation (65 Gt yr^{-1}), then accelerated by 18 percent until 2007, and remained constant thereafter. The average ice mass loss (7 Gt yr^{-1}) is dominated by ice dynamics (73 percent); the glacier's acceleration is dominated by surface mass balance (80 percent). Ice velocity apparently increases when the ocean surface is warmer, suggesting a linkage between ice dynamics and ocean temperature. Recent oceanographic observations by Rintoul et al. (2016) off the calving front of the Totten ice shelf showed that warm water (up to $-0.4\,^\circ$C) enters the cavity beneath the shelf through an 1,100-m-deep, 10-km-wide channel, facilitating basal melt.

6.3 Ice Shelves

With the rapid, ongoing breakup of the Ward Hut ice shelf (now smaller than 330 km^2) off northern Ellesmere Island, Arctic Canada, in the 2000s (Muller et al. 2008), ice shelves have become mainly a feature of Antarctica. The Filchner–Ronne shelf in the southern Weddell Sea has an area of approximately 422,000 km^2, a seaward-edge thickness of about 300 m, and a grounding line thickness of 1,600 m (Jenkins and Doake 1991; Scambos et al. 2007). The Ross Ice Shelf is the largest in the world, with an area of almost 473,000 km^2. Scambos et al. used MODIS data to map the surface morphology of the various ice sheets and reported the existence of eleven major ice shelves. The third largest is the Amery in East Antarctic (66,000 km^2).

Figure 6.10 illustrates the distribution of ice shelves around Antarctica and their areas. Selected satellite images are presented by Scambos et al. (2009a). Figure 6.11 illustrates the appearance of the edge of the Ross Ice Shelf in December 1996. It is about 600 km long and rises 15–50 m above the sea surface.

Depoorter et al. (2013) estimated the mass balance components for all ice shelves in Antarctica, using satellite measurements of calving and grounding-line flux, modeled ice shelf snow-accumulation rates, and a regional scaling that accounted for areas not surveyed. They obtained a total calving flux of 1,321 Gt yr^{-1} and a total basal mass balance

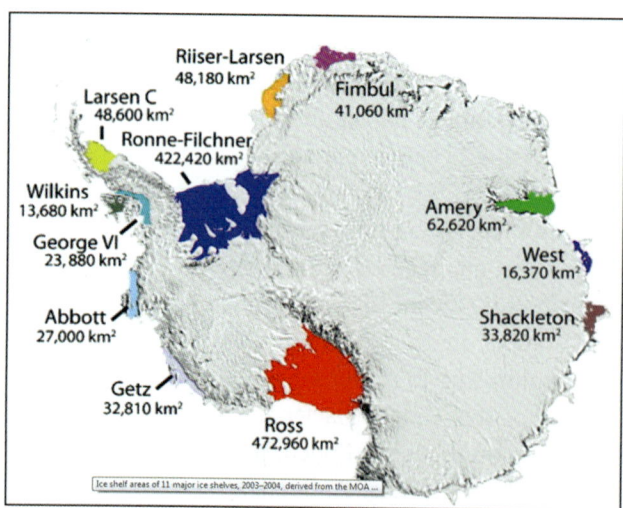

Figure 6.10 Map of the eleven major ice shelves and their areas in Antarctica.
Source: Scambos et al. 2007, 253, figure 8.

of $-1{,}454$ Gt yr^{-1}. Their result suggests that about half of the ice sheet surface mass gain is lost through oceanic erosion before reaching the ice front, and the calving flux is about 34 percent less than previous estimates.

Several mechanisms are thought to be involved in ice shelf breakup. Fürst et al. (2016) examined the "safety band" of Antarctic ice shelves – frontal areas that represent "passive shelf ice" and can be removed without dynamic implications. They showed that ice shelves in the Amundsen and Bellingshausen seas have limited or almost no "passive" portion, which implies that further retreat of current ice shelf fronts will have important dynamic consequences. This region is particularly vulnerable, for two reasons: Ice shelves have been

Figure 6.11 The edge of the Ross Ice Shelf, December 1996, from the *Nathaniel B. Palmer*.
Source: Michael van Woert, National Oceanographic and Atmospheric Administration, National Environmental Satellite, Data, and Information Service, Office of Research and Applications, www.photolib.noaa.gov/htmls/corp2399.htm.

thinning at high rates for two decades, and upstream grounded ice rests on a backward-sloping bed, a precondition for marine ice sheet instability. In contrast, Larsen C Ice Shelf, in the Weddell Sea, exhibits a large "passive" frontal area, suggesting that the July 2017 calving of a vast tabular iceberg will be unlikely to produce much dynamic change.

Ice shelves can be weakened by meltwater stored in ponds and crevasses that leads to fracturing and disintegration. This appears to have occurred with the rapid breakup of Larsen B in February 2002 and the Wilkins Ice Shelf in 2008 (Scambos et al. 2009b). However, Bell et al. (2017) have shown that the Nansen Ice Shelf has persistent active drainage networks of interconnected streams, ponds, and rivers that export a large fraction of the ice shelf's meltwater into the ocean, thereby preventing fracturing. Similar systems have been identified for the Larsen C, Ronne–Filchner, Ross, and Amery shelves. Kingslake et al. (2017) used aerial photographs from 1947 to 1960 and satellite images from 1973 onward to show that surface drainage has persisted for decades on at least twelve ice shelves. It has transported water on the Amery Ice Shelf as far as 120 km from grounded ice onto and across ice shelves, feeding vast melt ponds up to 80 km long and 43.5 km wide. On Shackleton Glacier, for example, water is transported up to 70 km from the edge of the East Antarctic plateau (1,350 m a.s.l.) onto the Ross ice shelf. In the Transantarctic Mountains, the Darwin, Nimrod, Lennox–King, and Liv glaciers all support surface drainage systems. Fifty percent of drainage systems originate within 3.6 km of blue ice areas and 50 percent originate within 8 km of exposed rock.

Another mechanism in ice stream and ice shelf retreat is the failure of approximately 100-m-high ice cliffs (Pollard et al. 2014). This kind of event, when combined with hydrofracture, can lead to rapid grounding-line retreat and may account for past collapses of the WAIS. A temporary slowing mechanism is the formation of floating melange ice (a mixture of sea ice, icebergs, and bergy bits) in narrow seaways that impede iceberg removal.

6.3.1 Melt

Several recent studies demonstrate that ice shelves, especially in the Antarctic Peninsula, are undergoing accelerating melt and disintegration. The most notable examples are the Larsen B and Wilkins shelves. The overall reduction in total ice shelf area around the peninsula during the last five decades has been estimated to exceed 28,000 km^2 (Cook and Vaughan 2010). Ice loss on the eastern side is mainly a result of warm air transported over the peninsula by the westerlies with associated descent and foehn warming. Ice shelf collapse is thought to result from hydrofracture of water-filled crevasses (Scambos et al. 2000). In the northern Antarctic Peninsula, surface warming began around 600 years ago (Sterken et al. 2012), and the rate of warming intensified during the last one hundred years (Mulvaney et al. 2012). The processes involved are still being investigated, but new work by Alley et al. (2016) suggests an important role for basal melt associated with warm Circumpolar Deep Water (CDW) upwelling onto the continental shelf. Hellmer and Olbers (1991) developed a two-dimensional model for the thermohaline circulation beneath an ice shelf.

Basal channels have a corresponding depression in the ice shelf surface and form lineations that are visible in MODIS and LandSat-8 imagery. Alley et al. (2016) used this imagery, together with airborne ice-penetrating radar and satellite laser altimetry spanning the period from 2002 to 2014, to map large (more than 1 km-wide) basal channels across all the ice shelves of Antarctica.

SUMMARY

The climate of Greenland is strongly determined by latitude and altitude. Cyclones enter Baffin Bay from North America, and many cross the ice sheet in the middle and upper troposphere. Others split at Cape Farewell and move into Baffin Bay or along the east coast. The southwest coast is tundra and the far north, polar desert. The west coast has warmed by 5 °C in winter since the 1870s, while spring and summer in southern Greenland show 1 °C cooling in the twentieth century. Mean monthly temperatures at Summit range between −10 and −48 °C. Mean annual accumulation is calculated to be 337 mm, with maximum amounts exceeding 2,000 mm observed in the southeast.

The Greenland ice sheet (GrIS) extends from 60° to 80° N and rises to 3,290 m. It contains a water volume equivalent to 7.5 m of sea level. A mega-canyon 800 m deep was identified in central Greenland and extends 750 km to the northwest. The GrIS has four altitudinal snow facies: the ablation facies, the soaked facies, the percolation facies, and the dry snow facies. In western Greenland, the equilibrium-line altitude is about 1,520 m and the dry snow zone is above 2,500 m.

Accumulation rates are approximately 0.1–0.3 m w.e. in the interior of Greenland, and 0.7–1 m w.e. near the northwest and southeast coasts. Sublimation from the surface and from drifting snow is between 40 and 60 Gt yr^{-1}. Surface melt monitored by passive microwave data shows irregular expansion since the 1980s. Maximum melt briefly affected 97 percent of the ice sheet in 2012 in response to summer temperature anomalies of 4–5 °C. Reduced Arctic sea ice in summer is facilitating blocking circulation patterns, which favor Greenland melt. Data from GRACE and Cryosat-2 have allowed mass balance changes to be determined. For 2011–2014, the GrIS lost an average of 269 Gt yr^{-1}. About 60 percent of the mass loss since 1991 is attributed to decreasing surface mass balance (SMB) and the remainder to discharge increase. Just four major glaciers have contributed half of the mass loss since 2000. Model simulations indicate that the SMB for 1960–1990 was positive, that it declined subsequently due to increased melt, and that 1920–1930 had comparable SMB to the 2000s.

Energy balance measurements at the Summit station in Greenland show that longwave loss in winter is offset by positive sensible heat flux. Net radiation is positive in summer, balanced by cooling due to sublimation and negative subsurface heat flux. Cloud cover increases the surface temperature at Summit by 6.9 °C annually.

Surface albedo values vary with aging, melt-caused grain size increase, and impurities. Mesoscale albedo variability in western Greenland is related mainly to the fractional coverage of distributed impurities. In the ablation zone, supraglacial streams flow into lakes, which either periodically drain into crevasses and moulins, or flow directly into moulins.

Meltwater ponds covered approximately 1.4 percent of an area in southwest Greenland in July 2012. In southern Greenland, meltwater is stored or routed via englacial aquifers. They cover 70,000 km^2 and hold a water volume of 140 Gt.

The Antarctic ice sheet (AIS) has an area of 14 million km^2 and a volume equivalent to 56.8 m of sea level. The larger East Antarctic ice sheet (EAIS) rests mostly on bedrock, while the marine-based West Antarctic ice sheet (WAIS) is mainly grounded well below sea level.

Twenty-five percent of the AIS is above 3,000 m elevation, and the Transantarctic Mountains that divide the EAIS and the WAIS rise above 4,500 m elevation. The climate in this area is extremely cold and arid. Mean temperatures at South Pole range from −29 °C in January to −59 °C in July. About 9–12 percent of the surface experiences some summer melting. Snow accumulation peaks around 1,200 m elevation. On an annual basis, the continent receives about 2,500 Gt of ice equivalent mass (7 mm of sea level). The snow facies differ from those in Greenland. The wet zone is found largely in the Antarctic Peninsula. One categorization has identified three types of microrelief on the WAIS: (1) depositional (dune fields), (2) redistributional (sastrugi, pits), and (3) erosional (glazed surfaces).

On the EAIS, megadunes are common and glaze covers 30 percent of the surface. Average annual integrated blowing snow sublimation is approximately 390 Gt.

Blue ice areas (BIAs) are areas where SMB is negative and sublimation due to katabatic winds is the main ablation process. They cover approximately 1.67 percent of the AIS. Four types are recognized: (1) the lee of a mountain, which is a barrier to snow drift; (2) valley glaciers with descending katabatic winds, which cause net erosion of the surface; (3) steep slopes with katabatic winds, which cause divergent snow transport; and (4) the lowest part of a glacier basin where katabatic winds remove snow. An energy budget comparison of a blue ice and a snow site suggests that the accumulation gradient due to the wind is more important than the ablation gradient in determining the extent of BIAs.

Mass loss of ice was pronounced in the Amundsen–Bellingshausen Sea sector from 1992 to 2006, accounting for 80 percent of the loss from the WAIS. Rapid glacier retreat rates are due to the retrograde bed elevation. The ice sheet mass changed by 65 Gt yr^{-1} during 1992–2011. In the same time interval, the ice mass of the Antarctic Peninsula decreased by 20 Gt yr^{-1} while the EAIS changed by only 14 ± 43 Gt yr^{-1}. The massive Totten Glacier in the EAIS has been identified as potentially unstable and "warm" ocean water is entering a large cavity beneath its ice shelf, promoting basal melt.

The Ward Hunt Ice Shelf of Ellesmere Island underwent rapid break up in the 2000s. There are eleven ice shelves around Antarctica. The Ross Sea ice shelf covers 473,000 km^2; in the Weddell Sea, the Filchner–Ronne Ice Shelf covers 422,000 km^2. Ice shelves can be weakened by meltwater stored in ponds and crevasses. Hydrofracture of water-filled crevasses appears to be responsible for the breakup of Larsen B. However, active surface drainage systems are reported on at least twelve ice shelves. They feed vast melt ponds on the Amery Ice Shelf. Basal channels in the ice shelves that transport upwelling Circumpolar Deep Water seems to be causing basal melt.

QUESTIONS

1. Compare the snow facies of Greenland and Antarctica.
2. Account for the pattern of accumulation over the Greenland ice sheet.
3. Which factors give rise to variations in ice sheet albedo?
4. What explains the expansion of surface melt on Greenland since the 1970s?
5. Trace the changes in surface mass balance and ice discharge since the 1990s.
6. Characterize the energy balance at Summit station in winter and summer.
7. Describe the physical characteristics of the Antarctic ice sheet.
8. Why is Antarctica so cold and dry?
9. Which factors give rise to blue ice areas?
10. Contrast and explain the changes in mass balance of the WAIS (including the Antarctic Peninsula) and the EAIS since the 1990s.
11. Why and how are ice shelves important features of Antarctica?
12. Which mechanisms are leading to ice shelf retreat and breakup?

References

Abdalati, W., and K. Steffen. 1997. "Snow Melt on the Greenland Ice Sheet as Derived from Passive Microwave Satellite Data." *Journal of Climate* 10: 165–75.

Aitken, A. R. A., et al. 2016. "Repeated Large-Scale Retreat and Advance of Totten Glacier Indicated by Inland Bed Erosion." *Nature* 533: 385–9.

Alley, K. E., et al. 2016. "Impacts of Warm Water on Antarctic Ice Shelf Stability through Basal Channel Formation." *Nature Geoscience* 9: 289–93.

Alley, R. B., et al. 2015. "Oceanic Forcing of Ice-Sheet Retreat: West Antarctica and More." *Annual Review of Earth and Planetary Science* 43: 207–31.

Bamber, J. L., et al. 2013. "Paleofluvial Mega-Canyon Beneath the Central Greenland Ice Sheet." *Science* 34: 997–9.

Bell, R. E., et al. 2017. "Antarctic Ice Shelf Potentially Stabilized by Export of Meltwater in Surface River." *Nature* 544: 344–8.

Benson, C. S. 1961. "Stratigraphic Studies in the Snow and Firn of the Greenland Ice Sheet." *Folia Geographica Danica* 9: 13–37.

Bintjana, R. 1999. "On the Glaciological, Meteorological, and Climatological Significance of Antarctic Blue Ice Areas." *Reviews of Geophysics* 37: 337–59.

Bintanja, R., and M. R. Van den Broeke. 1995. "The Surface Energy Balance of Antarctic Snow and Blue Ice." *Journal of Applied Meteorology* 34: 902–26.

Bøggild, C. E., et al. 2010. "The Ablation Zone in Northeast Greenland: Ice Types, Albedos and Impurities." *Journal of Glaciology* 56(195): 101–13.

Box, J. E. 2002. "Survey of Greenland Instrumental Temperature Records: 1873–2001." *International Journal of Climatology* 22: 1829–47.

Box, J. E., and K. Steffen. 2001. "Sublimation on the Greenland Ice Sheet from Automated Weather Station Observations." *Journal of Geophysical Research* 106(D24): 33965–81.

Bromwich, D. H., and T. R. Parish. 1998. "Meteorology of the Antarctic." In *Meteorology of the Southern Hemisphere*, edited by D. J. Karoly and D. G. Vincent, 175–200. Meteorology Monographs 27(49).

Bromwich, D. H., et al. 2001. "Modeled Precipitation Variability over the Greenland Ice Sheet." *Journal of Geophysical Research: Atmosphere* 106(D24): 33891–908.

Burgess, E. W., et al. 2010. "A Spatially Calibrated Model of Annual Accumulation Rate on the Greenland Ice Sheet (1958–2007)." *Journal of Geophysical Research: Earth Surface* 115: F02004.

Charalampidis, C. 2016. *Climatology and Firn Processes in the Lower Accumulation Area of the Greenland Ice Sheet*. Uppsala Dissertation 1372, Faculty of Science and Technology. Uppsala: Acta Universitatis Upsaliensis.

Chen, Q.-S., F. H. Bromwich, and L. Bai. 1997. "Precipitation over Greenland Retrieved by a Dynamic Method and Its Relation to Cyclonic Activity." *Journal of Climate* 10: 839–70.

Christie, F. D. W., et al. 2016. "Four-Decade Record pf Pervasive Grounding Line Retreat along the Bellingshausen Margin of West Antarctica." *Geophysical Research Letters* 43: 5741–9.

Chu, V. W. 2014. "Greenland Ice Sheet Hydrology: A Review." *Progress in Physical Geography* 38: 19–54.

Cook, A. J., and D. G. Vaughan. 2010. "Overview of Areal Changes of the Ice Shelves on the Antarctic Peninsula over the Past 50 Years." *Cryosphere* 4: 77–98.

Crary, A. P., and C. R. Wilson. 1961. "Formation of 'Blue' Glacier Ice by Horizontal Compressive Forces." *Journal of Glaciology* 3: 1045–50.

Depoorter, M. A., et al. 2013. "Calving Fluxes and Basal Melt Rates of Antarctic Ice Shelves." *Nature* 502: 89–92.

Domine, F., A.-S. Taillandier, and W. R. Simpson. 2007. "A Parameterization of the Specific Surface Area of Snow in Models of Snowpack Evolution, Based on 345 Measurements." *Journal of Geophysical Research* 112: F02031.

Enderlin, E. M., et al. 2014. "An Improved Mass Budget for the Greenland Ice Sheet." *Geophysical Research Letters* 41: 866–77.

Eterna, J., et al. 2009. "Higher Surface Mass Balance of the Greenland Ice Sheet Revealed by High-Resolution Climate Modeling." *Geophysical Research Letters* 36: L12501.

Fahnestock, M. A., et al. 2000. "Snow Megadune Fields on the East Antarctic Plateau: Extreme Atmosphere–Ice Interaction." *Geophysical Research Letters* 27: 3719–22.

Favier, V., et al. 2011. "An Updated and Quality Controlled Surface Mass Balance Dataset for Antarctica." *Cryosphere* 7: 583–97.

Fettweis, X., et al. 2016. "Reconstructions of the 1900–2015 Greenland Ice Sheet Surface Mass Balance Using the Regional Climate MAR Model." *Cryosphere Discussions*. doi: 10.5194/tc-2016-268.

Forster, R. R., et al. 2013. "Extensive Liquid Meltwater Storage in Firn within the Greenland Ice Sheet." *Nature Geoscience* 7: 95–8.

Frezotti, M., et al. 2002. "Snow Megadunes in Antarctica: Sedimentary Structure and Genesis." *Journal of Geophysical Research* 107(D18): ACL 1.1–1.12.

Fürst, J. J., et al. 2016. "The Safety Band of Antarctic Ice Shelves." *Nature Climate Change* 6: 479–82.

Grinsted, A., et al. 2003. "Dating Antarctic Blue Ice Areas Using a Novel Ice Flow Model." *Geophysical Research Letters* 30(19): 205, 1.1–1.5.

Hall, D. K., et al. 2012. "A Satellite-Derived Climate-Quality Data Record of the Clear-Sky Surface Temperature of the Greenland Ice Sheet." *Journal of Climate* 25: 4785–98.

Hamilton, R. A. 1958a. "The Meteorology of North Greenland during the Midsummer Period." *Quarterly Journal of the Royal Meteorological Society* 84: 142–58.

Hamilton, R. A. 1958b. "The Meteorology of North Greenland during the Midwinter Period." *Quarterly Journal of the Royal Meteorological Society* 84: 355–74.

Hanna, E., T. E. Cropper, and R. J. Hall. 2016. "Greenland Blocking Index 1851–2015: A Regional Climate Change Signal." *International Journal of Climatology* 36: 4847–61.

Haran, T., et al. 2013. "MEaSUREs Greenland Monthly Image Mosaics from MODIS." https://nsidc.org/data/nsidc-072

Harig, C., and F. J. Simons. 2016. "Ice Mass Loss in Greenland, the Gulf of Alaska, and the Canadian Archipelago: Seasonal Cycles and Decadal Trends." *Geophysical Research Letters*. doi: 10.1002/2016GL067759.

Hawley, R. L., et al. 2014. "Recent Accumulation Variability in Northwest Greenland from Ground-Penetrating Radar and Shallow Cores along the Greenland Inland Traverse." *Journal of Glaciology* 60: 375–82.

Hellmer, H. H., and D. J. Olbers. 1991. "On the Thermohaline Circulation beneath the Filchner–Ronne Ice Shelves." *Antarctic Science* 3: 433–42.

Hofer, S., et al. 2017. "Decreasing Cloud Cover Drives the Recent Mass Loss on the Greenland Ice Sheet." *Science Advances* 3: e1700584.

Hui, F.-M., et al. 2014. "Mapping Blue-Ice Areas in Antarctica Using ETM+ and MODIS Data." *Annals of Glaciology* 55(66): 129–37.

Janssens, I., and P. Huybrechts. 2000. "The Treatment of Meltwater Retention in Mass-Balance Parameterizations of the Greenland Ice Sheet." *Annals of Glaciology* 31: 133–40.

Jenkins, A., and C. S. M. Doake. 1991. "Ice–Ocean Interaction on Ronne Ice Shelf, Antarctica." *Journal of Geophysical Research* 96(C1): 791–813.

Kingslake, J., et al. 2017. "Widespread Movement of Meltwater onto and Across Antarctic Ice Shelves." *Nature* 544: 349–52.

Koenig, L. S., et al. 2014. "Initial In Situ Measurements of Perennial Meltwater Storage in the Greenland Firn Aquifer." *Geophysical Research Letters* 41: 81–5.

Koenig, L. S., et al. 2016. "Annual Greenland Accumulation Rates (2009–2012) from Airborne Snow Radar." *Cryosphere* 10: 1739–52.

Lenearts, J. T. M., et al. 2012. "Drifting Snow Climate of the Greenland Ice Sheet: A Study with a Regional Climate Model." *Cryosphere* 6: 891–9.

Li, X., et al. 2016. "Ice Flow Dynamics and Mass Loss of Totten Glacier, East Antarctica, from 1989 to 2015." *Geophysical Research Letters* 43: 6366–73.

Liu, J.-P., et al. 2016. "Has Arctic Sea Ice Loss Contributed to Increased Surface Melting of the Greenland Ice Sheet?" *Journal of Climate* 29: 3373–86.

MacGregor, J. A., et al. 2016. "A Synthesis of the Basal Thermal State of the Greenland Ice Sheet." *Journal of Geophysical Research: Earth Surface*. doi: 10.1002/2015JF003803.

McMillan, M., et al. 2016. "A High Resolution Record of Greenland Mass Balance." *Geophysical Research Letters* 43(10). doi: 10.1002/2016GL069666.

Merlivat, L., et al. 1973. "Tritium and Deuterium Content of the Snow in Groenland." *Earth and Planetary Science Letters* 19: 235–40.

Miller, N. B., et al. 2017. "Forcing and Responses of the Surface Energy Budget at Summit, Greenland." *Cryosphere* 11: 497–516.

Moon, K. R. 2012. *Investigations of the Dry Snow Zone of the Greenland Ice Sheet Using QuikSCAT*. MS thesis, Department of Electrical and Computer Engineering. Provo, Utah: Brigham Young University.

Morris, E. M., and D. J. Wingham. 2011. "The Effect of Fluctuations in Surface Density, Accumulation and Compaction on Elevation Change Rates along the EGIG Line, Central Greenland." *Journal of Glaciology* 57: 416–30.

Mosley-Thompson, E., et al. 2001. "Local to Regional-Scale Variability of Annual Net Accumulation on the Greenland Ice Sheet from PARCA Cores." *Journal of Geophysical Research* 106(33): 839–54.

Mote, T. L. 2014. *Greenland Daily Surface Melt 25 km EASE-Grid 2.0 Climate Data Record, 1979–2012*. [Digital media]. Boulder, CO: University of Colorado, National Snow and Ice Data Center.

Mote, T. L., and M. R. Anderson. 1995. "Variations in Snowpack Melt on the Greenland Ice Sheet Based on Passive-Microwave Measurements." *Journal of Glaciology* 41(137): 51–60.

Mottram, R. H. 2016. "Report." Cryolist 74.

Mouginot, J., et al. 2017. "Comprehensive Annual Ice Sheet Velocity Mapping Using Landsat-8, Sentinel-1, and RADARSAT-2 Data." *Remote Sensing* 9: 364–83.

Moussavi, M. S., et al. 2016. "Derivation and Validation of Supraglacial Lake Volumes on the Greenland Ice Sheet from High-Resolution Satellite Imagery." *Remote Sensing of the Environment* 183: 294–303.

Moustafa, S. E., et al. 2015. "Multi-Modal Albedo Distributions in the Ablation Area of the Southwestern Greenland Ice Sheet." *Cryosphere* 9: 905–23.

Muller, D. E., L. Copland, and D. Stern. 2008. "Examining Arctic Ice Shelves prior to the 2008 Breakup." *Eos* 89(49): 502–3.

Mulvaney, R., et al. 2012. "Recent Antarctic Peninsula Warming Relative to Holocene Climate and Ice-Shelf History." *Nature* 489: 141–4.

Nghiem, S. V., et al. 2012. "The Extreme Melt across the Greenland Ice Sheet in 2012." *Geophysical Research Letters* 39(20): L20502.

Noël, B., et al. 2016. "A Daily, 1-km Resolution Dataset of Downscaled Greenland Ice Sheet Surface Mass Balance (1958–2015)." *Cryosphere Discussions*. doi: 10.5194/tc-2016-145.

Nolin, A. W., and J. Dozier. 2000. "A Hyperspectral Method for Remotely Sensing the Grain Size of Snow." *Remote Sensing of the Environment* 74: 207–16.

Nolin, A. W., and J. Stroeve. 1997. "The Changing Albedo of the Greenland Ice Sheet: Implications for Climate Modeling." *Annals of Glaciology* 25: 51–7.

Ohmura, A., et al. 1999. "Precipitation, Accumulation and Mass Balance of the Greenland Ice Sheet." *Zeitschrift für Gletscherkunde und Glazialgeologie* 35: 1–20.

Overly, T. B., et al. 2016. "Greenland Annual Accumulation along the EGIG Line, 1959–2004, from ASIRAS Airborne Radar and Neutron-Probe Density Measurements." *Cryosphere* 10: 1679–94.

Palm, S. P., et al. 2011. "Satellite Remote Sensing of Blowing Snow Properties over Antarctica." *Journal of Geophysical Research: Atmospheres* 116(D16): D16123.

Palm, S. P., et al. 2017. "Blowing Snow Sublimation and Transport over Antarctica from 11 Years of CALIPSO Observations." *Cryosphere Discussions*. doi: 10.5194/tc-2017-45.

Palmer, S. J., et al. 2013. "Greenland Subglacial Lakes Detected by Radar." *Geophysical Research Letters* 40: 6154–9.

Pollard, D., R. M. DeConto, and R. B. Alley. 2014. "Potential Antarctic Ice Sheet Retreat Driven by Hydrofracturing and Ice Cliff Failure." *Earth and Planetary Science Letters* 412: 112–21.

Rignot, E., et al. 2008. "Recent Antarctic Ice Mass Loss from Radar Interferometry and Regional Climate Modelling." *Nature Geoscience* 1: 106–10.

Rignot, E., et al. 2014. "Widespread, Rapid Grounding Line Retreat of Pine Island, Thwaites, Smith, and Kohler Glaciers, West Antarctica, from 1992 to 2011." *Geophysical Research Letters* 421: 3502–9. doi: 10.1002/2014GL060140.

Rintoul, S. R., et al. 2016. "Ocean Heat Drives Rapid Basal Melt of the Totten Ice Shelf." *Science Advances* 2(12): e1601610.

Ryan, J. C., et al. 2016. "Attribution of Greenland's Ablating Ice Surfaces on Ice Sheet Albedo Using Unmanned Aerial Systems." *Cryosphere Discussions*. doi: 10.5194/tc-2016-204.

Scambos, T. A., et al. 2000. "The Link between Climate Warming and Breakup of Ice Shelves in the Antarctic Peninsula." *Journal of Glaciology* 46(154): 516–30.

Scambos, T. A., et al. 2007. "MODIS-Based Mosaic of Antarctica (MOA) Data Sets: Continent-wide Surface Morphology and Snow Grain Size." *Remote Sensing of the Environment* 111: 242–57.

Scambos, T., J. Bohlander, and B. Raup. 2009a. "Images of Antarctic Ice Shelves." https://nsidc.org/data/iceshelves_images/

Scambos, T., et al. 2009b. "Ice Shelf Disintegration by Plate Bending and Hydro-fracture: Satellite Observations and Model Results of the 2008 Wilkins Ice Shelf Break-ups." *Earth and Planetary Science Letters* 280: 51–60.

Scambos, T. A., et al. 2012. "Extent of Low-Accumulation 'Wind Glaze' Areas on the East Antarctic Plateau: Implications for Continental Ice Mass Balance." *Journal of Glaciology* 58(210): 633–47.

Scambos, T. A., et al. 2017a. "How Much, How Fast? A Science Review and Outlook for Research on the Instability of Antarctica's Thwaites Glacier in the 21st Century." *Global and Planetary Change* 152: 16–34.

Scambos, T. A., C. Shuman, and M. Fahnestock. 2017b. *Antarctic Megadunes, Wind Glaze, and Snow Roughness Variability in East Antarctica (and Mars): Polar Ice, Polar Climate, Polasr Change*. International Glaciological Society Symposium. Boulder, CO. Abstract 76A2607.

Scarchili, C., M. Frezzotti, and P. Grigioni. 2010. "Extraordinary Blowing Snow Transport Event in East Antarctica." *Climate Dynamics* 34: 1195–206.

Serreze, M. C., and R. G. Barry. 2014. *The Arctic Climate System*. 2nd ed. Cambridge: Cambridge University Press.

Shepherd, A., et al. 2012. "A Reconciled Estimate of Ice–Sheet Mass Balance." *Science* 338: 1183–9.

Shuman, C. A., K. Steffen, and C. R. Stearns. 2001. "A Dozen Years of Temperature Observations at Summit, Central Greenland Automatic Weather Stations 1987–99." *Journal of Applied Meteorology* 40(4): 741–52.

Simmonds, I. 1998. "The Climate of the Antarctic Region." In *Climates of the Southern Continents: Past, Present and Future*, edited by J. E. Hobbs, J. A. Lindesay, and H. A. Bridgman, 137–60. Chichester, UK: Wiley and Sons.

Sinisalo, A., and J. C. Moore. 2010. "Antarctic Blue Ice Areas: Towards Extracting Palaeoclimate Information." *Antarctic Science* 22: 99–115.

Smith, L. C., et al. 2015. "Efficient Meltwater Drainage through Supraglacial Streams and Rivers on the Southwest Greenland Ice Sheet." *Proceedings of the National Academy of Sciences* 112: 101–6.

Steffen, K., and J. E. Box. 2001. "Surface Climatology of the Ice Sheet: Greenland Climate Network 1995–1999." *Journal of Geophysical Research* 106(D24): 33951–64.

Sterken, M., et al. 2012. "Holocene Glacial and Climate History of Prince Gustav Channel, Northeastern Antarctic Peninsula." *Quaternary Science Review* 31: 93–111.

Takahashi, S., et al. 1992. "Bare Ice Fields Developed in the Inland." *Proceedings, NIPR Symposium on Polar Meteorology and Glaciology* 5: 128–39.

Tedesco, M., and A. I. Monaghan. 2009. "An Updated Antarctic Melt Record through 2009 and Its Linkages to High-Latitude and Tropical Climate Variability." *Geophysical Research Letters* 36: L18502.

Tedesco, M., et al. 2013. "Evidence and Analysis of 2012 Greenland Records from Spaceborne Observations: A Regional Climate Model and Reanalysis Data." *Cryosphere* 7: 615–30.

Tedesco, M., et al. 2015. "Greenland Ice Sheet." In *State of the Climate in 2014*, edited by A. Mekonnen, J. A. Renwick, and A. Sanchez-Lugo. *Bulletin of the American Meteorological Society* 96(7): S137–9.

Tedesco, M., et al. 2016. "The Darkening of the Greenland Ice Sheet: Trends, Drivers, and Projections (1981–2100)." *Cryosphere* 10: 477–96.

Tran, N., et al. 2008. "Snow Facies over Ice Sheets Derived from Envisat Active and Passive Observations." *IEEE Transactions on Geoscience and Remote Sensing* 46: 3694–708.

Turner, J., et al. 2014. "Antarctic Climate Change and the Environment: An Update." *Polar Record* 50: 237–59.

Van de Wal, R. S. W., et al. 2012. "Twenty-One Years of Mass Balance Observations along the K-Transect, West Greenland." *Earth System Science Data* 4(1): 31–5.

Van den Broeke, M. R., et al. 2008. "Partitioning of Melt Energy and Melt Water Fluxes in the Ablation Zone of the West Greenland Ice Sheet." *Cryosphere* 2(2): 179–89.

Van den Broeke, M. R., et al. 2016. "On the Recent Contribution of the Greenland Ice Sheet to Sea Level Change." *Cryosphere* 10: 1933–46.

Van Wyk de Vries, M., R. G. Bingham, and A. S. Hein. 2017. "A New Volcanic Province: An Inventory of Subglacial Volcanoes in West Antarctica." In *Exploration of Subsurface Antarctica: Uncovering Past Changes and Modern Processes*, edited by M. J. Siegert, S. S. R. Jamieson, and D. A. White. Special Publication 461. London: Geological Society.

Vaughan, D. G., et al. 1999. "Reassessment of Net Surface Mass Balance in Antarctica." *Journal of Climate* 12: 933–46.

Wang, Y.-F., et al. 2016. "A Comparison of Antarctic Ice Sheet Surface Mass Balance from Atmospheric Climate Models and In Situ Observations." *Journal of Climate* 29(14): 5317–37.

Winther, J. G., M. N. Jespersen, and G. E. Liston. 2001. "Blue-Ice Areas in Antarctica Derived from NOAA AVHRR Satellite Data." *Journal of Glaciology* 47(157): 325–34.

Oceanic Environments

7

This chapter examines the ocean environments of both polar regions – their hydrography, water masses, and currents. Global sea level changes are also described. Each polar sea is discussed, and then sea ice conditions and polynyas are examined.

7.1 Southern Ocean

The Southern Ocean was defined and its major characteristics described in Section 4.2. Here we look in more detail at its seasonal and geographical variability.

Some 99.7 percent of the area between latitudes 60° and 65° S is ocean. The greatest constriction in the Southern Ocean is at the Drake Passage, where the tip of South America is 1,100 km from the tip of the Antarctic Peninsula (Ostapoff 1965). The zonally averaged sea surface temperature (SST) is 0 °C at 61° S, while the equivalent temperature at 61° N is 4 °C higher. There are considerable departures from zonality, however. The Atlantic and Indian Ocean sectors are up to 2–3 °C warmer than the Pacific sector. The diurnal variation of SST in high latitudes is much less than 0.5 °C as a result of the large amount of cloud cover and persistent strong winds.

The surface salinity in the Southern Ocean south of the ocean Polar Front is generally constant around 33.9 psu, but local variations between 31 and 34.5 psu have been observed (Ostapoff 1965). The low value is caused by ice melt. The Polar Front (or Antarctic Convergence) generally coincides with a salinity minimum at 200 m depth of approximately 34.0–34.2 psu, while to the south there is a zonal ring of maximum values greater than 34.6 psu.

Tomczak and Godfrey (2003) provide a detailed account of Antarctic oceanography. Except near the continent, the waters of the Southern Ocean move eastward, forced by the southern westerlies. This is the only region of the planet where there is essentially unhindered circumglobal flow of ocean water.

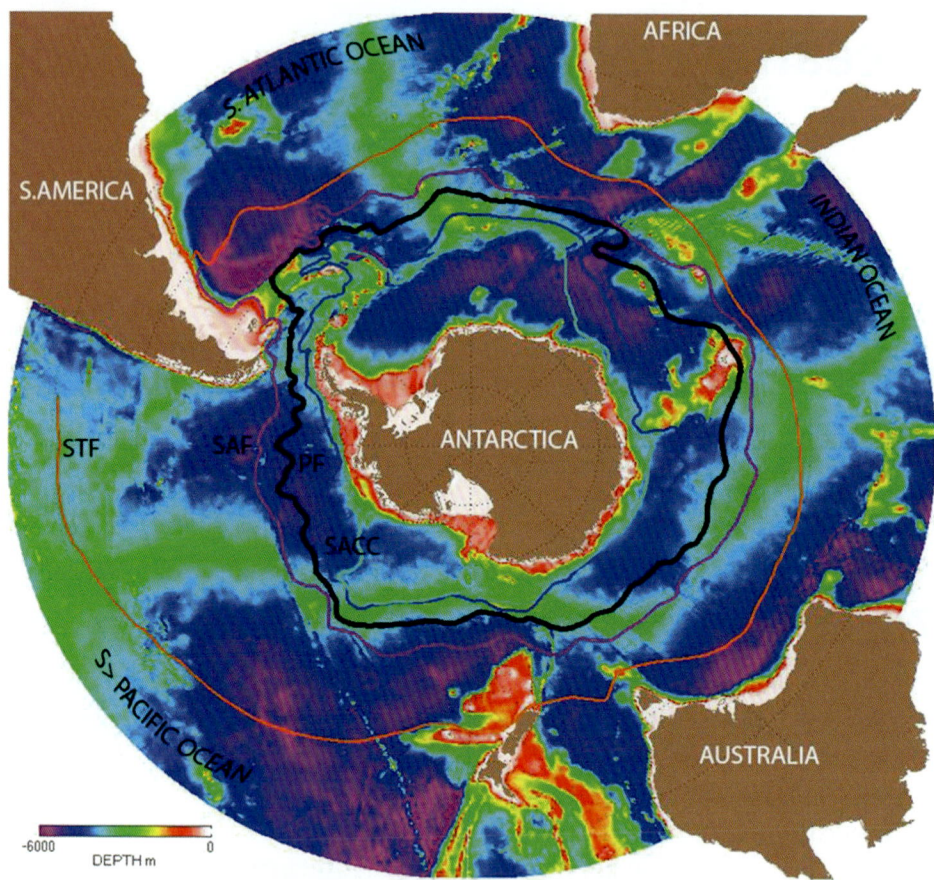

ANTARCTIC CIRCUMPOLAR CURRENT
SEAWATER DENSITY FRONTS (FROM ORSI et al. 1995),
AND *BATHYMETRY* OF THE SOUTHERN OCEAN (UP TO LATITUDE 25 S)

Figure 7.1 The Antarctic Circumpolar Current (black line) sea water density fronts (after Orsi et al. 1995) and bathymetry. SACC = Southern Antarctic Circumpolar Current Front, PF = Polar Front, SAF = Subantarctic Front, STF = Subtropical Front.
Source: Image from the GRACE mission. Courtesy of NASA/JPL-Caltech.
https://commons.wikimedia.org/w/index.php?curid=3526740.

Maximum velocities occur just north of the Polar Front, and in the Drake Passage the average maximum geostrophic velocity over 100 km distance is approximately 27 cm s^{-1}, with an estimated core velocity of 40–60 cm s^{-1} (Ostapoff 1965). At the bottom, velocities decrease to 5–10 cm s^{-1}.

The Antarctic Circumpolar Current (ACC) flowing continuously around the Antarctic continent is the strongest current in the world's oceans (Figure 7.1). It has a length of about 20,000 km (Rintoul 2000; Wyrtki 1960) and moves

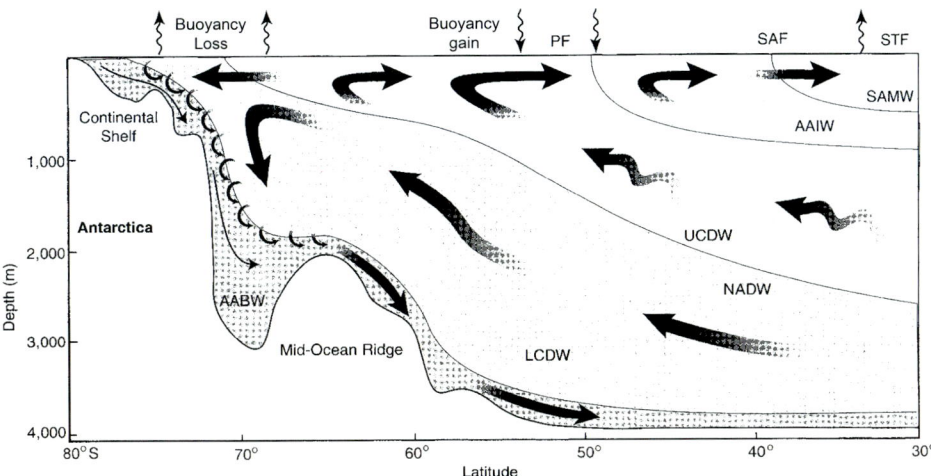

Figure 7.2 Schematic two-cell meridional overturning circulation in the Southern Ocean. An upper cell is primarily formed by northward Ekman transport and southward eddy transport in the UCDW layer. A lower cell is primarily driven by dense water formation near Antarctica.
Source: Speer et al. 2000, 3221, figure 8. Courtesy of American Meteorological Society.
© Copyright 2000 American Meteorological Society (AMS).

eastward throughout its depth from the surface down. Using three years of in situ velocity data (2006–2009) in the Drake Passage and twenty years of satellite altimetric observations of surface currents, Koenig et al. (2016) estimated the volume transport at 140 Sv (10^6 m^3 s^{-1}). However, Donahue et al. (2016) found a larger value of 173 Sv from moorings between 2007 and 2011 (see Section 4.2) This total is primarily baroclinic flow (127.7 Sv), with 45.6 Sv of barotropic origin. A barotropic flow is the same from the ocean surface to the bottom – that is, the flow is uniform with depth. Barotropic flow, by comparison, is in dynamical balance with the sea surface slope. The term "baroclinic" denotes the depth-dependent part of the flow. The baroclinic component of the flow results from the density distribution in the fluid, which varies due to differences in temperature and salinity. Hence, it is a function of the vertical shear. The Drake Passage is open to 2,000 m depth, which includes the Upper Circumpolar Deep Water (UCDW) layers.

Meridional sections across the ACC indicate an equatorward flow of surface water due to the westerly wind stress and poleward transport of subsurface and deep water (Figure 7.2). Mesoscale eddies between the surface Ekman layer and approximately 2,000 m depth transfer most of the ocean heat toward the Antarctic continent. The Upper Circumpolar Deep Water (UCDW) transports about 2 Sv southward, according to Wyrtki (1960). North Atlantic Deep Water (NADW) and

Lower Circumpolar Deep Water (LCDW) move southward below the UCDW. The bottom water flows northward (Figure 7.2). Observations show that winds and current almost coincide and that both the belt of maximum westerlies and the current show a shift to the south from about 40–45° S in the central South Atlantic to about 55–60° S in the eastern South Pacific.

The Antarctic Polar Front is located south of the ACC axis and south of the maximum westerlies. According to Wyrtki (1960), it can be either a divergent or a convergent phenomenon, depending on the strength and position of the westerly wind maximum. With weak westerlies, there is divergence to the north and convergence to the south of the main current that intensifies the Polar Front. If the westerlies are strong and displaced southward, there is convergence to the north and divergence to the south of the maximum westerlies, extending up to the ice edge. With the westerly wind belt in a northerly position, the Polar Front is in the range of divergent motion. Easterly winds develop near the continental margin and Antarctic divergence is displaced from the ice edge with poleward motion of surface water.

Kort (1964) notes that Antarctic Bottom Water (AABW) forms in autumn–winter when shelf water begins to cool and ice growth increases its salinity by brine drainage. Fofonoff (1956) demonstrated that shelf water sinks down the continental slope when the salinity is at least 34.51 psu. The continental shelf edge around Antarctica has a depth of up to 800 m. The regions where bottom water forms are the southwest and western Weddell Sea, the Ross Sea, the Shackleton shelf glacier, and near-shore areas of Princess Martha Coast (5° E to 20° W). Orsi et al. (1999) calculated that the rate of newly formed AABW sinking down the slope around Antarctica is 8–10 Sv.

The Southern Ocean is the only place where there is extensive direct upwelling of deep water to the sea surface. Dynamically, this movement is attributable to the open latitude band (56–63° S) of the Drake Passage. Antarctic Bottom Water (AABW) at 67° S, 30° W has a potential temperature of −0.88 °C and salinity of 34.64 psu (Johnson 2006). The net buoyancy gain shown in Figure 7.2 over much of the Southern Ocean is a result of both heat input and net precipitation.

The permanent thermocline that characterizes most of the world ocean is absent around Antarctica (Tomczak and Godfrey 2003). Density variations with depth are small, and the pressure gradient force is distributed more evenly through the water column. As a result, currents can extend to great depth.

Available hydrographic observations for the Southern Ocean were compiled by Orsi et al. (1995). The northern boundary of Subantarctic Surface Water delimits the Subantarctic Front (SAF). The sharp termination of the poleward extent of UCDW characteristics coincides with a frontal feature that separates the ACC from the Weddell Gyre. The traditional recognition of only three fronts is attributed by Sokolow and Rintoul (2009a, 2009b) to the emphasis placed on the Drake Passage. These researchers used fifteen years of sea surface height (SSH) data to map circumpolar ACC fronts. In twelve longitudinal sectors of 30°, regions of high gradient were defined as those that exceeded a threshold of 0.25 m per 100 km. Sokolow and Rintoul then found the SSH contours that most efficiently represented the high gradient regions. They showed that the ACC consists of multiple frontal filaments or jets that are aligned along particular streamlines throughout the circumpolar path of the current. The SSH value (approximate streamline) associated with each frontal branch was found to be nearly constant, both in time and around the circumpolar path. The frontal branches inferred by fitting SSH contours to fifteen years of SSH gradient maps agree very well with the front positions inferred from synoptic sections of water mass properties; see Figure 7.3.

Emery and Meincke (1986) provided a summary of global water masses. For the upper 500 m around Antarctica, they describe two components:

Subantarctic Surface Water (SASW): 3.2–15 °C, 34.0–35.5 psu
Antarctic Surface Water (AASW): −1.0 to 1.0 °C, 34.0–34.6 psu

Below 1,500 m, there are three components:

Circumpolar Deep Water (CDW): 0.1–2.0 °C, 34.62–34.73 psu
North Atlantic Deep Water (NADW): 1.5–4.0 °C, 34.8–35.0 psu
Antarctic Bottom Water (AABW): −0.9 to 1.7 °C, 34.64–34.72 psu

Giglio and Johnson (2016) used Argo float data north of 60° S for 2006–2013 to identify the Subantarctic and Polar Fronts in the Antarctic Circumpolar Current; there are two Subantarctic Fronts. Dynamic height contours can be used to identify all three fronts, which correspond to local maxima in vertical shear. The ACC fronts are associated with strong gradients in temperature and salinity. Based on potential temperature criteria, the Polar Front is farthest south, between 75° and 110° W at around 62° S, and around 61° S at 180° longitude. It is farthest north (48° S) at 75° E and at about 50° S from 30° W to 30° E. During 2006–2013, in the upper 2,000 m of the global ocean sampled by Argo floats, the southern hemisphere ocean was the main recipient of heat from global warming. Roemmich et al. (2015) have shown that ocean heat gain over the 0–2,000 m layer took place at a rate of 0.4–0.6 W m^{-2} during 2006–2013; 67–98 percent of this gain occurred in the southern hemisphere extratropical ocean.

Graham et al. (2012) use one hundred year simulations to show that the number and intensity of fronts is set largely by the bottom topography of the Southern Ocean. The number of fronts is reduced in regions where the path of the ACC is constricted or blocked by topography, as in the Drake Passage. Fronts

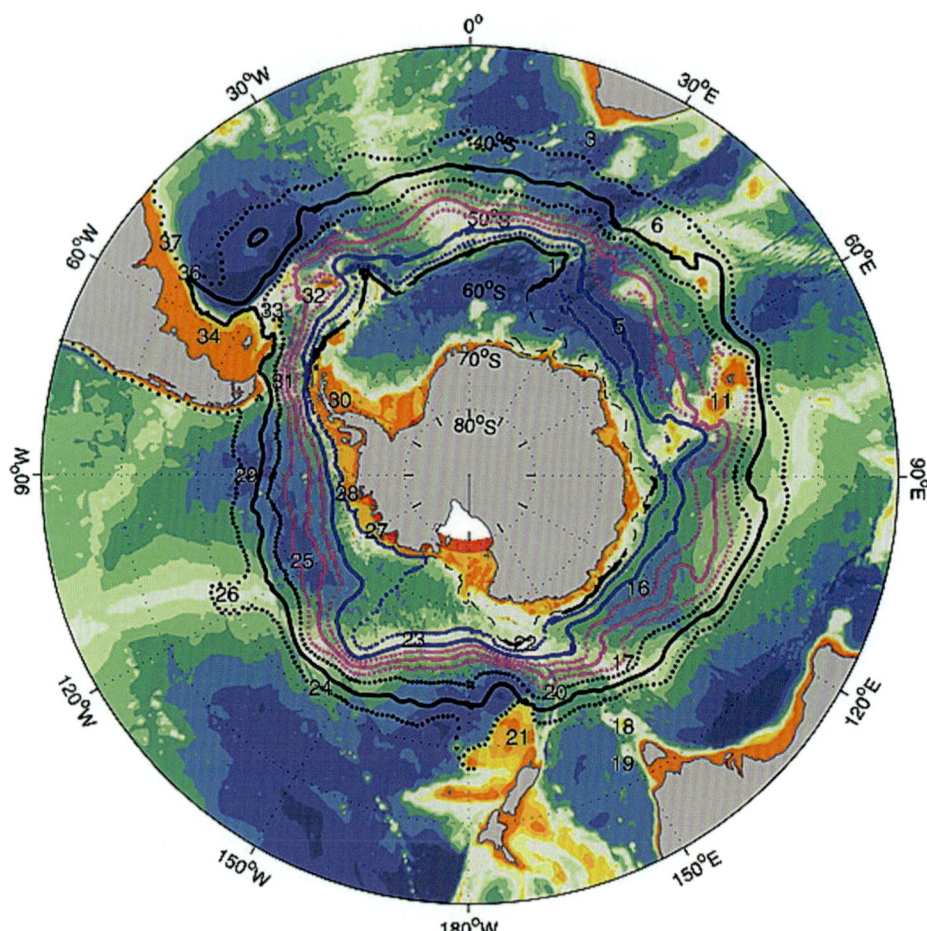

Figure 7.3 Mean ACC front positions mapped using sea surface height (SSH). The Southern Ocean fronts are color coded. The ACC fronts are plotted using local (estimated in 30° sectors) SSH labels. There are small discontinuities in the fronts owing to the positions of the fronts, mapped using local SSH labels, which are extended by 1° of longitude outside of the local sectors, and the frontal SSH labels change longitudinally. The path of the shelf boundary (SB), when obscured by sea ice and not visible in altimetry, is indicated by dashed line. The ACC fronts overlie the Southern Ocean bathymetry. STF = Subtropical Front; SAMW = Subantarctic Mode Water; AAIW = Antarctic Intermediate Water; AABW = Antarctic Bottom Water. Topographic and circulation features from west to east (starting from the prime meridian) are numbered as follows: 1, Weddle Front; 3, Agulhas Retroflection region; 5, Enderby Basin (Enderby Abyssal Plain); 6, Southwest Indian Ridge (SWIR); 11, Kerguelen Plateau; 16, Australian–Antarctic Basin; 17, Southeast Indian Ridge (SEIR); 18, South Tasman Rise; 19, Tasman Outflow; 20, Macquarie Ridge; 21, Campbell Plateau; 22, Mid-Ocean Ridge; 23, Pacific Antarctic Ridge; 24, Eltanin Fracture Zone System; 25, Amundsen Abyssal Plain; 26, East Pacific Ridge; 27, Getz Ice Shelf; 28, Abbot Ice Shelf; 29, Southeast Pacific Basin; 30, Antarctic Peninsula; 31, Drake Passage; 32, Scotia Sea; 33, Falkland Plateau; 34, Patagonian Shelf; 36, Brazil–Falkland Confluence Zone.

Source: Sokolow and Rintoul 2009a, C05018, figure 6. Courtesy of American Geophysical Union.

within the ACC are more barotropic and extend down to the ocean floor making them sensitive to the topography.

7.2 Ross Sea

The Ross Sea is a deep embayment in the Southern Ocean between Victoria Land in the west and Marie Byrd Land in the east. Its northern limit is the edge of the continental shelf. It covers about 960,000 km^2. The southern part is covered by the Ross Ice Shelf (see Section 6.3) and the sea is covered by sea ice for most of the year. The sea is generally less than 900 m depth; in the west, it is less than 300 m deep over wide areas. The ocean surface circulation is dominated by a wind-driven cyclonic (clockwise) gyre, which is forced partly by the east wind drift along the coastline that turns northward along the coast of Victoria Land. The gyre is accompanied by upwelling of deep water. The bottom water characteristics at 72° S, 165° W are potential temperature of −0.24 °C and salinity of 34.70 psu (Johnson 2006). The slow-moving Circumpolar Deep Water is a relatively warm, salty water mass that flows onto the continental shelf at certain locations in the eastern Ross Sea, influenced by the bottom topography. It has a temperature of 1–2 °C and a salinity between 34.62 and 34.73 psu.

Jacobs et al. (1970) report that the westerly current of CDW over the Ross Sea continental slope provides a dynamic barrier to northward thermohaline flow of dense Ross Sea Shelf Water (RSSW). Insufficient brine may be released by the freezing of sea ice to produce the major portion of RSSW. Ice Shelf Water (ISW), identified by a pronounced temperature minimum, has temperatures as low as −2 to −1 °C near its source at the base of the Ross Ice Shelf. Low-salinity Antarctic Bottom Water (AABW) forms during summer over the continental slope in the eastern Ross Sea from a mixture of CDW and ISW. Higher-salinity AABW is produced in summer in the western Ross Sea from a combination of RSSW and CDW.

7.3 Weddell Sea

The Weddell Sea is an embayment in the Antarctic continent that is delimited by the Antarctic Peninsula to the west and the coast of Coats Land to the east. It has an area of about 2.8 million km^2. The southern part is covered by the thick Ronne–Filchner Ice Shelf (see Section 6.3). The Antarctic continental shelf widens to 250 km along the Antarctic Peninsula and up to about 480 km along the southern edge of the Weddell Sea. Marking the edge of the continent, the break lies at an unusually great depth of about 500 m. The Weddell Sea is one of the few locations in the global ocean where deep bottom water forms and contributes to the global ocean thermohaline circulation (see Box 7.1). The wind-driven, clockwise, cyclonic Weddell Gyre gives rise to an outflow in the western part of the sea that transports sea ice and icebergs northward (Muench and Gordon 1995). It has a cold, low-salinity surface layer; at 68° S, 53° W the surface potential temperature is −1.85 °C and the salinity is 34.25 psu (Johnson 2006). This is separated by a thin, weak pycnocline from a thick layer of relatively warm and salty water, referred to as Weddell Deep Water, and a cold bottom layer. Based on data obtained from the western Weddell Sea during the austral winter 1992 US–Russian drifting ice station experiment, Muench and Gordon (1995) reported that the northward flow increased more than twice as much from south to north. A significant fraction of the northward transport was contained in a 300–500 m thick bottom layer of cold water.

Box 7.1 Thermohaline Circulation

The thermohaline circulation (THC) is the part of the large-scale ocean circulation that is driven by global density gradients created by fluxes of surface heat and freshwater (Figure 7.4). Density gradients are determined by gradients of water temperature and salt content. The THC is also known as the meridional overturning circulation (MOC) or the global conveyor belt, in recognition that these currents are responsible for the large-scale exchange of water masses in the ocean.

The thermohaline circulation is mainly driven by the formation of deep water masses in the northern North Atlantic and the Southern Ocean near Antarctica as a result of differences in water temperature and salinity. The absence of deep water formation in the Pacific Ocean may reflect a higher freshwater flux, making it more stable than the Atlantic Ocean. Atlantic Deep Water (ADW) flows southward from the Greenland Sea and eventually surfaces in the North Pacific, at which point some of it becomes the Indonesian through-flow. ADW intermingles with Antarctic Intermediate Water (AIW) and flows into the South Atlantic via Drake Passage. The Indonesian through-flow traverses the South Indian Ocean and mixes with returning AIW to form a northward flow to complete the ocean circulation. The entire circulation takes approximately 2,000 years.

Figure 7.4

The global thermohaline circulation. Blue paths represent deep-water currents, while red paths represent surface currents.
Source: Wikipedia, https://en.wikipedia.org/wiki/Thermohaline_circulation#/media/File:Thermohaline_Circulation_2.png. Robert Simmon, National Aeronautics and Space Administration. Minor modifications by Robert A. Rohde; released to the public domain by NASA Earth Observatory.

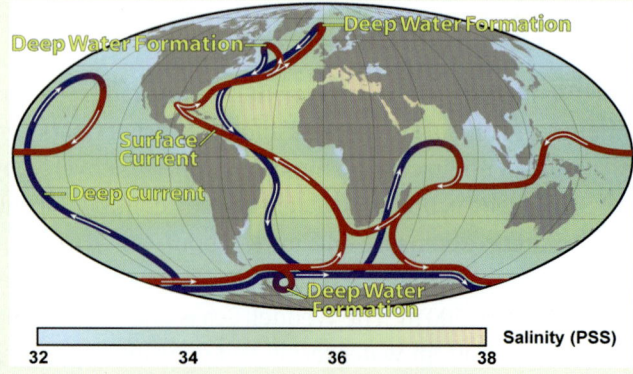

Table 7.1 Surface energy budget components in the Weddell Sea
(W m^{-2}, positive upward)

Surface	Sn	L↑	H	LE	Flux through ice
Sea ice: winter	0	20–40	−17	3	10
Sea ice: summer	−40	30–50	−5	3	–
Ice shelf: winter	0	30–50	16	0	1
Ice shelf: summer	−60	50–60	−12	2	1
Coastal polynya: winter	−2	100–130	240	80	
Coastal polynya: summer	−240	60–70	(30)	(20)	
Lead: winter	−5	70–110	190	70	
Lead: summer	−190	30–50	(20)	(14)	
Ocean: autumn	−100	60–70	27	30	

Winter is June to end of August. Summer is December to end of February.
Source: Launiainen and Vihma 1994, 412, table 5.

A significant phenomenon in the Weddell Sea is the sinking of water mass by the process of cabbeling. Cabbeling involves the formation of denser water by the mixing of two water masses with different temperatures and salinity; warm, salty water from mid-latitudes encounters cold, freshwater from melting sea ice and the effect of wind cooling near Antarctica. Offshore winds drive sea ice away from the coast, thereby leading to continual ice growth in coastal polynyas (see Section 7.9). An important associated process is brine extrusion during ice growth, which increases water salinity.

Launiainen and Vihma (1994) analyzed surface energy fluxes in the Weddell Sea from five drifting buoys during 1990–1992. The buoys drifted from about 73° S in the Weddell Sea northward to about 63° S, and then east–northeastward toward 25° to 0° W. Table 7.1 summarizes some of these researchers' results. Sensible heat fluxes were largest, 100–300 W m^{-2}, over leads and coastal polynyas that represented 5–7 percent of the area in winter. These fluxes approximately balanced the downward flux over the sea ice. Summer values of sensible heat there were small due to the small difference between air and sea surface temperatures. The mean annual heat loss from the Weddell Sea is estimated to be 20–30 W m^{-2}.

7.4 Arctic Ocean

The Arctic Ocean has limited contact with the Atlantic and Pacific oceans due to the presence of narrow passageways and shallow sills in this body of water. It is, therefore, designated as a Mediterranean sea (Tomczak and Godfrey 2003). The circulation of the Arctic Ocean, unlike that of the Southern Ocean, is driven by thermohaline forcing.

The seas that make up the Arctic Ocean are the Barents, Kara, Laptev, East Siberian, Chukchi, Beaufort, and Lincoln (Figure 7.5). The Barents Sea is the largest, with an area of 1.405 million km². Important seas adjacent to the Arctic Ocean are the Bering Sea and the Greenland Sea, as well as Baffin Bay. The Arctic Ocean has continental shelves covering 2.5 million km². About 40 percent of this area occurs in the interior shelves of the Kara Sea, Laptev Sea, East Siberian Sea, and Beaufort Sea (Williams and Carmack 2015), which are distinguished from the inflow and outflow shelves by their massive amounts of river runoff. In the mid-shelf region, wind and ice motion-induced surface stresses dominate mixing and circulation, resulting in high variability. Along the northern boundary, water is forced by upwelling- and downwelling-favorable surface stresses, which drive shelf-basin exchanges with cyclonic boundary currents of Atlantic and Pacific origin over the upper slope. Shelf-basin exchange is further modified by shelf-break morphology.

The Arctic Ocean is divided by the Lomonosov Ridge into two deep basins, the Canada Basin and the Eurasian Basin, which reach depths of 5449 m (Figure 7.6). The ridge extends across the Arctic from north of Greenland to the New Siberian Islands and in places is less than 400 m deep. A 1,700-km-wide opening exists to the North Atlantic along a large sill that runs from Greenland to Iceland, the Faroe Islands, and the Scotia bathymetric features. Approximate sill depths are 600 m in Denmark Strait (between Greenland and Iceland), 400 m between Iceland and the Faroe Islands, and 800 m between the Faroe Islands and Scotland.

The Arctic/sub-Arctic water masses were first identified by Emery and Meincke (1986). Updated information is provided by Talley et al. (2011, table S12.3):

Polar Surface Water (PSW): −1.5 to −1.9 °C, 31–34 psu
Bering Strait Water summer (Pacific Summer Water) (sBSW): −1.3 °C, 32–33 psu
Atlantic Water (AW): 200–1,000 m, 0–3 °C, >34.9 psu
Upper Polar Deep Water (uPDW): 1,000–1,700 m, −0.5 to 0 °C, 34.85–34.9 psu
Canada Basin Deep Water (CBDW): Below 1,700 m, −0.53 °C, 34.95 psu
Eurasian Basin Deep Water (EBDW): Below 1,700 m, −0.95 °C, 34.94 psu
Greenland Sea Deep Water (GSDW): Below 2,000 m, < −1.2 °C, 34.88–34.90 psu

Temperature and salinity data for the Arctic basin were assembled for a joint US–Russian atlas of Arctic oceanography for both summer and winter (Tanis and Timokhov 1997, 1998). Mean values are shown versus depth for the Canadian, Amundsen, and Nansen basins in Table 7.2.

Atlantic water penetrates into the Nansen basin at 300 m, where the average temperature is 1.7 °C and salinity shows a sharp increase. The salinity values at

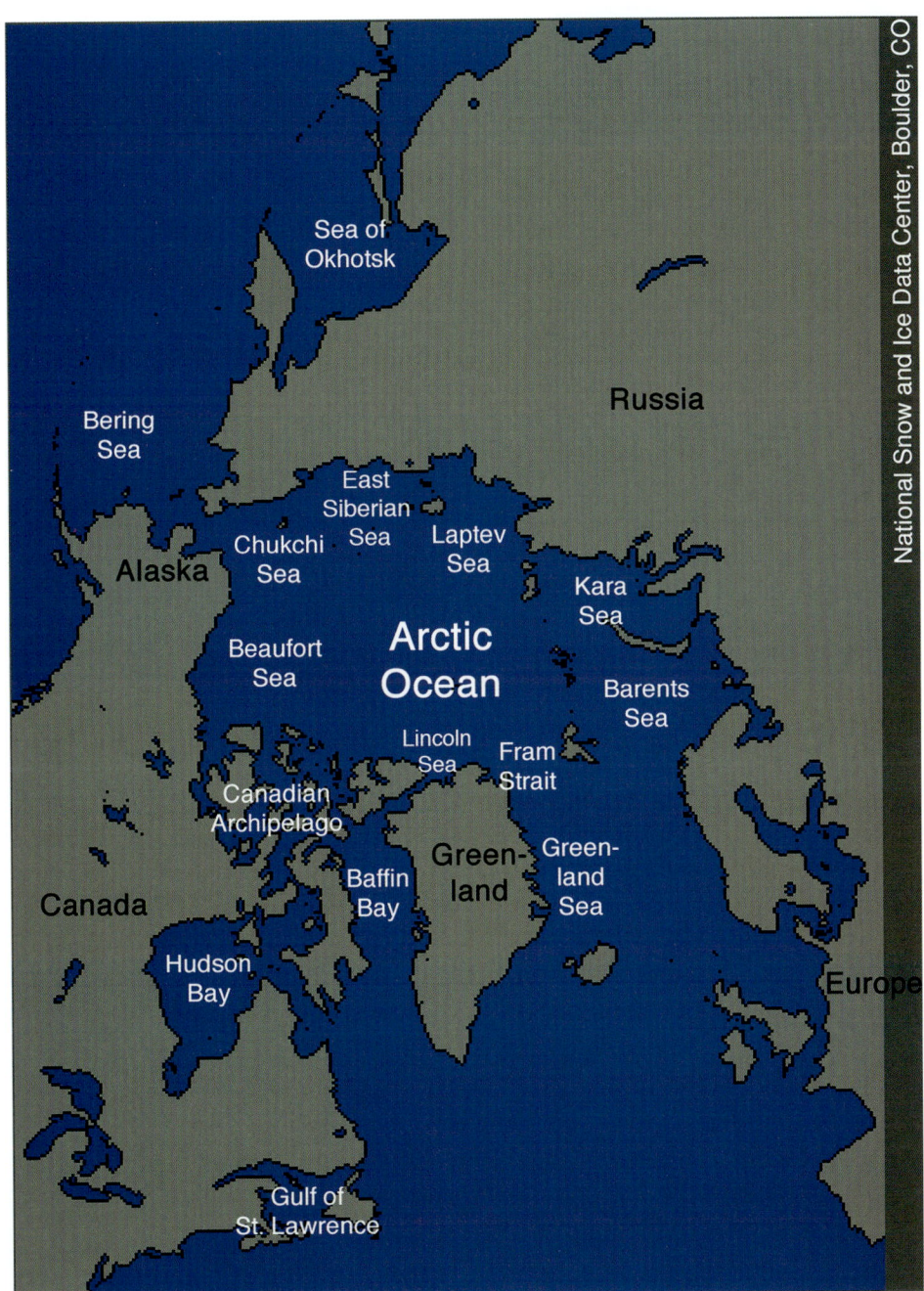

Figure 7.5 The Arctic seas.
Source: National Snow and Ice Data Center, University of Colorado, Boulder.

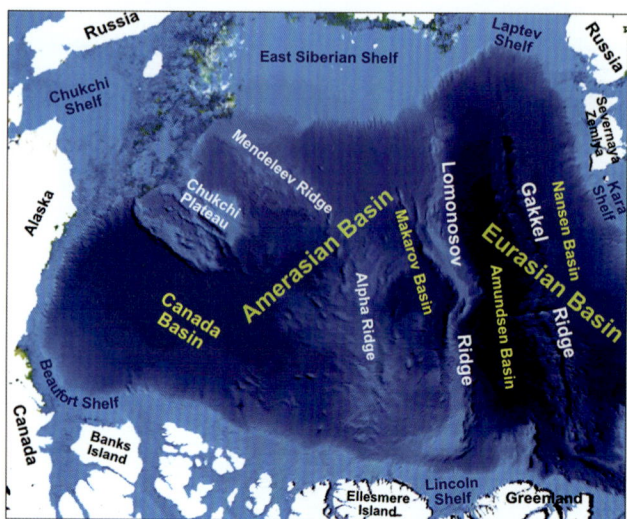

Figure 7.6 Arctic Basin bathymetry.
Source: Mike Norton, https://upload.wikimedia.org/wikipedia/commons/d/d4/Arctic_Ocean_b.

and below 2,000 m are remarkably uniform. The lowest water temperatures of −1.8 °C or less, and the highest salinities of approximately 34.9 psu, are found in winter in the northern Barents and Kara seas, the western Laptev Sea, and adjoining areas of the Nansen basin (see Table 7.2).

The Environmental Working Group (EWG) data set has enabled the study of long-term variability of the Arctic Ocean (Polyakov et al. 2004). Swift et al. (2005) used a special version of the EWG data, which allowed them to examine the variability in the upper layer, the Atlantic layer, and the Pacific waters (Rudels 2015). Similar strong and sudden changes as in the 1990s could be detected in the 1960s, which showed a strong increase in Atlantic water temperatures, and in the 1970s, which featured a salinity increase in the upper layer. The weakening and disappearance of the Pacific water in the central Arctic Ocean was found to occur in the mid-1980s, before the inflow of warm Atlantic water and the redistribution of the low-salinity shelf water into the Canadian Basin arose in the 1990s.

A cross-section of potential temperature and salinity in the Arctic Ocean is shown in Figure 7.7; the figure also highlights the bathymetric features. Potential temperature (θ) is used in oceanography to compare the temperature of water parcels at different pressure levels in the ocean where compressibility of the water plays a small role – namely, compression (expansion) causes a rise (fall) of temperature. Potential temperature is defined as the temperature of a water parcel that is moved adiabatically to a different pressure level. The adiabatic lapse rate in the ocean is approximately 0.1–0.2 °C km^{-1} (in contrast to the 9.8 °C km^{-1} rate for unsaturated air). In the ocean, θ is defined in reference to the sea surface, so it is always slightly lower than the actual temperature. It is a conservative property. Figure 7.7 illustrates the three main layers in the Arctic: (1) Polar Surface Water down to about 200 m, (2) intermediate water from about 200 to 800 m, and (3) deep and bottom waters below 800 m.

The surface circulation in the Arctic Ocean is driven mainly by the clockwise gyre that is a response to the winter ridge of high pressure over the western Arctic and the Beaufort Sea high in spring (Serreze and Barrett 2011). Ice circulates slowly around the gyre. On the Eurasian side, the Transpolar Drift Stream takes

Table 7.2 Mean temperature (T) and salinity (S, psu) data for (a) the Canadian, (b) Amundsen, and (c) Nansen basins (after Tanis and Timokhov 1997, 1998). Canadian Basin: Western data; Amundsen and Nansen basins: Russian data

Depth (m)	(a) T	S	(b) T	S	(c) T	S
5	−1.5	30.2	−1.6	30.3	−1.85	33.8
100	−1.4	32.4	−1.6	33.6	−1.1	34.35
300	0.0	34.6	0.8	34.8	1.7	34.9
1,000	0.4	34.9	−0.3	34.9	−0.2	34.9
2,000	−0.4	34.9	−0.7	34.9	−0.8	34.9
4,000			−0.7	34.95	−0.7	34.96

water and sea ice from the Eurasian coast to Fram Strait, where it forms the East Greenland Current. The Atlantic water that enters the Arctic via the Norwegian and Barents seas sinks beneath the surface polar water and flows counterclockwise around the basin below 250–300 m depth toward the Canadian Arctic archipelago. It has much higher temperature (0–0.5 °C) and salinity (34.9 psu) than the overlying Arctic Water (see Figure 4.10).

The Arctic Ocean has low surface salinity, which is approximately 27–30 psu over the Eurasian continental shelf and in the Beaufort Sea, for example. This is primarily caused by the immense quantity of river discharge in Eurasia (the Ob, Lena, and Yenisei rivers) and western North America (the Mackenzie River). The runoff during the October–September water year amounts to 404 $km^3 yr^{-1}$ for the Ob, 603 for the Yenisei, 525 for the Lena, and 333 for the Mackenzie (Serreze et al. 2003). Salinity values for water coming off the Eurasian river mouths drop to 20 psu or less over the coastal shelves. This decrease is important because in polar oceans density is primarily determined by salinity rather than by temperature. The low salinity of the Arctic is also partly a result of inflow of relatively freshwater from the North Pacific via Bering Strait, with this flow being equivalent to almost half of the river runoff. Halocline stratification is much stronger in the Amerasian Basin than in the Eurasian Basin, owing to additional inputs of low-salinity water from the Pacific via Bering Strait, and surface flow convergence under the atmospheric Beaufort High (Carmack et al. 2016). Pacific water enters the Arctic in the depth range 60–220 m.

A major feature of the Arctic Ocean since the 1990s has been a general warming trend (Seidov et al. 2015). Based on the World Ocean Database (WOD) that has been assembled by the National Oceanic and Atmospheric Administration (NOAA), differences of temperature between 2005–2010 and the coldest decade of 1975–1984 show that upper-ocean warming has occurred

Figure 7.7

Cross-section of (a) potential temperature and (b) salinity in the Arctic Ocean.

Source: Courtesy of Professor Lynne Talley, San Diego State University, and Elsevier. Talley et al. 2011, 418, figure 12.11a, b.

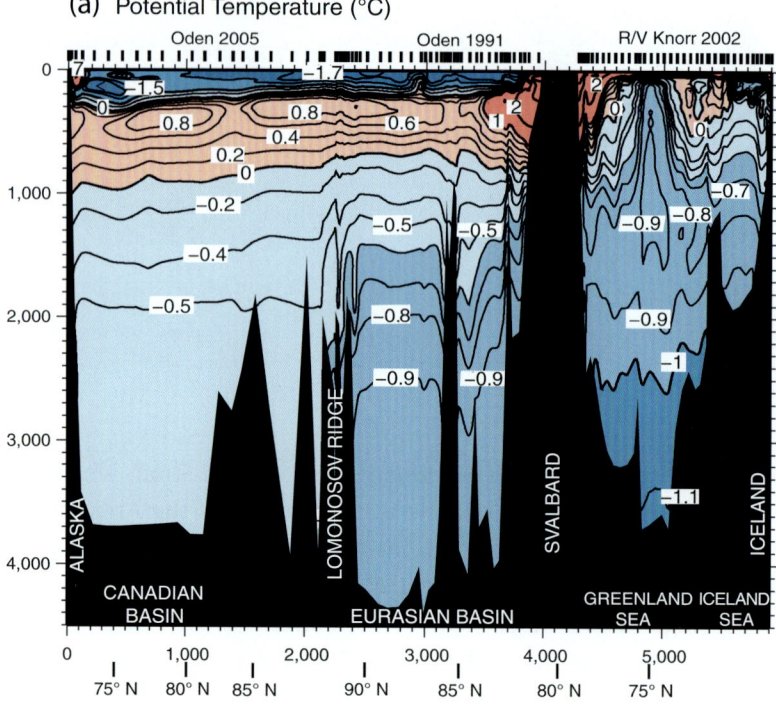

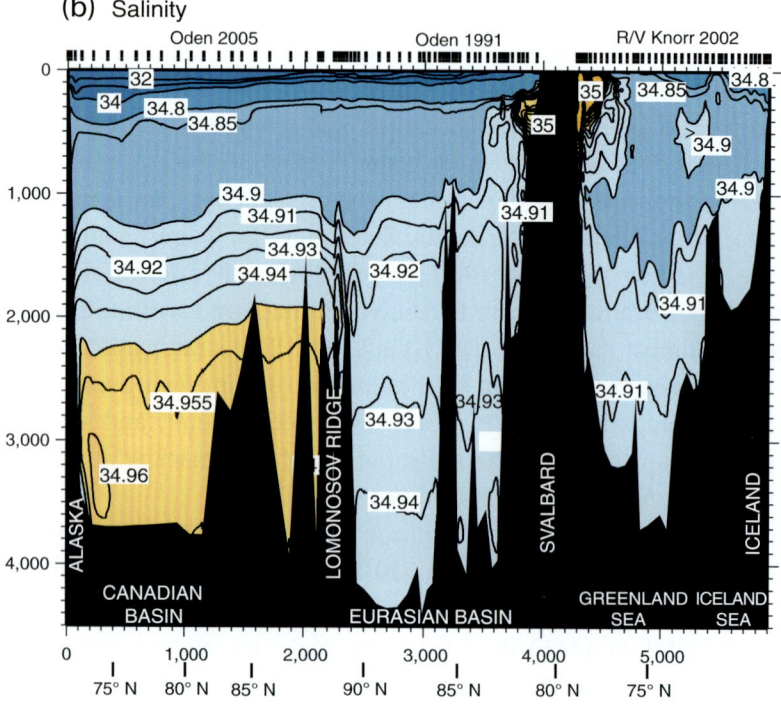

nearly everywhere, with the amplitudes being consistently high in the Greenland–Iceland–Norwegian Sea (GINS), Barents Sea, Eurasian Arctic, and western Arctic.

Levitus et al. (2000) found that heat content of the surface 0–3,000 m layer in the world ocean increased by approximately 2×10^{23} J between the mid-1950s and mid-1990s, which corresponds to a warming of 0.06 °C. Seidov et al. (2015) have shown that the heat content of the layer between 50 and 300 m depth in the GINS and the Arctic Ocean turned sharply positive, relative to the base climate of 1955–2006, after 2000. The temperature of the top 100–150 m layer in the Barents Sea has increased by approximately 4 °C since the late 1970s.

Rudels (2015) reports that north of the Laptev Sea, the temperature of the Atlantic water observed on the Nansen and Amundsen Basin Observational System (NABOS) cruise in 2002 was reduced considerably from the high values observed by Polarstern in the 1990s. However, the temperature measured at the NABOS moorings on the Laptev Sea slope began to rise in 2004, when a sudden increase in both the temperature and the thickness of the Atlantic layer was observed. The temperature has remained high ever since (Dmitrenko et al. 2008; Polyakov et al. 2005). Wells et al. (2013) reported, based on Argos data, that the heat storage in the northern North Atlantic increased during 1999–2010, especially between 60° and 70° N. Cold anomalies were observed during the same interval between 20° and 50° N.

7.5 Arctic Seas

The distribution of the Arctic seas around the basin is shown in Figure 7.5. These seas are discussed in turn in this section.

7.5.1 Barents Sea

The Barents Sea between Svalbard and Novaya Zemlya has an area of 1,420,000 km² and is rather shallow, with an average depth of about 225 m. It is linked to both the Norwegian Sea/North Atlantic and the Arctic Ocean, and extends to about 80° N. The northeastward Atlantic Current divides into three branches: the West Spitsbergen Current in eastern Fram Strait, flow into the Barents Sea, and the Norwegian Coastal Current (Pfirman et al. 1994). The warm, salty waters of the North Atlantic Current form a Polar Front, along with cold water from the north, that extends generally southeastward from Svalbard toward Novaya Zemlya (Loeng 1991). The southern part of the sea remains ice-free year-round to about

latitude 75° N. The loss of heat to the atmosphere, while Atlantic water traverses the Barents Sea, greatly modifies this water mass prior to its entry into
the Arctic Ocean proper. When passing through the Barents Sea, the Norwegian
Atlantic Current is strongly modified by cooling, mixing, and freezing during
winter, and all the Atlantic Water (AW) entering in the west is modified and
leaves the shelf toward the Arctic Ocean mostly with temperatures below 0 °C
(Ingvaldsen et al. 2004).

Loeng et al. (1997) and Ingvaldsen et al. (2004) have determined the transports
of water through the Barents Sea. There is an average inflow and outflow
transport of approximately 4.6 Sv in winter and 3.1 Sv in summer, of which the
through-flow of Atlantic water contributes 1.7 Sv in winter and 1.3 Sv in summer
across a section from 71.5° to 73.5° N (Ingvaldsen et al. 2004). A spring minimum, which is sometimes an outflow, is associated with northerly winds. Based
on data from moorings during October 1991 to September 1992, at 77.3° N, 62.9° E
to 78.8° N, 58.6° E, Schauer et al. (2002) have shown that the majority of the
Atlantic inflow entering from the Norwegian Sea leaves the northeastern Barents
Sea as cold, dense bottom water as a result of cooling and freezing. The flow
toward the Kara Sea was between 0.6 and 2.6 Sv during the period covered by
their analysis.

There is an outflow of cold Arctic Water just south of Bear Island – the Bear Island
Current – but its magnitude is unknown. The Atlantic domain shifts northward in
winter. This phenomenon is connected to coastal downwelling off Norway, which
moves the AW away from the coast. Arctic Water, northeast of the polar front,
appears to originate from convection due to sea ice formation, but some may be
advected from the northern Kara Sea and the Arctic Ocean (Pfirman et al. 1994).
It comprises three main components: the westward-flowing Persey Current near
78° N, the East Spitsbergen Current flowing southwestward out of the Arctic Ocean
between 45° and 30° E, and the strong Hopen–Bjørnaya Current flowing southward
along the eastern flank of the Spitsbergen Banks to about 74.5° N.

From autumn to early summer, the region north of the Polar Front is ice
covered. The ice melts as summer progresses and the margin shifts northward.
In mid- to late summer, this melt forms a 20-m-thick surface layer of relatively
warm (due to radiative heating) freshwater above the cold AW. Here, warm,
saline Atlantic-derived water is found at depths less than 75 m.

7.5.2 Kara Sea

The Kara Sea is located north of Russia between Novaya Zemlya and Franz Josef
Land in the west and Severnaya Zemlya in the east. It has an approximate area of

880,000 km^2 and an average depth of only 110 m. The central regions have a large number of island groups. About 82 percent of the Kara Sea occupies part of the Siberian shelf; accordingly, about 40 percent of it is less than 50 m deep. The water is very cold (-1.4 °C) and typically remains frozen for nine months of the year.

The Kara Sea receives large amounts of discharge from the Ob (approximately 400 km^3) and Yenisei (630 km^3) rivers, so the salinity in summer off the mouths of the Ob and Yenisei is only 10–12 psu. Over the rest of the sea, the salinity is 25 psu in winter and 22 psu in summer. The northern Kara Sea is influenced by deep Atlantic water, which penetrates the Kara Sea from the Arctic Basin via the deep St. Anna and Voronin troughs. According to Schauer et al. (2002), in 1996, warm Atlantic water, with temperatures up to 3 °C, was located in the western St. Anna Trough at 200 m depth.

7.5.3 Laptev Sea

The Laptev Sea off northern Siberia is bounded in the west by the Taimyr Peninsula and Severnaya Zemlya and in the east by the New Siberian Islands and Kotelny Island. The Siberian mainland is dissected by several large gulfs and bays. The sea has an area of 660,000 km^2 and a mean depth of about 50 m. The southern and southeastern areas, comprising 45 percent of the total area, have water depths ranging from 10 to 50 m. A dramatic incline, beginning at 100 m depth and ending at the 3,000 m level, divides the sea into the northern and southern parts along the parallel of the Vil'kitsky Strait. Assuming the 200-m isobath to be the shelf boundary, the shelf zones make up 72 percent of the area of the Laptev Sea (Timokhov 1994).

The runoff to the Laptev Sea from five major rivers – Khatanga, Anabar, Lena, Olenek, and Yana – is about 767 km^3. The summer salinity off the mouth of the Lena River is only 5–10 psu, but in the northern parts it rises to 28 psu. Except in August–September, this sea remains largely ice covered. In winter, water temperature varies from -1.4 °C in the eastern sea, up to -0.8 °C in the northwestern sector. In summer, the southwestern upper 15-m layer is warmed to a temperature of 5–7 °C and river runoff near the coast results in surface temperatures of 8–10 °C (Timokhov 1994). Temperatures increase to 1° in the southeastern part, but remain about -1° in the northern areas. Surface water motion is generally cyclonic, with eastward flow along the coasts.

The Laptev Sea is one of the core areas for ice production in the Arctic Ocean, with a distinct connection to Transpolar Drift characteristics. It transports large quantities of sediment that are incorporated into the ice due to the processes that form frazil ice (a suspension of ice crystals). Ice production in the Laptev Sea was mapped with the Moderate-Resolution Imaging Spectroradiometer (MODIS) by Preusser et al. (2016), which is an especially valuable source of data in this region, given the narrow and elongated flow leads close to the fast-ice edge. These researchers' results showed that polynyas in the Laptev Sea contribute at least 7 percent to the total potential sea-ice production in Arctic polynyas.

The ice drift and export in the Laptev Sea has been studied by Alexandrov et al. (2000). According to these researchers, during an "average year," sea ice is exported from the Laptev Sea through its northern and eastern boundaries, with maximum and minimum export occurring in February and August, respectively. The winter ice outflow from the Laptev Sea has been shown to vary between 251,000 km^2 (1984–1985) and 732,000 km^2 (1988–1989), with the average being 483,000 km^2. Sea ice is exported into the East Siberian Sea mostly in summer, the volume discharged having a mean value of 69,000 km^2. Out of the seventeen summers investigated by Alexandrov et al., twelve of them were characterized by sea ice import from the Arctic Ocean into the Laptev Sea through its northern boundary.

7.5.4 East Siberian Sea

The East Siberian Sea lies between the New Siberian Islands in the west and Wrangel Island in the east. Three straits connect it to the Laptev Sea. The East Siberian Sea has an area of 910,000 km^2 and a mean depth of about 50 m. It is only 10–20 m deep in the west and central parts and 30–40 m deep in the east. Assuming the 200-m isobath to be the shelf boundary, the shelf zones make up 96 percent of the area of the Siberian Sea (Timokhov 1994). The 50-m isobath passes almost parallel to the mainland shore at an average distance of 550–650 km from it.

Near the western boundary, there is the large New Siberian Island archipelago, which is divided into three main groups: the Lyakhovsky islands, the Anjou islands, and the De Long islands. For most of the year, the sea is ice covered. The water temperature at the surface in the winter decreases from the southwest to northeast and is near freezing. The salinity in winter has a tendency to increase from the southwest, rising from 17–18 psu in the western part to 32–33 psu in the northeastern part. In summer, the water temperature at the surface decreases northward, from 5 to 7° C in the coastal zone to −1 to −1.5 °C in the northern part.

Based on cruise data from September 2000, Semiletov et al. (2005) identified two distinct water masses in the southern part of the sea. West of about 160° E, there is freshwater flux from the Lena River; temperatures average 2.6 °C and salinity 22 psu. In the eastern part, there is Pacific inflow, with average temperatures of 0.6 °C and 30 psu salinity.

7.5.5 Chukchi Sea

The Chukchi Sea lies between Wrangel Island and Point Barrow, Alaska. Its southern boundary is the Arctic Circle. It has an area of close to 600,000 km^2 and a mean depth of 70 m. Slightly more than 55 percent of this sea is less than 50 m deep. Pacific water flows into the Chukchi via Bering Strait. Northward transport through Bering Strait is seasonal, averaging 1.0 Sv during April–September and 0.6 Sv during October–March (Woodgate et al. 2005a, 2005b). The mean inflow is opposed to the prevailing winds and is forced primarily by a slope in sea level from the North Pacific to the Arctic Ocean. The inflow branches into pathways following Herald Canyon, Central Channel (71° N, 175° E, between Herald and Hannah shoals), and the Alaskan coast (Spall 2007). The northward flow is at its maximum in summer. More than half of the transport exits the Chukchi Sea via Barrow Canyon. The seasonal cycle of salinity in the southern and central Chukchi Sea is dominated by advection through Bering Strait, while local atmospheric forcing and brine rejection are more important north of Herald and Hanna Shoals and in Barrow Canyon (off Point Barrow).

In late summer, the Chukchi Sea exhibits a two-layer water column with well-mixed surface and bottom layers separated by a strong pycnocline (Ladd et al. 2016). During winter, a combination of surface cooling, wind mixing from storms, and brine rejection from ice formation vertically mix the water column. Atlantic Water upwelling events, attributed to easterly winds, occur frequently in Barrow Canyon. During some upwelling events, AW with potential temperature greater than -1 °C and salinity greater than 33.6 psu has been observed to upwell from deeper than 200 m in the Arctic Basin onto the Chukchi Shelf via Barrow Canyon.

A winter polynya that occurs along the Alaskan coast between Cape Lisburne and Point Barrow is the largest in the western Arctic (see Section 7.9). Observations from 1990–1991 and 1991–1992 show the temperature in the central Chukchi Sea is at the freezing mark from late in the year through early summer. Winter water in the Chukchi has a salinity of about 32.5 psu, and the salinity of summer water is 34 psu or greater (Spall 2007). Advection through Bering Strait is important for the large-scale timing of ice melt in the Chukchi Sea.

7.5.6 Beaufort Sea

The Beaufort Sea is located between a line north of Point Barrow in the west and Banks Island and the southwestern edge of Prince Patrick Island in the east. It has an area of about 475,000 km^2 and a mean depth of around 1,000 m. The continental shelf is narrow near Point Barrow but is more than 140 km wide off the mouth of the Mackenzie River. The surface circulation over the deeper portions of the Beaufort Sea is dominated by the Beaufort gyre, which moves clockwise over the Canada Basin. The pack ice drifts westward in the Beaufort gyre but the subsurface flow along the shelf slope is eastward. Pickart (2004) has shown that the eastward current is concentrated in an approximately 20-km-wide jet along the shelf break. This jet has three distinct seasonal configurations: (1) in late spring to late summer, cold, winter-transformed Bering Sea water is advected eastward; (2) from about mid-summer to early autumn, a surface-intensified current advects predominantly Bering summer water (for brief periods in August–September, the local winds come from the west); and (3) from mid-autumn to mid-spring, under easterly winds, the jet transports upwelled Atlantic water. The upper 100 m of this jet has a temperature of about $-1.4\,°C$ in summer and $-1.8\,°C$ in winter. It overlies Pacific water that enters via Bering Strait. Below is a layer of warm Atlantic water at $0–1\,°C$ that has circulated around the Arctic Ocean. The bottom water has a temperature of -0.4 to $-0.8\,°C$.

The inflow into the Beaufort Sea is about $420\ km^3$. Data on the Mackenzie River inflow are provided by Macdonald et al. (1989). About 70 percent (by volume) of the shelf water in autumn 1986 was due to river runoff and 30 percent was due to ice melt. Hence the shelf here is largely estuarine in character. Freshwater is removed by winter ice growth early after freeze-up. Then, as the ice thickens, removal of water from the shelf by flushing takes over and lasts approximately 150 days.

7.5.7 Lincoln Sea

The Lincoln Sea stretches from northern Ellesmere Island in the west to Cape Morris Jessup, northern Greenland, in the east. The northern limit is defined as the great circle between those two headlands. It has an area of $64,000\ km^2$ and a mean depth of approximately 250 m. The sea is ice covered throughout the year, with the thickest sea ice found in the Arctic Ocean, up to 15 m thick in places. Haas et al. (2006) measured multiyear ice thicknesses of 3.9–4.2 m using airborne electromagnetic induction (EMI) measurements. Although some of the ice is exported via Nares Strait, the majority is transported east of Greenland via Fram Strait.

Water depths in the Lincoln Sea range from 100 to 300 m. The water has three distinct properties. First, in the inner part of the Lincoln Sea shelf, the temperature and salinity increase from the surface, where they are about $-1.5\,°C$ and 32.4 psu, to about 400 m, and then remain constant to the seafloor. Second, the water over the outer part of the shelf, including the slope, has attributes similar to those in the Canadian Basin, which are not unlike those of waters from the Pacific. Third, the waters north of the slope have characteristics matching those of the Eurasian Basin.

Based on year-round current measurements between 1989 and 1994, Newton and Sotirin (1997) showed that there is an eastward undercurrent, confined to the continental slope, with a width of about 50 km and speeds of 5–9 cm s^{-1}. Temperature and salinity characteristics of this undercurrent are similar to those of Canadian Basin waters, suggesting the existence of a boundary current system that is continuous along the continental slope north of Alaska and the Canadian Arctic archipelago.

7.5.8 Greenland Sea

The Greenland Sea extends east of Greenland to the Svalbard archipelago, north to Fram Strait and the Arctic Ocean, and southeast and south to the Norwegian Sea, Iceland, and Denmark Strait. It has an area of 1.2 million km^2. The average depth is 1,450 m, with the maximum depth being 4,850 m.

The average surface water temperature for the Greenland Sea in winter is -1 °C or less in the north and 1–2 °C in the south; the corresponding summer temperatures are about 0 and 6 °C, respectively. The surface water salinity is 33.0–34.0 psu in the east and less than 32.0 psu in the western parts. Waters of the north-flowing North Atlantic Current (NAC) sink in the Arctic Ocean. Some return south in the cold East Greenland Current (EGC), an important part of the Atlantic conveyor belt that flows along the east coast of Greenland and transports Arctic pack ice southward from October to August. In the eastern part of the Greenland Sea is the warm Spitsbergen Current, an offshoot of the NAC. These two currents form a counter-clockwise gyre. The transport of Atlantic water, which has a temperature warmer than 1 °C, accounts for about half of the northward flow (Rudels 2015; Schauer et al. 2008). The northward flow in the two West Spitsbergen Current branches is estimated to be 1.8 Sv, with 1.3 Sv of Atlantic water warmer than 2 °C in the eastern branch, and 4.9 Sv, with 1.7 Sv of Atlantic water in the off-shore branch, giving a total northward flow of 6.6 Sv (Beszczynska-Möller et al. 2011). The southward flow in the East Greenland Current is larger, 8.6 Sv, giving a net outflow of 2.0 Sv. The EGC flows from Fram Strait (79° N) to Cape Farewell, transporting large amounts of Arctic ice (see Section 7.3). From a mooring at 75° N in 1994–1995, Woodgate et al. (1999) found an annual mean transport of 21 Sv (across 14–9° W), with the transport varying from 11 Sv in summer to 37 Sv in winter. The flow comprises a seasonally varying wind-driven component (two thirds) and a more consistent thermohaline component (one third).

Offshore from the East Greenland Current, south of 64° N, is relatively saline (34.9–35.0 psu) and warm (4–6 °C) water in the Irminger Current. This branches from the North Atlantic Current at about 26° W and is located above the western slope of the oceanic Reykjanes Ridge (63.5° N, 23° W), southwest of Iceland (Gyory et al. n.d.). Transport estimates for the Irminger Current are in the range of 8–11 Sv (Tomczak and Godfrey 2003, 244).

The Greenland Sea, as a result of strong surface cooling and wind stress, undergoes a large amount of mixing. The surface mixed layer has a depth of 40–50 m (Blindheim and Østerhus 2005). Convective events in the central Greenland Sea in the 1990s extended down to between 1,200 and 2,000 m.

In winter, a large area north of Iceland, between Greenland and Jan Mayen, known as the West Ice, is covered by continuous ice. An ice area, the Odden, covering as much as 330,000 km² in most years, extended eastward from the main East Greenland ice edge in the vicinity of 72–74° N during the winters of 1966–1972 (the Great Salinity Anomaly), and in 1982, 1986, 1989, 1997, and 1998. However, this feature has rarely developed since 2000 (Rogers and Hung 2008). Its formation is associated with the presence of very cold polar surface water in the Jan Mayen Current, high pressure in high latitudes of the North Atlantic, a negative NAO, and anomalous westerly winds. Air temperature and downward longwave flux anomalies in the preceding autumn are unusually low in advance of a winter Odden ice cover, while heat fluxes are weakly positive.

7.5.9 Canadian Arctic Archipelago

The Canadian Arctic archipelago (CAA) consists of three major straits running from east to west – Nares Strait, Jones Sound, and Lancaster Sound – which all open to Baffin Bay. These straits provide a small net outflow from the Arctic Ocean. The shallowest sill is 125 m in Barrow Strait between Cornwallis Island and Somerset Island.

Steele et al. (1996) computed the geostrophic transports between different parts of the Arctic Ocean, and estimated that the outflow through the CAA was 0.56 Sv. For Nares Strait, a mooring array indicated a southward flow of 0.9 Sv (Münchow et al. 2006), but this value has subsequently been lowered. Rudels (2015) suggests that the total transport southward through the passages of the CAA is between 1.4 and 1.8 Sv.

Pack ice is prevalent for much of the year in the CAA, especially in the north and west. The channels in the southern and eastern parts usually clear by late summer. In 2016, the Northwest Passage became totally ice free during that season.

7.5.10 Adjacent Seas of the North Atlantic

Baffin Bay and Davis Strait

The passages connecting the Arctic Ocean to the North Atlantic exit into Baffin Bay, Davis Strait, and the Labrador Sea. Baffin Bay lies between Greenland to the east, Baffin Island to the west, and Devon and Ellesmere islands to the northwest. Smith Sound in the northern part is occupied by the North Water polynya (see Section 7.9). Baffin Bay has an area of almost 890,000 km^2, an average depth of 760 m, and a maximum depth of 2,136 m. The West Greenland Current flows northward from Cape Farewell and recirculates to the west off northwest Greenland. This flow, together with Arctic water flowing south through Nares Strait and Smith Sound, joined by flow out of Lancaster Sound, drives the southward-flowing Baffin Current. The Baffin Current carries sea ice and Greenland icebergs southward into the Labrador Current and to the Grand Banks off Newfoundland. Davis Strait has a sill depth of only 350–550 m.

Labrador Sea

The Labrador Sea is located east of the coast of Labrador, south of 60° N to a line between Newfoundland and Cape Farewell. It has an area of 840,000 km^2, an average depth of approximately 1,900 m, and a maximum depth of 4,300 m. The surface water temperature varies between -1 °C in winter, when two thirds of the sea is ice covered, and 5–6 °C in summer.

The water formed in the central Labrador Sea produces Labrador Sea Water, which is lighter than the North Atlantic Deep Water below it. The densest NADW is formed by mixing of water from the East Greenland Current in the Denmark Strait (between Iceland and Greenland). The Labrador Sea Water circulates around the subpolar gyre many times and mixes with other polar water masses on its journey. NADW flows at depth from the Labrador Sea to Antarctica (Tomczak and Godfrey 2003, 257). It forms by deep convection events in about six out of ten winters, giving rise to water with temperatures of 3.0–3.6 °C and salinities of 34.86–34.96 psu.

Hudson Bay

Hudson Bay is connected to the North Atlantic by Hudson Strait and to the Arctic Ocean by Foxe Basin. It has an area of 1.2 million km^2, an average depth of only 100 m, and a maximum depth of 270 m. It is ice covered from mid-December to early June, but surface water temperatures reach 8–9 °C in summer in the western part of the bay. Hudson Bay functions essentially as an estuary. There is a slow cyclonic gyre in the water. River runoff is large – approximately 700 km^3 – so the salinity is low; this raises the freezing point and facilitates ice growth.

7.5.11 Marginal Seas of the North Pacific

Bering Sea

The Bering Sea is bounded by Bering Strait on the north, Alaska on the east, the Russian Far East on the west, and the Alaska Peninsula and the Aleutian Islands to the south. It has an area of 2.3 million km^2. The sea is basically divided into a shallow (less than 200 m) Siberian–Alaskan shelf area in the east and north and a deep (3,500–3,800 m) basin in the south and west. There is a cyclonic gyre in the Bering Sea basin, with the south-flowing Kamchatka Current in the west and the north-flowing Bering Slope Current in the east (Stabeno et al. 1999). The Alaska Stream enters the Bering Sea from the North Pacific through gaps in the Aleutian Island chain. Flow over the east Bering Sea shelf is generally first to the northwest, and then northward toward Bering Strait.

Sea ice begins to form over the shelves and is advected southward from Bering Strait in November. Ice forms in polynyas (see Section 7.9) on the leeward side of islands and coasts. It reaches its maximum extent in mid-March, covering about one third to one half of the Bering Sea, and begins to retreat in April.

Sea of Okhotsk

The Sea of Okhotsk is a marginal sea of the northwest Pacific Ocean. It is located between the Kamchatka Peninsula on the east, the Kuril Islands on the southeast, Hokkaido to the south, the island of Sakhalin along the west, and the eastern Siberian coast along the west and north. It has an area of almost 1.6 million km^2, an average depth of 860 m, and a maximum depth of 3,370 m. There is a broad, shallow shelf (less than 200 m) in the north; the deep Kuril Basin is found in the south. A cyclonic gyre is present in the Sea of Okhotsk, with a boundary current flowing southward along the coast of Sakhalin Island.

The Sea of Okhotsk receives a large amount of runoff from the Amur River, so it has low surface salinity, which in turn facilitates freezing. Hence there is sea ice cover from October–November to June, or locally July. Coastal polynyas generate dense water on the shelves, which is transported southward in the Sakhalin Current into the Kuril Basin (Gladyshev et al. 2001). The sea surface temperature reaches 8–12 °C in summer, and the salinity drops to 32.5 psu during the same season. The southwestern part is warmed by waters from the Sea of Japan and the eastern part by Pacific Ocean inflow.

7.6 Global Sea Level

As discussed in Chapter 2, global sea level has varied by more than 120 m over the last million years. Ice losses from glaciers and ice sheets are making increasing contributions to global sea level rise. However, to put those contributions in context, we must briefly review the components that are involved.

The first is thermal expansion due to ocean warming. This warming involves a density decrease and, therefore, a volume increase (steric sea level rise). The majority of this warming and expansion currently takes place in the upper 700 m. This contribution has increased to 1.1 mm yr^{-1} for 1993–2010 (Church and Clark 2013). The large heat capacity of the ocean means that there is considerable delay before the full effects of surface warming are felt throughout the ocean depth. As a result, the ocean will not achieve equilibrium and global average sea level will continue to rise for centuries after atmospheric greenhouse gas (GHG) concentrations have stabilized.

A second major contribution to global sea level is the mass loss of land ice. There were increasing contributions for 1993–2010 from Greenland (0.33 mm yr^{-1}) and Antarctica (0.27 mm yr^{-1}). Glacier contributions, including those from independent ice bodies around Greenland, increased from 0.69 mm yr^{-1} for 1901–1990 to 0.86 mm yr^{-1} for 1993–2008. Terrestrial storage in reservoirs and dams was 0.38 mm yr^{-1} for 1993–2010. The total global average land ice storage of 2.8 mm yr^{-1} compares with an altimetric observation of 3.2 mm yr^{-1}.

It has proved difficult to determine sea level trends for the Arctic due to the paucity of tide gauges and the limited extent of open water available for satellite altimetry. Recent analysis of both data sources by Svendsen et al. (2016), however, has provided a reconstruction that shows that the Arctic mean sea level trend between 68° and 82° N was around 1.5 mm ± 0.3 mm yr^{-1} for the period 1950–2010, in good agreement with the global mean value of 1.8 ± 0.3 mm yr^{-1} over the same period (Church and White 2011).

7.7 Sea Ice

Sea ice cover is a key element of polar regions that plays a critical role in global and regional climate and oceanic processes. There are three major types of sea ice: first-year ice (FYI), multiyear ice (MYI) and (land)fast ice, each of which has different physical characteristics, as described in this section. A guide to sea ice information services around the world has recently been issues by the World Meteorological Organiztion (2017) that includes descriptions of ice types.

Ice begins to form in the ocean when the surface cools to about -1.8 °C for average ocean salinity (34.5 psu). Ice floats because it has a density of about 917 kg m^{-3} at 0 °C compared with a density of 1,000 kg m^{-3} for liquid water. Freshwater has a maximum density at 3.98 °C, but for every 5 psu increase in salinity, the freezing point decreases by 0.28 °C. The temperature of maximum density disappears when the salinity exceeds 24.7 psu. Cooling makes the surface water denser, setting up convection. However, the whole water column does not have to cool to freezing before ice can form, only the upper layer above the level of density maximum – the pycnocline. In the Arctic, the pycnocline is found at 50–150 m depth. Sea ice has two phases: salt-free ice and liquid brine (Ackley 1996). At a growing ice interface, most of the salts are rejected. Brine, gas, and solid salts are usually trapped at sub-grain boundaries within a lattice of essentially pure ice (Timco and Weeks 2010). First-year sea ice has a typical salinity in the range of 4–6 psu.

Initially, a suspension of tiny ice crystals, called frazil or grease ice, forms in the surface water. In calm conditions, the frazil crystals freeze together to form a continuous thin sheet of transparent ice, called nilas. Water molecules freeze onto the bottom of the ice in a process known as congelation, and the nilas thickens, turning first gray and then white. Congelation ice has a columnar structure. The ice thickens by the extension of ice platelets below; salt is rejected and descends in salty plumes. Brine inclusions are trapped between the platelets as they thicken.

Waves can maintain a suspension of frazil crystals and lead to the growth of small cakes of slush. These grow by accretion and eventually form pancake ice. At the sea ice margin, the pancakes are only a few centimeters in diameter, but in the interior of the pack ice they can be 3–5 m in diameter and 50–70 cm thick. These coalesce into floes and ultimately form a sheet, although rafting may occur locally, increasing the thickness by two to three times. Young ice is more than 30 cm thick, and first-year ice may reach 1.5–2 m in thickness.

Thickness estimates in the Arctic from the Cryosat 2 altimeter and for thin ice from the L-band (1.4 GHz) sensor on the Soil Moisture and Ocean Salinity (SMOS) satellite show mean thickness increases, respectively, from 1.46 and 0.45 m in November 2015 to 1.90 and 0.47 m in April 2016. SMOS thickness peaked at 0.58 m in December 2015 (Ricker et al. 2017).

7.7.1 Sea Ice Edge Location

The sea ice edge is commonly defined by the 15 percent ice concentration line. A transition zone 100–200 km wide – the marginal ice zone –lies between the ice edge and the boundary of ice having a concentration coverage of more than 80 percent. This zone is typically wider around Antarctica than in the Arctic.

Bitz et al. (2005) performed a modeling study of the mean position and seasonal range of the ice edges in the North Atlantic, North Pacific, and South Atlantic sector of the Southern Ocean. The departure of the wintertime ice edge (here, the limit of 50 percent concentration in ice coverage) from a symmetrical ring around either pole was shown to depend primarily on coastline shape, ice motion, and the melt rate at the ice–ocean interface. At any location, the principal drivers of the oceanic heat flux that melts sea ice are absorbed solar radiation and the convergence of heat transported by ocean currents. In regions where the ice edge extends relatively far equatorward, absorbed solar radiation is the largest component of the ocean energy budget, and the large seasonal range of insolation causes the ice edge to traverse a large distance. In contrast, at relatively high latitudes, the ocean heat flux convergence is the largest component. It has a small annual range, so the ice edge there traverses a much smaller distance.

Holland and Kimura (2016) analyzed the ice concentration budgets at both poles using AMRS-E brightness temperature data for 2003–2010. They derived a climatology of the ice concentration budget to enable observational decomposition of the seasonal dynamic and thermodynamic changes in ice cover. In both hemispheres, the spring ice loss was dominated by ice melting. In other seasons, ice divergence maintained freezing in the inner pack while advection caused melting at the ice edge, as ice was transported beyond the region where it could be sustained by thermodynamic processes. Mechanical redistribution by wind stress provided an important sink of ice concentration in the central Arctic and around the Antarctic coastline.

On synoptic time scales, ice edge retreat in the Barents Sea has been shown to respond to intense intrusions of moist air (maximum at 900 hPa) into the Arctic during autumn and winter through their impact on local temperature and sea ice concentration (Woods and Caballero 2016). Woods and Caballero found that the vertical structure of the warming associated with moist intrusions is amplified in the lower layers, corresponding to a transition of local conditions from a "cold clear" state with a strong inversion to a "warm opaque" state with a weaker inversion. Composite analyses of winter intrusions indicate surface downward longwave flux anomalies of up to 30 W m^{-2} and surface air temperature anomalies of up to 4.5 °C. In the marginal sea ice zone (75–80° N, 20–80° E), the passage of an intrusion also causes a reduction of 6 percent in sea ice concentration, which persists for many days after the intrusion has passed. There is a positive trend in the number of intrusion events crossing 70° N during December and January that can explain roughly 45 percent of the surface air temperature and 30 percent of the sea ice concentration trends observed in the Barents Sea during the past two decades.

Table 7.3 Relative contributions (percent) of ocean heat sources in selected MIZs in the CCSM2 (Bitz et al. 2005)

Region	Absorbed solar radiation	OHFC
Barents Sea	37	63
Greenland Sea	40	60
Bering Sea	64	36
Sea of Okhotsk	67	33
Labrador Sea	80	20
South Atlantic	83	17

Tsubouchi et al. (2012) estimated Arctic Ocean boundary fluxes through the four main gateways (Bering Strait, Davis Strait, Fram Strait, and Barents Sea) for 32 days in summer 2005. They found the following transport-weighted mean properties for water entering the Arctic: potential temperature of 4.49 °C and salinity of 34.50 psu. For water leaving the Arctic, including sea ice, the corresponding properties were 0.25 °C and 33.81 psu. Hence, the net effect was to freshen and cool the inflows by 0.69 psu in salinity and 4.23 °C, respectively. The net heat flux (including sea ice) was 189 ± 37 TW, representing a loss from the ocean to the atmosphere.

The main sources of ocean heating are solar absorption in summer and the ocean heat flux convergence (OFHC) year-round. The relative importance of these terms averaged over the marginal ice zones (MIZ) is shown in Table 7.3. The average OHFC near the ice edge is shown to be 65 W m^{-2} south of 62° N and 100 W m^{-2} north of 72° N. It is approximately 200 W m^{-2} in the ice-free Norwegian Sea as a result of the North Atlantic Current, compared to 40 W m^{-2} in the mostly ice-covered Bering Sea. Ice is maintained in the East Greenland Sea by the advection of ice and cold water in the East Greenland Current.

In the southern hemisphere in the model, autumn and winter growth rates are highest next to the continent in coastal polynya-like features. The ice edge advances rapidly in autumn, as the growth rates away from the coast are much higher then than in winter. In summer, there is no melt at the top of the ice due to snow cover and because melt ponds are virtually absent. Basal and lateral melt is important in that season.

7.7.2 Arctic Sea Ice

Arctic sea ice is currently made up of about 80 percent FYI and 20 percent MYI, a reversal from the proportions that prevailed up until the early 1990s (Arctic

Climatology Project 2000, Sea Ice Atlas). FYI has a higher fraction of level ice, whereas MYI has more ridges and hummocks. In the Arctic in winter, ice divergence accounts for half as much ice volume change as ice growth. In summer, basal plus lateral melt exceeds melt at the top surface in the Arctic Basin, except where continental climates have a strong influence on the sea-ice cover, as in the Canadian archipelago. At SHEBA in 1997–1998, Perovich et al. (1999) reported approximately 100 cm of basal melt and only 30 cm of top melt.

The ice edge pattern in the northern hemisphere in summer is relatively zonal, while in winter it varies greatly with longitude. The location of the wintertime ice edge in the northern hemisphere depends on the ice dynamics, which advect ice toward the periphery of the ice pack, and basal and lateral melt, which is confined to a narrow band near the ice edge.

Arctic sea ice circulates slowly clockwise around the Beaufort gyre and moves from the Asiatic coast across the pole in the Transpolar Drift Stream (TPDS) to exit the Arctic via the Fram Strait in the East Greenland Current. Kwok et al. (2004) summarized ice export estimates for 1978–2002 and examined, over a nine-year record, the associated variability in the time-varying upward-looking sonar (ULS) thickness distributions in Fram Strait. The average annual ice area flux over the period was 866,000 km^2 yr^{-1}. Between the 1980s and 1990s, the decadal difference in the net exported ice area was approximately 400,000 km^2, or roughly half the annual average.

Using thickness estimates from ULS moorings, Kwok et al. (2004) estimated the average annual ice volume flux (1991–1999) to be 2,218 km^3 yr^{-1} (0.07 Sv). Over the ULS ice thickness data set, there was an overall decrease of 0.45 m in the mean ice thickness and a decrease of 0.23 m over the winter months (December through March). Correspondingly, the mode of the MY ice thickness exhibited an overall decrease of 0.55 m and a winter decrease of 0.42 m.

Krumpen et al. (2016) analyzed a data set of ground-based and airborne electromagnetic ice thickness measurements collected during summers between 2001 and 2012. The primary source of the surveyed sea ice leaving Fram Strait is the Laptev Sea; the age of this ice decreased from three to two years between 1990 and 2012. The thickness data consistently show a general thinning of sea ice over the last decade, with a decrease in modal thickness of second-year and multiyear ice. The mean thickness decreased from 2.58 m in 2001 to 2.17 m in 2012. Overall, Krumpen et al. found a positive trend in the monthly Fram Strait area flux, noting that it has increased by 25 percent since the 1960s.

The most prominent atmospheric driver of anomalous sea-ice motion across Fram Strait is an east–west dipole pattern of sea level pressure (SLP) anomalies with centers of action located over the Barents Sea and Greenland (Tsukernik et al. 2009) The association between the SLP dipole pattern and Fram Strait ice

Table 7.4 Surface feature height statistics for (a) 20 cm and (b) 80 cm thresholds in the central Arctic and Beaufort–Chukchi Sea (Petty et al. 2016)

	Central Arctic (a)	Beaufort–Chukchi	Central Arctic (b)	Beaufort–Chukchi
	Mean (m)	Mode (m)	Mean (m)	Mode (m)
FYI	1.03	(a) 0.45	0.97	0.45
MYI	1.35	0.45	1.10	0.45
All	2.09	1.65	1.96	1.45

motion is maximized at 0-lag, persists year-round, and is strongest on time scales of 10–60 days.

Gudkovich (1961) showed that there are two different oceanic circulation patterns in the Arctic Ocean. First, an anticyclonic regime exists where the area of the Beaufort gyre increases and the area of the cyclonic Laptev gyre shrinks; the TPDS originates from the Laptev, East Siberian, and Chukchi seas and transports ice toward the Greenland Sea. Second, a cyclonic regime is observed, with a contraction of the Beaufort gyre occurring simultaneously with an expansion of the Laptev gyre; the TPDS slows down and its entrance shifts toward the Beaufort Sea. Proshutinsky and Johnson (1997) found that these regimes alternate every five to seven years.

A surface feature characterization of ice in the central Arctic and Beaufort–Chukchi Sea has recently been performed using airborne IceBridge data for 2009–2014 (Petty et al. 2016). Elevation threshold values of 20 and 80 cm above a derived level ice surface were used, which means that the former included snow sastrugi features as well as ice deformation features. The results demonstrated predominantly higher surface features (1 m or greater) in the central Arctic region, mainly north of Greenland and the Canadian archipelago, and predominantly lower features (1 m or less) in the Beaufort–Chukchi Sea region. Feature heights were markedly higher (1.5–1.7 m or greater) along the coast of Greenland. Table 7.4 shows the overall results for 2009–2014. Regression of surface feature height with total ice thickness showed a mean correlation of 0.72, based on a square-root relationship.

Kapsch et al. (2016) have analyzed the role of downwelling longwave (LWD) and shortwave (SWD) radiation on summer sea ice in the Arctic. They found that positive LWD anomalies in spring and early summer have significant impact on the September ice extent, whereas winter anomalies produce only a small effect. Positive anomalies in spring and early summer initiate an earlier melt onset, thereby triggering several feedback mechanisms that amplify melt during the succeeding months. Realistic positive SWD anomalies appear to be important only if they occur after the melt has started and the albedo is significantly reduced. Simultaneous positive LWD and negative SWD anomalies during cloudy conditions during spring have a significant impact on summer sea ice, while summer clouds have almost no effect.

The relation between Arctic sea ice and clouds is seasonally dependent. It was examined during 2006–2008 by Kay and Gettelman (2009). No cloud response to sea-ice loss was found in summer, but low clouds did form over newly open water during early autumn. This seasonal variation in the cloud response to sea-ice loss can be explained by near-surface static stability and air–sea temperature gradients. During summer, temperature inversions and weak air–sea temperature gradients limit atmosphere–ocean coupling. In contrast, relatively low static stability and strong air–sea gradients during early autumn permit upward turbulent fluxes of heat and increase low cloud formation over newly open water.

This work was extended by Taylor et al. (2013) using Cloud-Aerosol Lidar with Orthogonal Polarization (CALIOP) data, CALIPSO CloudSat Cloud Profiling Radar (CPR), Clouds and Earth's Radiant Energy System (CERES), and AQUA MODIS in a cloud property vertical profile merging process for July 2006 through June 2010. The covariances between Arctic low cloud properties and sea ice concentration were quantified. Smaller average cloud fraction and liquid water were found where there was more sea ice. The largest-magnitude cloud–sea ice covariance occurred between 500 and 1,200 m when the lower tropospheric stability ($\theta_{850\ hPa} - \theta_{SFC}$) was between 16 and 24 K. Increased lower tropospheric stability was associated with decreases in low cloud fraction, cloud liquid water, cloud ice water, and cloud total water. The covariance between low cloud properties and sea ice was found to be largest in autumn. Regionally, the Laptev, Chukchi, and Beaufort seas exhibited the largest regional covariance between cloud properties and sea ice in summer and autumn; the Barents and Kara Sea regions exhibited the largest covariance in winter. Cloud properties were found overall to vary more between two atmospheric regimes than with sea ice concentration. However, the covariance between the liquid water path (LWP) and the sea ice concentration in autumn was of similar magnitude to the average LWP differences between the stable and highly stable regimes.

Arctic sea-ice predictability has been shown to involve the effects of both persistence and re-emergence of anomalies. A review of predictability was presented by Guemas et al. (2016). Persistence of sea ice area (SIA) was shown to have a characteristic e-folding time scale that varies seasonally from two to five months, based on data for 1978–2008 (Blanchard-Wrigglesworth et al. 2011). July–August SIA was significantly correlated with that in September.

Guemas et al. also estimated the persistence time scale of sea ice thickness (SIT) as approximately one year in the central Arctic and a few months in the seasonal ice zone. The ocean provides important predictability. Based on observations for 1864–1998, Vinje (2001) identified a strong link between the Atlantic water temperatures in the southern Norwegian Sea and the sea ice extent (SIE) two to three years later in the Barents and Kara seas, via the warm advection by the North Atlantic Current.

Re-emergence of SIA anomalies has been shown to be a further factor in persistence. Blanchard-Wrigglesworth et al. (2011) highlighted two different mechanisms for the re-emergence of SIA anomalies on time scales from a few months up to one year. One mechanism explains the re-emergence from the melt season to the growth season due to the persistence of SST anomalies. A negative (positive) SIA anomaly in spring is associated with a positive (negative) SST anomaly along the sea ice edge, which favors a negative (positive) SIA anomaly when the sea ice cover returns during the next autumn. A different mechanism explains the re-emergence from the growth season to the melt season as a result of the persistence of SIT anomalies. A positive (negative) SIA anomaly in the growth season is associated with an early (late) date of freeze-up, locally creating a positive (negative) SIT anomaly that slows down (accelerates) the sea-ice retreat during the next spring, and is therefore associated with a local positive (negative) SIA anomaly. Bushuk and Giannakis (2017) extended this work by showing the roles of SIT to sea ice concentration (SIC) in growth-to-melt season re-emergence and of SST and SLP to SIC in melt-to-growth season re-emergence.

The Sea Ice Outlook (SIO) has been prepared since 2008 by many research groups. Hamilton and Stroeve (2016) analyzed the performance of more than 400 predictions from SIO's first eight years, testing for differences in ensemble skill across years, months, and five types of methods: heuristic, statistical, mixed, and ice–ocean or ice–ocean–atmosphere models. Their results highlighted a pattern of easy and difficult years, corresponding roughly to the distinction between climate and weather. Difficult years, in which most predictions were far from the observed extent, tended to have large positive or negative excursions from the overall downward trends. In contrast to these large interannual effects, ensemble improvement from June to July and August was modest. Among method types, predictions based on statistics and ice–ocean–atmosphere modeling more closely matched the actual effects. Thinning ice that is sensitive to summer weather, complicating prediction, reflects the current transitional era between a past Arctic cool enough to retain much thick, resistant multiyear ice, and a warmed future Arctic where little ice remains at the end of summer.

7.7.3 Landfast Ice

Landfast ice comprises two main components – bottom fast ice where grounded pressure ridges anchor the ice in shallow water (less than approximately 18 m depth) and attached floating ice over deep water. Off northern Alaska, floating fast ice covers most of the shelf out to the 18-m isobath, but inward of the 2-m isobaths the ice is grounded on the seafloor. This boundary is marked by a grounded pressure ridge (stamukhi). Ridging is caused by wind and wave action in the early winter (Barry 1993; Barry et al. 1979; Mahoney et al. 2007). Druck-enmiller et al. (2012) have reported that landfast ice thickness in spring off Barrow during 2008–2012, as measured along trails constructed by Iñupiat hunters, had modal thickness values of 1.5–1.6 m. Mahoney et al. (2015) have reported on the landfast ice mass balance station operated off Barrow by the University of Alaska since 2000; this station has measured changes in the growth and melt of landfast ice. The authors conclude that the mean annual maximum ice thickness of 1.5 m from 2000 to 2015 was significantly thinner than the thicknesses of around 1.8 m commonly reported during the 1970s – a difference attributed to shorter, warmer winters.

In Home Bay, Baffin Island, the edge of the fast ice can be over water that is 180 m deep some 70 km offshore (Jacobs et al. 1975). In the Kara Sea, Volkov et al. (2002) discerned two basic mechanisms of fast ice formation. First, grounded pressure ridges stabilize the ice, facilitating fast-ice growth in shallow regions (less than 25 m depth). The spatial extent of this ice is limited by the thickness of the pressure ridge and the ocean depth. Second, further fast-ice growth may occur as ice floes drift onshore and attach themselves onto the coast or fast-ice edge. This second mechanism is the main formation mechanism in the northeastern Kara Sea. Olason (2016) notes that offshore islands will prevent the ice drift under offshore winds, allowing fast ice to form over deep water. A crucial element is the formation of static arches in granular materials passing through openings and converging channels. It has been shown that the strength of the arch depends critically on the uniaxial compressive strength of the material. Thus, ice arches will form in channels and narrow passages.

According to 10-day sea ice charts for the region during 1933–2006 (Mahoney et al. 2008), fast ice forms first off the Taimyr Peninsula (the Severozemelsky region) around December 11. It remains mostly unaltered until the summer breakup, which occurs from July 11 to August 11.

Russian scientists have long recognized "ice massifs" in the Eurasian shelf seas. Massifs, formed by fast ice, appear and persist in regions with shallow depths and irregular coastlines. Timokhov (1994) noted that in the eastern parts of the Kara and Laptev seas, as well as to the west of the East Siberian Sea, fast ice extends over hundreds of kilometers from the shore, forming the basis for the Severnaya Zemlya, Yana, and New Siberian ice massifs. Each ice massif has its own respective flaw polynya. For example, in the Kara Sea, the Anderma and Yamal polynyas correspond to the Novaya Zemlya massif, and the Ob-Yenisei and Severnaya Zemlya polynyas correspond to the northern Kara massif. In the Laptev Sea, opposite the Taimyr massif, there is an extensive Siberian polynya.

7.7.4 Sea Ice Leads

Sea ice leads are important climatologically, oceanographically, and geophysically. Leads (open and refrozen) account for about half of the turbulent heat transfer to the atmosphere in winter, even though they represent only about 1 percent of the sea ice area (Maykut 1978).

A five-year climatology of leads in the western Arctic was derived by Miles and Barry (1998) using Defense Meteorological Satellite Program (DMSP) thermal and visible band imagery with 2.7- and 0.6-km resolution, respectively. In these images, leads of 200–300 m width are detectable. The occurrence (density) and orientation of leads are derived from gridded maps made at 10-day intervals. In the Miles and Barry climatology, relative lead densities are observed to be highest in early winter, decreasing 20 percent from November through April. The highest densities are observed in the central Canada Basin, and the lowest are in the East Siberian Sea. Preferred lead orientations are identified as generally north–south in the Beaufort Sea sector and east–west in the East Siberian Sea sector, with transitional orientations in the intermediate area. The spatial patterns of mean ice divergence and lead density are in general correspondence, with the highest values found in the Beaufort Sea and the lowest in the East Siberian Sea. The strength of the association between the lead density and divergence is indicated by the correlation coefficient ($r = 0.68$); the r values for November–January and February–April are 0.53 and 0.71, respectively. The circular correlation of lead orientation with the shear is 0.81 for NDJ and 0.75 for FMA. The mean angle of $90.7°$ for NDJ and $89.1°$ for FMA indicates a strong association between the two orientations: In theory, fractures in ice are expected to form orthogonally to the direction of the principal stress. The preeminent geometric feature of the lead distributions is a characteristic rectilinear pattern, with an intersection angle of about $30°$, in accordance with theory (Erlingsson 1991). This angle appears to be consistent throughout the range of scales observable on the images, from kilometers to hundreds of kilometers.

Willmes and Heinemann (2016) used the thermal infrared (IR) data from MODIS for January to April 2003–2015 over the entire Arctic Ocean to determine lead frequencies and regional characteristics on a daily basis. They found that the marginal ice zones in Fram Strait and the Barents Sea are the primary regions for lead activity. There are also distinct patterns of predominant fracture zones in the Beaufort Sea and along the shelf-breaks, mainly in the Siberian sector along the flaw polynyas in the Kara, Laptev, and East Siberian seas (Figure 7.8). Note that some areas of leads appear to be related to shoals – for example, Hanna Shoal in Figure 7.8. Wang et al. (2016) established that wintertime lead area fraction during the last three decades has not undergone significant trends. However, a substantial positive trend in lead area fraction was found in summer, located where sea ice concentration is already low.

Energy Budgets

Lindsay and Makshtas (2003) assembled energy balance data over the Arctic pack ice using measurements from Soviet North Pole stations; selected data are shown in Table 7.5. Solar downwelling peaks in June, whereas net radiation reaches its maximum in July. The turbulent energy flux terms are small throughout the year. About 60 W m^{-2} is available for ice melting in June and July.

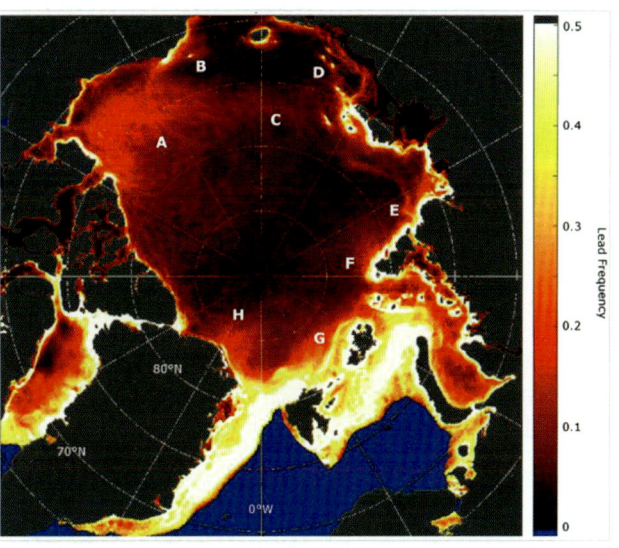

Figure 7.8 Average lead frequency in the pan-Arctic, January to April, 2003–2015. A cutoff value of 0.5 is applied. A = Beaufort Sea, B = Hanna Shoal, C = band between the Beaufort Sea and New Siberian Islands, D = two unknown lead hot spots in the East Siberian Sea, E = Vilkitsky canyon outflow region, F = fracture zone east of Severnaya Zemlya, G = elongated region with high lead frequency northwest of Franz Josef Land, H = enhanced lead activity north of Greenland.
Source: Willmes and Heinemann 2016, 9, figure 5. Courtesy of MDPI.

7.7.5 Arctic Sea Ice Trends

Records of sea ice conditions are highly variable in their source (ship, aircraft reconnaissance, satellite remote sensing, and submarine sonar), descriptors (extent, concentration, thickness), regional extent, and time interval covered. Recently, Walsh et al. (2017) have assembled an Arctic database from 1850 to the present. In doing so, they combined information from six sources prior to the availability of satellite passive microwave data in 1979:

1. The W. Dehn collection of sea ice charts for the Alaskan region for 1953–1986
2. The Russian Arctic and Antarctic Research Institute database, spanning 1933–2006 and covering the Eurasian Arctic, including the Chukchi Sea

Table 7.5 Monthly mean energy budgets over Arctic pack ice (W m^{-2}) (from Lindsay and Makshtas 2003, 416–17, table 4.1)

	January	April	June	July	September
S↓	0	146	308	231	44
L↓	164	188	291	304	262
Rn	−27	−4	55	69	−8
H	10	0	−2	2	2
LE	1	0	−10	−6	−4
Bottom flux	15	12	2	0	−2
Heat storage	−1	0	18	31	−14
Available for melting	0	0	27	34	1

3. The National Research Council of Canada sea ice data spanning about 1815–2000 and covering the eastern Canadian waters – Baffin Bay, Davis Strait, the Labrador Sea, and the Gulf of St. Lawrence
4. The historical sea ice charts of the Danish Meteorological Institute (DMI), spanning 1894–1956
5. The Arctic Climate System Study (ACSYS) database of North Atlantic ice edge positions, 1750–2002
6. Whaling ship reports for the Alaskan region from the Bockstoce collection, 1849–1914

The coverage of the DMI charts includes all of the Atlantic sector and the Pacific sector as far south as the Bering Sea.

The time series of March and September ice extent from the Walsh et al. (2017) database is shown in Figure 7.9. These authors have also demonstrated that the decrease of pan-Arctic sea-ice extent from the 1920s to the 1940s is most apparent in the summer months. However, the recent decrease is apparent in all seasons and is the only systematic excursion from the mean that is prominent in all seasons. Winter ice reached a record low of 14.42 million km^2 in March 2017, surpassing previous records in March 2015 and 2016.

A long-term perspective on the recent sea ice decline in late summer has been provided by Kinnard et al. (2011) using four types of high-resolution terrestrial proxy records from the circum-Arctic region and ocean cores. They conclude that while extensive uncertainties remain, especially before the sixteenth century, both the duration and the magnitude of the current decline in sea ice seem to be unprecedented for the past 1,450 years. Enhanced advection of warm Atlantic

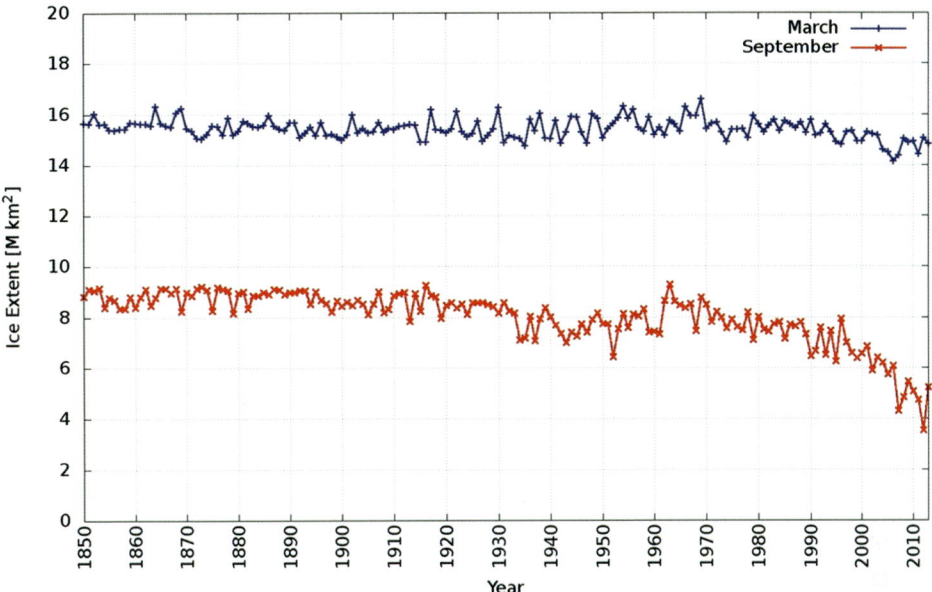

Figure 7.9 March and September Arctic ice extent from 1850.
Source: Walsh et al. 2017, 100, figure 8. Courtesy of American Geographical Society.

water to the Arctic seems to be the main factor driving the decline of sea ice extent on multidecadal time scales. This process is also observed during the recent sea ice decline (Spielhagen et al. 2011). Paradoxically, Kinnard et al. also report an interval from the late fifteenth to the early seventeenth century, during the Little Ice Age, when sea ice extent also decreased. They attribute this to enhanced southerly advection of warm air into the Arctic.

Observations in the Bering Strait were made in summers 1778 and 1779 by British navigators James Cook and Charles Clerke. The ice edge was encountered at 70.7° and 70.3° N in the respective summers. Stern (2016) reported that these limits were essentially maintained until the 1990s, when northward retreat of summer ice in the Chukchi Sea became increasingly evident.

Regional trends have also been examined by Walsh et al. (2017). For example, increased sea ice cover relative to preceding decades was noted in the Greenland Sea from the 1960s to the early 1970s. This expansion of the ice cover, apparent in both September and March, coincided with the Great Salinity Anomaly (see Box 4.3) that migrated through the East Greenland waters in the 1960s and circulated around the North Atlantic over the next decade (Dickson et al. 1988). The outstanding feature of both the Barents Sea and Greenland Sea time series is the decrease in sea ice cover during the last few decades. This decrease took both regions to period-of-record lows in March and, in at least one or two of the years since 2000, to record lows in September.

The Canadian archipelago saw frequent light ice years from the 1880s to the early 1910s, and then from the 1940s through the 1950s. Ice cover in this region increased from about 1960 to 1980, consistent with a cooling in this region during these decades. However, since the 1990s, the region has seen an unprecedented retreat of sea ice during the warm season.

The Beaufort and Chukchi seas are dominated by the recent retreat of summer sea ice. While both seas continue to be completely ice covered in March, the September sea ice coverage has decreased so precipitously that the extent in the past few years has been less than half the extent in the 1970s and 1980s. The Chukchi Sea, in particular, has been nearly ice free in autumn since September 2007. The Bering Sea ice, which is seasonal, does not show a significant reduction over the post-1850 period, in contrast to the Beaufort and Chukchi seas.

The most consistent sea ice concentration record is that provided by passive microwave remote sensing since late 1978 (Meier et al. 2014b). This record began with the Scanning Multichannel Microwave Radiometer (SMMR) on NASA's Nimbus 7 satellite, which was succeeded in 1987 by the Special Sensor Microwave Imager (SSM/I) on DMSP satellites until 2008, and by the Special Sensor Microwave Imager/Sounder (SSMIS) from 2008 to the present. The 1980s showed mostly interannual variability, but this was followed by progressive decline, especially in summer ice extent in the 2000s (Serreze and Stroeve 2015). The average decrease was 13 percent decade^{-1} in September, compared with less than 3 percent decade^{-1} in March (see Figure 7.9). The average ice area (1979–2013) for March was 15.5×10^6 km^2, and that for September 6.4×10^6 km^2 (Meier et al. 2014b).

A record September sea ice concentration minimum of 4.3 million km^2 was set in 2007 (Stroeve et al. 2012). This record was broken in September 2012 (Figure 7.10), when the ice shrank to 3.4 million km^2 (Figure 7.11), in part as a result of the export of multiyear ice through Fram Strait and in part due to the influence of a North Pacific storm in early August that broke up the already thin ice. September 2017 featured the eighth lowest sea ice extent on record.

The age of the sea ice has decreased greatly since the 1980s. Figure 7.12 shows that the proportion of FYI increased from about 55 percent in 1985 to 70 percent in 2016, while ice aged four years or greater has vanished.

The atmospheric response to sea ice loss during 1979–2009 has been analyzed by Smith et al. (2017) using the Hadley Centre GEM3 coupled model. They found that a weak low sets up over the Arctic in summer and autumn, which also leads to warming in the North Atlantic Ocean. Increased Antarctic sea ice over the same time interval drives a poleward shift of the southern hemisphere midlatitude jet, especially in the cold season.

The relative contributions of greenhouse gas forcing and internal atmospheric variability to summer sea ice loss have recently been studied by Ding et al. (2017). Using model calculations, they showed that internal variability has dominated the Arctic summer circulation trend and may be responsible for about 40 percent of the overall decline in September sea ice since 1979. The tendency for a stronger anticyclonic (clockwise) circulation over Greenland and northeastern Canada in June, July, and August has increased the downward longwave radiation above the ice, with warming and moistening of the lower troposphere also occurring. These changes are followed by negative sea ice anomalies in September. The source of the internal variability is indicated by a relationship between SST variability in the tropical Pacific and annual mean atmospheric circulation over the Arctic, centered over Greenland, that is a driver of the sea ice trend.

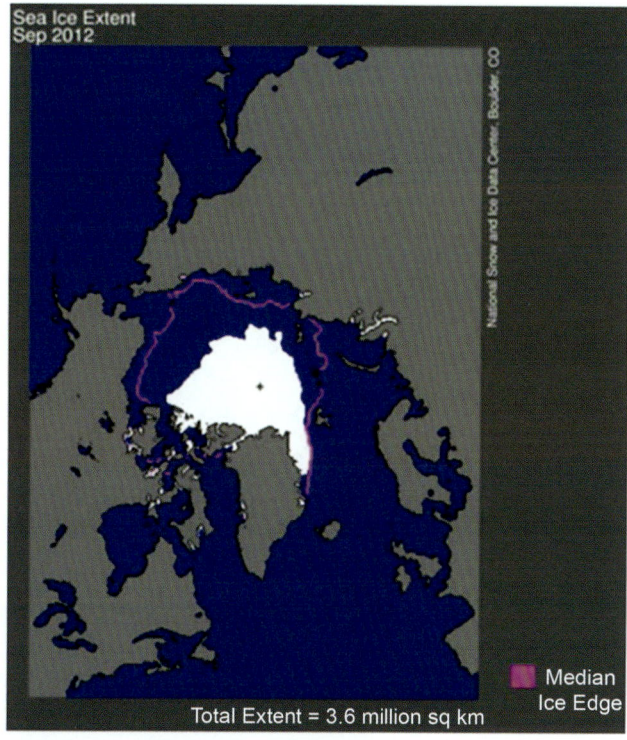

Figure 7.10 Trends in Arctic sea ice extent in September, 1979–2017.
Source: National Snow and Ice Data Center.

Changes in regional sea ice extent in the Arctic during autumn and early winter 1979–2014 have been analyzed by Chen et al. (2016). The largest negative trends (approximately -20 percent decade^{-1}) were found during autumn in the Beaufort Sea, the Barents–Kara Sea, and the Laptev–East Siberian Sea. During early winter, the largest trends in sea ice extent were found in the regions of Hudson Bay and the Barents–Kara Sea, around -10 percent decade^{-1}. Sea ice losses in the Beaufort Sea and the Barents–Kara Sea were associated with a cooling over Eurasia, but in the former region the circulation anomaly was reminiscent of a Rossby wave train across the North Pacific, whereas in the latter area the pattern projected onto the negative phase of the Arctic Oscillation.

Analysis of September open water fraction in the Pacific and Atlantic sectors for 1979–2014 by Goldstein et al. (2016) has suggested the development of a

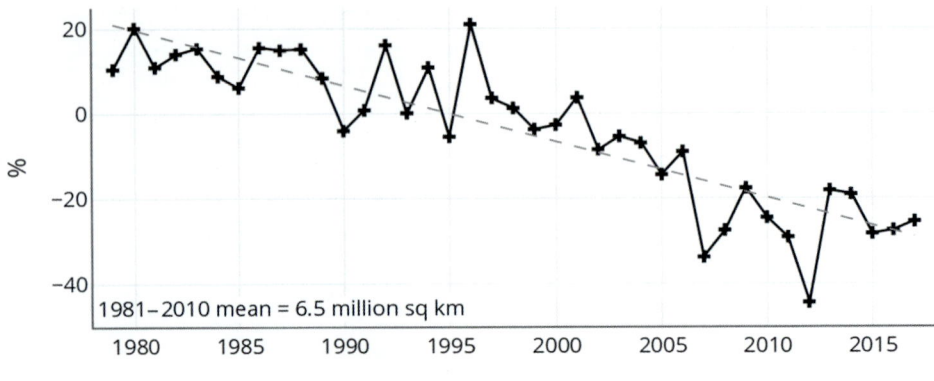

Figure 7.11 Arctic sea ice extent for the record minimum in September 2012. The purple line is the 1981–2010 median position.
Source: National Snow and Ice Data Center, Sea Ice Index.

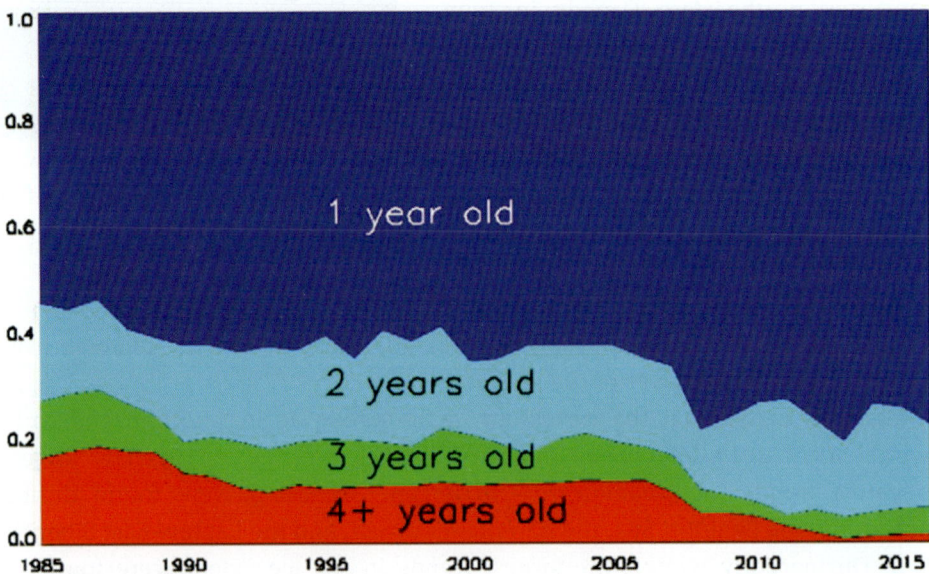

Figure 7.12 Time series of sea ice age coverage, 1985–2016. The coverages are presented as fractions, or percent, of the total sea ice areal coverage.
Source: M. Tschudi, NOAA Arctic Report Card, 2016; Perovich et al. 2016, figure 4.3c.

statistically significant shift in the mean and an increase in the variance around 1988 and another breakpoint around 2007 in the Pacific sector. Breakpoints in the Atlantic sector record of open water were also evident in 1988 and 2007, but were more weakly significant. The breakpoints appeared to be associated with concomitant shifts in average ice age, and tended to lead to changes in Arctic circulation regimes.

The characteristics of changes in ice conditions during 1979–2014 were very different in the Atlantic and Pacific sectors (divided by the 100° E and 100° W meridians) of the Arctic (Lynch et al. 2016). The trend in September open water fraction was 5.2 percent decade^{-1} in the Arctic north of 70° N, only 2.1 percent decade^{-1} in the Atlantic sector, but 9.1 percent decade^{-1} in the Pacific sector. The per-decade trend reached 17.2 percent in the East Siberian Sea, 16.0 percent in the Chukchi Sea, and 12.0 percent in the Beaufort Sea. Anomalies in the Pacific sector ice cover can be partially compensated for by anomalies of opposite sign in the Atlantic sector. An assessment of linkages between summer atmospheric patterns and sectoral anomalies in the area of maximum open water north of 70° N demonstrates that there is asymmetry in the mechanisms.

Years with low ice extent and high open water fraction are uniformly associated with positive 925-hPa temperature anomalies and southerly flow in both the Atlantic and Pacific sectors. However, years with high ice extent and low open water fraction in both sectors reveal two dominant mechanisms. Some years with anomalously low maximum open water fraction are associated with negative temperature anomalies and southerly transport – a cool summer pattern that allows ice to persist over larger areas. In contrast, other low-open-water years are characterized by a mechanism, whereby – even when melting – ice cover is continually replenished by advection from the north.

According to Alexiev et al. (2016), increase of surface air temperature (SAT) in the marine Arctic (the part covered with sea ice in winter) was closely related to reduction of sea ice extent (SIE) in summer for 1980–2014. Based on this finding, anomalies of Arctic September SIE were reconstructed from the beginning of twentieth century using a linear regression. The reconstructed SIE shows a substantial decrease in the 1930–1940s, with a minimum occurring in 1936, although the decrease was only a half of the decline in the record year of 2012.

Recent increases in Atlantic water inflow into the eastern Eurasian Basin (north of Severnaya Zemlya) have been reported by Polyakov et al (2017). Since 2003, enhanced release of oceanic heat has reduced winter sea-ice formation at a rate now comparable to losses from atmospheric thermodynamic forcing. Polyakov et al. note that release of 1 W m^{-2} over the year causes a sea ice loss of 10 cm; upward flux through the pycnocline averaged 12 W m^{-2} for winter 2013–2014 and 7.5 W m^{-2} for winter 2014–2015. These are equivalent to 54- and 40-cm reductions in ice growth, respectively, in the east Eurasian Basin. The researchers term this process the "atlantification" of the Eurasian Basin.

Relations between sea ice and tropical convection as expressed in the Madden Julian Oscillation (MJO) were investigated by Henderson et al. (2014). Anomalies in daily change in sea ice concentration were isolated for all phases of the Real-Time Multivariate MJO index during both summer (May–July) and winter (November–January) months. The relationship between sea ice concentration and the MJO included three aspects. First, the MJO projects onto the Arctic atmosphere in both winter and summer, as shown by statistically significant "wavy patterns" at 500 hPa, with a variety of wave numbers, and consistent anomaly sign changes in composites of surface and mid-tropospheric atmospheric fields for different MJO phases. In November–January, height anomalies in MJO phase 2 (convection over the Indian Ocean) resemble positive AO polarity while height anomalies in phases 6 and 7 (convection over the Western Pacific) resemble negative AO polarity. Second, the MJO modulates Arctic sea ice in both summer and winter seasons, with the region of greatest variability shifting with the migration of the ice margin poleward (equatorward) during the summer (winter) period. This variability is supported by corresponding anomalies in surface wind and temperature. Third, the MJO modulates Arctic sea ice regionally, often resulting in dipole-shaped patterns of variability between anomaly centers in the Barents and Greenland seas in January.

For the southern Beaufort Sea and Amundsen Gulf, Galley et al. (2008) analyzed Canadian Ice Service charts for 1980–2004. In summer, a trend toward increased old sea ice concentration occurred near the mouth of the Amundsen Gulf, with a trend toward decreasing summer first-year sea ice farther west. In winter, increasingly the thick first-year sea ice extent appears to be replacing young sea ice within the flaw lead system in the region. The dynamically driven breakup of sea ice in spring in the Amundsen Gulf is a highly variable event, taking between two and twenty-two weeks to completely remove ice from the gulf. The timing and duration of the open water season depend upon the extent and timing of old ice influx. Freeze-up occurs very quickly, proceeding from west to east with little temporal variability.

Ice in the Eurasian Arctic has generally decreased since 1933, based on data from Soviet and Russia ice charts (Mahoney et al. 2008). The retreat has not been continuous, however, with the data showing two periods of retreat separated by a partial recovery between the mid-1950s and mid-1980s. The charts, in combination with air temperature records, suggest that the retreat in recent years is pan-Arctic wide and year-round in some regions, whereas the retreat in the early to mid-twentieth century was confined to summer and autumn in the Russian Arctic.

Rothrock et al. (1999) showed changes in Arctic ice thickness by comparing submarine sonar ice draft data from 1958 through 1976 to measurements from the 1990s. Their results indicate that there was thinning at every point of comparison between 1993 and 1997 with similar data acquired between 1958 and 1976; the mean ice draft at the end of the melt season decreased by about 1.3 m (40 percent) in most of the deep water portion of the Arctic Ocean, from 3.1 m in 1958–1976 to 1.8 m in the 1990s.

Kwok and Rothrock (2009; Kwok et al. 2009) have reported that within the data release area of declassified submarine sonar measurements (covering 38 percent of the Arctic Ocean), the overall mean winter ice thickness of 3.64 m in 1980 can be compared to a 1.89 m mean during the last winter (2008–2009) of the ICESat record – an astonishing decrease of 1.75 m in thickness. Between 1975 and 2000, the steepest rate of decrease was 0.08 m yr^{-1} in 1990, compared to a slightly higher winter/summer rate of 0.10/0.20 m yr^{-1} in the five-year ICESat record (2003–2008).

There have been two recent expeditions to the Arctic to measure ice and snow cover thickness. Haas et al. (2017) made in situ measurements at ten sampling sites in the Lincoln Sea between Ellesmere Island and 87.1° N in April 2017. Mean and modal total ice thicknesses ranged between 2 and 3.4 m and between 1.8 and 2.9 m, respectively. Coincident snow thicknesses ranged between 0.3 and 0.47 m (mean) and 0.1 and 0.5 m (mode). There was excellent agreement with the snow climatology published by Warren et al. (1999) and with published long-term ice thinning rates.

Merkouriadi et al. (2017) and Gallet et al. (2017) investigated the physical properties of first-year (FYI) and second-year (SYI) ice in the Atlantic sector of the Arctic Ocean, during the Norwegian young sea ICE (N-ICE2015) expedition (January–June 2015). Snow depth was 41 ± 19 cm in January and 56 ± 17 cm in February, which is significantly greater than Warren et al. (1999) described for this region. The snow water equivalent was 14.5 cm over FYI and 19 cm over SYI. For April–June overall, the snow thickness was about 20 cm greater than the climatology for SYI, with an average of 55 ± 27 cm and 32 ± 20 cm for FYI.

Sea ice thickness and volume were determined by Kwok and Rothrock (2009) using ICESat data from 2003 to 2008. They found a greater than 42 percent decrease in MYI coverage since 2005, with a remarkable thinning of approximately 0.6 m in MYI thickness over four years. In contrast, the average thickness of the seasonal ice in midwinter (approximately 2 m), which covered more than two thirds of the Arctic Ocean in 2007, exhibited a negligible trend. Total MYI volume in the winter has experienced a net loss of 6,300 km^3 (more than 40 percent) in the four years since 2005, while the FYI cover gained volume owing to increased overall area coverage. The Arctic has a maximum ice volume of 16,400 km^3 in the spring.

Landy et al. (2017) used ICESat and Cryosat 2 data to analyze ice thickness in the eastern Canadian Arctic during 2003–2016. The mean seasonal growth rate was 23 cm month^{-1} from November to April. In Hudson Bay, the ice was 40 cm thicker in the east than in the northwest, whereas in Baffin Bay it was 20 cm thicker in the west than in the east. The April thickness reached 2.12 m in Foxe Basin; it was 1.67 m in eastern Hudson Bay and 1.25 m in the northwest part.

There is extensive and persistent fast ice in the waters of the Canadian Arctic archipelago. Howell et al. (2016) analyzed trends observed at the Cambridge Bay, Resolute, Eureka, and Alert sites during 1957–2014, representing some of the Arctic's longest records of landfast ice thickness. Observed end-of-winter (maximum) trends of landfast ice thickness were statistically significant at Cambridge Bay (-4.31 ± 1.4 cm decade^{-1}), Eureka (-4.65 ± 1.7 cm decade^{-1}), and Alert (-4.44 ± 1.6 cm decade^{-1}), but not at Resolute. Over the more than fifty-year record, the ice thinned by approximately 0.24–0.26 m at Cambridge Bay, Eureka, and Alert, with essentially negligible change occurring at Resolute. Although statistically significant warming in spring and fall was present at all sites, only low correlations between temperature and maximum ice thickness were present; snow depth was found to be more strongly associated with the negative ice thickness trends.

Russian fast ice thickness measurements were reported by Frolov et al. (2005). They show a positive trend during 1940–1973 of $+0.35$ cm yr^{-1} and a negative trend of -0.52 cm yr^{-1} during 1973–2000.

When Smedsrud et al. (2016) analyzed a new time series from 1935 to 2014 of Fram Strait sea ice area export, they found that the long-term annual mean export is about 880,000 km^2, representing 10 percent of the sea ice–covered area inside the basin. There was large interannual and multidecadal variability, and no long-term trend, over the past eighty years. However, the last decade has witnessed increased ice export, with several years having annual ice exports that exceeded 1 million km^2. Since 1979, annual export has increased by about 6 percent decade^{-1}, due to higher southward ice drift speeds caused by stronger southward geostrophic winds, a phenomenon largely explained by increasing surface pressure over Greenland. Spring and summer area export increased by 11 percent decade^{-1}. In contrast, the 1950–1970 period had relatively low export during spring and summer, and mid-September sea ice extent was consistently higher during these decades than both before and after them. Thus export anomalies during spring have a clear influence on the following September sea ice extent in general, and for the recent decade, the export may be partly responsible for the accelerating decline in Arctic sea ice extent. Ice export during winter will generally result in new ice growth and contributes to thinning inside the Arctic Basin. Increased ice export during summer or spring will, in contrast, contribute directly to open water farther north and a reduced summer sea ice extent through the ice–albedo feedback. The relatively low spring and summer export from 1950 to 1970 is, therefore, consistent with a higher mid-September sea ice extent for those years.

7.8 Antarctic Sea Ice

There are distinct differences between sea ice characteristics in the two polar regions. Nearly all ice in the Southern Ocean is seasonal, and its structure differs from that in the Arctic in significant ways. Antarctic sea ice extent retreats to a minimum of 3.1×10^6 km^2 in February, which is just 17 percent of the maximum extent of 18.5×10^6 km^2 in September (Parkinson 2014). The extent of pack ice and of the Marginal Ice Zone (MIZ) has been shown to depend on the passive microwave algorithm that is used.

Stroeve et al. (2016) compared the results from the NASA Team and Bootstrap algorithms. The annual values (10^6 km^2) for all Antarctica are as follows:

	NASA team	Bootstrap
MIZ	3.83	2.54
Polynya	0.59	0.39
Pack ice	6.49	8.53

Thus, applying the same thresholds for both sea ice algorithms results in a MIZ from the NASA Team algorithm that is, on average, twice as large as in the Bootstrap algorithm and that contains considerably more broken ice within the consolidated pack ice.

There are rather sparse data available on ice thickness in the Antarctic. Two decades of data compiled by the SCAR Antarctic Sea Ice Processes and Climate (ASPeCt) program, totaling more than 23,000 observations, gave a mean thickness of all ice as 0.87 ± 0.91 m, compared with a level-ice thickness of 0.62 m (Worby et al. 2008). Kurtz and Markus (2012) used satellite laser altimetry data from NASA's ICESat combined with passive microwave measurements to analyze basin-wide changes in Antarctic sea ice thickness and volume over a five-year period from 2003 to 2008. The ICESat data for 2003–2005 and the shipboard measurements collected during the ASPeCt program show good agreement in spring:

	ICESat mean (m)	Ships mean (m)
Spring (October–November)	0.79	0.73
Summer (February–March)	0.58	0.35

The thickest ice resides in the western Weddell Sea, the Bellingshausen and Amundsen seas, the western Ross Sea, and surrounding the Antarctic coastline. The thinnest ice is found in the eastern Weddell Sea, the eastern Ross Sea, portions of the Indian and Pacific Oceans, and toward the northern edge of the sea ice. The ICESat record shows that the 2003–2008 mean ice volume reached a minimum of 3,357 km^3 in the summer, grew to 8,125 km^3 in the autumn, and reached its maximum of 11,111 km^3 in the spring.

Strong wind and wave interactions can significantly increase sea ice thickness by rafting and ridging (Lewis et al. 2011) as can heavy snow loading, flooding, and freezing at the top (Massom et al. 2001). Nghiem et al. (2016) employed QuikSCAT data for 1999–2009 to determine ice trajectories in the Southern Ocean. Their work shows that sea ice, grown earlier in the ice season, drifts northward away from the Antarctic continent, forming a circumpolar frontal ice zone (FIZ) behind the ice edge. In the circumpolar sea ice zone adjacent to the sea ice edge, the scatterometer data exhibit a band of strong radar backscatter, which is consistent with the signature of older, thicker, and rougher sea ice with more snow cover in the FIZ. The formation of this band is attributable to a longer exposure to wind and wave actions, and thickening over time by ice growth and greater snow accumulation. This band of sea ice is up to 1,000 km wide and serves as a "Great Shield," encapsulating and protecting younger and thinner ice in the interior ice pack. Three age classes of sea ice can be distinguished from the backscatter signatures: rough older ice, older ice, and younger ice. Additionally, permanent ice (ice shelf and fast ice) as well as melt on ice can be identified. Figure 7.13 shows the distributions of these ice classes near the maximum of each season for 1999–2009. In all years, the Antarctic sea ice cover was totally surrounded by a FIZ of rough, older ice. The outer ice edge of the FIZ is close to the coast in regions such as the Somov Sea, D'Urville Sea, and Mawson Sea, or far away from the coast – particularly in the Lazarev Sea, where the FIZ may extend all the way to Bouvet Island (Figure 7.13).

Nghiem et al. (2016) have shown that in general, the sea ice edge is determined by the presence of warm water and is typically enclosed by the −1.0 °C SST isotherm in each year. In the interior sea ice region behind the FIZ, persistent katabatic winds force the opening, production, and advection of ice. In the newly opened or ice divergence areas, protected behind the FIZ, the young ice can have a high growth rate. This process is enhanced by the very cold air that is advected off the continent and ice shelves. The isotherm is in the proximity of the southern ACC front, as delineated by Kim and Orsi (2014).

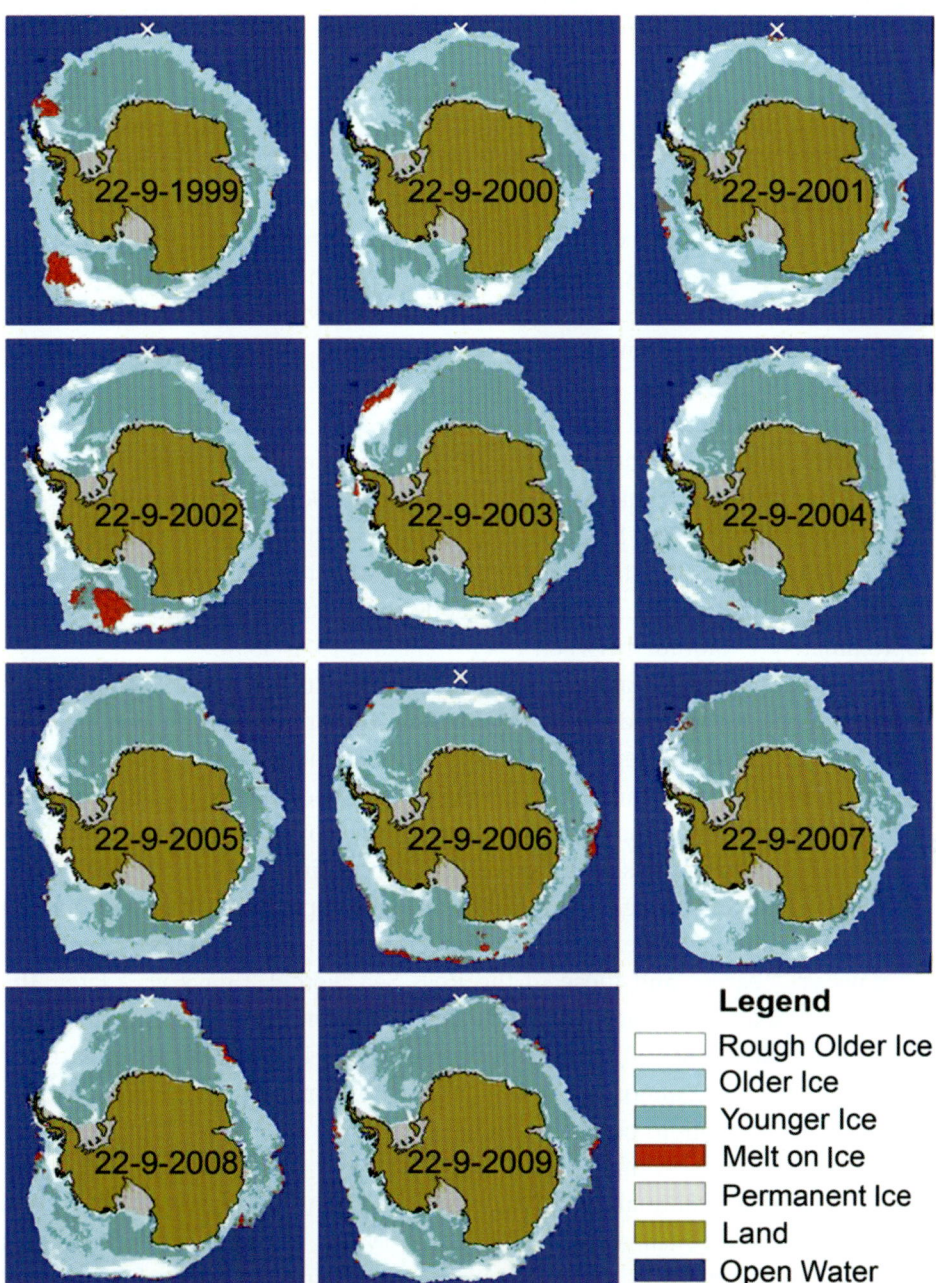

Figure 7.13 Synoptic classes of Antarctic sea ice around the September equinox in 1999–2009. The location of Bouvet Island is marked with the white cross.
Source: Nghiem et al. 2016, 286, figure 4.

7.8.1 Sea Ice Trends in the Antarctic

Recently, sea ice reported in ships' log books from the Heroic Age of Antarctic exploration (1897–1917) have been analyzed by Edinburgh and Day (2016). They found that, while in most sectors the ice extent was comparable to today, in the Weddell Sea the edge was 1.0–1.7° farther north at that time. A proxy sea ice record from 1702 for the Amundsen–Ross Sea sector indicates opposite trends in winter sea ice extent in the Bellingshausen–Weddell Sea, where this extent has declined over time, and in the Amundsen–Ross Sea, where it expanded northward by 1° latitude during the twentieth century (Thomas and Abram 2016). Maximum extent in this region was observed during the mid-1990s.

Whaling ship records of inferred summer ice edge location from the 1930s to 1980s were analyzed by de la Mare (2009). The data suggest that there was a zonal mean southward migration of the summer (October–March) ice edge of 2.4° latitude in the 1970s–1980s compared to the 1930s–1950s. King and Harangozo (1998) used coastal station temperature data to infer a southward migration of the autumn and winter sea ice edge along the western Antarctic Peninsula of approximately 1° latitude between 1945–1954 and 1973–1994.

Satellite passive microwave records for the Southern Ocean for 1979–2014 from the NOAA/NSIDC Goddard-merged climate data record of monthly mean sea ice concentration product (Meier et al. 2014a) show an increase in sea ice cover that is most pronounced in the summer months of December–April (Hobbs et al. 2016). This trend is dominated by increased sea ice coverage in the western Ross Sea, which is offset by a strong decrease in the Bellingshausen and Amundsen seas. The trends in sea ice areal coverage are accompanied by related trends in yearly duration. The retreat dates in the Amundsen–Bellingshausen Sea were 1.2 days yr^{-1} earlier and those in the Ross Sea were 1.2 days yr^{-1} later; the corresponding advance dates were 1.9 days yr^{-1} later and 1.3 days yr^{-1} earlier, respectively. November 2016 witnessed a record low ice concentration, with the Amundsen Sea being almost ice free. The lower extents of sea ice coverage have continued and remain to be explained.

Re-examination of Nimbus 1, 2, and 3 satellite data from the 1960s has identified large interannual variations in sea ice extent (Gallagher al. 2014). The September 1964 ice mean area was a record 19.7×10^6 km^2, while in August 1966 the maximum sea ice extent fell to 15.9×10^6 km^2.

The atmosphere is thought to be the primary driver of sea ice trends. Cyclonic flow around the Amundsen Sea low drives warm poleward winds into the Antarctic Peninsula–Bellingshausen Sea region, along with a cold equatorward wind over the Ross Sea, with clear implications for the dipole in sea ice trends between these two regions (Hosking et al. 2013; Turner et al. 2015). However, the linkages of the Southern Annular Mode (SAM) and the Amundsen Sea low are complex. A positive SAM trend has increased the southern westerlies but it is unclear how this change has affected sea ice extent. The ocean also has an essential role in explaining the seasonality of the trend patterns. Mixed-layer feedback processes between sea ice and ocean have had a role in modulating the sea ice trends, and there appears to be a spatial dependence on where these processes are important. Nevertheless, the record length is short, and Stroeve et al. (2016) point out that the sea ice increase over the last 36 years remains within the range of intrinsic internal variability.

Meehl et al. (2016) have examined the relationship between the sea ice expansion between 2000 and 2014 and tropical Pacific climatic conditions. They found that the Interdecadal Pacific Oscillation, an internally generated mode of climate variability, transitioned from positive to negative in the late 1990s, with an average cooling of tropical Pacific sea surface temperatures. Sea-level pressure and 850-hPa wind changes near Antarctica since 2000 have been conducive to expanding Antarctic sea ice extent, particularly in the Ross Sea region in all seasons, involving a deepening of the Amundsen Sea low. These atmospheric circulation changes are mainly driven by precipitation and convective heating anomalies related to the Interdecadal Pacific Oscillation in the equatorial eastern Pacific, with additional contributions from convective heating anomalies in the Southwest Pacific convergence zone and tropical Atlantic regions.

Cerrone et al. (2017) examined the roles of the SAM, the Semiannual Oscillation (SAO), the Pacific–South American (PSA) teleconnection, and the zonal wave number 3 (ZW3) mode on the variability of sea ice concentration for 1982–2013. Most of the sea ice temporal variability was concentrated in the two- to four-year time range associated with the constructive superposition of the PSA and ZW3 patterns. Interannual variations were related to the SAM and SAO patterns. The two-year signal resulted from the superposition of the positive phases of SAM, ZW3, and PSA patterns. The 2.7-year signal combined positive oscillations of the PSA and ZW3, and the four-year signal featured superposed positive oscillations of the SAM and ZW3 patterns and the negative phase of the PSA. This four-year signal is apparent only after 2000, whereas the other two signals are noted throughout the record.

Comiso et al. (2017) analyzed an improved passive microwave record of sea extent and surface temperature for November 1978–December 2015. They found that there was a strong correlation, with a one-month lag in surface temperature, measured at -0.96 during the growth season and -0.98 during the melt season. There was only a weak relationship between ice extent and the SAM index of circulation. The record ice extent of 20 million km^2 in 2014 was also shown to display a high sensitivity to surface temperature. Surprisingly, Antarctic sea ice extent fell to a record minimum in October 2016 and reached an all-time low of 2.28 million km^2 on March 3, 2017.

Turner et al. (2017) reported that during the spring months of September to November 2016, the Antarctic sea ice extent decreased by 6.82 million km^2. They observed that ice retreat in the Weddell Sea took place rapidly via strong northerly flow with poleward heat fluxes, after an early maximum ice extent in late August. The Amundsen Sea low was at record strength in September. Rapid ice retreat occurred in the Ross Sea in November, when there was record high surface pressure, with the SAM at its most negative for that month since 1968.

Stuecker et al. (2017) showed that the extreme El Niño event that peaked in December to February 2015–2016 contributed to pronounced extratropical southern hemisphere SST and sea ice extent anomalies in the eastern Ross, Amundsen, and Bellingshausen seas that persisted in part until the 2016 austral spring. A second factor was internal variability of the SAM, which promoted the exceptional low sea ice extent in November–December 2016.

7.9 Polynyas

A polynya (a Russian term) is an ocean area that is largely ice free and that is surrounded by sea ice or sea ice and land. It may form by either of two processes: as a sensible-heat polynya or as a latent-heat polynya. The former, which is thermodynamically driven, typically occurs when warm water upwells keeping the surface water temperature at or above freezing. This reduces ice production and may stop it entirely. A sensible-heat polynya forms in the open ocean and the upwelling is accounted for by the bottom topography. In contrast, a latent-heat polynya is an open water region between a barrier and the ice pack, where the ice is driven away from the coast, an ice shelf, a grounded iceberg, or landfast ice, by offshore winds or ocean currents. New ice forms in the open water, which is then herded downwind toward the first-year pack ice, where the new ice is consolidated onto the pack. Ice growth leads to latent heat release as well as brine expulsion. This process increases ocean salinity, causing the higher-density water to sink. Through this mechanism, latent-heat polynyas in the coastal regions of Antarctica serve as a major source of the world's bottom waters.

Some polynyas are hybrids of the sensible-heat and latent-heat types. For example, the Barrow Coastal Polynya is considered to be a wind-driven hybrid polynya, with both sensible heat from upwelling Atlantic Water and wind-driven divergence caused by the northeasterly wind (Hirona et al. 2016).

Barber and Massom (2007) provide summary tables of the physical characteristics of many Arctic and Antarctic polynyas, and a map of Arctic polynyas (Figure 7.14). There are large polynyas along the Siberian coast and many smaller ones in the Canadian Arctic archipelago.

Preusser et al. (2016) have analyzed MODIS data for sixteen circum-Arctic polynyas for November to March from 2002–2003 to 2014–2015. All polynya regions combined covered an average thin-ice area of 184×10^3 km^2 and created an average total wintertime accumulated ice production of about 1,444 km^3. The main contributors (53 percent) were the Kara Sea region, the North Water polynya, and scattered smaller polynyas in the Canadian Arctic archipelago. The mean thin-ice thickness was estimated to be 13 cm.

Tamura et al. (2008) estimated that about 10 percent of sea ice production in the southern hemisphere is accounted for by coastal polynyas. Mean values of annual cumulative sea-ice production for the four major Antarctic coastal polynyas for 1992–2001 were as follows: Ross Sea, 390 km^3; Darnley, 181 km^3; Mertz, 120 km^3; and Shackleton, 110 km^3. The total amount for thirteen polynyas was 1,410 km^3.

The coastal polynya area around Antarctica during June, July, August, and September (wintertime) 1992–2008 was estimated from SSM/I data to be 245,000 km^2 (Kern 2009). The polynyas along East Antarctica (60–160° E) accounted for about 40 percent of the total; the most persistent were located along the Lars–Christensen Coast (LCC), Prydz Bay, the western Davis Sea, Mertz Glacier, and in the Ross Sea along the Ross Ice Shelf and in Terra Nova Bay. The polynya at the LCC was observed on 110 ± 5 days during winters 1992–2008 and covered an average area of 2,400 km^2 on more than 90 days.

Dale et al. (2016) examined the relationship between wind strength and sea ice concentration anomalies in the Ross Sea polynya in winter (April–October) 2001–2014. Persistent weak winds near the edge of the Ross Ice Shelf were generally associated with positive SIC anomalies in the Ross Sea polynya. Conversely, negative SIC anomalies in this area occurred during persistent strong southerly winds. Strong winds caused significant advection of sea ice in the region. The sea motion anomalies indicated the production of new ice by thermodynamic growth.

According to Barber and Massom (2007), the Ross Sea Polynya (RSP) is the largest in the Antarctic, with a winter area of around 20,000 km^2. Two smaller polynyas are located in the western part of the Ross Sea: the Terra Nova Bay

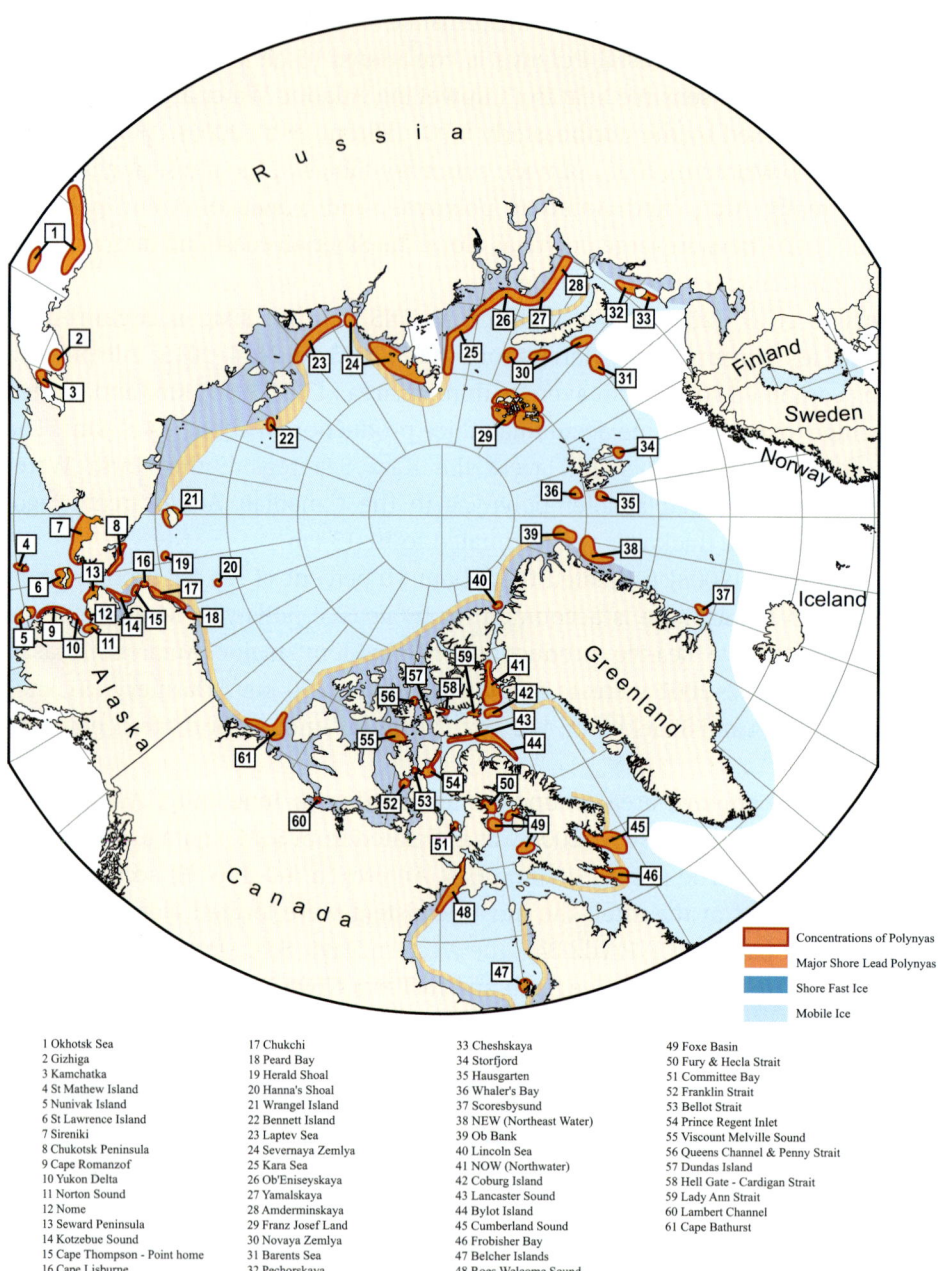

Figure 7.14 Distribution of polynyas in the Arctic (Barber and Massom 2007).
Source: Smith, W. O., Jr., and D. G. Barber, eds. *Polynyas: Windows to the World*. Amsterdam: Elsevier Oceanography Series, Vol. 74, p. 9, figure 1.

Polynya (TNBP), with a mean area of 1,300 km^2 and maxima up to 5,000 km^2, and the McMurdo Sound Polynya (MSP), with an area about two thirds of the TNBP (Hollands and Dierking 2016). The TNBP, which is oriented east–west, is bounded by the Drygalski Ice Tongue in the south and by the Campbell Ice Tongue in the north. It is maintained during the winter season by 25–40 m s^{-1} katabatic winds that are channeled by glacial valleys and flow off the ice sheet over the adjacent sea ice (Bromwich and Kurtz 1984). There is a period of maximum efficiency in sea ice production from July to November.

Estimates of ice production and dense water formation in global polynyas during nine winters have been made by Oshima et al. (2016) based on Advanced Microwave Scanning Radiometer for EOS (AMSR-E) passive microwave data. They found that ice production rate is high in Antarctic coastal polynyas, in contrast to Arctic coastal polynyas. This is consistent with the formation of dense Antarctic Bottom Water (AABW). The Ross Ice Shelf polynya has by far the highest ice production (253 km^3) in the southern hemisphere. The Cape Darnley polynya (65–69° E) is the second highest production area (127 km^3). Ten other polynyas have ice productions ranging from 27 to 83 km^3.

The Okhotsk Northwestern polynya exhibits the highest ice production (400 km^3) in the northern hemisphere, and the resultant dense water formation leads to overturning in the North Pacific. The next largest polynyas are the North Water polynya, with 152 km^3, and the Anadye–St. Lawrence Island polynya, with 140 km^3; another six polynyas range in size from 15 to 71 km^3. Most of the ice production in northern hemisphere polynyas occurs in autumn.

Like leads, polynyas are a source of heat and moisture to the atmosphere. Thus, they modify the weather in surrounding areas.

The North Water (NOW) in northern Baffin Bay, between Greenland and Ellesmere Island, covers an area of 85,000 km^2 in spring. It forms south of an ice bridge across northern Smith Sound. Three recurring polynyas are recognized within the NOW region: the Smith Sound, Lady Ann Strait, and Lancaster Sound polynyas (Steffen 1985). Eventually, the three separate polynyas become contiguous in the early spring, forming the NOW polynya. Barber et al. (2001a, 2001b) showed that the North Water is maintained by both latent-heat and sensible-heat processes. Airborne remote sensing was carried out over the NOW in winters 1978–1979 and 1980–1981 by Steffen and Ohmura (1985) and Steffen and Lewis (1988). Gray–white ice with estimated thickness of 15–30 cm was the dominant surface type. Sea surface temperatures increased from west to east by 10–15 °C across northern Baffin Bay. The highest SSTs were off Cape Alexander in Smith Sound, where the range of −1 to −15 °C was 20 °C higher than the SSTs over the fast ice. Cells of upwelling warm (Atlantic) water were common off West Greenland and near Wolstenholme Island, the Carey Islands, and Smith Sound.

Temperatures in these cells in December–January were most frequently in the range of -1.0 to $-0.8\,°C$. The energy budget of the NOW in January is -56 W m^{-2} for Rn, -28 W m^{-2} for H, and -23 W m^{-2} for LE, with a residual of 222 W m^{-2} (Steffen and Ohmura 1985). For the six winter months, the subsurface heat supply from the water is 173.5 W m^{-2}. The supply of heat by refreezing is no more than 35 W m^{-2}, so the sensible heat extracted from the water is about 139 W m^{-2}.

For the Kara Sea, Kern et al. (2005) estimated the average polynya area was 21.2×10^3 km^2 for the winters (January–April) of 1996–1997 to 2000–2001, being as large as 32.0×10^3 km^2 in 1999–2000 and smaller than 12×10^3 km^2 in 1998–1999. The modeled cumulative winter ice-volume flux out of the Kara Sea varied between 100 and 350 km^3 yr^{-1}. Bareiss and Görgen (2005) showed that in November–June of 1979–1980 to 2001–2002, the mean area of the West New Siberian polynya in the southeast Laptev Sea averaged 4,000 km^2 and had a mean duration of 14 days, while the Annabar–Lena polynya averaged 3,000 km^2 and had a mean duration of 22 days. The mean cumulative areas of the two were $1,713 \times 10^3$ km^3 and $1,152 \times 10^3$ km^2, respectively, associated with a mean frequency of 12.4 polynya events during November–June in all investigated regions of the Laptev Sea.

Winter polynya areas to the northeast of Svalbard have long been known as Whaler's Bay. Their persistence and re-emergence between 2011 and 2014 was investigated by Ivanov et al. (2016). Recent increased seasonality of Arctic sea ice cover enables an enhanced influence of oceanic heat on sea ice and on the heat transported by Atlantic Water. The "memory" of ice-depleted conditions in summer is transferred to the autumn through excess heat content in the upper mixed layer, which in turn transfers this "memory" via thinner and younger ice to midwinter conditions. This thinner ice facilitates the formation of polynyas and leads. Thermohaline convection-induced upward heat flux from the Atlantic layer retards ice formation, either keeping ice thin or blocking ice formation entirely.

An open ocean polynya was observed in the Weddell Sea near Maud Rise during three austral winters, 1974–1976. It was identified in Electrically Scanning Microwave Radiometer (ESMR) imagery, but has not recurred since that time. Holland (2001) explains this feature through a mechanism by which modest variations in the large-scale oceanic flow past the Maud Rise seamount caused a horizontal cyclonic eddy to be shed from its northeast flank. The shed eddy transmitted a divergent Ekman stress into the sea ice, leading to a crescent-shaped opening in the pack. Thermodynamic interaction with the atmosphere further enhanced the opening by inducing oceanic convection. The Maud Rise polynya has not re-formed, probably as a result of enhanced ocean stratification due to freshening but had not recurred until September 2017 when an area of 80,000 km^2 unexpectedly opened up. The causes are uncertain.

SUMMARY

Almost all of the area between latitudes 60° and 65° S is ocean. Except near Antarctica, the waters of the Southern Ocean move eastward at 40–60 cm s^{-1}, forced by the westerlies. The Antarctic Circumpolar Current (ACC) has a volume transport of approximately 173 Sv, three fourths of which is baroclinic. The Antarctic Polar Front (APF) is located south of the ACC axis and the maximum westerlies. Antarctic Bottom Water forms in autumn and winter when shelf waters sink down the continental slope at a rate of 8–10 Sv, mainly in the southwest Weddell and Ross seas. Rather than three fronts (as at Drake Passage), the circumpolar ACC comprises multiple frontal filaments. The Subantarctic Surface Water (SASW) and Antarctic Surface Water (AASW) water masses are found in the upper 500 m around Antarctica. Below 1,500 m there are three water masses: Circumpolar Deep Water (CDW), North Atlantic Deep Water (NADW), and Antarctic Bottom Water (AABW). The number and intensity of fronts are determined largely by the bathymetry. The number is reduced where the ACC's path is constricted. Two sub-Antarctic fronts and a Polar Front have strong temperature and salinity gradients.

The Ross Sea embayment covers approximately 960,000 km^2. The southern part is the Ross Ice Shelf; the remainder has sea ice for much of the year. The surface ocean circulation has a wind-driven cyclonic (clockwise) gyre, accompanied by upwelling deep water. The Weddell Sea, east of the Antarctic Peninsula, covers approximately 2.8 million km^2. In the south is the Ronne–Filchner Ice Shelf. The wind-driven cyclonic gyre transports sea ice northward in the western part. Offshore winds lead to the formation of coastal polynyas. Surface energy budget analysis has shown that sensible heat fluxes from the 5–7 percent of the area that comprises leads or coastal polynyas largely balance downward fluxes over the sea ice.

The Arctic Ocean is considered to be a Mediterranean sea due to its relative isolation. Continental shelves cover 2.5 million km^2 and receive massive amounts of river runoff. The Arctic Ocean is divided by the Lomonosov Ridge into the Canadian and Eurasian basins. Water masses in the upper 500 m are the Atlantic Subarctic Upper Water (ASUW) and Pacific Subarctic Upper Water (PSUW). Below 1,500 m are deep and abyssal waters and Arctic Bottom Water (ABW).

Surface circulation and ice drift in the Arctic Ocean reflect a clockwise gyre driven by the Beaufort Sea high pressure. On the Eurasian side, the Transpolar Drift Stream takes water and sea ice from the Eurasian coast to Fram Strait, where it forms the East Greenland Current. Atlantic water enters the Norwegian and

Barents seas, sinks, and flows counterclockwise around the Arctic Basin. The Arctic has low surface salinity owing to the immense river runoff and fresh water entering this ocean via Bering Strait. Water temperatures have risen since the 1950s, especially in the entryways. The upper Barents Sea has warmed 4 °C since the 1970s.

Many diverse seas surround the Arctic Ocean. The rather shallow Barents Sea (1.4 million km^2) has a main branch of the warm, salty North Atlantic Current that forms a Polar Front with cold water from the north. The Barents Sea is ice free year-round to about 75° N. The Kara Sea (888,000 km^2) is part of the Siberian shelf, with nearly half of it having a depth less than 50 m. The sea is very cold and frozen for eight months of the year. Runoff from the Ob and Yenisei rivers causes very low salinity.

The Laptev Sea (660,000 km^2) has a mean depth of approximately 50 m. It is a major region of winter sea ice growth and export. Runoff from the Lena and four other major rivers causes low salinity. Three channels connect the Laptev Sea to the East Siberian Sea (910,000 km^2). The East Siberian Sea is nearly all continental shelf, with depths less than 50 m, and has sea ice cover for most of the year. The Chukchi Sea (approximately 600,000 km^2) has a mean depth of 70 m. Pacific water enters via Bering Strait, forced by a slope in sea level. Northward flow is stronger in summer. Upwelling events of Atlantic Water occur in Barrow Canyon. A large winter polynya forms along the Alaskan coast. The Beaufort Sea (475,000 km^2) has a mean depth of approximately 1,000 m. The continental shelf is narrow in the west but wide off the Mackenzie delta. The surface circulation is clockwise around the Beaufort gyre, but an eastward subsurface jet occurs along the shelf break. The upper water temperature is approximately −1.4 °C in summer and −1.8 °C in winter. The Lincoln Sea (64,000 km^2) is covered all year with the thickest ice in the Arctic. There is a narrow, eastward undercurrent along the continental slope, suggesting a link to the undercurrent off Alaska.

The Greenland Sea (1.2 million km^2) is linked to the Arctic Ocean via Fram Strait. It has an average depth of 1,450 m and features deep troughs. North-flowing North Atlantic Current (NAC) water sinks in the Arctic, and part of it returns south as the East Greenland Current (EGC), which transports Arctic pack ice south from October to August. The eastern part of the sea is home to the warm West Spitsbergen Current, which, with the NAC, forms a counterclockwise gyre. The EGC has a southward transport of 8.6 Sv with a net outflow of 2 Sv. At 75° N, a mooring showed a mean transport of 11 Sv in summer and 37 Sv in winter, with two thirds of this transport being wind driven and one third attributable to thermohaline circulation. Deep convection events were reported in the central Greenland Sea in the 1990s.

The Canadian Arctic archipelago (CAA) has three main straits with shallow sills that open to Baffin Bay. Total outflow of Arctic water through the CAA is approximately 1.5 Sv. Sea ice persists for much of the year in the north and west, whereas channels in the east and south usually clear by late summer.

Adjacent seas of the North Atlantic are Baffin Bay, Davis Strait, Labrador Sea, and Hudson Bay. Baffin Bay (approximately 890,000 km^2) has an average depth of 760 m. The West Greenland Current flows northward before recirculating off northwest Greenland and joining Arctic water from Nares Strait to form the Baffin Current, which transports sea ice and Greenland icebergs southward to Newfoundland. Davis Strait links Baffin Bay to the Labrador Sea (840,000 km^2). Two thirds of the sea is ice covered in winter. North Atlantic Deep Water forms by deep convection in about six out of ten winters and flows at depth to Antarctica. NADW is overlain by lighter Labrador Sea Water. Hudson Bay (approximately 1.2 million km^2) is linked to the North Atlantic by Hudson Strait. It has an average depth of approximately 100 m and is ice covered from mid-December to early June. There is a slow cyclonic gyre in this area, and the large runoff makes it function as an estuary.

Marginal seas of the North Pacific are the Bering Sea and the Sea of Okhotsk. The former has an area of 2.3 million km^2, with a shallow shelf in the east and north and a deep basin in the south and west. There is a cyclonic gyre with the south-flowing Kamchatka Current in the west. The Alaska stream of Pacific water enters through gaps in the Aleutian Island chain. Sea ice in November is advected south from polynyas and by mid-March covers one third to one half of the sea. The Sea of Okhotsk (approximately 1.6 million km^2) has an average depth of 860 m, with a broad shelf in the north. There is a cyclonic gyre and southward current along Sakhalin Island. Runoff from the Amur River gives this sea low salinity, facilitating freezing from October–November to June.

Ocean warming and expansion caused 1.1 mm of annual sea level rise from 1992 to 2010. Over this period, Greenland produced 0.33 mm of sea level rise, Antarctica 0.27 mm, and terrestrial storage 0.38 mm. Glacier melt accounted for 0.86 mm of this amount for 1993–2008.

There are three categories of sea ice: first-year ice (FYI), multiyear ice (MYI), and (land)fast ice. The sea begins to freeze at around −1.8 °C with average ocean salinity. Only the layer above the pycnocline has to cool to enable freezing. Growing sea ice rejects brine, which drains out. In calm conditions, frazil crystals form on the surface and freeze into a sheet of nilas. Congelation ice grows on the base and the ice thickens, turning first gray and then white. Waves can lead to the formation of cakes of slush and pancakes that can be several meters in diameter in the interior of the pack. FYI can reach 1.5–2 m in thickness. The sea ice edge is defined by the 15 percent concentration line. The 100- to 200-km-wide marginal ice zone (MIZ) extends to concentrations of 80 percent.

Arctic ice is currently about 60 percent FYI and 40 percent MYI, a reversal from the pattern in the 1980s. In winter, ice divergence accounts for half as much volume change as ice growth. In summer, basal and lateral melt exceed that at the top surface.

Arctic ice circulates around the Beaufort gyre and moves from the Asian coast to Fram Strait, where it exits the basin. For 1978–2002, the annual ice area export was 866,000 km^2 and the annual ice volume flux was 2,218 km^3 (0.07 Sv). There has been a 0.45 m decrease in mean ice thickness in Fram Strait.

The Arctic Ocean has two oceanic circulation patterns that alternate every five to seven years. One is anticyclonic, where the Beaufort gyre expands and that in the Laptev Sea shrinks. The other is cyclonic, with the reverse tendency.

Cloud cover does not respond to sea ice loss in summer, but low clouds form over open water in autumn. The Laptev, Chukchi, and Beaufort seas have the largest cloud–sea ice covariance in summer–autumn; for the Barents–Kara Sea, the maximum occurs in winter.

Persistence of sea ice characteristics affects predictability. July–August sea ice area is correlated with that in September. Re-emergence of sea ice anomalies also occurs from the melt to the growth season, and vice versa; these phenomena are related to SST anomalies and sea ice thickness anomalies, respectively. Analysis of sea ice predictions has highlighted the role of "easy" and "difficult" years as reflecting climate versus weather. Predictions based on statistics and coupled models have shown the best performance in predicting sea ice extent.

Landfast ice comprises bottom fast ice in 2-m water depths as well as attached floating ice. Off northern Alaska, the latter extends to the 18-m isobaths. In parts of the Kara, Laptev, and East Siberian seas, fast ice extends hundreds of kilometers offshore. It provides a basis for ice massifs, each of which is associated with a flaw polynya.

Sea ice leads form in only approximately 1 percent of the ice area, but account for about half of the turbulent heat transfer to the atmosphere in winter. Lead densities are highest in early winter, decreasing 20 percent by April in the western Arctic. Preferred orientations are north–south in the Beaufort Sea and east–west in the East Siberian Sea. There is a strong correlation between divergence and lead density, and between shear and lead orientation. Leads show a rectilinear pattern, with an intersection angle of approximately 30°. MODIS IR data for January to April 2003–2015 over the entire Arctic have shown that the main lead activity occurs in the MIZ of Fram Strait and the Barents Sea.

Major fracture zones are in the Beaufort Sea and along flaw polynyas off Siberia.

Energy balance data for Arctic pack ice show small turbulent fluxes and about 60 W m^{-2} available for melting in June–July.

Arctic sea ice trends from a data compilation from 1850 to the present show that a decrease in sea ice extent from the 1920s to the 1940s was apparent in summer, whereas the recent decline has affected sea ice extent in all seasons. This decline appears unprecedented for the past 1,450 years. The East Greenland Sea had more sea ice during the 1960s and 1970s, coincident with the Great Salinity Anomaly. The CAA and the Chukchi Sea have seen major recent declines in summer and autumn ice cover.

The passive microwave record since 1978 is the most consistent source of information on sea ice extent in the Arctic. In September, the average decrease in ice extent was 13 percent decade^{-1}, compared with less than 3 percent decade^{-1} in March. September 2012 saw a record minimum of 4.3 million km^2 of Arctic sea ice.

Internal variability seems to account for approximately 40 percent of the September sea ice decline since 1979. It involves Pacific SSTs and atmospheric circulation over Greenland. September open water fractions in the Pacific and Atlantic sectors over 1979–2014 show breakpoints around 1988 and 2007. Increases in open water have been much greater in the Pacific sector than in the Atlantic, with the largest increases occurring in the Chukchi and East Siberian seas. Since 2003, there has been increased Atlantic water inflow into the eastern Eurasian Basin, with oceanic heat reducing winter sea ice growth.

Links exist between the tropical Madden Julian Oscillation (MJO) and mid-tropospheric AO circulation and both winter and summer sea ice concentrations.

Ice in the Eurasian Arctic has generally decreased since 1933, but experienced a partial recovery from the mid-1950s to the mid-1980s.

Arctic ice thickness for the 1990s shows a 40 percent thinning compared with 1958–1976 data. Winter ice thinned by 1.75 m from 1980 to 2008–2009. MYI coverage decreased 42 percent from 2005 to 2008, and thinned by 0.6 m during this span. Landfast ice measurements in the CAA from 1957 to 2014 have shown thinning of 0.25 m (except at Resolute) that reflects changes in snow cover.

A record of Fram Strait ice export for 1935–2014 shows no long-term trend in ice export, except for an increase in the 2000s due to stronger northerly winds.

Antarctic ice is nearly all seasonal. Its mean extent occurs between February and September and ranges from 3 to 18.5 million km^2. Two decades of observations show a mean thickness value of 0.87 m. Wind and wave interaction lead to rafting, while the snow load depresses the ice, flooding the surface, which freezes on top. Ice drifts northward and forms a circumpolar frontal ice zone, up to 1,000 km wide, that is older, thicker, and rougher. It protects the younger ice in the interior of the pack.

Ship's logs from 1897 to 1917 suggest ice extent comparable to now, except in the Weddell Sea where ice coverage reached 1.0–1.7° farther north. Whaling records indicate the summer ice edge in the 1970s and 1980s was 2.4° south of that in the 1930s. Satellite data for 1979–2014 show increasing ice cover in the summer. Increases in the western Ross Sea have been offset by decreases in the Bellingshausen–Amundsen seas. Changes in the tropical Pacific since 2000 and deepening of the Amundsen Sea low have affected the extent of ice in the Ross Sea. An improved passive microwave record (November 1978–December 2015) shows strong negative correlations between sea ice extent and one-month lagged surface temperature in the growth season, and similar positive correlations in the melt season.

Polynyas are areas of largely ice-free water surrounded by sea ice or by ice and a coastline. They may form by upwelling warm water (sensible-heat polynya) or by winds driving ice away from a coast or other barrier. New ice forms in the open water, releasing latent heat, and is pushed downwind until it is consolidated onto the pack ice. Sixteen Arctic polynyas were shown to generate about 144 km^3 of sea ice. Thirteen Antarctic polynyas produced 1,410 km^3 of sea ice and accounted for approximately 10 percent of southern hemisphere sea ice production. The winter coastal polynyas of Antarctica cover approximately 245,000 km^2. The Ross Sea polynya is approximately 20,000 km^2 and is forced by strong southerly winds. The largest polynyas (Ross Sea and Cape Darnley) also form dense Antarctic Bottom Water.

The northwestern Okhotsk polynya has the highest ice production in the northern hemisphere. The North Water (NOW) polynya in northern Baffin Bay (85,000 km^2 in spring) is maintained by both latent heat and sensible heat from upwelling Atlantic water off West Greenland. Polynyas in the Laptev Sea have been shown to have a mean duration of 14 days, with twelve events noted between November and June.

In 1974–1976, an open ocean polynya was observed by ESMR close to Maud Rise. It has not re-formed since then.

QUESTIONS

1. Describe the main features of the Antarctic Circumpolar Current.
2. Compare the Ross and Weddell seas.
3. What are the major differences between the Arctic Ocean and the Southern Ocean?
4. How does river runoff affect the Arctic Ocean?
5. Compare the Arctic seas north of Eurasia with those north of North America and Greenland.
6. Compare the East Greenland Sea and Baffin Bay–Davis Strait.

7. What are the components of global sea level rise?
8. Compare sea ice growth processes in the Arctic and the Antarctic.
9. How has Arctic sea ice changed since the 1980s?
10. How do landfast ice and pack ice differ?
11. What roles do sea ice leads and polynyas play in the ice balance and energy balance?
12. Describe the characteristics of polynyas and what maintains them.

References

Ackley, S. F. 1996. "Sea Ice." In *Encyclopedia of Applied Physics*. Vol. 17, 81–103. New York: VCH Publishers.

Alexandrov, V. Y., et al. 2000. "Sea Ice Circulation in the Laptev Sea and Ice Export to the Arctic Ocean: Results from Satellite Remote Sensing and Numerical Modeling." *Journal of Geophysical Research* 105(C5): 17143–59.

Alexiev, G. A., N. Glok, and A. Smirnov. 2016. "On Assessment of the Relationship between Changes of Sea Ice Extent and Climate in the Arctic." *International Journal of Climatology* 36: 3407–12.

Arctic Climatology Project. 2000. *Environmental Working Group Joint U.S.–Russian Sea Ice Atlas*, edited by F. Tanis and V. Smolyanitsky. [Digital media.] Boulder, CO: National Snow and Ice Data Center.

Barber, D., et al. 2001a. "Physical Processes within the North Water (NOW) Polynya." *Atmosphere–Ocean* 39: 163–6.

Barber, D., et al. 2001b. "Sea-Ice and Meteorological Conditions in Northern Baffin Bay and the North Water (NOW) Polynya between 1979 and 1996." *Atmosphere–Ocean* 39: 343–59.

Barber, D. G., and R. A. Massom. 2007. "The Role of Sea Ice in Arctic and Antarctic Polynyas." In *Polynyas: Windows to the World*. Vol. 74, edited by W. O. Smith, Jr., and D. G. Barber, 1–54. Amsterdam: Elsevier Oceanography Series.

Bareiss, J., and K. Görgen. 2005. "Spatial and Temporal Variability of Sea Ice in the Laptev Sea: Analyses and Review of Satellite Passive-Microwave Data and Model Results, 1979 to 2002." *Global and Planetary Change* 48: 2854.

Barry, R. G. 1993. "Canada's Cold Seas." In *Canada's Cold Environments*, edited by H. M. French and O. Slaymaker, 29–61. Montreal and Kingston: McGill- Queen's University Press.

Barry, R. G., R. E., Moritz, and J. C. Rogers. 1979. "The Fast Ice Regimes of the Beaufort and Chukchi Sea Coasts, Alaska." *Cold Regions Science and Technology* 1: 129–52.

Beszczynska-Möller, A., et al. 2011. "A Synthesis of Exchanges through the Main Oceanic Gateways to the Arctic Ocean." *Oceanography* 24: 82–99.

Bitz, C. M., et al. 2005. "Maintenance of the Sea Ice Edge." *Journal of Climate* 18: 2903–21.

Blanchard-Wrigglesworth, E., et al. 2011. "Persistence and Inherent Predictability of Arctic Sea Ice in a GCM Ensemble and Observations." *Journal of Climate* 24: 231–50.

Blindheim, J., and S. Østerhus. 2005. "The Nordic Seas: Main Oceanographic Features." In *The Nordic Seas: An Integrated Perspective*, edited by H. Drange et al., 11–38. Geophysics Monograph 158. Washington, DC: American Geophysical Union.

Bromwich, D. H., and D. D. Kurtz. 1984. "Katabatic Wind Forcing of the Terra Nova Bay Polynya." *Journal of Geophysical Research* 89(C3): 3561–72.

Bushuk, M., and D. Giannakis. 2017. "The Seasonality and Interannual Variability of Arctic Sea Ice Reemergence." *Journal of Climate* 30: 4657–76.

Carmack, E. C., et al. 2016. "Freshwater and Its Role in the Arctic Marine System: Sources, Disposition, Storage, Export, and Physical and Biogeochemical Consequences in the Arctic and Global Oceans." *Journal of Geophysical Research: Biogeoscience* 121: 675–717.

Cerrone, D., et al. 2017. "Dominant Covarying Climate Signals in the Southern Ocean and Antarctic Sea Ice Influence during the Last Three Decades." *Journal of Climate* 30: 3055–77.

Chen, H. W., R. B. Alley, and F. Zhang. 2016. "Interannual Arctic Sea Ice Variability and Associated Winter Weather Patterns: A Regional Perspective for 1979–2014." *Journal of Geophysical Research: Atmospheres* 121. doi: 10.1002/2016JD024769.

Church, J. A., and P. U. Clark. 2013. "Sea Level Change." In *Climate Change 2013: The Physical Science Basis. Contribution of Working Group I to the Fifth Assessment Report of the Intergovernmental Panel on Climate Change*, edited by T. F. Stocker et al., 1137–216. Cambridge: Cambridge University Press.

Church, J. A., and N. J. White. 2011. "Sea-Level Rise from the Late 19th to the Early 21st Century." *Surveys in Geophysics* 32: 585–602.

Comiso, J. C., et al. 2017. "Positive Trend in the Antarctic Sea Ice Cover and Associated Changes in Surface Temperature." *Journal of Climate* 30: 2251–67.

Dale, E. R., et al. 2016. "Atmospheric Forcing of Sea Ice Anomalies in the Ross Sea Polynya Region." *Cryosphere Discussions*. doi: 10.5194/tc-2016-89.

de la Mare, W. K. 2009. "Changes in Antarctic Sea-Ice Extent from Direct Historical Observations and Whaling Records." *Climate Change* 92: 461–93.

Dickson, R. R., et al. 1988. "The 'Great Salinity Anomaly' in the Northern North Atlantic 1968–1982." *Progress in Oceanography* 20(2): 103–51.

Ding, Q.-H., et al. 2017. "Influence of High-Latitude Atmospheric Circulation Changes on Summertime Arctic Sea Ice." *Nature Climate Change* 7: 289–95.

Dmitrenko, I. A., et al. 2008. "Towards a Warmer Arctic Ocean: Spreading of the Early 21st Century Atlantic Water Warm Anomaly along the Eurasian Basin Margins." *Journal of Geophysical Research* 113: C05023.

Donahue, K. A., et al. 2016. "Mean Antarctic Circumpolar Current Transport Measured in Drake Passage." *Geophysical Research Letters* 43: 760–7.

Druckenmiller, M. L., et al. 2012. "Trails to the Whale: Reflections of Change and Choice on an Iñupiat Icescape at Barrow, Alaska. *Polar Geography* 35: 5–29.

Edinburgh, T., and J. J. Day. 2016. "Estimating the Extent of Antarctic Summer Sea Ice during the Heroic Age of Exploration." *Cryosphere Discussions*. doi: 10.5194/tc-2016-90.

Emery, W. J., and J. Meincke. 1986. "Global Water Masses: Summary and Review." *Oceanologica Acta* 9: 383–91.

Erlingsson, B. 1991. "The Propagation of Characteristics in Sea Ice Deformation Fields." *Annals of Glaciology* 15: 73–80.

Fofonoff, N. P. 1956. "Some Properties of Sea Water Influencing the Formation of Antarctic Bottom Water." *Deep-Sea Research* 4: 32–5.

Frolov, I. V., et al. 2005. "Landfast Ice and Polynyas of the Arctic seas." In *Remote Sensing of Sea Ice in the Northern Sea Route: Studies and Applications*, edited by O. M. Johannessen et al., 58–9. Chichester, UK: Springer.

Gallagher, D. W., G. G. Campbell, and W. N. Meier. 2014. "Anomalous Variability in Antarctic Sea Ice Extents during the 1960s with the Use of Nimbus Data." *IEEE Journal on Selected Topics* 7: 881–7.

Gallet, J.-C., et al. 2017. "Spring Snow Conditions on Arctic Sea Ice North of Svalbard, during the Norwegian Young Sea ICE (N-ICE2015) Expedition." *Journal of Geophysical Research: Atmospheres* 122. doi: 10.1002/2016JD026035.

Galley, R. J., et al. 2008. "Spatial and Temporal Variability of Sea Ice in the Southern Beaufort Sea and Amundsen Gulf: 1980–2004." *Journal of Geophysical Research* 113: C05S9.

Giglio, D., and G. C. Johnson. 2016. "Subantarctic and Polar Fronts of the Antarctic Circumpolar Current and Southern Ocean Heat and Freshwater Content Variability: A View from Argo." *Journal of Physical Oceanography* 46: 749–68.

Gladyshev, S., et al. 2001. "Distribution, Formation, and Seasonal Variability of Okhotsk Sea Mode Water." *Journal of Geophysical Research* 108: C3186.

Goldstein, M. A., et al. 2016. "Abrupt Transitions in Arctic Open Water Area." *Cryosphere Discussions*. doi: 10.5194/tc-2016-108.

Graham, R. M., et al. 2012. "Southern Ocean Fronts: Controlled by Wind or Topography?" *Journal of Geophysical Research: Oceans* 117: C08018.

Gudkovich, Z. M. 1961. "Relation of the Ice Drift in the Arctic Basin to Ice Conditions in the Soviet Arctic Seas." *Trudy Okeanograficheskiy Komitet Akademy Nauk USSR* 11: 14–21.

Guemas, V., et al. 2016. "A Review on Arctic Sea-Ice Predictability and Prediction on Seasonal to Decadal Time-Scales." *Quarterly Journal of the Royal Meteorological Society* 142: 546–61.

Gyory, J., A. J. Mariano, and E. H. Ryan. n.d. "The Irminger Current: Ocean Surface Currents." http://oceancurrents.rsmas.miami.edu/atlantic/irminger.html

Haas, C., et al. 2017. "Ice and Snow Thickness Variability and Change in the High Arctic Ocean Observed by In Situ Measurements." *Geophysical Research Letters*. doi: 10.1002/2017GL075434.

Haas, C., S. Hendricks, and M. Doble. 2006. "Comparison of the Sea Ice Thickness Distribution in the Lincoln Sea and Adjacent Arctic Ocean in 2004 and 2005." *Annals of Glaciology* 44: 247–52.

Hamilton, L. C., and J. Stroeve. 2016. "400 Predictions: The SEARCH Sea Ice Outlook 2008–2015." *Polar Geography* 39(4): 274–87.

Henderson, G. R., B. S. Barrett, and D. M. LaFleur. 2014. "Arctic Sea Ice and the Madden-Julian Oscillation (MJO)." *Climate Dynamics* 43: 2185–96.

Hirano, D., et al. 2016. "A Wind-Driven, Hybrid Latent and Sensible Heat Coastal Polynya off Barrow, Alaska." *Journal of Geophysical Research: Oceans* 121: 980–97.

Hobbs, W. R., et al. 2016. "A Review of Recent Changes in Southern Ocean Sea Ice, Their Drivers and Forcings." *Global and Planetary Change* 1343: 228–50.

Holland, D. M. 2001. "Explaining the Weddell Polynya: A Large Ocean Eddy Shed at Maud Rise." *Science* 292: 1694–700.

Holland, P. R., and N. Kimura. 2016. "Observed Concentration Budgets of Arctic and Antarctic Sea Ice." *Journal of Climate* 29(14): 5241–9.

Hollands, T., and W. Dierking. 2016. "Dynamics of the Terra Nova Bay Polynya: The Potential of Multi-Sensor Satellite Observations." *Remote Sensing of the Environment* 187: 30–48.

Hosking, J. S., et al. 2013. "The Influence of the Amundsen–Bellingshausen Seas Low on the Climate of West Antarctica and Its Representation in Coupled Climate Model Simulations." *Journal of Climate* 26: 6633–48.

Howell, S. E. L., et al. 2016. "Landfast Ice Thickness in the Canadian Arctic Archipelago from Observations and Models." *Cryosphere Discussions*. doi: 10.5194/tc-2016-71.

Ingvaldsen, R. B., L. Asplin, and H. Loeng. 2004. "The Seasonal Cycle in the Atlantic Transport to the Barents Sea during the Years 1997–2001." *Continental Shelf Research* 24: 1015–32.

Ivanov, V., et al. 2016. "Arctic Ocean Heat Impact on Regional Ice Decay: A Suggested Positive Feedback." *Journal of Physical Oceanography* 46: 1437–56.

Jacobs, J. D., R. G. Barry, and R. L. Weaver. 1975. "Fast Ice Characteristics with Special Reference to the Eastern Canadian Arctic." *Polar Record* 17: 521–36.

Jacobs, S. S., A. F. Amos, and P. M. Bruchhausen. 1970. "Ross Sea Oceanography and Antarctic Bottom Water Formation." *Deep-Sea Research* 17: 935–62.

Johnson, G. C. 2006. "Quantifying Antarctic Bottom Water and North Atlantic Deep Water Volumes." *Journal of Geophysical Research: Oceans* 113(C5): C05027.

Kapsch, M. L., et al. 2016. "The Effect of Downwelling Longwave and Shortwave Radiation on Arctic Summer Sea Ice." *Journal of Climate* 29: 1143–59.

Kay, J. E., and A. Gettelman. 2009. "Cloud Influence on and Response to Seasonal Arctic Sea Ice Loss." *Journal of Geophysical Research* 114: D1820.

Kern, S. 2009. "Wintertime Antarctic Coastal Polynya Area: 1992–2008." *Geophysical Research Letters* 36: L14501.

Kern, S., et al. 2005. "A Comprehensive View of Kara Sea Polynya Dynamics, Sea-Ice Compactness and Export from Model and Remote Sensing Data." *Geophysical Research Letters* 32(15): L1550.

Kim, Y. S., and A. H. Orsi. 2014. "On the Variability of Antarctic Circumpolar Current Fronts Inferred from 1992–2011 Altimetry." *Journal of Physical Oceanography* 44: 3054–71.

King, J. C., and S. A. Harangozo. 1998. "Climate Change in the Western Antarctic Peninsula since 1945: Observations and Possible Causes." *Annals of Glaciology* 27 (27): 571–5.

Kinnard, C., et al. 2011. "Reconstructed Changes in Arctic Sea Ice over the Past 1,450 Years." *Nature* 479: 509–12.

Koenig, Z., et al. 2016. "Anatomy of the Antarctic Circumpolar Current Volume Transports through Drake Passage." *Journal of Geophysical Research: Oceans* 121: 2572–95.

Kort, V. G. 1964. "Antarctic Oceanography." In *Research in Geophysics*. Vol. 2, edited by H. Odishaw, 309–33. Cambridge, MA: MIT Press.

Krumpen, T., et al. 2016. "Recent Summer Sea Ice Thickness Surveys in Fram Strait and Associated Ice Volume Fluxes." *Cryosphere* 10: 523–34.

Kurtz, N. T., and T. Markus. 2012. "Satellite Observations of Antarctic Sea Ice Thickness and Volume." *Journal of Geophysical Research* 117: C08025.

Kwok, R., G. F. Cunningham, and S. S. Pang. 2004. "Fram Strait Sea Ice Outflow." *Journal of Geophysical Research: Oceans* 109: C01009.

Kwok, R., et al. 2009. "Thinning and Volume Loss of the Arctic Ocean Sea Ice Cover: 2003–2008." *Journal of Geophysical Research* 114: C07005.

Kwok, R., and D. A. Rothrock. 2009. "Decline in Arctic Sea Ice Thickness from Submarine and ICESat Records: 1958–2008." *Geophysical Research Letters* 36: L15501.

Ladd, C., et al. 2016. "Winter Water Properties and the Chukchi Polynya." *Journal of Geophysical Research: Oceans*. doi: 10.1002/2016JC011918.

Landy, J. C., et al. 2017. "Sea Ice Thickness in the Eastern Canadian Arctic: Hudson Bay Complex and Baffin Bay." *Remote Sensing of the Environment* 200: 281–94.

Launiainen, J., and T. Vihma. 1994. "On the Surface Heat Fluxes in the Weddell Sea." In *The Polar Oceans and Their Role in Shaping the Global Environment*, edited by O. M. Johannessen, R. D. Muench, and J. E. Overland, 398–419. Geophysics Monograph 85. Washington, DC: American Geophysical Union.

Levitus, S., et al. 2000. "Warming of the World Ocean." *Science* 287: 2225–9.

Lewis, M. J., et al. 2011. "Sea Ice and Snow Cover Characteristics during the Winter–Spring Transition in the Bellingshausen Sea: An Overview of SIMBA 2007." *Deep-Sea Research* 58(9–10): 1019–38.

Lindsay, R. W., and A. P. Makshtas. 2003. "Air–Sea Interactions in the Presence of the Arctic Pack Ice." In *Arctic Environment Variability in the Context of Global Change*, edited by L. P. Bobylev, K. Ya. Kondratyev, and O. M. Johannesen, 416–17. Chichester, UK: Praxis Publishing.

Loeng, H. 1991. "Features of the Physical Oceanographic Conditions of the Barents Sea." *Polar Research* 10: 5–18.

Loeng, H., V., Ozhigin, and B. Adlandsvick. 1997. "Water Fluxes through the Barents Sea." *Journal of Marine Science* 54: 310–17.

Lynch, A. H., et al. 2016. "Linkages between Arctic Summer Circulation Regimes and Regional Sea Ice Anomalies." *Journal of Geophysical Research: Atmospheres* 121: 7868–80.

Macdonald, R. W., et al. 1989. "Composition and Modification of Water Masses in the Mackenzie Shelf Estuary." *Journal of Geophysical Research* 94(C12): 18057–70.

Mahoney, A., H. Eicken, and S. Hendricks. 2015. "Tracking a Newly Predominant Ice Type: SIZONet Observations of First-Year Ice Thickness North of Alaska." ARCUS. arctic-observing-open-science-meeting/18-november-2015.

Mahoney, A., et al. 2007. "Alaska Landfast Sea Ice: Links with Bathymetry and Atmospheric Circulation." *Journal of Geophysical Research: Oceans* 112(C2): C02001.

Mahoney, A., et al. 2008. "Observed Sea Ice Extent in the Russian Arctic, 1933–2006." *Journal of Geophysical Research: Oceans* 113: C11005.

Massom, R. A., et al. 2001. "Snow on Antarctic Sea Ice." *Reviews of Geophysics* 39: 413–45.

Maykut, G. A. 1978. "Energy Exchange over Young Sea Ice in the Central Arctic." *Journal of Geophysical Research* 83: 3646–58.

Meehl, G. A., et al. 2016. "Antarctic Sea-Ice Expansion between 2000 and 2014 Driven by Tropical Pacific Decadal Climate Variability." *Nature Geoscience* 9: 590–5.

Meier, W. N., et al. 2014a. "Verification of a New NOAA/NSIDC Passive Microwave Sea-Ice Concentration Climate Record." *Polar Research* 33: 21004.

Meier, W. N., et al. 2014b. "Arctic Sea Ice in Transformation: A Review of Recent Observed Changes and Impacts on Biology and Human Activity." *Reviews of Geophysics* 52: 185–217.

Merkouriadi, I., et al. 2017. "Winter Snow Conditions on Arctic Sea Ice North of Svalbard during the Norwegian Young Sea ICE (N-ICE2015) Expedition." *Journal of Geophysical Research: Atmospheres* 122. doi: 10.1002/2017JD026753.

Miles, M. W., and R. G. Barry. 1998. "A 5-Year Satellite Climatology of Winter Sea Ice Leads in the Western Arctic." *Journal of Geophysical Research* 103(10): 21723–34.

Muench, R. D., and A. L. Gordon. 1995. "Circulation and Transport of Water along the Western Weddell Sea Margin." *Journal of Geophysical Research* 100(C9): 18503–15.

Münchow, A., H. Melling, and K. K. Falkner. 2006. "An Observational Estimate of Volume and Freshwater Flux Leaving the Arctic Ocean through Nares Strait." *Journal of Physical Oceanography* 36: 2025–41.

Newton, J. L., and B. J. Sotirin. 1997. "Boundary Undercurrent and Water Mass Changes in the Lincoln Sea." *Journal of Geophysical Research: Oceans* 102: 3393–403.

Nghiem, S. V., et al. 2016. "Geophysical Constraints on the Antarctic Sea Ice Cover." *Remote Sensing of the Environment* 181: 281–92.

Olason, A. 2016. "A Dynamical Model of Kara Sea Land-Fast Ice." *Journal of Geophysical Research: Oceans* 232: 3141–58.

Orsi, A. G., G. C. Johnson, and A. L. Bullister. 1999. "Circulation, Mixing, and Production of Antarctic Bottom Water." *Progress in Oceanography* 109: 43–55.

Orsi, A. H., T. Whitworth, and W. D. Nowlin. 1995. "On the Meridional Extent and Fronts of the Antarctic Circumpolar Current." *Deep-Sea Research* 42: 641–73.

Oshima, K. I., S. Nihashi, and K. Iwamoto. 2016. "Global View of Sea-Ice Production in Polynyas and Its Linkage to Dense/Bottom Water Formation." *Geoscience Letters* 3: 13.

Ostapoff, F. 1965. "Antarctic Oceanography." In *Biogeography and Ecology in Antarctica*, edited by J. Mieghem et al., 97–126. Dordrecht: Springer Science.

Parkinson, C. L. 2014. "Global Sea Ice Coverage from Satellite Data: Annual Cycle and 35-yr Trends." *Journal of Climate* 27(24): 9377–82.

Perovich, D. K., et al. 1999. "Year on Ice Gives Climate Insights." *Eos* 80: 485–6.

Perovich, D. K., et al. 2016. "Sea Ice in Arctic Report Card 2016." www.arctic.noaa.gov/Report-Card

Petty, A. A., et al. 2016. "Characterizing Arctic Sea Ice Topography Using High-Resolution IceBridge Data." *Cryosphere* 10: 1161–79.

Pfirman, S. L., D. Bauch, and T. Gammelsrod. 1994. "The Northern Barents Sea: Water Mass Distribution and Modification." In *The Polar Oceans and Their Role in Shaping the Global Environment*, edited by O. M. Johannessen, R. D. Muench, and J. E. Overland, 77–94. Geophysics Monograph 85. Washington, DC: American Geophysical Union.

Pickart, R. S. 2004. "Shelfbreak Circulation in the Alaskan Beaufort Sea: Mean Structure and Variability." *Journal of Geophysical Research: Oceans* 109(C4): C04024.

Polyakov, I. V., et al. 2004. "Variability of the Intermediate Atlantic Water of the Arctic Ocean over the Last 100 Years." *Journal of Climate* 17: 4485–94.

Polyakov, I. V., et al. 2005. "One More Step towards a Warmer Arctic." *Geophysical Research Letters* 32: L17065.

Polyakov, I. V., et al. 2017. "Greater Role for Atlantic Inflows on Sea-Ice Loss in the Eurasian Basin of the Arctic Ocean." *Science* 356: 285–91.

Preusser, A., et al. 2016. "Circumpolar Polynya Regions and Ice Production in the Arctic: Results from MODIS Thermal Infrared Imagery for 2002/2003 to 2014/2015 with a Regional Focus on the Laptev Sea." *Cryosphere Discussions*. doi: 10.5194/tc-2016-133.

Proshutinsky, A., and M. A. Johnson. 1997. "Two Circulation Regimes of the Wind-Driven Arctic Ocean." *Journal of Geophysical Research* 102: 12493–514.

Ricker, R., et al. 2017. "A Weekly Arctic Sea-Ice Thickness Data Record from Merged CryoSat-2 and SMOS Satellite Data." *Cryosphere Discussions*. doi: 10.5194/tc-201.

Rintoul, S. R. 2000. "Southern Ocean Currents and Climate." *Papers and Proceedings of the Royal Society of Tasmania* 133: 41–50.

Roemmich, D., et al. 2015. "Unabated Planetary Warming and Its Ocean Structure since 2006." *Nature Climate Change* 5: 240–5.

Rogers, J. C., and M.-P. Hung. 2008. "The Odden Ice Feature of the Greenland Sea and Its Association with Atmospheric Pressure, Wind, and Surface Flux Variability from Reanalyses." *Geophysical Research Letters* 35: L08504.

Rothrock, D. A., Y. Yu, and G. A. Maykut. 1999. "Thinning of the Arctic Sea-Ice Cover." *Geophysical Research Letters* 26(23): 3469–72.

Rudels, B. 2015. "Arctic Ocean Circulation, Processes and Water Masses: A Description of Observations and Ideas with Focus on the Period Prior to the International Polar Year 2007–2009." *Progress in Oceanography* 132: 22–67.

Schauer, U., et al. 2002. "Atlantic Water Flow through the Barents and Kara Seas." *Deep-Sea Research* 49(12): 2281–98.

Schauer, U., et al. 2008. "Variation of Measured Heat flow through the Fram Strait between 1997–2006." In *Arctic–Subarctic Ocean Fluxes*, edited by R. R. Dickson, J. Meincke, and P. Rhines, 65–85. Dordrecht: Springer.

Seidov, D., et al. 2015. "Oceanography North of 60°N from World Ocean Database." *Progress in Oceanography* 132: 153–73.

Semiletov, I., et al. 2005. "The East Siberian Sea as a Transition Zone between Pacific-Derived Waters and Arctic Shelf Waters." *Geophysical Research Letters* 32(10): GLO 224920.

Serreze, M. C., and A. P. Barrett. 2011. "Characteristics of the Beaufort Sea High." *Journal of Climate* 24: 159–82.

Serreze, M. C., et al. 2003. "The Large-Scale Hydro-climatology of the Terrestrial Arctic Drainage System." *Journal of Geophysical Research* 108(D2): 8160.

Serreze, M. C., and J. Stroeve. 2015. "Arctic Sea Ice Trends, Variability and Implications for Seasonal Forecasting." *Philosophical Transactions of the Royal Society* A373: 2140159.

Smedsrud, L. H., et al. 2017. "Fram Strait Sea Ice Export Variability and September Arctic Sea Ice Extent over the Last 80 Years." *Cryosphere* 11: 65–79.

Smith, D. G., et al. 2017. "Atmospheric Response to Arctic and Antarctic Sea Ice: The Importance of Ocean–Atmosphere Coupling and the Background State." *Journal of Climate* 30: 4547–65.

Sokolow, S., and S. R. Rintoul. 2009a. "Circumpolar Structure and Distribution of the Antarctic Circumpolar Current Fronts: 1. Mean Circumpolar Paths." *Journal of Geophysical Research: Oceans* 114: C005208.

Sokolow, S. and S. R. Rintoul. 2009b. "Circumpolar Structure and Distribution of the Antarctic Circumpolar Current Fronts: 2. Variability and Relationship to Sea Surface Height." *Journal of Geophysical Research: Oceans* 114: C11019.

Spall, M. A. 2007. "Circulation and Water Mass Transformation in a Model of the Chukchi Sea." *Journal of Geophysical Research: Oceans* 112: C05025.

Speer, K., S. R. Rintoul, and B. Sloyan. 2000. "The Diabatic Deacon Cell." *Journal of Physical Oceanography* 30: 3212–22.

Spielhagen, R. F., et al. 2011. "Enhanced Modern Heat Transfer to the Arctic by Warm Atlantic Water." *Science* 331: 450–3.

Stabeno, P. J., J. D. Schumacher, and K. Otahni. 1999. "The Physical Oceanography of the Bering Sea." In *Dynamics of the Bering Sea*, edited by T. R. Loughlin, and K. Ohtani, 1–28. Fairbanks, AK: University of Alaska Sea Grant.

Steele, M., et al. 1996. "A Simple Model Study of the Arctic Ocean Freshwater Balance, 1979–1985." *Journal of Geophysical Research* 101: 20833–48.

Steffen, K. 1985. "Warm Water Cells in the North Water, Northern Baffin Bay during Winter." *Journal of Geophysical Research* 90(C5): 9129–36.

Steffen, K., and J. E. Lewis. 1988. "Surface Temperatures and Sea Ice Typing for Northern Baffin Bay." *International Journal of Remote Sensing* 9: 409–22.

Steffen, K., and A. Ohmura. 1985. "Heat Exchange and Surface Conditions in North Water, Northern Baffin Bay." *Annals of Glaciology* 6: 178–81.

Stern, H. L. 2016. "Polar Maps: Captain Cook and the Earliest Historical Charts of the Ice Edge in the Chukchi Sea." *Polar Geography* 39: 220–7.

Stroeve, J. C., et al. 2012. "The Arctic's Rapidly Shrinking Sea Ice Cover: A Research Synthesis." *Climatic Change* 110(3–4): 1005–27.

Stroeve, J. C., et al. 2016. "Mapping and Assessing Variability in the Antarctic Marginal Ice Zone, Pack Ice and Coastal Polynyas in Two Sea Ice Algorithms with Implications on Breeding Success of Snow Petrels." *Cryosphere* 10(4): 1823–43.

Stuecker, M. F., C. M. Bitz, and K. C. Armour. 2017. "Conditions Leading to the Unprecedented Low Antarctic Sea Ice Extent during the 2016 Austral Spring Season." *Geophysical Research Letters*. doi: 10.1002/2017GL074691.

Svendsen, P. L., O. B. Andersen, and A. A. Nielsen. 2016. "Stable Reconstruction of Arctic Sea Level for the 1950–2010 Period." *Journal of Geophysical Research: Oceans*. doi: 10.1002/2016JC011685.

Swift, K., et al. 2005. "Long-Term Variability of Arctic Ocean Waters: Evidence from Reanalysis of the EWG Data Set." *Journal of Geophysical Research* 110: C03012.

Talley, L., et al. 2011. *Descriptive Physical Oceanography*. 6th ed. New York: Elsevier.

Tamura, T., K. I. Ohshina, and S. Nihashimm. 2008. "Mapping of Sea Ice Production for Antarctic Coastal Polynyas." *Geophysical Research Letters* 35(7): L07606.

Tanis, F., and I. Timokhov, eds. 1997. *Joint U.S.–Russian Atlas for the Arctic Ocean: Oceanographic Atlas for the Winter Period.* [Digital media.] Boulder, CO: University of Colorado, National Snow and Ice Data Center.

Tanis, F., and I. Timokhov, eds. 1998. *Joint U.S.–Russian Atlas for the Arctic Ocean: Oceanographic Atlas for the Summer Period.* [Digital media.] Boulder, CO: University of Colorado, National Snow and Ice Data Center.

Taylor, P. C., et al. 2013. "Covariance between Arctic Sea Ice and Clouds within Atmospheric State Regimes at the Satellite Footprint Level." *Journal of Geophysical Research: Atmospheres* 120: 12656–78.

Thomas, E. R., and N. Abram. 2016. "Ice Core Reconstruction of Sea Ice Change in the Amundsen–Ross Seas since 1702 AD." *Geophysical Research Letters* 43(7). doi: 10.1002/2016GL06813.

Timco, G. W., and W. F. Weeks. 2010. "A Review of the Engineering Properties of Sea Ice." *Cold Regions Science and Technology* 60(2): 107–29.

Timokhov, L. A. 1994. "Regional Characteristics of the Laptev and the East Siberian Seas: Climate, Topography, Ice Phases, Thermohaline Regime, Circulation." *Reports on Polar Research* 144: 15–31.

Tomczak, M., and J. S. Godfrey. 2003. *Regional Oceanography: An Introduction.* 2nd ed. Delhi: Daya Publishing House.

Tsubouchi, T., et al. 2012. "The Arctic Ocean in Summer: A Quasi-Synoptic Inverse Estimate of Boundary Fluxes and Water Mass Transformation." *Journal of Geophysical Research* 117(C1): C01024.

Tsukernik, M., et al. 2009. "Atmospheric Forcing of Fram Strait Sea Ice Export: A Closer Look." *Climate Dynamics* 35: 1349–60.

Turner, J., et al. 2015. "Antarctic Sea Ice Increase Consistent with Intrinsic Variability of the Amundsen Sea Low." *Climate Dynamics* 46: 2391–402.

Turner, J., et al. 2017. "Unprecedented Springtime Retreat of Antarctic Sea Ice in 2016." *Geophysical Research Letters* 44: 6868–75.

Vinje, T. 2001. "Anomalies and Trends of Sea Ice Extent and Atmospheric Circulation in the Nordic Seas during the period 1864–1998." *Journal of Climate* 14: 255–67.

Volkov, V. A., et al. 2002. *Polar Seas Oceanography: An Integrated Case Study of the Kara Sea.* London: Springer.

Walsh, J. E., et al. 2017. "A Database for Depicting Arctic Sea Ice Variations Back to 1850." *Geography Review* 107: 89–107.

Wang, Q., et al. 2016. "Sea Ice Leads in the Arctic Ocean: Model Assessment, Interannual Variability and Trends." *Geophysical Research Letters* 43: 7019–27.

Warren, S. G., et al. 1999. "Snow Depth on the Arctic Sea Ice." *Journal of Climate* 12: 1814–29.

Wells, N. C., N. Couldrey, and V. O. Ivchenko. 2013. "Comparison of North Atlantic Heat Storage Estimated during the ARGOS Period (1999–2010)." *Ocean Science Discussions* 10: 2363–98.

Williams, W. J., and E. C. Carmack. 2015. "The 'Interior' Shelves of the Arctic Ocean: Physical Oceanographic Setting, Climatology and Effects of Sea-Ice Retreat on Cross-Shelf Exchange." *Progress in Oceanography* 139: 24–41.

Willmes, S., and G. Heinemann. 2016. "Sea-Ice Wintertime Lead Frequencies and Regional Characteristics in the Arctic, 2003–2015." *Remote Sensing* 8: 4–19.

Woodgate, R. A., K. Aagaard, and T. J. Weingartner. 2005a. "Monthly Temperature, Salinity, and Transport Variability of the Bering Strait through Flow." *Geophysical Research Letters* 32: L04601.

Woodgate, R. A., et al. 2005b. "A Year in the Physical Oceanography of the Chukchi Sea: Moored Measurements from Autumn 1990–1991." *Deep-Sea Research* 52: 3116–49.

Woodgate, R. A., E. Fahrbach, and G. Rohardt. 1999. "Structure and Transport of the East Greenland Current at 75°N from Moored Current Meters." *Journal of Geophysical Research* 104: 18059–72.

Woods, C., and R. Caballero. 2016. "The Role of Moist Intrusions in Winter Arctic Warming and Sea Ice Decline." *Journal of Climate* 29(12): 4473–85.

Worby, A. P., et al. 2008. "Thickness Distribution of Antarctic Sea Ice." *Journal of Geophysical Research* 113: C05S92.

World Meteorological Organization. 2017. *Sea-Ice Information Services in the World*. 3rd ed. WMO No. 574. Geneva: World Meteorological Organization.

Wyrtki, K. W. 1960. "The Antarctic Circumpolar Current and the Antarctic Polar Front." *Deutsche Hydrographische Zeitschrift* 13(4): 153–74.

The Third Pole

The extensive area of high elevations in Central Asia, notably the Tibetan Plateau, averaging about 4,500 m, and adjacent mountain ranges, has given rise to the concept of a Third Pole (Schild 2008) where environmental conditions – climate, biota, permafrost, and glaciers – resemble those in the Arctic. The Central Asian part includes the Hindu Kush, Pamir, Tien Shan, and Kun Lun ranges (Figure 8.1). The southern section includes the Karakorum–Himalayan mountain system and Tibet, spanning an area of more than 4.3 million km^2 in Afghanistan, Bangladesh, Bhutan, China, India, Myanmar, Nepal, and Pakistan. It has fourteen peaks rising above 8,000 m and the headwaters of ten major river systems (Indus, Ganges, Brahmaputra, Irrawaddy, Salween, Mekong, Yangtze, Yellow, Tarim, and Amu Darya). Based on these features, this area has been called "the roof of the world." It also has 36,800 glaciers and ice caps covering 49,870 km^2 (Yao et al. 2004) and 1,200 lakes (larger than 1 km^2) with an area of 47,000 km^2 (Chen et al. 2015).

The Himalaya are 2,500 km long, running from 34–36° N at 27° E in the northwest to 27–28° N at 90° E. They comprise the Greater Himalaya in the north, with peaks averaging 6,000 m and rising above 8,500 m; the Lesser Himalaya in the center; and the lower Siwalik in the south. The total width of the Himalaya is 250–400 km. The heavily glacierized Karakorum ranges lie to the west of the Himalaya; they are 500 km long and 200 km wide. Farther west are the southwest–northeast Kun Lun mountains.

The northern mountain ranges of Central Asia are the 2,000-km-long Tien Shan extending from 67° to 95° E between 40° and 45° N, the east–west Pamir–Alai of southern Uzbekistan–Kyrgyzstan, and the north–south Pamir of Tajikistan. The central Tien Shan rise to 7,400 m and have extensive ice cover. The western Tien Shan include the Alatau ranges that rise to approximately 5,000 m. The southwestern Pamir is heavily glacierized between 2,500 and 5,000–6,000 m (Figure 8.2), including the 75-km-long Fedtchenko glacier, while the northwest has alpine relief. The southeastern Pamir is a high upland with peaks reaching 5,000–5,500 m.

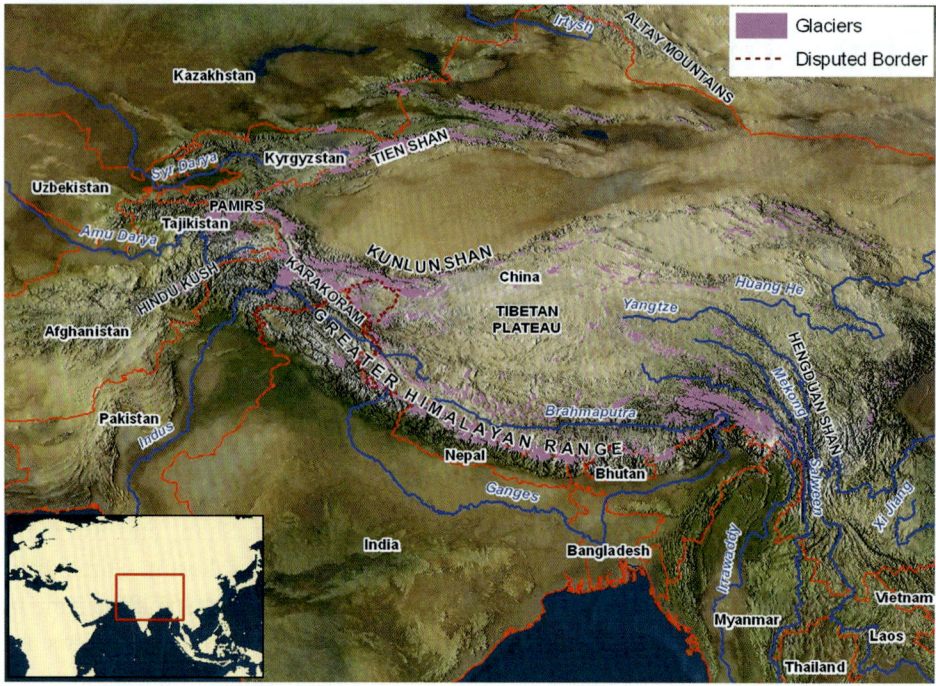

Figure 8.1 The mountain ranges, river systems, and glaciers of High Asia, the Central Asian region that constitutes the Third Pole.
Source: USAID. The CHARIS project at the National Snow and Ice Data Center, Boulder, CO. https://nsidc.org/charis/project-summary/.

Figure 8.2 The upper part of the Abramov Glacier in the Pamir.
Source: R. G. Barry.

The Third Pole differs from the polar regions proper in its lack of a polar night and day and in the fact that its climatic conditions are strongly determined by altitudinal controls, rather than latitude. Continentality is a further factor that locally gives rise to aridity. Its vegetation is primarily steppe grassland, rather than tundra; aridity plays a major role in this distribution. The Tibetan Plateau has many internal mountain ranges with snow and ice cover (Figure 8.3).

This chapter focuses first on the climatic characteristics of the region and then on its cryospheric components – snow cover, glaciers, and frozen ground. As with the two

main polar regions, the Third Pole is undergoing rapid environmental changes, the consequences of which are also described, as
they are of great significance for the peoples
of the area.

8.1 Climate

8.1.1 Circulation

The atmospheric circulation over the
Tibetan Plateau is strongly influenced by
the topography, which has a major dynamic
influence on it. In winter, the mid-latitude
westerlies are split by the Tibetan Plateau

Figure 8.3 A typical section of the Tibetan Plateau with
mountain ranges in the distance.
Source: R. G. Barry.

up to at least 600 hPa. The subtropical westerly jet stream at 300 hPa is anchored
south of the Himalaya over northern India from November through May, while to
the north of the plateau the polar front jet may vary widely in its latitudinal
location. As the spring season advances, the jet stream begins to shift northward.
In summer, heating of the plateau surface transfers sensible heat to the lower
atmosphere, which supports a shallow thermal low at 600 hPa in the afternoon
and early evening (Yanai and Li 1994). The ground–air temperature difference
over the western plateau reaches 10 °C around noon in June and July. Over the
eastern plateau in summer latent heat also contributes to atmospheric warming.
The thermal low leads to low-level inflow and active cumulus convection over
the central plateau that reaches a maximum around 1800 LST (Local Sidereal
Time). At upper levels (200 hPa), there is an anticyclonic circulation in summer
and the equatorward thermal gradient (a reversal of the poleward pattern of
winter) supports the 150-mb easterly tropical jet stream at 15° N over India.

In the November to April period, when the subtropical westerly jet stream is
south of the Himalaya, "western disturbances" from the Middle East travel
eastward, bringing snowfall and gales to the Hindu Kush and Karakorum. The
snow line is about 1,500 m in the west. This regime ends in May as the jet stream
shifts poleward. The south Asian monsoon begins to affect the Nepal Himalaya in
June, with the precipitation extending to the northwest as the summer advances.
Monsoon rains occur in spells referred to as "bursts" lasting about ten days with
intervening drier periods, called "breaks." The precipitation is mainly associated
with monsoon depressions from the Bay of Bengal that move westward, steered
by the upper easterlies. Seasonal amounts decrease westward, where the monsoon season is shorter, and also northward along the valleys. The annual total

precipitation in the central Himalaya shows maximum amounts at the foot of the Siwalik ranges and at 2,000–2,400 m in the Lesser Himalaya. In eastern Nepal–Sikkim, totals of up to 3,400 mm occur around 1,800 m elevation, decreasing to 500–300 mm at 4,000–5,000 m (Barry 2008, 368–78). However, in the Annapurna Range (approximately 28.4° N, 84.3° E), a local station network operated during 1999–2001 showed annual totals of 5,000 mm at around 3,000 m altitude on the southern slopes, dropping to 1,100 mm in the rain shadow on the north side of the main range (Putkonen 2004).

The Central Asian mountains are under the influence of the Siberian high-pressure cell during September to April. Its presence is a result of the isolation of the region and the low temperatures formed over the snow cover. Its central pressure is typically 1,030–1,050 hPa and it is located about 40° N, 95° E. Cold highs can move into the area southeastward from Scandinavia and the Kara Sea and southward from the Taimyr Peninsula.

Classification of cyclone and anticyclone trajectories over Central Asia for 1942–1951 was provided by Bugaev et al. (1962). Table 8.1a summarizes the monthly frequencies for cyclones, showing that most cyclones affect the region in the cold season. Table 8.1b gives the corresponding information for anticyclones. Four other types did not exceed three cases in any month over the ten years of the Bugaev et al. series. Type 1, entering from the northwest, is dominant in October–November; type 2, entering from the southwest, is of similar frequency to type 1 in January, February, and April.

In summer, northeastern Asia is affected by westerlies, but cyclones are infrequent. The average number of westerly invasions during 1934–1944 was two per month in December to February, but five per month in June, July, and August. The frequency of thermal depressions was five per month in July and August. The frequency of cold air invasions from the north averaged two per month for January to June, 1935–1944, but reached five per month in July and August in the same time span. Cold air invasions from the northwest were more frequent, averaging four or five per month from November to March, nine per month in June, and more than six in the other months.

The precipitation regime in the Alai range at the Abramov glacier (3,780 m) shows a spring maximum and a nearly dry July–September, with a mean annual total (1968–1988) of 727 mm. Mean monthly temperatures range from −14 to 6 °C. In the central Tien Shan, winters are dry and there is a clear summer maximum. In the Ak–Shirak mountains, the annual precipitation is only about 300 mm, with 60 percent falling in June–September. The western Pamir has about 229 days with precipitation annually, compared with only 165 days in the eastern Pamir. The western Pamir is influenced by spring cyclones that move eastward from Iran. For the Pamir and Tien Shan, Getker (1985) determined

Table 8.1 Frequency of (a) cyclone and (b) anticyclone trajectories over Central Asia, 1942–1951 (Bugaev et al. 1962, table 145 and table 151)

(a)

Type	J	F	M	A	M	J	J	A	S	O	N	D	Year
1	6	6	1	4	4	–	1	1	–	2	1	2	28
2	19	17	16	17	9	10	3	3	1	13	11	17	136
3	12	8	20	17	4	2	–	–	1	3	2	71	
4	–	2	2	2	2	–	–	1	–	–	3	2	21
5	1	–	3	3	8	–	4	1	3	1	–	2	26

Type 1: Move eastward along about 45° N from 40° to 75° E, sometimes becoming semi-stationary.
Type 2: SW–NE track from 32° N, 35° E to 48° N, 80° E. Only one in five reaches Western Siberia; the rest occur over north Central Asia or Kazakhstan.
Type 3: Moves eastward near 45° N from 35° to 80° E. The tracks bend south around the Pamir–Alai and Tien Shan.
Type 4: Moves parallel to type 2 but a few degrees farther north.
Type 5: Northernmost track from approximately 50° N, 45° E to 48° N, 70° E with anticyclonic curvature.

(b)

Type	J	F	M	A	M	J	J	A	S	O	N	D	Year
1	9	15	14	14	17	13	9	3	14	30	24	19	181
2	11	13	3	11	4	4	–	2	–	2	2	6	58

Type 1: Enters from the northwest, moves latitudinally, and moves out to the northeast.
Type 2: Moves over Central Asia from southwest to northeast.

generalized amounts of solid precipitation (50 percent probability); amounts were 440 and 300 mm, respectively. Corresponding mean maximum snow water equivalent (SWE) values were 350 and 180 mm.

Getker (1985) also analyzed precipitation and SWE gradients in the Pamir–Alai and western and central Tien Shan using a glaciological approach (see Barry 2008, 383–5). Annual and solid precipitation amounts are about 50 percent greater in the Pamir than in the Tien Shan. On macroslopes in peripheral basins with a spring–summer maximum, annual precipitation increases from 500–600 mm at low elevations to 1,500–1,600 mm in higher zones. In interior basins, the increase is from 350–500 mm to 800–1,000 m. In peripheral basins with a winter–spring maximum, the increase is from 100–300 to 500–1,000 mm. In interior basins with a winter–spring maximum, the increase is from 350–400 to 800–1,200 mm. Maximum SWE on the Pamir firn plateau is double that at the equilibrium-line altitude (ELA) and more than ten times that in the lowlands

(Getker 1985). Annual precipitation increases from 300 mm at 3,000 m elevation on the Khan Tengry massif in the central Tien Shan to 900 mm at 5,000 m and above (Aizen et al. 1997).

8.1.2 Energy Budget and Temperature

Energy budgets were measured at fourteen stations on the Tibetan Plateau in 1997 (Xu and Haginoya 2001). Net radiation (Rn) exceeded 100 W m^{-2} at 4,500–4,700 m elevation in summer, due to low levels of water vapor content and high levels of solar radiation. Sensible heat (H) reached a maximum in May–June with values of 54–76 W m^{-2}, while latent heat (LE) peaked in July–August with 41–83 W m^{-2}. At Lhasa (3,650 m), Rn in July was 170 W m^{-2}, H was 39. and LE was 71 W m^{-2}. In winter, Rn at Lhasa was −8 W m^{-2}. In winter, net radiation was generally about 25 W m^{-2} (Xu et al. 2010).

The climatology of temperature over the plateau has been analyzed by Frauenfeld et al. (2005) using data from 161 stations and the ERA-40 (European Centre for Medium-Range Weather Forecasts reanalysis) data for the period September 1957 to December 2000. The temperature fields in these data are considered representative of the mean elevation of 2.5 × 2.5-degree grid cells. Annual mean temperatures across virtually all of the plateau are less than 0 °C as a result of its high elevation. The western part, with elevations generally above 5,000 m, is characterized by means around −10 °C. Temperatures on the eastern part (around 3,000–4,000 m elevation) are higher, with those at lower elevations surrounding the plateau being much higher. Means of approximately 15 °C are found north and east of the plateau and values greater than 25 °C are found to the south.

The spring and autumn climatologies closely resemble the annual mean pattern. In contrast, during winter, the plateau is characterized by temperatures around −25 °C in the higher west and −15 °C in the lower east (Figure 8.4). In the summer season, temperatures over the entire plateau rise above 0 °C. The east warms up to 5–10 °C, and high elevations of the west warm up to 0–5 °C.

8.1.3 Precipitation

Precipitation over the Tibetan Plateau decreases steadily from southeast (500–600 mm) to northwest (50–100 mm) (Xu and Haginoya 2001), as shown in Figure 8.5. However, in 1998 in the southeast, totals were 700–1,000 mm at stations located between 28.5° and 31.4° N, 94.3° and 100° E, at around 2,600–3,300 m elevation (Xu et al. 2005).

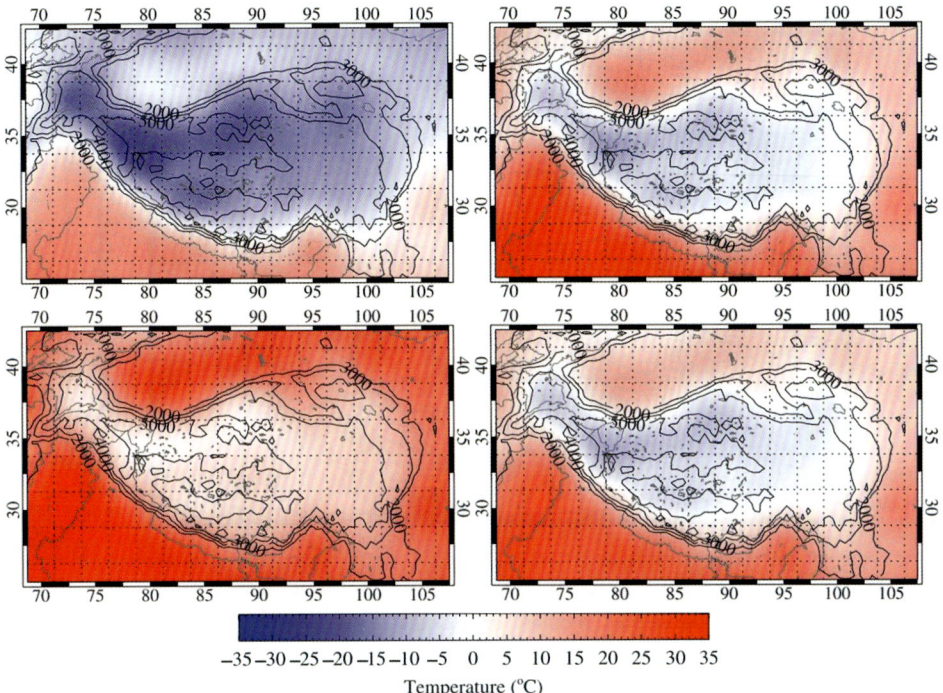

Figure 8.4 Seasonal temperature climatology of the Tibetan Plateau as depicted by the 2,000–5,000 m digital elevation model (DEM) height contours: (top left) winter, (top right) spring, (bottom left) summer, and (bottom right) autumn.
Source: Frauenfeld et al. 2005, figure 7. Courtesy of American Geophysical Union.

Precipitation over the Tibetan Plateau is mostly convectional from large cumulonimbus cells, and hail is not uncommon. Geostationary satellite data for 1989–1994 show two periods of maximum diurnal cloud activity (Fujinami and Yasunari 2001). The first, in March–April, affects most of the plateau, especially the southern part (30° N, 90° E) and a region from 35° N, 80° E to 31° N, 102° E. This activity is related to cold air in upper westerly troughs and daytime surface heating. In June, warm, humid air from the South Asian monsoon initiates diurnal cloud activity in the southeast; this gradually moves toward 30° N, 86° E, where activity is concentrated during the monsoon season. Mountain ranges within the plateau experience orographic uplift influences.

Analysis of the spatial distribution of ∂O^{18} and ∂D (deuterium) stable isotopes along a southwest–northeast transect across the plateau in July 1986 showed that whereas south of the Himalaya, moisture comes from the Indian monsoon circulation, between the Himalaya and Tanggula Mountains its source is the Bay of Bengal and northern India (Tian et al. 2001). This moisture is transported

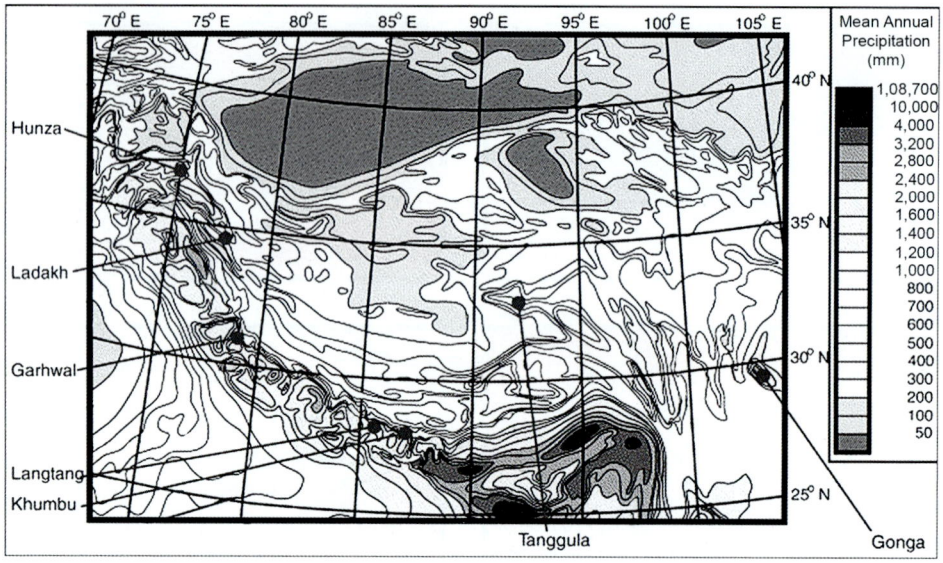

Figure 8.5 Mean annual precipitation (mm) over the Tibetan Plateau.
Source: Barry 2008, 414, figure 5.24; modified after Owen et al. 2006.

by southerly flow, reaching the plateau through valleys like the Brahmaputra in the eastern Himalaya. In the western Himalaya and southwest Tibetan Plateau, the moisture source is the Indian Ocean and Indian peninsula. Monsoon air is unable to cross the Tanggula Mountains. North of the Tanggula Mountains, more than half of the moisture is from recycling over inland Asia and the remainder from local convection. Also, Huang and Shen (1986), based on an analysis of atmospheric water vapor flux and flux divergence, showed that over Qinghai (northeastern Tibetan Plateau), moisture is transported from north to south and from west to east in May–June. In late July–August, it comes from the south.

   Li (2017) analyzed the daily regimes in summer over the Tibetan Plateau using hourly data for 2007–2013 for 100 stations. He showed that over the southeast, there are large amounts, high frequency, and strong intensity of rainfall, whereas frequency and intensity are low in the west and north. Over the northeastern plateau and the Yurlung Zangbo River valley, there is low frequency but high intensity of rainfall, along with a large proportion of strong events. The southern slopes of the Himalaya have high-frequency, low-intensity rainfall. Precipitation over the middle reaches of the Yurlung Zangbo River valley, however, is high intensity and is closely related to mesoscale convective systems. Analysis of the diurnal phase and duration of precipitation identified "late afternoon/short duration" and "nighttime/long duration" modes. There is a correlation of

0.5 between water vapor content and precipitation frequency, and the mean duration is correlated with water vapor at 0.65.

Tibet has seasonal snow cover, but amounts
are generally small. For example, Naqu (31.4° N,
91.9° E, 4,500 m altitude) had a maximum SWE
of 20 mm during 1994–2000 (Sato 2001) and
seasonal freezing of the ground penetrated to
1.6–2 m. As a consequence of the greater aridity,
the snow line is 400–500 m higher on the north
side of the Nangpa Pass (28.1° N, 85.6° E, 5,500

Table 8.2 Estimates of seasonal latent heat flux (W m^{-2}) averaged over the Tibetan Plateau (from J. Peng et al. 2016a)

Season	HOLARS	LandFlux EVAL
Winter	9.3	8.1
Spring	21.4	21.7
Summer	47.3	45.7
Autumn	25.2	21.8

m) than on the south side. Li (1983; see Shi et al. 2008b, 24–260) provided data
on the number of snow days for 1951–1980. Over most of the plateau there are
about 100 snow days annually. Based on Scanning Multichannel Microwave
Radiometer (SMMR) data (1978–1987), only 59 percent of the plateau is snow
covered in winter. Further information is given later in this chapter.

J. Peng et al. (2016a) performed a detailed cross-comparison of existing latent
heat flux products over the Tibetan Plateau. Table 8.2 compares the seasonal
averages for two products. Results show that the spatial pattern from the High
Resolution Land Atmosphere Surface Parameters from Space (HOLAPS) evapotranspiration (ET) demonstrator data set is very similar to the LandFlux-EVAL
data set (a benchmark ET product from the Global Energy and Water Cycle
Experiment). Similar to precipitation, there is decreasing ET from the southeast
to northwest over the plateau, particularly in spring and summer. For sensible
heat, the opposite pattern is observed.

8.1.4 Climate Change

There are ninety-seven stations located above 2,000 m above sea level (a.s.l.) on
the Tibetan Plateau. The longest records at five stations began before the 1930s,
but most records date from the mid-1950s. Analyses by Liu and Chen (2000)
of the temperature series show that the main portion of the plateau has experienced statistically significant warming since the mid-1950s, especially in winter,
but the recent warming in the central and eastern plateau did not reach the
level of the 1940s warm period until the late 1990s. Compared with the northern
hemisphere average, the warming of the plateau occurred several decades
earlier. The linear rates of temperature increase during 1955–1996 were about
0.16 °C decade^{-1} for the annual mean and 0.32 °C decade^{-1} for the winter mean,
which exceeded those for the northern hemisphere. There was also a tendency
for the warming trend to increase with elevation on the plateau and its

surrounding areas. This suggests that the Tibetan Plateau is one of the most sensitive areas responding to global climate change. In an analysis of data from ninety stations, Wang et al. (2008) found warming over the whole Tibetan Plateau increased at a rate of 0.36 °C decade^{-1} during 1961–2007, which is double the previous estimate made by Liu and Chen (2000) based on data for 1955–1996.

Existing observations indicate that the temperature is rising at a higher rate in Nepal and Chinese regions of the Himalaya compared with rest of the Himalaya (Gautam et al. 2013). Diodato et al. (2011) have shown that in the last few decades the Himalayan and Tibetan Plateau region has warmed at a faster rate than it did in the last century. An increase of 0.5 °C in annual average maximum temperature occurred over 1971–2005 compared to 1901–1960. According to Dash et al. (2007), the western Indian Himalaya saw a 0.9 °C rise in annual average maximum temperature over 1901–2003. They observed that much of this trend was related to increases after 1972. Immerzeel (2008) reported a basin-wide warming trend similar to the global average (0.6 °C/100 years for the 1901–2002 gridded data set) for the Brahmaputra basin in the eastern Indian Himalaya and Tibetan Plateau.

According to analysis performed by Cai et al. (2017), the entire Tibetan Plateau had a warming trend of 0.5 °C decade^{-1} for 1961–2014 based on data for eighty-six stations – a finding confirmed by moderate-resolution imaging spectroradi-ometer (MODIS) daytime and nighttime land surface temperature trends at corresponding pixels, but with a lower warming trend seen for the entire Tibetan Plateau of 0.27 or 0.25 °C decade^{-1}, respectively. There was, however, a cooling trend for central regions of 0.48 or 0.18 °C decade^{-1} in terms of daytime or nighttime mean trends. Higher elevations showed a continuous warming trend, especially in winter. For the entire Tibetan Plateau, seasonal rates were winter, 0.59 °C decade^{-1}; summer, 0.53 °C decade^{-1}; autumn, 0.47 °C decade^{-1}; and spring, 0.41 °C decade^{-1}. Further work by Liu and Yan (2017) using 139 stations in and around the plateau for 1961–2012 confirmed the greater warming at higher elevations. Based on data from seventy-three stations around 2-km elevation, forty-six stations around 3 km, and sixteen stations around 4 km, they found that the warming rates of mean annual air temperature were 0.316, 0.331, and 0.359 °C decade^{-1}, respectively. These values were also about 0.1 °C decade^{-1} higher than the global values. Seasonally, above 2 km elevation, winter showed the strongest warming, with a rate of 0.468 °C decade^{-1}, while spring had the weakest warming. Minimum temperatures had the greatest warming rate of 0.426 °C decade^{-1}, about 1.6 times that of maximum temperatures. In winter, minimum temperatures in the 4.5–5 km zone increased by about 0.83 °C decade^{-1}.

Pepin et al. (2015) reviewed the mechanisms involved in elevation-dependent warming. They described snow–albedo feedbacks, cloud cover change, water vapor modulation of longwave heating, reduced windiness, and increased absorbing aerosols, but concluded there were insufficient observations in Tibet to make any determination of causes.

A declining trend of monsoon precipitation in the western Indian Himalaya and an increasing trend in the eastern Indian Himalaya have been observed, whereas increasing precipitation and stream flow are noted in many parts of the Tibetan Plateau (Gautam et al. 2013).

The Corrected (for gauge undercatch)–China Meteorological Administration (CMA) and Asian Precipitation – Highly Resolved Observational Data Integration Towards Evaluation of Water Resources (APHRODITE) estimates for the Tibetan Plateau generally show decreasing trends in summer precipitation and increasing trends in spring and winter precipitation during 1961–2007 at both the basin and the plateau scales (Tong et al. 2014). However, the Corrected-CMA estimates show larger values in trends and more cases with significance than the APHRO-DITE estimates, suggesting the effects of the undercatch corrections incorporated in the CMA data on the precipitation trends. There are extensive regions with an increasing tendency relative to precipitation throughout the central, southeastern, and northern plateau in the Corrected-CMA data. The increasing trends in the southeastern part of the plateau (20–50 mm decade^{-1}) are generally greater than those in the northern region (0–10 mm decade^{-1}). The largest increasing rates (20–100 mm decade^{-1}) are located around 95° E, 30° N. Areas with a decreasing precipitation tendency mostly include the eastern edge of plateau, the upstream section of the Yangtze and Yellow rivers, Qiandam Basin, and the upstream section of the Yarlung Zangbo. Using the 0.5 × 0.5 degree APHRODITE data and seven other gridded precipitation data sets (gauge and satellite), Song et al. (2016) found moderate increases in the inner and northeastern Tibetan Plateau and northwest Xinjiang, and obvious decreases in the southeastern plateau. However, in the Himalaya and Karakorum, there are large discrepancies among different data sets, with the Global Precipitation Climatology Project (GPCP) and APHRODITE precipitation data sets showing significant decreases in precipitation along the Himalaya and other data sets exhibiting strong spatial heterogeneity or slight variations.

In a review study, Yang (2017) reported that between 1984 and 2006 the monsoon-affected region of the southern and eastern Tibetan Plateau received less precipitation, more evaporation, less soil moisture, and less runoff, resulting in a general shrinkage of lakes in this region. However, W. Zhang et al. (2017) reported a 13 percent increase in precipitation in southeastern Tibet during May 1979–2014, attributed to earlier monsoon onset. In central, northern, and western regions, which are dominated by westerly air circulation, Yang (2017) found that there was more precipitation, evaporation, soil moisture, and runoff, which together with more glacier melt resulted in the general expansion of lakes. Increased atmospheric precipitable water content is seen as a cause of the precipitation increase; estimates of this water content increase range from 0.12 to 0.21 mm decade^{-1}. As a result of warming and increasing glacier melt, the lake area on the Tibetan Plateau has increased over the last thirty-two years. In a global mapping study using 30-m resolution Landsat imagery since 1984, Pekel et al. (2016) determined that the water area on the plateau had increased by 390 km^2 yr^{-1}. resulting in a gain of 8,300 km^2 (20 percent).

In a review of studies for Central Asia (central and inner Tian Shan and Pamir), Unger-Shayesteh et al. (2013) reported that most studies agree general warming trends have accelerated since the 1970s, but vary with regard to seasonal changes and the magnitude of the warming. The reported mean annual temperature trend ranges between -0.1 and $+0.6\,°C$ decade^{-1}. There is evidence that the warming has been stronger in the cold season and is associated with an increase of minimum (monthly) temperatures. With regard to precipitation changes, no consistent picture about precipitation increase or decrease could be drawn from the reviewed studies.

A study by Chen et al. (2011) analyzed the temporal precipitation variations in arid Central Asia (ACA) during 1930–2009 using monthly gridded precipitation from the Climatic Research Unit. The results show that the annual precipitation in this westerly circulation–dominated arid region has generally increased during the past eighty years, with an increasing trend (0.7 mm decade^{-1}) in winter. The precipitation variations also differ regionally, such that ACA can be divided into five distinct subregions: West Kazakhstan (I), East Kazakhstan (II), Central Asia Plains (III), Kyrgyzstan (IV), and Iran Plateau (V). The annual precipitation falls fairly evenly in all seasons in the two northern subregions (regions I and II, approximately north of 45° N), whereas precipitation falls mainly in winter and spring (accounting for up to 80 percent of the annual total) in the three southern subregions. The annual precipitation has increased in all subregions, except the southwestern ACA (subregion V), during the past eighty years. A significant increase in precipitation has occurred in subregions I and III. The long-term trends in annual precipitation in all subregions are determined mainly by trends in winter precipitation.

Hu et al. (2017) examined the variations of annual precipitation over Central Asia (from Uzbekistan to Xinjiang) during the periods 1901–2013, 1951–2013, and 1979–2013 using the latest version of Global Precipitation Climatology Centre (GPCC) full data reanalysis version 7. They found a sharp decline during 1901–1944, followed by an increase until the 1980s, and fluctuations thereafter.

During 1979–2013, the mountainous area showed a greater increasing trend than the entire region. There was an increasing trend in the annual precipitation in Xinjiang and decreasing trends over the five countries of Central Asia during 1951–2013.

A monthly record of gridded temperature and precipitation data has recently been compiled for Central Asia (35.28–50.25° N, 50.40–91.98°E) by Zhou et al. (2017b). Their analysis, which focused on 1951–2010, identified 113 out of 369 stations as having statistically significant increases of mean annual air temperature in the 1970s, centered on the years 1976–1977, but found no significant changes in precipitation.

The Tibetan Plateau is the source of the Yangtze and Yellow rivers and is known as the "water tower" of Asia. Runoff of the upper streams of seven rivers has been analyzed from 1960 to the early 1990s: Yellow River, Dadu River, Yalong River, Jinsha River (upper stream of the Yangtze River), Lancang (Mekong) River, Nujiang (Salween) River, and Yarlung Zangbo (Brahmaputra) River (Shi et al. 2008b, 292–7). A synthetic runoff estimate for the seven rivers showed that between the high-water years of 1962–1966 and the low-water years of 1969–1973, total annual runoff was reduced by 22 percent. After the 1970s, runoff increased slowly. Runoff from the Yellow River differed from runoff from the other rivers in that it increased noticeably from the 1970s to 1980s. The upper catchments of the Yellow River, with an area of 121,972 km^2 and an altitudinal range from 2,550 to 6,280 m, have mean annual precipitation ranging from 300 to 700 mm, 73 percent of which falls during June–September. Average runoff is 174 mm (Shi et al. 2008b, 296).

Wang and Yang (2015) analyzed runoff changes over the Tibetan Plateau during 1960–2000. They found an increasing trend of annual runoff in the upstream sections of the Nujiang River, Lancang River, and Qilian Mountains, dominated by the increase of base flow; and a decreasing trend of annual runoff in the upper reaches of the Yarlung Zangbo River, Yellow River, and Yangtze River, dominated by the reduction of overland flow. Change in the amount of runoff was mainly due to change in precipitation.

Rising temperature accelerates glacier melting and increases the summer overland flow. In addition, rising temperature may reduce overland flow and increase the base flow due to changes in the active layer, which leads to the increase of soil water storage capacity, and subsequently changes the pattern of runoff generation. Qiu (2012) reported that G. Wang in Chengdu studied water entering the upper Yangtze River and found that the amount has fallen by 15 percent over the past four decades, despite a 15 percent increase in glacial melt and increased rainfall over the same period. The cause appears to be a deeper active layer (more than 60 cm) that allows the soil to hold more water. Wang also reported that the area of alpine wetland and high-vegetation-cover alpine meadow had decreased by 37 and 16 percent, respectively, which would reduce runoff.

In the wet southeastern part of Tibet, the upper catchment of the Yarlang Tsangpo (Brahmaputra) River had annual mean precipitation of 663 mm and runoff of approximately 435 mm for 1962–2002, according to Li et al. (2013). The basin is 0.2 million km^2 in area, with a mean altitude of 4,500 m. However, the alpine zone has less than 300 mm precipitation while the downstream subtropical forest receives more than 2,000 mm.

8.2 Cryosphere

The cryosphere in Central Asia comprises glaciers, snow cover, and perennially or seasonally frozen ground. These are discussed in turn in this section, together with their temporal changes.

8.2.1 Glaciers

The identification and inventory of glaciers in western China are detailed in Shi (2008a) in the first Chinese Glacier Inventory (CGI-1) for each of fourteen mountain ranges. The glacier data were based on digitized topographic maps produced from aerial photographs between the 1960s and 1980s. Table 8.3 summarizes the salient points.

The most extensive mountain range systems are the Kunlun and Hengduan, while the largest glacier areas are in the Himalaya and Karakorum. The Karakorum have the most glacier cover, with 62 percent of the mountains being covered. The Siachen glacier on the north slope of the Karakorum is the largest nonpolar glacier, with a length of 72 km. The mean glacier coverage for all mountain areas in China is just 2.5 percent.

Guo et al. (2015) reported on the second inventory of glaciers in China (CGI-2) based on satellite images. They listed the data by river basin rather than mountain ranges, but provided comparable data to that offered by Shi et al. (2008a). The glacier area for 42,370 glaciers in western China is 43,087 km^2 in GCI-2, compared with 52,043 km^2 in CGI-1. Direct comparison of the two inventories is not possible, for several reasons. Notably, the cloud-covered monsoon region of southeast Tibet (with a glacier area of 8,753 km^2) was not included in CGI-2. Including those glaciers in a subsequent analysis gives a revised CGI-2 total of 48,000 km^2 (W.-O. Guo, personal communication, 2017). The glacier area for the Tibetan Plateau is given as 7,269 km^2 in GCI-2, compared with 7,836 km^2 in GCI-1. The largest difference is for the Ganges drainage basin: 18,100 km^2 in

Table 8.3 Glaciers in western China by mountain range (from Shi 2008a, table 3.1) and the Randolph Glacier Inventory. Note that several mountain ranges overlap into other national territories that are listed in the far right column

Mountain system	Area (km²)	Number of glaciers	Glacier area (km²)	Glacier volume (km³)	Total glacier area (km²)
Altai	26,800	403	280	16	1,163@
Sawir	4,400	21	13	1	
Tianshan	21,900	9,035	9,225	1,011	West 9,531@ East 2,854@
Pamir	23,800	1,289	2,696	249	10,233@
Karakorum	26,600	1,363	6,263	692	22,862@
Kunlun	478,106	7,697	12,267	1,283	West 8,153@ East 3,251@
Altun	56,300	235	275	16	
Qilian	132,500	5,827	1,931	93	1,637@
Qiangtan	6,822	958	1,802	162	2,581*
Tanggula	141,300	1,530	2,213	184	2,213*
Gangdise	158,300	3,554	1,760	81	1,760*
Nyainquentangha	110,600	7,080	10,700	1,003	9,120*
Inner Tibet					7,923@
South and East Tibet					3,873@
Hengduan	356,300	1,725	1,579	97	4,382@
Himalaya	202,500	6,472	8,418	712	West 7,868@ Central 5,447@ East 4,904@
Hindukush					2,933@
Total	2,373,300	46,377	59,425	5,600	

* Yao et al. (2012).
@ Randolph Glacier Inventory version 6

CGI-1 versus 7,880 km² in CGI-2. It is probable that CGI-1 misidentified permanent snow banks as glaciers.

Another glacier inventory for Central Asia was recently published by Nuimura et al. (2015). The Glacier Area Mapping for Discharge from the Asian Mountains (GAMDAM) inventory is based on Landsat data for 1999–2003. Glacier areas by river basin are shown in Table 8.4. GAMDAM emphasizes the predominance of the southern watersheds and Qinghai-Tibet.

Ye et al. (2017a, 2017b) worked out the glacier area on the Tibetan Plateau (i.e., TPG1976, TPG2001, TPG2013) based on three mosaics by Landsat and

Table 8.4 Glacier areas by river basin in Central Asia (Nuimura et al. 2015)

Watershed	Glacier area (km^2)
Amu Darya	3,154
Indus	26,018
Ganges	10,621
Brahmaputra	17,419
Irrawaddy	73
Salween	2,198
Mekong	586
Yangtze	2,441
Yellow	189
Tarim Interior	2,768
Qinghai–Tibetan Interior	10,000
Total	75,466

HJ-1A/HJ-1B satellite images from the mid-1970s. Only debris-free ice was delineated and analyzed in this study. In the 1970s, glaciers covered 1.7 percent of the Tibetan Plateau study area as a whole. Glacier area was 44,366 ± 2,827 km^2 in the 1970s, 42,210 ± 1,621 km^2 in 2001, and 41,137 ± 1,616 km^2 in 2013. Compared with previous inventories of High Mountain Asia, area differences ranged from −19.6 percent (TPG1976 minus CGI-1) to −3.6 percent (TPG2013 minus CGI-2) to −1.1 percent (TPG2001 minus CGI-2), while the area difference for TPG2001 minus the GAMDAM Glacier Inventory (GGI; Nuimura et al. 2015) was +10.4 percent (Ye et al. 2017a). GAMDAM has less glacierized area than TGP2001 over the entire Tibetan Plateau. Nuimura et al. (2015) acknowledged that "our exclusion of steep headwalls that are unaffected by glacier mass balance potentially discounts glaciers located on steep ground, resulting in an underestimation of total ice volume and median elevations in the GGI." The GAMDAM inventory tends to have areas smaller than those measured in conformance with Global Land Ice Measurements from Space (GLIMS) guidelines.

Earl and Gardner (2016) used Landsat 5 and 7 to map glaciers in northern Asia, including the Altai Mountains. They reported a total of 2,446 glaciers (mean area = 0.475 km^2), with a total area of 1,163 km^2.

The entire Himalaya in India, Bhutan, and Nepal contain 43,208 glaciers, according to the GLIMS data, with a glacier area of about 47,310 km^2; the whole Karakorum Mountains have a glacier area of 16,000 km^2 (Dyurgerov and Meier 2005). The majority of the glacier area is distributed between 5,000 and 6,000 m a.s.l., with about half as much between 4,000 and 5,000 m a.s.l.

The snowline altitude rises from 2,800 m in the Altai (49° N) to 5,400 m on the north slope of the Kunlun Mountains (37° N) and 5,800 m over the Tibetan Plateau (34° N). On the north slope of Mt. Everest (Qomalungma), it reaches 6,000 m.

Ice core records from three locations in Tibet – the Guliya in the western Kun Lun (31.3° N, 81.5° E, 6,200 m), Puruogangri (33.9° N, 89.1° E, 5,860 m), and Dunde in the Qilian Shan (38.1° N, 96.4° E, 5,325 m) – have rather similar annual net mass balances, averaging 0.22 m water equivalent (w.e.) for Guliya in the far northwestern Tibetan Plateau, 0.35 m w.e. for Puruogangri in the central Tibetan Plateau, and 0.39 m w.e. for Dunde. The Dunde ice cap ranges from 94 to 167 m in thickness, and the area was found to have a 10-m temperature at the summit of −7.3 °C (Shi et al. 2008b, 387). The ELA is at around 4,900 m (Wu and Thompson 1988). The Guliya ice cap has an area of 376 km², an average thickness of 200 m, and a maximum of 350 m. Its 10-m temperature at the summit was −19.4 °C (Thompson et al. 1995). The net balance on Dasuopu glacier at 7,200 m on Mt. Xixibangma in the central Himalaya (28.3° N, 85.7° E) shows a different pattern, with an annual accumulation rate of 0.7 m w.e. The 10-m ice temperature at the summit was −14 °C (Shi et al. 2008b, 385) and the snowline was at 6,200 m.

Neckel et al. (2014) used ICESat lidar data to determine glacier mass loss during 2003–2009 over the Tibetan Plateau and surroundings. The most negative mass budgets of −0.77 m w.e. yr^{-1} were found for the Qilian Shan and eastern Kunlun, while a mass gain of +0.37 m w.e. yr^{-1} was found in the westerly-dominated north–central part of the Tibetan Plateau.

Rasmussen (2013) analyzed the sensitivity of Central Asian glaciers to temperature by using the US National Centers for Environmental Prediction (NCEP)/US National Center for Atmospheric Research (NCAR) reanalysis database for 1948–2010. Rasmussen's positive degree-day model of annual glacier-wide summer surface balance used upper-air temperatures at NCEP/NCAR grid points at altitudes representative of each of six glaciers between 40° and 45° N, 71° and 87° E in the Tien Shan and two others near 50° N, 87° E. Sensitivity to temperature change ranged between −0.5 and −0.2 m w.e. yr^{-1} $°C^{-1}$ for these glaciers. Precipitation, however, was not a good estimator of the interannual variation of winter surface balance.

8.2.2 Glacier Changes

There are relatively few records of glacier changes in Tibet and the Himalaya–Karakorum ranges. Moreover, a caveat applies in regard to such data: These records tend to be from small to medium-sized ice bodies at lower altitudes that are more easily accessible, but may not be representative of the regional pattern.

Glacial retreat on the Tibetan Plateau and surrounding regions has occurred since the 1960s, but has intensified in the past decade, according to Yao et al. (2007). The amount of glacial retreat is relatively small in the continental interior of the plateau and increases toward the margins, with the greatest retreat around the more maritime edges. The glacial retreat has caused an increase of more than 5.5 percent in river runoff from the plateau, with larger increases in the Tarim River basin. Glacial retreat has also caused rising lake levels in the areas with large coverage of glaciers, such as the Nam Co Lake and Selin Co Lake areas. Rising lake levels are devastating grasslands and villages near the lakes.

Ye et al. (2016, 2017a) analyzed glacier changes on the Tibetan Plateau using glacier coverage data (TPG1976, TPG2001, TPG2013) from time series of Landsat satellite images since the 1970s. Compared with previous inventories of High Mountain Asia (CGI-1 and CGI-2), there was a wide range of differences in glacier area, making for uncertainties in the results. In the 1970s, glaciers covered 1.7 percent of the study area as a whole. Glacier area on the Tibetan Plateau decreased from 44,366 km^2 in the 1970s to 42,210 km^2 in 2001 (4.9 percent), and decreased again to 41,137 km^2 in 2013 (2.6 percent). Glaciers in the external catchments tended to shrink more rapidly between the mid-1970s and approximately 2013, by 8.7 percent (0.24 percent yr^{-1}), than those in the interior basins (6.2 percent or 0.17 percent yr^{-1}). Glacier shrinkage rate was four times greater in the marginal (external) drainage basins of the southeast Tibetan Plateau (the Mekong catchment) than in the interior drainage basins of the northwest plateau (the Tarim catchment) from 1976 to 2013.

ICESat/GLAS data from 2003 to 2009 were used to calculate glacier surface elevation changes for each mountain and basin on Tibetan Plateau. The overall average rate of glacier surface elevation change was -0.24 m yr^{-1}. For external catchments, glacier mass change totaled -12.67 Gt yr^{-1} and that for internal basins totaled -2.59 Gt yr^{-1}. The glacier thinning rate was more marked in the marginal (external) drainage basins of the south and east Tibetan Plateau than in the interior drainage basins of the north and west. Values ranged from -0.85 m yr^{-1} in the central Himalaya to -0.61 m yr^{-1} in the Mt. Nyqingtangula area in central Tibet, to -0.38 m yr^{-1} in the Tanggula mountains in central Tibet, and to a near zero balance in the Pamir.

Ye et al. (2006a) quantified glacier variations in the Mt. Naimona balance in the western Himalaya by integrating glacier coverage data from Advanced Space-borne Thermal Emission and Reflection Radiometer (ASTER) and Landsat imagery at four different times: 1976, 1990, 1999, and 2003. She showed that glaciers in the region both retreated and advanced during this period of twenty-eight years; however, retreat dominated. The variations for glaciers in the western Himalayan region were dramatic compared with those in other regions in Central Asia. From 1976 to 2003, glacier area decreased from 84.41 to 77.29 km^2. By sequential images, it was shown that glacier areas shrank, on average, by 0.17, 0.19, and 0.77 km^2 yr^{-1} during the periods 1976–1990, 1990–1999, and 1999–2003, respectively. Thus, it appears that glacier retreat has accelerated.

Shangguan et al. (2004) analyzed changes in glaciers at the head of Yurungkax River (centered at 35.7° N, 81° E) in the heavily glaciated west Kunlun Mountains by using aerial photos (1970), as well as Landsat Thematic Mapper (TM) (1989) and Enhanced Thematic Mapper Plus (ETM+) (2001) images. A comparative analysis performed for glacier length/area variations since 1970 showed that the prevailing characteristic of glacier variation has been ice wastage, although changes in glacier area were very small in this region. Results indicate that a small enlargement of ice extent during 1970–1989 was followed by a reduction of more than 0.5 percent during 1989–2001. The enlargement of glaciers during 1970–1989 might have been caused by the decrease in air temperature and the increase in precipitation during the 1960s, while the glacier shrinkage during 1989–2001 might be a reaction to increasing air temperature (discussed earlier).

Ye et al. (2006b) described quantitative measurements of glacier variations in the Mt. Geladandong region of central Tibet using geographic information system (GIS) and remote sensing technologies. Data from Landsat images at three different times (1973–1976, 1992, and 2002) were compared with glacier areas digitized from a topographic map based on aerial photographs taken in 1969. The analysis showed that while some glaciers have advanced during the past thirty years, others have retreated. The area of retreat is much larger than the area of advance. The total glacier area decreased from 889 km^2 in 1969 to 847 km^2 in 2002, a reduction of almost 43 km^2 (4.8 percent). The variation of glacier area in the Mt. Geladandong region is not as large as that in other regions within the Tibetan Plateau. These glacier areas decreased 4.7 km^2 (0.68 km^2 yr^{-1}) during 1969–1976, 15.4 km^2 (0.96 km^2 yr^{-1}) during 1976–1992, and 22.4 km^2 (2.24 km^2 yr^{-1}) during 1992–2002, indicating recent accelerated retreat. The recession rates of glacier termini also increased. It is likely that the increase in summer air temperature is the major reason for glacier shrinkage in the Mt. Geladandong region of central Tibet.

Geodetic estimates of mass changes in the Karakorum revealed balanced budgets or possibly slight mass gains since about the year 2000. Indications for longer-term stability exist but no mass budget analyses are available before 2000. Bolch et al. (2016) have shown that glaciers in the Hunza River basin (Central Karakorum) have, on average, been in balance since the 1970s. Heterogeneous behavior and frequent surge activities were also characteristic for the period before 2000.

Bashir et al. (2017) reexamined the issue of the "Karakorum anomaly" of positive glacier mass balance using twelve stations located in 34–35° N, 72–76° E; three stations were at 600–800 m and the rest at 1,400–2,300 m. The instrumental records show that all locations exhibit similar trends in hydroclimatic change at the regional scale. Increases in water vapor, cloudiness, and precipitation and decreases in net radiation, near-surface wind speed, and potential evapotranspiration have resulted in a positive, annual hydrologic mass balance.

In summer, increased precipitation, decreased discharge, and diminished potential evapotranspiration result in a positive mass balance that is stored as glacier ice. In spring, the combination of decreased precipitation and increased runoff indicates there is a negative mass balance. Given that spring season flows in the Indus River are mainly due to snow melt and summer season contributions are mainly due to glacier melt, it can be assumed that the snowfields are more vulnerable to early melt in the spring, whereas the glaciers are more protected from melting in the summer, especially when monsoon moisture helps to nourish the glacier upper catchments.

Gautam et al. (2013) reported that Scherler et al. (2011) analyzed 286 glaciers from the Hindu Kush, Karakorum, western Indian Himalaya, Tibetan Plateau, west Kunlun Shan, and southern central Himalaya (Nepal, Bhutan, Sikkim, Uttarakhand, and Himanchal) through satellite images from 2000 to 2008. They found 58 percent of sampled glaciers in the westerlies-influenced Karakorum region were either stable or slowly advancing, while more than 65 percent of glaciers in the monsoon-influenced regions are retreating, associated with decreasing monsoon precipitation (discussed earlier in this chapter); several heavily debris-covered glaciers appeared to be stable. Spatially, they found a higher concentration of retreating glaciers (79 percent) in the western Indian Himalaya and in the northern central Himalaya and west Kunlun Shan (86 percent), where debris-free glaciers are dominant. By comparison, 65 and 73 percent of sampled glaciers were found to be retreating at relatively slower rates, respectively, in the Nepal–Bhutan Himalaya and Hindukush, where debris cover is common.

Racoviteanu et al. (2015) investigated spatial patterns in glacier characteristics and area changes since 1962 in the eastern Himalaya–Nepal (Arun and Tamor basins) and India (Teesta basin in Sikkim) based on various satellite imagery. They compared glacier surface area changes from 1962 to 2000–2006 and dependence of those changes on glacier topography and climate on the eastern side of the topographic barrier (Sikkim) versus the western side (Nepal). Glacier mapping from 2000 Landsat and ASTER yielded 1,463 km^2 total glacierized area, of which 569 km^2 was located in Sikkim and 488 km^2 in eastern Nepal. Supraglacial debris covered 11 percent of the total glacierized area, and supraglacial lakes covered about 5.8 percent of the debris-covered glacier area alone. Glacier area loss (from 1962 to 2000) was 0.50 percent yr^{-1}, with little difference in the losses seen in Nepal (0.53 percent yr^{-1}) versus Sikkim (0.44 percent yr^{-1}). Glacier area change was controlled mostly by glacier area, elevation, and altitudinal range, and, to a smaller extent, by slope and aspect. In the Kanchenjunga–Sikkim area, Racoviteanu et al. estimated a glacier area loss of 0.23 percent yr^{-1} from 1962 to 2006. Clean glaciers exhibited more area loss, on average, from 1962 to 2006 (34 percent) than debris-covered glaciers (22 percent). Glaciers in this region of the Himalaya are shrinking at similar rates to those reported for the last decades in other parts of the Himalaya.

Almost all glaciers in the eastern Himalaya are showing retreat. Landsat imagery of Bhutan shows a 23.3 percent area loss for glaciers between 1980 and 2010, with losses mostly observed below 5,600 m a.s.l., and a greater area loss occurring for clean glaciers (Bajracharya et al. 2014). Kääb et al. (2012) computed a 2003–2008 specific mass balance of −0.26 to −0.34 m w.e. yr^{-1} (for different density scenarios for snow and ice) for eastern Nepal and Bhutan using laser altimetry.

Using declassified Corona satellite imagery and ASTER data, Maurer et al. (2016) quantified surface lowering, ice volume change, and geodetic mass balance during 1974–2006 for glaciers in the eastern Himalaya, centered on the Bhutan–China border. The wide range of glacier types allowed for the first mass balance comparison between clean, debris, and lake-terminating (calving) glaciers in the region. Measured glaciers showed significant ice loss, with an estimated mean annual geodetic mass balance of −0.13 m w.e. yr^{-1} for ten clean-ice glaciers, −0.19 m w.e. yr^{-1} for five debris-covered glaciers, −0.28 m w.e. yr^{-1} for six calving glaciers, and −0.17 m w.e. yr^{-1} for all glaciers combined. While there was no information on debris cover depth, in this region it enhances melt, in contrast to the situation for the Kachenjunga–Sikkim area (discussed earlier). Contrasting hypsometries along with melt pond, ice cliff, and englacial conduit mechanisms resulted in statistically similar mass balance values for both clean-ice and debris-covered glacier groups. Calving glaciers accounted for 18 percent (66 km^2) of the glacierized area, yet contributed 30 percent (−0.7 km^3) to the total ice volume loss.

In the Rongbuk catchments on the northern slope of the Himalaya, elevation changes of glacier surfaces were investigated by comparing a digital elevation model (DEM) generated from the 2006 Advanced Land Observing Satellite (ALOS)/Panchromatic Remote-Sensing Instrument for Stereo Mapping (PRISM) stereo-pair images with the base DEM derived from the 1:50,000 topographic maps from 1974 (Ye et al. 2015). The average elevation change rate of glacier surfaces in Rongbuk catchment was estimated at 0.47 m yr^{-1} between 1974 and 2006. Such surface lowering rates varied significantly with glaciers and altitudes. Notably, the debris-covered ice thinned much more rapidly than the exposed ice at higher altitudes. Overall, glaciers in Rongbuk catchment lost mass of 0.06 Gt yr^{-1} during 1974–2006. Compared to hydrological measurements in 2010 at the Mt. Qomolangma station, glacier imbalance constituted more than 50 percent of the Rongbuk runoff (Ye et al. 2015).

Yao et al. (2012) reported on the glacier status on the Tibetan Plateau and surroundings, identifying the retreat of 82 glaciers, area reduction of 7,090 glaciers, and mass-balance change of 15 glaciers. The 7,090 glaciers had an area of approximately 13,363.5 km^2 in the 1970s and approximately 12,130.7 km^2 in the 2000s. The Himalaya show the most extreme glacial shrinkage. In the southeast, the glacial area was reduced at a rate of 0.57 percent yr^{-1} during the 1970s through 2000s. The smallest rate of glacier contraction was observed in the eastern Pamir region, which had an area-reduction rate of 0.07 percent yr^{-1}. The most negative mass balances occurred along the Himalaya, ranging from $-1,100$ to -760 mm yr^{-1}. The least negative mass balance was in the interior of the plateau. In the northeast, the mass balance became positive ($+250$ mm yr^{-1}). Mass-balance measurements of the Qiyi (39.3° N, 97.7° E), Xiaodongkemadi (33.0° N, 92.1° E), and Kangwure (28.5° N, 85.8° E) glaciers date from 1975, 1989, and 1992 respectively. The Qiyi and Xiaodongkemadi glaciers had a positive mass balance until the early 1990s, but then the mass balance turned increasingly negative. The Kangwure Glacier mass balance has been increasingly negative since 1992.

Kraaijenbrink et al. (2017) calculated the impact of a global 1.5 °C warming on High Asian glaciers (from the Tien Shan to the Himalaya). Their work indicates that this global warming will lead to a warming of 2.1 °C in the High Asian region, and that 64 percent of the ice mass stored in the glaciers at present will remain by the end of the century. By comparison, ensemble projections for three representative concentration paths – RCP 4.5, RCP 6.0, and RCP 8.5 – suggest that projected mass losses of 49, 51, and 64 percent, respectively, will occur by the end of the century.

Wu et al. (2017) studied the temperate glaciers of the Kangri Karpo Mountains of southeast Tibet, an especially humid area. They analyzed elevation changes using geodetic methods with DEMs derived from topographic maps (1980), the Shuttle Radar Topography Mission (SRTM) (2000), and TerraSAR-X/TanDEM-X (2014). Glacier area and length changes were derived from topographical maps and Landsat TM/ETM+/Operational Land Imager (OLI) images between 1980 and 2015. Wu et al. determined that the Kangri Karpo Mountains contain 1,166 glaciers, which had an area of 2,048 km^2 in 2015. Ice cover in the area decreased by 680 km^2 (24.9 percent) or 0.71 percent yr^{-1} from 1980 to 2015. However, nine glaciers advanced over this period. Glaciers with area of 788 km^2 in the region, as derived from DEM differencing, have experienced a mean mass deficit of 0.46 m w.e. yr^{-1} over 1980–2014.

Two ice cores have been retrieved from approximately 5,800 m a.s.l. at Mt. Nyainqêntanglha and Mt. Geladaindong in the southern and central Tibetan Plateau (Kang et al. 2015). The combined tracer analysis of tritium (^3H), ^{210}Pb (lead), and mercury, as well as other chemical records, provided evidence supporting the view that the sites had not seen net ice accumulation since at least the 1950s and 1980s, respectively. These results imply an annual ice loss rate of more than several hundred millimeters w.e. over the past thirty to sixty years. Both mass balance modeling at the sites and in situ data from the nearby glaciers confirmed a continuous mass loss in the region due to dramatic warming in recent decades. The findings suggest that the loss of accumulation area of glaciers is a possibility from the southern to central Tibetan Plateau at high elevations, probably up to about 5,800 m a.s.l.

For Central Asia, most studies have confirmed that glaciers in the Tien Shan and the Pamir continue to retreat and to shrink, though little is known about mass and volume changes. The detected changes in glacier area are regionally heterogeneous, with stronger glacier area losses at the northern and eastern margins of the Tian Shan mountain system (Unger-Shayesteh et al. 2013). In some ranges of the northern Tien Shan (the Zailiyskiy and Kungey Alatoo), glaciers experienced a large relative area reduction of about 30 percent from 1955 to 1999, according to Bolch (2007). Changes in the extent of 335 glaciers in the eastern Terskey–Alatoo, inner Tien Shan, between the end of the Little Ice Age (LIA; mid-nineteenth century), 1990, and 2003 have been estimated through the delineation of glacier outlines and the LIA moraine positions on the Landsat TM and ASTER imagery for 1990 and 2003, respectively (Kutuzov and Shahgedanova 2009). By 2003, the glacier surface area had decreased by 19 percent (76 km^2) of the LIA value. Mapping of 109 glaciers revealed that glacier surface area decreased by 12.6 percent of the 1965 value between 1965 and 2003. Detailed mapping of ten glaciers using historical maps and aerial photographs from the 1943–1977 period has enabled glacier extent variations over the twentieth century to be identified with a higher temporal resolution. Glacial retreat was slow in the early twentieth century, but increased considerably between 1943 and 1956, and then again after 1977. The post-1990 period has been marked by the most rapid glacier retreat since the end of the LIA.

Narama et al. (2010) used Corona data from 1970, Landsat data from 2000, and ALOS data from 2007 to study glaciers in four regions in the Tien Shan Mountains: Pskem, Ili–Kungöy, At-Bashy, and southeast Fergana. Pskem, to the northeast of Tashkent (the Ugam, Maidantal, and Pskem ranges, and the western part of the Talas range), has 525 glaciers; Ili–Kungöy (the Ili Ala-Too and Kungöy Ala-Too, north of Lake Ysyk-Köl) has 735 glaciers; the At-Bashy range (southeast Kyrgyzstan) has 192 glaciers: and the southeast Fergana range (Kyrgyzstan) has 306 glaciers. Between about 1970 and 2000, the glacier area decreased by 19 percent in the Pskem region, 12 percent in the Ili–Kungöy region, 12 percent in the At-Bashy region, and 9 percent in the southeast Fergana region. From 2000 to 2007, glacier area shrank by a further 5 percent in the Pskem region, 4 percent in the Ili–Kungöy region, and 4 percent in the At-Bashy region, but remained stable in the southeast Fergana region. Hence, the most dramatic glacier shrinkage has occurred in the outer ranges of the Tien Shan Mountains. Recent glacier area loss is a result of rising summer temperatures. There have been smaller area losses for orographically closed basins in the eastern Tien Shan.

In the entirety of the Altai mountains, there are 1,281 glaciers that cover 1,191 km^2 (Y. Zhang et al. 2017). Changes determined by a mass balance model, forced by a regional climate model, for 1990–2011 show a mean mass loss for the region of −0.69 mm yr^{-1}, with 81 percent of glaciers showing negative mass balance. Glaciers in the western part, where temperatures rose and precipitation decreased, have shown the largest losses.

For seventy glaciers in the middle Chinese Tien Shan, a moderate area loss of 13 percent was detected for the period from 1963 to 2000 (Li et al. 2006). The annual percentage of area change for glaciers in the Chinese Tianshan since 1960 has been 0.31 percent, according to Wang et al. (2011). Temperature and precipitation at fourteen meteorological stations in the region displayed a marked positive tendency from 1960 to 2009, increasing at rates of 0.34 °C decade^{-1} and 11 mm decade^{-1}, respectively. The temperature in the dry seasons (from November to March) increased rapidly, at a rate of 0.46 °C decade^{-1}, but the precipitation increased only 2.3 mm decade^{-1}. While the temperature in the wet seasons (from April to October) increased at a rate of 0.25 °C decade^{-1}, the precipitation increased at 8.7 mm decade^{-1}.

Unger-Shayesteh et al. (2013) attribute the bigger changes in the outer Central Asian mountain ranges to three factors: (1) the higher ice-mass turnover due to higher precipitation amounts and hence higher sensitivity of the glaciers to climate change; (2) lower average elevation of the glacier termini; and (3) with increasing continentality toward the inner and eastern parts of the Tien Shan, the predominance of summer-accumulation–type glaciers. Such glaciers exhibit a smaller mass turnover and reduced ablation due to fresh snow falling in the warm season and the associated increase in surface albedo.

The few glaciers with long-term mass balance data in Central Asia (Tuyuksu, Golubin, Karabatkak, Molodezhniy, and Urumqi No. 1) all showed increasingly negative balances from the mid-1970s to 2010. Aizen et al. (2007) found accelerated area reduction and thinning rates for the Akshiirak glaciers, with absolute annual area-reduction rates in 1977–2003 being more than double those in 1943–1977.

The contribution of annual glacier ice melt in the Himalaya to stream flow is a question that has recently been addressed. Glacier meltwater along two rivers in two watersheds, situated in the monsoon-influenced part of the Nepal Himalaya, was studied by Racoviteanu et al. (2013). Glacier ice melt was positively correlated with the basin glacierized area and contributed 58.3 percent to annual flow in the small Langtang Khola watershed in the Trishuli basin (43.5 percent glacierized area) and 21.2 percent to annual flow in the Hinku watershed in Dudh Kosi basin (34.7 percent glacierized area). Of this amount, 17.7 and 4.1 percent of stream flow, respectively, were due to the contribution of debris-covered glaciers in Langtang Khola and Hinku. The contribution of glacier ice melt to measured discharge decreased substantially toward lowland locations in both study areas, falling to 9.5 percent of the stream flow measured at Betrawati (600 m) and 7.4 percent at Rabuwa Bazaar (470 m), about 50 km from the glacier termini. The glacier ice melt contribution decreased to 4.5 percent of annual discharge farther downstream in the Trishuli basin (at 325 m, about 75 km from the glacier termini).

Ye et al. (2009) investigated changes in glaciers and supraglacier lakes in the Mt. Qomolangma region on the northern slopes of the middle Himalaya on the Tibetan Plateau. Glaciers in this region have both retreated and advanced in the past 35 years, with retreat dominating. The glacier retreat area was 3.23 km^2 (0.75 km^2 yr^{-1}) during 1974–1976, 8.68 km^2 (0.36 km^2 yr^{-1}) during 1976–1992, 1.44 km^2 (0.12 km^2 yr^{-1}) during 1992–2000, 1.14 km^2 (0.22 km^2 yr^{-1}) during 2000–2003, and 0.52 km^2 (0.07 km^2 yr^{-1}) during 2003–2008. Smaller glaciers retreated more rapidly than larger ones. At the same time, supraglacier lakes on the debris terminus of the Rongbuk Glacier were enlarged dramatically, from 0.05 km^2 in 1974 to 0.71 km^2 in 2008 – more than thirteen times larger the baseline area. In addition, glacier changes showed spatial differences; for example, the glacier retreat rate was the fastest at glacier termini between 5,400 and 5,700 m a.s.l. (Ye et al. 2009).

The question of how much water melting glaciers contribute to lake changes under a warmer climate remains unanswered for local lakes, as these bodies of water are fed by four sources: rainfall, snow melt, frozen ground melt, and glacier melt. As part of studies addressing the impact of glacier-melting effects on lakes in high-altitude closed basins of the western Himalaya, Ye et al. analyzed changes in glaciers and lakes in the Yamzhog Yumco basin in southern Tibet (Ye et al. 2007) and in the Mapam Yumco basin in the western Tibet (Ye et al. 2008), by means of GIS and remote sensing. Glacier and lake variations in the Yamzhog Yumco basin were studied by integrating series of spatial data from topographic maps and Landsat images at three different times: 1980, 1988–1990, and 2000. The results indicate that the total glacier area decreased from 218 km^2 in 1980 to 215 km^2 in 2000 (1.5 percent). Glacier recession rates were clearly higher in the 1990s than the 1980s due to the warmer climate. The total lake area decreased by about 67 km^2 during 1980–1990 and increased by 32 km^2 during 1990–2000. Changes in the lake area in the basin occurred rapidly, most likely caused primarily by the change in precipitation and evaporation in the basin, and secondarily by the increased water supply from melting glaciers.

Mapam Yumco basin (more simply called Mapam basin: 7,786 km^2) is a complex closed basin in southwestern Tibet located above 4,500 m a.s.l. Landsat images in 1990, 1999, and 2000 and ASTER images in 2003 were used to evaluate changes in this area. The total glacier and lake areas in Mapam basin in 1974 were 108 and 782 km^2, respectively, occupying 1.4 and 10.0 percent of the basin area. The glacier area decreased monotonically during the period, with the rate of glacier change accelerating from -0.20 km^2 yr^{-1} (1973–1990) to -0.32 km^2 yr^{-1} (1990–1999), and -0.36 km^2 yr^{-1} (1999–2003).

Lake changes in Mapam, however, were more complicated (Ye et al. 2008). The total lake area decreased from 1974 to 1999, but then slightly increased from 1999 to 2003. Lake area decreased by 22.87 km^2 from 1974 to 1990 (with lake decrease amounting to 27.24 km^2 in some areas and lake growth to 4.37 km^2 in others), decreased by 13.92 km^2 from 1990 to 1999 (with lake decrease of 19.33 km^2 in some regions and lake growth of 5.41 km^2 in others), and then increased by 2.63 km^2 from 1999 to 2003 (with lake decrease in some areas amounting to 8.94 km^2 and lake growth in others to 11.57 km^2). The change rate of the total lake area first became more negative, accelerating from -1.43 km^2 yr^{-1} (1973–1990) to -1.55 km^2 yr^{-1} (1990–1999), but then became positive, $+0.66$ km^2 yr^{-1} (1999–2003). The lake area enlargement in the last period could not compensate for the lake area reduction in the earlier periods, a similar change to that seen in the Yamzhog basin (Ye et al. 2007). The change rate in lake area recession became more negative, changing from -1.70 km^2 yr^{-1} (1974) to -2.15 km^2 yr^{-1} (1990) to -2.24 km^2 yr^{-1} (2003), while for the expanded areas the change rate increased from 0.27 to 0.60 to 2.89 km^2 yr^{-1} over the same years. Overall, although the total glacier area retreat (7.53 km^2) was much smaller than the total lake area reduction (34.16 km^2), the glacier change by percentage (-7.0 percent) of the total was more significant than the lake change (-4.4 percent).

Ye et al. (2008) showed that both glacier and lake areas in the Mapam basin decreased from 1974 to 2003. Lake area reduction by 34.16 km^2 was significant due to the large decrease in imbalances between precipitation and evaporation (P − E), similar to the trend seen in the Yamzhog basin. Increased melting of glaciers and frozen ground in the basin also contributed to the lake enlargement in 1974–2003. However, the total glacier recession by 7.53 km^2 did not supply enough water for lake growth to compensate for lake reduction in the same periods. In the Mapam basin, both enlargement and reduction of lakes accelerated in 1974–2003, which might be an indicator of an accelerated water cycle process over the Tibetan Plateau forced by the warming climate (Ye et al. 2008).

8.2.3 Snow Cover

Station data on seasonal snow cover over the Tibetan Plateau, which began to be collected in 1951, are relatively sparse. Since 1978, coverage has been improved through the collection of passive microwave data. Shi (2008b) provided a map of the annual mean number of snow cover days, which range from 60 to more than 150 across most of the plateau. Snowfall amounts are typically low and average daily snow depths are only 5–15 cm over most of the plateau. Maximum and minimum average snow depths, based on SMMR data for 1978–1987, over the plateau are 21 and 10 cm, respectively. About 60 percent of the plateau has winter snow cover. The Qaidan basin, Yarlang Zanbo river valley, and the northern plateau especially lack snow cover, based on station data (Li 1994). The winter snow depth is largest in southeast Tibet, where it reaches 30–40 cm. The largest snow depths (50–60 cm) in Central Asia are in the Altai Mountains. In the Tianshan, Pamir, and Karakorum mountains, depths average 40–50 cm; in the Kunlun, they average 20–30 cm and in the Qilian, only 10–20 cm.

The annual cycle of snow cover on the Tibetan Plateau begins in mid-September and lasts until June, with the maximum occurring in January–February (Shi 2008b). Analysis of station snow data for 1951–1997 over northwestern China by Li (1999) does not show any trends related to this cycle.

Huang et al. (2017) used daily MODIS snow products for 2001–2014 to analyze snow cover over the Tibetan Plateau. The annual average perennial snow cover area (SCA) on the Tibetan Plateau is approximately 16 percent over the entire year and 23 percent during the hydrologic year (early October to late April). The monthly SCA reaches a maximum in February (approximately 30 percent) and a minimum in August (2.7 percent). Huang et al. found that the SCA tended to increase at elevations below 2,000 m a.s.l. in the northeastern Qaidam basin, whereas it decreased at higher elevations. The SCA also exhibited a mean decrease over the entire plateau. Snow cover days (SCD) tended to decrease over approximately 60 percent of the plateau and increase in 38.5 percent. The SWE tended to decrease in almost 33 percent of the area and increase over 20 percent. Decreased snowfall and increased rainfall and temperature were identified as the main reasons for the decrease in SCD and SWE. Rainfall increased over 72 percent of the plateau in 2001–2014, especially in the north and in the northern Himalaya. At high elevations, snow cover decreased with rising air temperatures.

Giao et al. (2012) analyzed snow cover in eastern Tibet (Lhasa River basin, Niyang River basin, and Changdu region) for 1979–2005 using MODIS data. At lower-elevation sites, the length of the snow-free season increased. In contrast, at higher elevations, it decreased. The shorter snow season at lower elevation is attributed to an increase in air temperatures, whereas at higher elevations an increase in precipitation appears to compensate for the increase in air temperature.

Wang et al. (2017) analyzed SCA and SWE over the Tibetan Plateau for 1979–2006 using passive microwave data. They found a significant decreasing trend in the western part for both SCA and SWE in summer and autumn and in the southern part for SCA in all four seasons. A significant increasing trend was identified in the central–eastern part for SCA in autumn, winter, and spring and in the eastern and far western parts for SWE in winter and spring. An increase in air temperature was accompanied by SCA decrease in the southern and western parts, but by SCA and SWE increase in the central–eastern part.

For Central Asia, only a few studies have investigated changes in seasonal snow cover. They suggest a decrease in maximum snow depth and a reduction in snow cover duration (Unger-Shayesteh et al. 2013). Zhou et al. (2017a) used Advanced Very High Resolution Radiometer (AVHRR) 1-km data for 1986–2008 and MODIS data for 2000–2009 to study snow cover in northern Central Asia. This area was defined as including Central Asian Republics, Kazakh Steppe, Aral–Caspian Desert, Tarim, Siberian Altai–Sayan, Mongolian Altai, Western Tien Shan, Northern Tien Shan, Issyk Kul, Inner Tien Shan, Eastern Tien Shan, Central Tien Shan, Western Pamir, Pamiro–Alai, Central Pamir, and Eastern Pamir. The area-weighted mean long-term SCD for the whole area was 95.2 ± 65.7 days. High-elevation mountainous areas (above 3,000 m) in Altai, Tien Shan, and Pamir, which accounted for about 2.8 percent of the total area in Central Asia, had more than 240 SCD. Area-weighted mean SCD in the whole region did not exhibit significant trends from 1986 to 2008. However, low-elevation areas (below 2,000 m) in Central Tien Shan and Eastern Tien Shan, as well as mid-elevation areas (1,000–3,000 m) in the western Tien Shan, Pamiro–Alai, and western Pamir, experienced increasing SCD, associated with both earlier onset of snow cover and later melting. Decrease of SCD was observed in mountainous areas of the Altai, Tien Shan, and Pamir.

8.2.4 Permafrost

Permafrost occupies about 3.5 million km^2 of the area of Central Asia, according to Zhao et al. (2008). The Tibetan Plateau is underlain by discontinuous permafrost (50–90 percent of the area underlain by permafrost) in the northern part, and by either sporadic permafrost (10–50 percent of the area underlain by permafrost) or isolated permafrost (less than 10 percent of the area) in the south. Zhou et al. (2000), however, identify predominantly continuous permafrost (70–90 percent), widespread island permafrost (30–70 percent), and sparse island

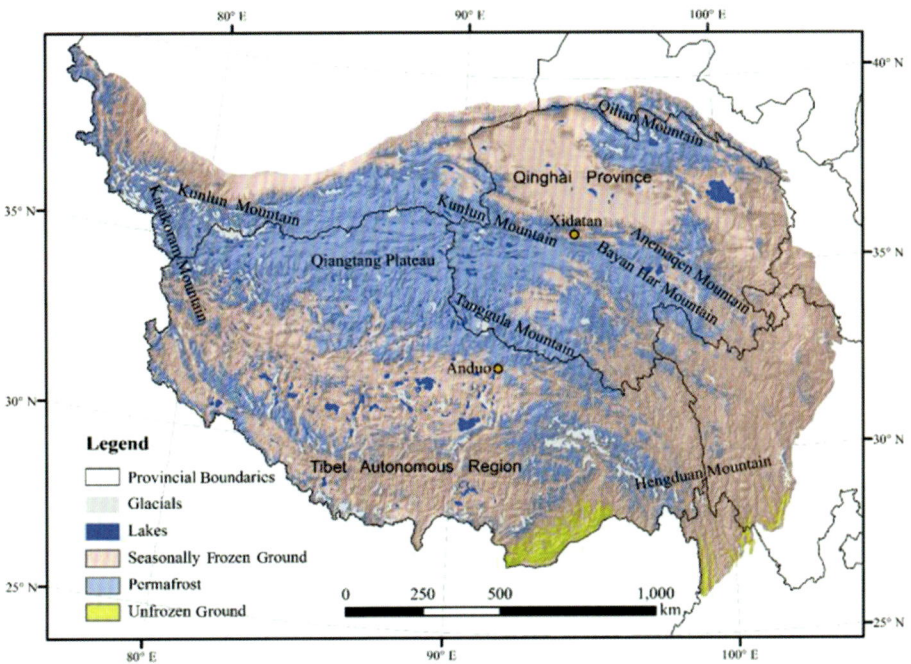

Figure 8.6 Distribution of permafrost and seasonally frozen ground on the Tibetan Plateau. Source: Zou et al. 2017, figure 2.

permafrost (less than 30 percent). The total permafrost area is 1.06 million km^2, or 40 percent of the plateau (Zou et al. 2016) (Figure 8.6). Cheng and Wu (2007) estimate that approximately 36.6 percent of the permafrost is in the discontinuous zone, 46.6 percent in the sporadic zone, and 16.8 percent in the isolated patch zone. The 1,100-km-long Qinghai–Xizhang railroad, which opened in 2006 and traverses from Golmud to Lhasa, has specially engineered embankments to avoid permafrost degradation, since it crosses 550 km underlain by permafrost, half of it "warm" and more than one third of it ice-rich permafrost, which is susceptible to thawing.

The lower altitudinal limit of permafrost in the north is at about 4,200 m a.s.l. Most of the permafrost is warm and thin, as well as ice-poor due to the overall aridity. Seasonally frozen ground occupies 1.46 million km^2, or 56 percent, of the plateau (Zou et al. 2016).

Permafrost is present in Central Asia in the Qinghai–Tibetan Plateau, Tien Shan, Pamir, and Mongolia (Zhao et al. 2010). Monitoring of the ground thermal regime in these regions over the past several decades has shown that the permafrost has been undergoing significant changes caused by climate warming.

Observations of ground temperature at the top of permafrost (TTOP) at a Kunlun Pass site showed a warming trend from 1996 to 2001 of about 0.3 °C. Measurements in a 30-m bore hole at the Cryosphere Research Station on the Qinghai–Tibetan Plateau (35.7° N, 94.1° E) indicate that ground temperatures at 12–20 m rose by 0.15–0.36 °C between 2001 and 2003 (Zhao et al. 2008). In the vicinity of the southern lower altitudinal limit of permafrost around Amdo–Liangdaohe, sporadic permafrost decreased in area by 36 percent and the lower altitudinal limit of permafrost rose by 50–80 m from the mid-1960s to the mid-1990s.

During the International Polar Year 2007–2008, measured mean annual ground temperature (MAGT) at a depth of 6 m ranged from −3.2 to 0.2 °C on the Qinghai–Tibetan Plateau and the active-layer thickness (ALT) varied between 105 and 322 cm at different sites. Ground temperatures at the bottom of the active layer have warmed, on average, by 0.06 °C yr^{-1} over the past decade. The increase in permafrost temperatures in the northern Tien Shan from 1974 to 2009 ranged from 0.3 to 0.6 °C. At present, measured permafrost temperatures vary from −0.5 to −0.1 °C. The ALT increased from 3.2–4 m in the 1970s to a maximum of 5.2 m between 1995 and 2009.

Cheng and Wu (2007) demonstrated, using long-term temperature measurements, that the lower altitudinal limit of permafrost rose by 25 m in the northern plateau during the 1970s–2000s and by 50–80 m in the south from 1982 to 2001. The thickness of the active layer increased by 0.15–0.50 m and ground temperature at a depth of 6 m rose by about 0.1–0.3 °C between 1996 and 2001.

Permafrost degradation is one of the main factors responsible for a dropping groundwater table at the source areas of the Yangtze River and Yellow River, which in turn results in lowering lake water levels, drying swamps, and shrinking grasslands. Ran et al. (2017) have evaluated the reduction of permafrost stability over the plateau from the 1960s to the 2000s. They estimated decadal mean annual air temperatures (MAATs) by integrating remote sensing–based estimates of mean annual land surface temperatures, Leaf Area Index, and fractional snow cover values, as well as decadal mean MAATs taken at 152 weather stations. They found a continuous rise of approximately 0.04 °C yr^{-1} in the decadal mean MAAT values over the past half-century. The area of stability that became degraded was approximately 154 × 10^4 km^2, equal to 88 percent of the permafrost area in the 1960s. The stability of 75 percent of the extremely stable permafrost, 90 percent of the stable permafrost, 90 percent of the sub-stable permafrost, 92 percent of the transitional permafrost, and 33 percent of the unstable permafrost has been reduced to lower levels of stability. Approximately 49 percent of the unstable permafrost and 96 percent of the extremely unstable permafrost have degraded to seasonally frozen ground. The mean elevations of the extremely stable, stable, sub-stable, transitional, unstable, and extremely unstable permafrost areas increased by 88, 97, 155, 185, 161, and 250 m, respectively. The degradation mainly occurred from the 1960s to the 1970s and from the 1990s to the 2000s.

X.-Q. Peng et al. (2016b) analyzed observations from more than 800 stations with gridded mean monthly air temperature data across China from 1951. The results show that MAAT increased statistically significantly, by 0.29 °C decade^{-1} from 1967 to 2013, with greater warming occurring on the Tibetan Plateau. Changes of mean annual area extent of seasonal soil freeze–thaw state decreased significantly for completely frozen ground, while the area extents of partially frozen ground and unfrozen ground both increased.

SUMMARY

The Third Pole comprises the 4,500-m-high Tibetan Plateau, the mountain ranges of Central Asia to the north, and the Karakorum–Himalaya to the south. The name denotes the cold, arid climates and the extensive areas of snow, ice, and frozen ground found this region. This area also contains the headwaters of ten major river systems.

Atmospheric circulation is strongly affected by the topography. In winter, the westerlies are split by the Tibetan Plateau and the subtropical jet is anchored south of the Himalaya. In summer, the Tibetan Plateau has a shallow thermal low and a high-level anticyclone. The Asian monsoon affects the Himalaya and the southeastern Tibetan Plateau in July through August.

Net radiation on the Tibetan Plateau is more than 100 W m^{-2} in summer. Sensible heat peaks in May–June (50–75 W m^{-2}), while latent heat is 40–80 W m^{-2} in July–August.

Winter temperatures average -25 °C in the higher western Tibetan Plateau and -15 °C in the east. In summer, the temperature in the east reaches 5–10 °C and that in the west 0–5 °C. Precipitation decreases from 500–600 mm in the southeast to 50–100 mm in the northwest. Convective precipitation is dominant over the Tibetan Plateau.

The moisture source south of the Himalaya is the Indian monsoon, while northward to the Tanggula Mountains moisture comes from the Bay of Bengal and northern India. North of the Tanggula, half of the moisture is recycled over inland Asia and the rest is from convection. Tibet has small amounts of seasonal snow cover. There are about 100 snow days annually and maximum SWE at Naqu is 20 mm.

Most station records on the Tibetan Plateau begin in the 1950s. Warming began several decades earlier in this area than in the northern hemisphere as a whole.

Over 1955–1996, mean winter temperature rose by 0.32 °C decade^{-1}. Over 1961–2007, the rate was 0.36 °C decade^{-1}. In 1961–2014, the Tibetan Plateau warmed at 0.59 °C in winter, 0.53 °C in summer, 0.47 °C in autumn, and 0.41 °C in spring. Warming is higher at elevations around 4 km but the cause is uncertain.

The Himalaya show decreasing precipitation in the west and increased precipitation in the east. Different data sets for the Tibetan Plateau show conflicting results. However, over 1961–2007, the southeastern plateau received less precipitation and the inner and northeastern areas more precipitation. Lake area on the Tibetan Plateau has increased by 20 percent since 1984 due to more precipitation and glacier melt.

Data for Central Asia indicate warming trends since the 1970s that are greater in the cold season and for minimum temperatures. Precipitation in arid Central Asia has increased since 1930, mainly in winter. The mountain regions of Central Asia showed increases for 1979–2013, and Xinjiang for 1951–2013.

Runoff data from seven major rivers on the Tibetan Plateau show that between 1962–1966 and 1969–1973, runoff was reduced 22 percent, but then increased slowly beginning in the 1970s. Runoff from the Yellow River increased from the 1970s to 1980s. For 1960–2000, runoff decreased in the upper Yellow and Yangtze rivers and increased in the Qilian Mountains.

The cryosphere in Central Asia comprises glaciers, snow cover, and frozen ground. The total glacier area is approximately 113,000 km^2, with 47,000 km^2 in the Himalaya and 16,600 km^2 in the Karakorum. Most of the glacier area is between 5,000 and 6,000 m. The snowline is at 5,800 m over the Tibetan Plateau. The Guliya ice cap in northwest Tibet has a mass balance of 0.22 m w.e., the Dunde ice cap in the Qilian Shan has a mass balance of 0.39 m w.e., and at Dasuopu in the central Himalaya the mass balance is 0.7 m w.e. For 2002–2009, the most negative annual mass balance was −0.77 m w.e. in the Qilian Shan and eastern Kunlun, while there was a mass gain of 0.27 m w.e. in the north–central Tibetan Plateau. Glaciers have retreated on the Tibetan Plateau and surroundings since the 1960s, especially in the more maritime edges, with this trend intensifying in the last decade. Mass changes in the Karakorum have been in balance or slightly positive since about 2000. Over 2000–2008, 58 percent of Karakorum glaciers were stable or slowly advancing, while 65 percent of glaciers in the monsoon-dominated Himalaya were retreating.

In Sikkim and part of Nepal, supraglacial debris in the 2000s covered 11 percent of the glacierized area, and supraglacial lakes covered almost 6 percent of the debris cover. The overall glacier area loss rate from 1962 to 2000 was 0.5 percent yr^{-1}, while in Sikkim it was 0.23 percent yr^{-1} from 1962 to 2006. Glaciers in Bhutan lost 23 percent of their area between 1980 and 2010. In Tibet and surroundings, the area covered by 7,000 glaciers decreased from approximately 13,363 km^2 in the 1970s to 12,130 km^2 in the 2000s. Area decrease was most extreme in the southeast Himalaya (0.57 percent yr^{-1}) from the 1970s to 2000s, and least evident in the eastern Pamir (0.07 percent yr^{-1}).

Glaciers are shrinking in the Tien Shan and Pamir. The glacier area in the northern Tien Shan decreased by 30 percent between 1955 and 1990. In the Tien Shan region, glacier area decrease between 1970 and 2007 ranged from 9 percent in southeast Fergana to 24 percent in Pskem, with this retreat being attributed to rising summer temperatures. The bigger changes in the outer Central Asian ranges are attributed to (1) higher ice-mass turnover due to higher precipitation, (2) lower elevation of glacier termini, and (3) summer accumulation in the inner and eastern Tien Shan, which reduces ablation.

Glacier melt contributes 58 percent to the annual flow in a small Himalayan watershed and 32 percent to another flow in the Dudh Kosi basin. About 40 km from the termini, those values drop to 9 and 7 percent, respectively.

Snow cover days (SCD) range from 60 to 150 days across most of the plateau. Average daily snow depths are 5–15 cm. Maximum and minimum snow depths, based on SMMR data, are 21 and 10 cm, respectively. About 60 percent of the plateau has snow cover. In the Tien Shan, Pamir, and Karakorum, snow depths average 40–50 cm; they are 50–60 cm in the Altai. The average annual perennial snow cover area (SCA) is 23 percent of the total area for October–April. In 2001–2014, snow cover days decreased for more than 60 percent of the plateau; snow water equivalent (SWE) decreased over about 33 percent of this area and increased over 20 percent. In Central Asia, maximum snow depth has decreased, as has snow cover duration. For all of northern Central Asia, the area-weighted mean snow cover duration is 95 days.

The Tibetan Plateau is underlain by continuous permafrost in the north and discontinuous or isolated permafrost in the south, occupying 1 million km^2, while Central Asia has 3.5 million km^2 of permafrost. The lower altitudinal limit for permafrost is 4,200 m in the north. Most permafrost in the Third Pole is thin, warm, and ice-poor. Seasonally frozen ground affects 56 percent of the Tibetan Plateau. In response to climate warming, temperature at the top of the permafrost table in Kunlun Pass rose 0.3 °C between 1996 and 2003. Near the southern lower altitude limit, the sporadic permafrost area decreased by 37 percent from the mid-1960s to the mid-1990s and the altitudinal limit rose by 50–80 m. In 2007–2008, mean annual ground temperature at 6 m depth on the plateau was between −3.2 and 0.2 °C. The active-layer thickness varied from 105 to 322 cm. Permafrost temperatures in the northern Tien Shan rose 0.3–0.6 °C from 1974 to 2009. The lower altitudinal limit on the Tibetan Plateau rose by 25 m during the 1970s to 2000s and by 50–80 m in the south during 1982–2001.

QUESTIONS

1. In what ways does the Third Pole resemble and differ from the Arctic and the Antarctic?
2. How do the Tibetan Plateau–Himalaya affect the seasonal air circulation in the region?
3. Describe the principal features of the distributions of air temperature and precipitation over the Tibetan Plateau.
4. Describe the characteristics of seasonal snow cover in Tibet and its trends.
5. Describe the temperature changes over Tibet since the mid-1950s.
6. Which precipitation changes over the Tibetan Plateau are fairly well established?
7. Compare the degree of glacierization of the major mountain ranges of the Third Pole.
8. Discuss some of the regional variations in glacier area across the Third Pole.
9. Describe the contribution of glacier melt to stream flow in the Himalaya.
10. Which features of permafrost in Tibet are unusual?
11. Describe the major recent changes in permafrost conditions in the Third Pole region.

References

Aizen, V. B., et al. 1997. "Glacier Regime of the Higher Tien Shan Mountains, Pobeda–Khan Tengry Massif." *Journal of Glaciology* 43(145): 503–21.

Aizen, V. B., et al. 2007. "Glacier Changes in the Tien Shan as Determined from Topographic and Remotely Sensed Data." *Global and Planetary Change* 56: 328–40.

Bajracharya, S. R., S. B. Maharjan, and F. Shrestha. 2014. "The Status and Decadal Change of Glaciers in Bhutan from the 1980s to 2010 Based on Satellite Data." *Annals of Glaciology* 55: 159–66.

Barry, R. G. 2008. *Mountain Weather and Climate*. 3rd ed. Cambridge: Cambridge University Press.

Bashir, T., et al. 2017. "A Hydrometeorological Perspective on the Karakoram Anomaly Using Unique Valley-Based Synoptic Weather Observations." *Geophysical Research Letters*. doi: 10.1002/2017GL075284.

Bolch, T. 2007. "Climate Change and Glacier Retreat in Northern Tien Shan (Kazakhstan/Kyrgyzstan) Using Remote Sensing Data." *Global and Planetary Change* 56: 1–12.

Bolch, T., et al. 2016. "Glaciers in the Hunza Catchment (Karakoram) Are in Balance since the 1970s." *Cryosphere Discussions*. doi: 10.5194/tc-2016-197.

Bugaev, V. A., et al. 1962. *Synoptic Processes of Central Asia*. [Translation of Russian original, Uzbek Academy of Sciences, USSR, 1957). Geneva: World Meteorological Organization.

Cai, D., et al. 2017. "Spatiotemporal Temperature Variability over the Tibetan Plateau: Altitudinal Dependence Associated with the Global Warming Hiatus." *Journal of Climate* 30: 969–84.

Chen, D., et al. 2015. "Assessment of Past, Present and Future Environmental Changes on the Tibetan Plateau." *Chinese Science Bulletin* 60: 3025–35.

Chen, F., et al. 2011. "Spatiotemporal Precipitation Variations in the Arid Central Asia in the Context of Global Warming." *Science China Earth Sciences* 54(12): 1812–21.

Cheng, G.-D., and T.-H. Wu. 2007. "Responses of Permafrost to Climate Change and Their Environmental Significance, Qinghai–Tibet Plateau." *Journal of Geophysical Research* 112: F02S03.

Dash, S. K., et al. 2007. "Some Evidence of Climate Change in Twentieth-Century India." *Climate Change* 85: 299–321.

Diodato, N., G. Bellocchi, and G. Tartari. 2011. "How Do Himalayan Areas Respond to Global Warming?" *International Journal of Climatology* 32: 975–82.

Dyurgerov, M. B., and M. F. Meier. 2005. *Glaciers and the Changing Earth System: A 2004 Snapshot.* Occasional Paper No. 58. Boulder, CO: Institute of Arctic and Alpine Research, University of Colorado.

Earl, L., and A. S. Gardner. 2016. "A Satellite-Derived Glacier Inventory for North Asia." *Annals of Glaciology* 57(71): 50–60.

Frauenfeld, O., T. Zhang, and M. C. Serreze. 2005. "Climate Change and Variability Using European Centre for Medium-Range Weather Forecasts Reanalysis (ERA-40) Temperatures on the Tibetan Plateau." *Journal of Geophysical Research: Atmospheres* 110. doi: 10.1029/2004JD005230.

Fujinami, H., and T. Yasunari. 2001. "The Seasonal and Intraseasonal Variability of Diurnal Cloud Activity over the Tibetan Plateau." *Journal of the Meteorological Society of Japan* 79(6): 1207–27.

Gautam, M. R., G. R. Timilsina, and K. Acharya. 2013. *Climate Change in the Himalayas: Current State of Knowledge.* Policy Research Working Paper No. WPS 6516. Washington, DC: World Bank.

Getker, M. I. 1985. Snow Resources of the Mountain Regions of Middle Asia. [In Russian]. Avtoreferat. DSc. Dissertation, Institute of Geography, Academy of Sciences, USSR, Moscow.

Giao, J., et al. 2012. "Spatiotemporal Distribution of Snow in Eastern Tibet and the Response to Climate Change." *Remote Sensing of the Environment* 121: 1–9.

Guo, W.-Q., et al. 2015. "The Second Chinese Glacier Inventory: Data, Methods and Results." *Journal of Glaciology* 61(226): 357–72.

Hu, Z.-Y., et al. 2017. "Variations and Changes of Annual Precipitation in Central Asia over the Last Century." *International Journal of Climatology* 37. doi: 10.1002/joc.4988.

Huang, F.-J., and R.-J. Shen. 1986. "The Source of Water Vapor and Its Distribution over the Qinghai–Xizang Plateau during the Period of Summer Monsoon." In *Proceedings of the International Symposium on the Qinghai-Xizang Plateau and Mountain Meteorology,* 596–603. Beijing: Science Press/Boston, MA: American Meteorological Society.

Huang, X.-D., et al. 2017. "Impact of Climate and Elevation on Snow Cover Using Integrated Remote Sensing Snow Products in Tibetan Plateau." *Remote Sensing of the Environment* 190: 274–88.

Immerzeel, W. 2008. "Historical Trends and Future Predictions of Climate Variability in the Brahmaputra Basin." *International Journal of Climatology* 28: 243–54.

Kääb, A., et al. 2012. "Contrasting Patterns of Early Twenty-First-Century Glacier Mass Change in the Himalayas." *Nature* 488: 495–8.

Kang, S., et al. 2015. "Dramatic Loss of Glacier Accumulation Area on the Tibetan Plateau Revealed by Ice Core Tritium and Mercury Records." *Cryosphere* 9: 1213–22.

Kraaijenbrink, P. D. A., et al. 2017. "Impact of a Global Temperature Rise of 1.5 Degrees Celsius on Asia's Glaciers." *Nature* 549: 257–63.

Kutuzov, S., and M. Shahgedanova. 2009. "Glacier Retreat and Climatic Variability in the Eastern Terskey–Alatoo, Inner Tien Shan between the Middle of the 19th Century and Beginning of the 21st Century." *Global and Planetary Change* 69: 59–70.

Li, B. L., et al. 2006. "Glacier Change over the Past Four Decades in the Middle Chinese Tien Shan." *Journal of Glaciology* 178(52): 425–32.

Li, F.-P., et al. 2013. "The Impact of Climate Change on Runoff in the Southeastern Tibetan Plateau." *Journal of Hydrology* 505: 188–201.

Li, J. 2017. "Hourly Station-Based Precipitation Characteristics over the Tibetan Plateau." *International Journal of Climatology*. doi: 10.1002/joc.5281.

Li, P.-J. 1983. "Distribution of Snow Cover in China." [In Chinese]. *Journal of Glaciology and Geocryology* 5: 9–18.

Li, P.-J. 1994. "Dynamic Characteristics of Snow Cover in West China: Snow and Ice Cover Interactions with the Atmosphere and Ecosystems." *International Association of Hydrological Sciences Publications* 223: 141–52.

Li, P.-J. 1999. "Variation of Snow Water Resources in Northwestern China, 1951–1997." *Science in China, Series D* 42: 72–9.

Liu, X., and B. Chen. 2000. "Climatic Warming in the Tibetan Plateau during Recent Decades." *International Journal of Climatology* 20(14): 1729–42.

Liu, X., and L.-B. Yan. 2017. "Elevation-Dependent Climate Change in the Tibetan Plateau." In *Climate Science: Oxford Research Encyclopedia*. doi: 10.1093/acrefore/9780190228620.013.593.

Maurer, J. M., S. B. Rupper, and J. M. Schaefer. 2016. "Quantifying Ice Loss in the Eastern Himalayas since 1974 Using Declassified Spy Satellite Imagery." *Cryosphere* 10: 2203–15.

Narama, C., et al. 2010. "Spatial Variability of Recent Glacier Area Changes in the Tien Shan Mountains, Central Asia, Using Corona ($\sim$1970), Landsat ($\sim$2000), and ALOS ($\sim$2007) Satellite Data." *Global and Planetary Change* 71: 42–54.

Neckel, N., et al. 2014. "Glacier Mass Changes on the Tibetan Plateau 2003–2009 Derived from ICESat Laser Altimetry Measurements." *Environmental Research Letters* 9: 014009.

Nuimura, T., et al. 2015. "The GAMDAM Glacier Inventory: A Quality-Controlled Inventory of Asian Glaciers." *Cryosphere* 9: 849–64.

Owen, L. A., et al. 2006. "Climatic and Topographic Controls on the Style and Timing of Late Quaternary Glaciation throughout Tibet and the Himalaya Defined by 10Be Cosmogenic Radionuclide Surface Exposure Dating." *Quaternary Science Reviews* 24 (12–13): 1391–411.

Pekel, J.-F., et al. 2016. "High-Resolution Mapping of Global Surface Water and Its Long-Term Changes." *Nature* 540(7633): 418–22.

Peng, J., et al. 2016. "Comparison of Satellite-Based Evapotranspiration Estimates over the Tibetan Plateau." *Hydrology and Earth-System Sciences* 20: 3167–82.

Peng, X.-Q., et al. 2016. "Response of Changes in Seasonal Soil Freeze/Thaw State to Climate Change from 1950 to 2010 across China." *Journal of Geophysical Research: Earth Surface.* doi: 10.1002/2016JF003876/.

Pepin, N., et al. 2015. "Elevation-Dependent Warming in Mountain Regions of the World." *Nature Climate Change* 5(5): 424–30.

Putkonen, J. K. 2004. "Continuous Snow and Rain Data at 500–400 m near Annapurna, Nepal, 199–2001." *Arctic and Antarctic Alpine Research* 36: 244–8.

Qiu, J. 2012. "Thawing Permafrost Reduces River Runoff." *Nature News.* doi: 10.1038/nature.2012.9749.

Racoviteanu, A. E., R. Armstrong, and M. W. Williams. 2013. "Evaluation of an Ice Ablation Model to Estimate the Contribution of Melting Glacier Ice to Annual Discharge in the Nepal Himalaya." *Water Resources Research* 49: 5117–33.

Racoviteanu, A. E., et al. 2015. "Spatial Patterns in Glacier Characteristics and Area Changes from 1962 to 2006 in the Kanchenjunga–Sikkim Area, Eastern Himalaya." *Cryosphere* 9: 505–23.

Ran, Y.-H., X. Li, and G.-D. Cheng. 2017. "Climate Warming Has Led to the Degradation of Permafrost Stability in the Past Half Century over the Qinghai–Tibet Plateau." *Cryosphere Discussions.* https://doi.org/10.5194/tc-2017120.

Rasmussen, L. A. 2013. "Meteorological Controls on Glacier Mass Balance in High Asia." *Annals of Glaciology* 54(63): 352–9.

Sato, T. 2001. "Spatial and Temporal Variations of Frozen Ground and Snow Cover in the Eastern Part of the Tibetan Plateau." *Journal of the Meteorological Society of Japan* 79: 519–34.

Scherler, D., B. Bookhagen, and M. R. Strecker. 2011. "Spatially Variable Response of Himalayan Glaciers to Climate Change Affected by Debris Cover." *Nature Geoscience* 4: 156–9.

Schild, A. 2008. "ICIMOD's Position on Climate Change and Mountain Systems." *Mountain Research and Development* 28: 328–31.

Shangguan, D.-H., et al. 2004. "Glacier Changes at the Head of Yurunkax River in the West Kunlun Mountains in the Past 32 Years." *Acta Geographica Sinica* 59(6): 852–62.

Shi, Y.-F., ed. 2008a. *Concise Glacier Inventory of China.* Shanghai: Popular Science Press.

Shi, Y.-F., ed. 2008b. *Glaciers and Related Environments in China.* Beijing: Science Press.

Song, Ch.-Q., et al. 2016. "Precipitation Variability in High Mountain Asia from Multiple Datasets and Implication for Water Balance Analysis in Large Lake Basins." *Global and Planetary Change* 145: 20.

Thompson, L. G., et al. 1995. "A 1000-Year Ice Core Climate Record from the Guliya Ice Cap, China: Its Relationship to Global Climate Variability." *Annals of Glaciology* 21: 175–81.

Tian, L., et al. 2001. "Tibetan Plateau Summer Monsoon Northward Extent Revealed by Measurements of Water Stable Isotopes." *Journal of Geophysical Research* 106(D22): 28081–8.

Tong, K., et al. 2014. "Tibetan Plateau Precipitation as Depicted by Gauge Observations, Reanalyses and Satellite Retrievals." *International Journal of Climatology* 34(2): 265–85.

Unger-Shayesteh, K., et al. 2013. "What Do We Know about Past Changes in the Water Cycle of Central Asian Headwaters? A Review." *Global and Planetary Change* 110: 4–25.

Wang, B., et al. 2008. "Tibetan Plateau Warming and Precipitation Changes in East Asia." *Geophysical Research Letters* 35: L14702. doi: 10.1029/2008GL034330.

Wang, Sh., et al. 2011. "Glacier Area Variation and Climate Change in the Chinese Tianshan Mountains since 1960." *Journal of Geographical Sciences* 21: 63–73.

Wang, Y., and D. Yang. 2015. *Impact of Cryosphere Hydrological Changes on the River Runoff in the Tibetan Plateau.* American Geophysical Union, Fall Meeting 2015, abstract #C33E-0873.

Wang, Z.-B., R.-G. Wu, and G. Huang. 2017. "Low-Frequency Snow Changes over the Tibetan Plateau." *International Journal of Climatology.* doi: 10.1002/joc.5221.

Wu, R.-P., et al. 2017. "Recent Glacier Mass Balance and Area Changes in the Kangri Karpo Mountain Derived from Multi-sources of DEMs and Glacier Inventories." *Cryosphere Discussions.* https://doi.org/10.5194/tc-2017-153.

Wu, X.-D., and L. G. Thompson. 1988. "A 40-Year Record in an Ice Core from the Dunde Ice Cap, China." *Annals of Glaciology* 10: 221.

Xu, J.-Q., and S. Haginoya. 2001. "An Estimation of Heat and Water Balances on the Tibetan Plateau." *Journal of the Meteorological Society of Japan* 79(1B): 485–504.

Yanai, M., and C. Li. 1994. "Mechanisms of Heating and the Boundary Layer of the Tibetan Plateau." *Monthly Weather Review* 102: 3305–21.

Yang, K. 2017. "Observed Regional Climate Change in Tibet over the Last Decades." In *Oxford Research Encyclopedia of Climate Science.* doi: 10.1093/acrefore/9780190228620.013.587.

Yao, T., et al. 2004. "Recent Glacial Retreat in High Asia in China and Its Impact on Water Resources in Northwest China." *Science in China, Series D: Earth Science* 47: 1065–75.

Yao, T., et al. 2012. "Different Glacier Status with Atmospheric Circulations in Tibetan Plateau and Surroundings." *Nature Climate Change* 2: 663–7.

Yao, T.-D., et al. 2007. "Recent Glacial Retreat and Its Impact on Hydrological Processes on the Tibetan Plateau, China, and Surrounding Regions." *Arctic and Antarctic Alpine Research* 39(4): 642–50.

Ye, Q.-H., T. D. Yao, and R. Naruse. 2008. "Glacier and Lake Variations in the Mapam Yumco Basin, Western Himalayas of the Tibetan Plateau, from 1974 to 2003 Using Remote Sensing and GIS Technologies." *Journal of Glaciology* 54(188): 933–5.

Ye, Q.-H., et al. 2006a. "Glacier Variations in the Mt. Naimona'Nyi Region, Western Himalayas, in the Last Three Decades." *Annals of Glaciology* 43: 385–9.

Ye, Q.-H., et al. 2006b. "Monitoring Glacier Variations on Geladandong Mountain, Central Tibetan Plateau, from 1969 to 2002 Using Remote-Sensing and GIS Technologies." *Journal of Glaciology* 52(179): 537–45.

Ye, Q.-H., et al. 2007. "Glacier and Lake Variations in the Yamzhog Yumco Basin in the Last Two Decades Using Remote Sensing and GIS Technologies." *Journal of Glaciology* 53(183): 673–6.

Ye, Q.-H., et al. 2009. "Monitoring Glacier and Supra-Glacier Lakes from Space in Mt. Qomolangma Region of the Himalayas on the Tibetan Plateau in China." *Journal of Mountain Science* 6: 101–6.

Ye, Q.-H., et al. 2015. "Glacier Mass Changes in Rongbuk Catchment on Mt. Qomolangma from 1974 to 2006 Based on Topographic Maps and ALOS PRISM Data." *Journal of Hydrology* 530: 273–80.

Ye, Q.-H., et al. 2016. "A Review on the Research of Glacier Changes on the Tibetan Plateau by Remote Sensing Technologies." *Journal of Geo-Information Science* 18(7): 920–30.

Ye, Q.-H., et al. 2017a. "Glacier Changes on the Tibetan Plateau Derived from Landsat Imagery: Mid-1970s – 2000–13." *Journal of Glaciology* 63(238): 273–87.

Ye, Q.-H., et al. 2017b. *Glacier Changes and Its Spatial Differences over the Tibetan Plateau since the 1970s.* Abstract 76A2634. Symposium on Polar Ice, Polar Climate, and Polar Change, International Glaciological Society, Boulder, CO.

Zhang, W., T. Zhou, and L. Zhang (2017). "Wetting and Greening Tibetan Plateau in Early Summer in Recent Decades." *Journal of Geophysical Research: Atmospheres* 122. doi: 10.1002/2017JD026468.

Zhang, Y., et al. 2017. "Glacier Mass Balance and Its Potential Impacts in the Altai Mountains over the Period 1990–2011." *Journal of Hydrology* 553: 662–77.

Zhao, L., et al. 2008. "Regional Changes of Permafrost in Central Asia." In *9th International Permafrost Conference, Fairbanks, Alaska.* Vol. 1, edited by D. L. Kane and K. M. Hinkel, 2061–9. Fairbanks, AK: Institute of Northern Engineering.

Zhao, L., et al. 2010. "Thermal State of Permafrost and Active Layer in Central Asia during the International Polar Year." *Permafrost and Periglacial Processes* 21: 198–207.

Zhou, H., E. Aizen, and V. Aizen. 2017a. "Seasonal Snow Cover Regime and Historical Change in Central Asia from 1986 to 2008." *Global and Planetary Change* 148: 192–216.

Zhou, H., E. Aizen, and V. Aizen. 2017b. "Constructing a Long-Term Monthly Climate Data Set in Central Asia." *International Journal of Climatology.* doi: 10.1002/joc.5259.

Zhou, Y., et al. 2000. *Geocryology in China.* Beijing: Science Press.

Zou, D.-F., et al. 2017 "A New Map of the Permafrost Distribution on the Tibetan Plateau." *Cryosphere* 11: 2527–42.

9 Future Environments in the Polar Regions

As we have seen, polar regions are undergoing rapid and accelerating changes in environmental conditions. Arctic sea ice is shrinking, glaciers and ice caps are retreating worldwide, mass loss from the Greenland and Antarctic ice sheets is increasing, permafrost is thawing, and the tundra is greening. Driving these changes are global warming and polar amplification, modulated by natural variability in the oceans and atmosphere. Projections for the twenty-first century are based on simulations with multiple coupled models of the atmosphere, ocean, sea ice, and biosphere system. These projections are discussed in this chapter, following a brief survey of climate forcings.

9.1 Greenhouse Gas and Aerosol Forcings

The annual increase in CO_2 concentration in the atmosphere rose from a base of 316 ppm in 1959 at rate of 0.94 ppm yr^{-1} in 1959, from 369.5 ppm in 2000 at a rate of 1.62 ppm yr^{-1}, and from 400.8 ppm in 2015 at a rate of 3.05 ppm yr^{-1}. Figure 9.1 illustrates the rise in the CO_2 mixing ratio at Barrow, Alaska, from 1975 to 2015. A record daily maximum of 412 ppm was reached in April 2017 at the National Oceanic and Atmospheric Administration's (NOAA) station on Mauna Loa, Hawaii. This increase constitutes a unique spike in the history of the atmosphere, according to Glikson (2016). During the Paleocene–Eocene Thermal Maximum (PETM; see Chapter 2), carbon release into the atmosphere was about an order of magnitude less than the present amount. The rate of mean annual temperature rise since 1750 CE is also an order of magnitude greater than during the PETM and almost as great as that observed in the Eemian Interglacial, 1,250,000 years ago.

Representative concentration pathways (RCPs) are four greenhouse gas (GHG) concentration trajectories adopted by the Intergovernmental Panel on Climate Change (IPCC) for its Fifth Assessment report in 2013. These RCPS, which are described in Box 9.1, are widely used in climate model simulations.

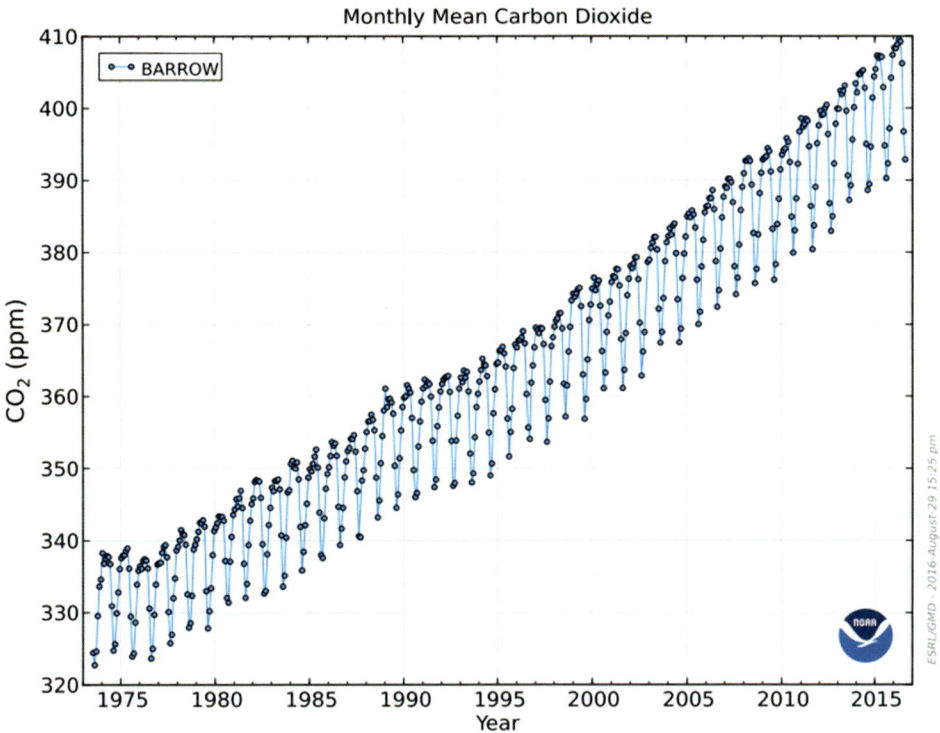

Figure 9.1 The CO_2 emission ratio increase at Barrow, Alaska, 1975–2015.
Source: National Oceanic and Atmospheric Administration, Dr. Pieter Tans.

In RCP 8.5, the concentration of atmospheric CO_2 reaches 936 ppm in 2100 – more than double its present value of 400 ppm. The projected increases in global surface temperature by the end of the twenty-first century based on Earth System Models are 1.0 °C (with a 95 percent confidence interval of 0.3–1.7 °C) and 3.7 °C (2.6–4.8 °C) for RCPs 2.6 and 8.5, respectively (IPCC 2013).

Friedrich et al. (2016) have demonstrated that the global climate sensitivity to CO_2 doubling is highly dependent on the background climate state, based on analysis of a 784-ka record of sea surface temperature (SST) and a transient paleoclimate model simulation. Their work shows that the SAT response to a CO_2 doubling amounts to 1.78 °C per CO_2 doubling for cold phases and 4.88 °C per CO_2 doubling for warm phases. Applying the RCP 8.5 forcing scenario until the year 2100, the anthropogenic forcing results in a global mean surface air temperature (SAT) anomaly of 5.86 °C by 2100. The uncertainties in the sensitivity and the ocean's heat uptake efficiency result in a likely range of 4.78–7.36 °C for the global mean SAT anomaly.

Box 9.1 Representative Concentration Pathways

RCP 2.6, RCP 4.5, RCP 6, and RCP 8.5 are named after a possible range of values for radiative forcing in the year 2100 relative to pre-industrial values: +2.6, +4.5, +6.0, and +8.5 W^{-2}, respectively (Moss et al. 2008).

RCPs 2.6 and 8.5, which specify future emissions of greenhouse gases and aerosols, are two extreme scenarios. The former is a strong mitigation scenario involving major replacement of fossil fuel use with renewable energy and nuclear power, as well as the implementation of new technologies for carbon capture and storage. This emission pathway is representative of scenarios in the literature that lead to very low greenhouse gas concentration levels by 2100. In essence, RCP 2.6 is a "peak-and-decline" scenario. In contrast, RCP 8.5 is a high-emission scenario that assumes rapid population growth, along with modest improvements in the efficiency of energy usage, and thus ongoing high demand for energy from fossil fuels. It is characterized by increasing greenhouse gas emissions over time, representative of scenarios in the literature that lead to high greenhouse gas concentration levels. RCP 4.5 is a stabilization scenario in which total radiative forcing is stabilized shortly after 2100, without overshooting the long-run radiative forcing target level. RCP 6 is a stabilization scenario in which total radiative forcing is stabilized shortly after 2100, without overshoot, through the application of a range of technologies and strategies for reducing greenhouse gas emissions.

An overview of these RCPs may be found in G. P. Wayne "The Beginner's Guide to Representative Concentration Pathways" (www.skepticalscience.com/rcp.php).

There have been several recent attempts to quantify the separate contributions to observed Arctic temperature change from greenhouse gases and other anthropogenic and natural forcing agents. Chylek et al. (2014) analyzed data for annual mean SAT north of 64° N from 1900 to 2012. They found that GHGs and the Atlantic Meridional Oscillation (AMO) were able to account for as much as 86.6 percent of the observed Arctic temperature variance. The positive phase of the Atlantic Multidecadal Oscillation (AMO) was found to govern the inflow of warm, saline water into the West Eurasian shelf seas. There the sea ice extent in the Barents–Kara Sea is uniquely exposed to open ocean conditions. Anthropogenic warming rates were 0.27 C° decade^{-1} from 1955 to 2012 and 0.31 C° decade^{-1} from 1985 to 2012. These results suggest that only about half of the recent (1985–2012) Arctic warming (0.31 °C decade^{-1}, out of 0.64 °C decade^{-1}) may be due to anthropogenic causes.

Najab et al. (2015) quantified the separate contributions to observed Arctic land temperature change from GHGs and other anthropogenic and natural forcing agents over the period 1913–2012. GHGs alone would have warmed the Arctic by 3 °C (2–4 °C) over the past century, but their effect has been offset by 1.8 °C

(1.3–2.2 °C) of cooling induced by aerosol forcing. The offset in the Arctic is substantially greater than the fraction of GHG-induced warming that has been offset by these forcings globally. The calculated net warming of 1.2 °C is very close to the observed warming. Natural forcing did not contribute to the observed long-term warming in a discernible way. It should be noted that Arctic temperatures will rise by 5 °C when global average warming is 2 °C.

Global warming, it should be noted, is largely due to an increase in *nighttime* temperature. Daily minimum temperatures have increased at twice the rate of daytime temperatures since 1950 (roughly 1.0 versus 0.5 °C).

Perlwitz et al. (2015) have examined the trends in 1,000–500 hPa warming since 1979. The main factors responsible for Arctic tropospheric warming are recent decadal fluctuations and long-term changes in SSTs, both located outside the Arctic. Arctic sea ice decline is the largest contributor to near-surface Arctic temperature increases, but accounts for only about 20 percent of the increase.

The Antarctic springtime stratospheric ozone hole was discovered by the British Antarctic Survey from data obtained with a ground-based instrument at Halley Bay, Antarctica (see Section 4.1 and Box 9.2). The October ozone loss was

Box 9.2 The Ozone Hole

In 1974, Mario Molina and Sherwood Rowland proposed that manufactured chlorofluorocarbons used as refrigerants and in aerosol sprays could destroy ozone, analogous to the suggestion of Paul Crutzen (1973) for nitrous oxide. McElroy et al. (1986) reached a similar conclusion for bromine, which is a stronger catalyst of ozone destruction than chlorine. In 1995, Crutzen, Molina, and Rowland were awarded the Nobel Prize in Chemistry for their work on stratospheric ozone. Susan Solomon (1999) proposed that chemical reactions on polar stratospheric clouds (PSCs) in the cold Antarctic stratosphere cause a massive seasonal increase in the amount of chlorine present in active, ozone-destroying forms. PSCs form only when temperatures fall to −80 °C in early spring. In such conditions, the ice crystals in the clouds provide a suitable surface for conversion of unreactive chlorine compounds into reactive chlorine compounds, which readily deplete ozone.

The United States banned aerosol spray cans in 1978, but further action was delayed by resistance from the chemical industry and Reagan-era opposition to regulation. In 1985, however, twenty nations signed the Vienna Convention for the Protection of the Ozone Layer, which established a framework for negotiating international regulations on ozone-depleting substances. That same year, the discovery of the Antarctic ozone hole over Halley Bay was announced by Farman et al. (1985), causing a revival in public attention to the issue.

In 1987, representatives from forty-three nations signed the Montreal Protocol. The goal of this agreement was to reduce CFC production by 50 percent over twelve years. In 1989, this reduction was

Box 9.2 (continued)

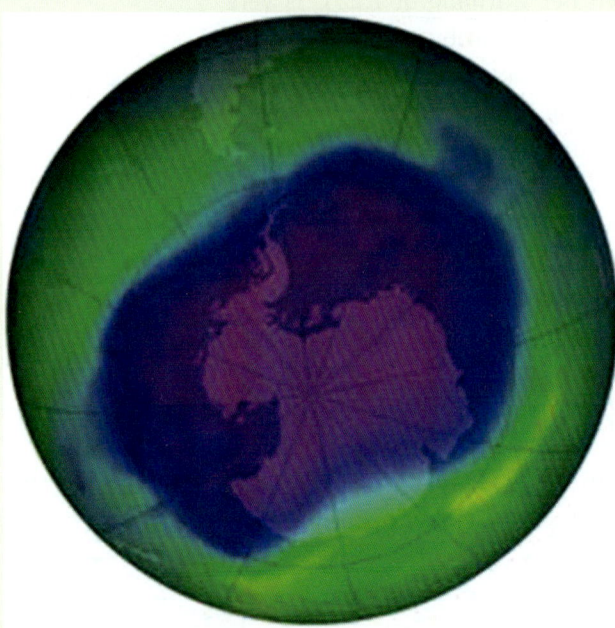

Figure 9.2 Record ozone hole of 29.5 million km² over Antarctica on September 24, 2006.
Source: National Aeronautics and Space Administration, www.nasa.gov/vision/earth/lookingatearth/ozone_record.html.

tightened to 100 percent in ten years. Meanwhile, the US halocarbon industry began to shift its position, although resistance persisted in Europe for several years.

Gradually, CFCs became replaced by the less damaging hydrochlorofluorocarbons (HCFCs). Hydrofluorocarbons (HFCs) were also used to replace CFCs. HFCs, which do not contain either chlorine or bromine, do not contribute at all to ozone depletion, although they are potent greenhouse gases.

The ozone hole forms each year in the austral spring (September–November), when there is a sharp decline (up to 60 percent) in the total ozone over most of Antarctica. During the cold dark Antarctic winter, polar stratospheric clouds form when temperatures drop below −78 °C. These ice clouds are responsible for chemical changes that promote production of chemically active chlorine and bromine.

When sunlight returns to the Antarctic in the spring, the chlorine and bromine activation leads to rapid ozone loss, which then results in the ozone hole over the continent.

Figure 9.2 shows the record extent of the Antarctic ozone hole on September 24, 2006. Although some ozone depletion also occurs in the Arctic during the boreal spring (March–May), wintertime temperatures in the stratosphere are not persistently low for as many weeks, which results in less ozone depletion in this region.

first reported in 1985. Aircraft and satellite measurements confirmed that the springtime ozone loss was a continent-wide feature. The ozone loss is related to halogen (chlorine)-catalyzed chemical destruction, which takes place following spring sunrise in the Antarctic. The chlorine is derived from human-made chlorofluorocarbons (CFCs) that have migrated to the stratosphere and then been broken down by solar ultraviolet radiation, freeing chlorine.

International agreements to regulate the production of chlorine compounds, first adopted in 1987, have begun to reverse the trend in ozone hole strength, but full recovery is not expected to occur until about 2050. In the meantime, ozone loss is promoting cooling of the lower stratosphere, which in turn increases the likelihood of formation of PSCs.

Previdi and Polyani (2017) used a chemistry–climate model to demonstrate that the decrease of oxygen-depleting substances resulting from the Montreal Protocol will lead to a substantial decrease in Antarctic surface mass balance over the period 2006–2065, relative to a hypothetic scenario in which the Montreal Protocol is not implemented. This decrease is predicted to produce an additional 25 mm of global sea level rise by the year 2065, relative to the present day. However, this additional sea level rise is more than offset by a reduction in ocean thermal expansion.

9.2 Sea Ice

Projections of the rate of sea ice loss in the Arctic during this century are one of the most critical issues for climate science. Regional patterns of Arctic sea ice loss in July–October during the twenty-first century have been modeled using the Coupled Model Intercomparison Project, phase 5 (CMIP5) ensemble (forty-two models) by Laliberté et al. (2016). They found that ice-free conditions will occur first on the Eurasian side of the Arctic. Regions along the northern sea route will be more reliably ice free than regions along the Northwest Passage, the transpolar sea route, and the Canadian Arctic archipelago, which will retain substantial sea ice cover past mid-century. Overall, ice-free conditions in the Arctic will likely be confined to September for several decades to come in many regions. Currently, shipping activities are driven by commercial concerns, with little regard being given to trends in ice extent.

Overland and Wang (2013) have reviewed approaches to predicting the timing of Arctic summer sea ice loss by considering three strategies: (1) extrapolation of sea ice volume data, (2) assumptions of several more-rapid loss events like those observed in 2007 and 2012, and (3) climate model projections. Time horizons for a nearly sea ice–free summer for these three approaches are roughly 2020 or earlier, 2030 ± 10 years, and 2040 or later. Jahn et al. (2016) have shown that internal variability alone leads to a prediction uncertainty of about two decades, while scenario uncertainty between the strong (RCP 8.5) and medium (RCP 4.5) forcing scenarios adds at least another five years to the time horizon. Walsh (2008) provided maps of March and September Arctic sea ice extent from 1980–2000 out to 2070–2090 based on the mean of fourteen IPCC models using the A1b scenario (Figure 9.3).

Screen and Williamson (2017) have assessed the likelihood of an ice-free Arctic given global warming of 1.5 and 2 °C. They determined that a summer ice-free

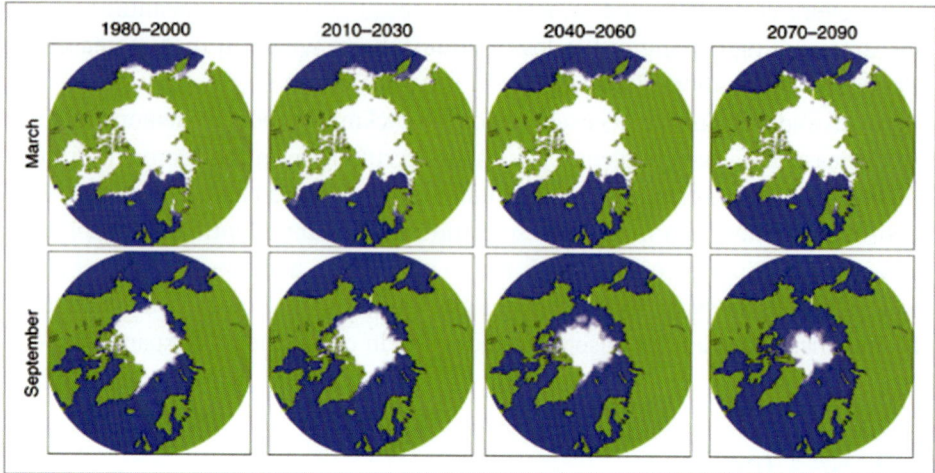

Figure 9.3 Maps of present and projected (2040–2060) sea ice cover for March (top) and September (bottom) from the Max Planck Institute (Hamburg) model, which displays the median ice coverage of fourteen models used in the Intergovernmental Panel on Climate Change Fourth Assessment. Blue shading shows the model's present-day ice coverage; white shows mean ice coverage for 2040–2060 in A1B scenario simulations.
Source: Walsh 2008, S19, figure 14.
Courtesy of Ecological Society of America.

Arctic is virtually certain to be avoided if the 1.5 °C target is achieved, whereas this scenario has a 40 percent likelihood of occurring with a 2 °C warming.

Notz and Stroeve (2016) have demonstrated that sea ice decline in September is linearly related to cumulative CO_2 emissions. The observed linear relationship implies a sustained loss of 3 ± 0.3 m^2 of September sea-ice area per metric ton of CO_2 emission during 1953–2015. Based on this sensitivity, Arctic sea ice will be lost throughout September with an additional 1,000 Gt of CO_2 emissions. If the current emission of 35 Gt CO_2 per year continues, the limit of 1,000 Gt will be reached before mid-century. However, internal variability causes an uncertainty of around 20 years for the first year of a near-complete loss of Arctic sea ice (less than 1 million km^2).

Observations of the Barents Sea ice in winter indicate that it has been reduced to less than one third of the pre-satellite era mean (Onarheim 2017). Projections from a large ensemble simulation show that a winter ice-free Barents Sea will occur for the first time during 2061–2088.

The fastest decrease and greatest variability of sea ice extent during 1983–2013 are in the Beaufort, Chukchi, and East Siberian seas (Letterly et al. 2016). Thinner summer ice in these areas is more susceptible to changes in winter cloud cover, heat and moisture advection, and surface winds. Using two climate reanalyses and satellite data, researchers have shown that increased wintertime surface cloud forcing contributed to the 2007 summer sea ice minimum. An analysis over the period 1983–2013 revealed that cloud forcing anomalies in the East Siberian and Kara Seas precondition the ice pack and explain 25 percent of the variance in late summer sea ice concentration. This finding was supported by Moderate-Resolution Imaging Spectroradiometer (MODIS) cloud cover anomalies, which explain up to 45 percent of the variance in sea ice concentration.

The open water duration in the Arctic has been analyzed by Barnhart et al. (2015), who created maps for the period 1920–2100 using daily output from a thirty-member initial-condition ensemble of business-as-usual climate simulations. The majority of the Arctic nearshore regions will begin leaving the range of internal variability in 2040. Models suggest that ice will cover coastal regions for only half of the year by 2070.

Andry et al. (2017) have examined the strength of the surface albedo feedback in the Arctic for simulations of the future climate (CMIP5 RCP 8.5). Their analysis shows a distinct peak in this feedback around 2100. This maximum is linked to increased seasonality in sea ice cover as sea ice recedes. Sea ice retreat during spring (April–June) turns out to be the dominant factor affecting the strength of the annual surface albedo feedback in the Arctic. Hence, changes in sea ice seasonality and the associated fluctuations in surface albedo feedback strength will exert a time-varying effect on Arctic amplification during the projected warming over the next century. Andry et al.'s results also show that the annual minimum will occur progressively earlier. Around the year 2150, average Arctic sea ice thickness is predicted to be reduced below a threshold of about 0.5 m, spring sea ice extent decreased sharply, and winters too mild to allow the formation of even a thin layer of seasonal ice.

Changes in shipping activity in the Canadian Arctic during 1990–2015 in relation to declining sea ice extent have been analyzed by Pizzolato et al. (2016). Statistically significant increases in shipping activity have been observed in the Hudson Strait (450–550 km traveled per year, or 4.5–5.5 transit equivalents annually), the Beaufort Sea (50–450 km traveled per year, or 0.5–5.5 transit equivalents annually), Baffin Bay (50–350 km traveled per year, or 0.5–3.5 transit equivalents annually), and regions in the southern route of the Northwest Passage (50–250 km traveled per year, or 0.5–2.5 transit equivalents annually). Increases in shipping activity are significantly correlated with reductions in sea ice concentration in regions of the Beaufort Sea, western Parry Channel, western Baffin Bay, and Foxe Basin. Changes in multiyear ice-dominant regions in the

Canadian Arctic appear to more strongly influence changes in shipping activity compared with seasonal sea ice regions.

Melia et al. (2016) simulated the impact of sea ice loss during this century on trans-Arctic shipping using CMIP5 climate models. By mid-century for standard open-water vessels, the frequency of navigable periods doubles, with routes across the central Arctic becoming available. European routes to Asia typically open ten days earlier via the Arctic than alternatives by mid-century, and thirteen days earlier by late century, while North American routes become available four days earlier. Future greenhouse gas emissions have a larger impact by late century: The likelihood of the shipping season lasting four to eight months in RCP 8.5 is double that in RCP 2.6, but both models show substantial interannual variability. Ice-strengthened vessels will likely be able to make Arctic transits for ten to twelve months annually by late century.

The integral role of sea ice in the life of Arctic indigenous peoples has been amply illustrated by Gearheard et al. (2013) in a work that incorporates the perceptions of native residents and details their everyday experiences in utilizing its resources. This analysis demonstrates the sensitivity of these peoples' relationship to the sea ice conditions at Barrow, Alaska; Clyde River; and Baffin Island and Qaanaaq, northwest Greenland. The projected loss of summer ice will greatly impact these communities in the coming decades.

9.3 The Arctic Links to Mid-Latitudes

The relationships between mid-latitude climate and Arctic conditions have received considerable attention in recent years. Ayarzagüena and Screen (2016) analyzed the possible future relation between Arctic sea ice loss and the severity of cold air outbreaks in mid-latitudes. Applying changing temperature thresholds relating to climate conditions of the time, they found that cold air outbreaks do not change in frequency or duration in response to projected sea ice loss. However, they do become less severe, mainly due to advection of warmed polar air, since the dynamics associated with their occurrence are largely not affected. Cold air outbreaks weaken even in mid-latitude regions where the winter mean temperature decreases in response to Arctic sea ice loss.

Cohen (2016) analyzed the role of Arctic sea ice decline on mid-latitude climatic anomalies. This analysis revealed that observed trends in hemispheric circulation over the period of Arctic amplification since the 1980s more closely resemble the variability associated with Arctic boundary forcings than that linked to tropical forcing. Furthermore, analysis of intraseasonal temperature variability showed that the cooling in mid-latitude winter temperatures has been accompanied by an increase in temperature variability and not a decrease, popularly referred to as "weather whiplash."

Vihma (2014) reviewed the effects of Arctic sea ice loss on mid-latitude weather and climate. The reduction in sea ice has increased the heat flux from the ocean to atmosphere in autumn and early winter. This has locally increased air temperature, moisture, and cloud cover and reduced the static stability in the lower troposphere. Studies based on observations, atmospheric reanalyses, and model experiments suggest that the sea ice decline, together with increased snow cover in Eurasia, favors circulation patterns resembling the negative phase of the North Atlantic Oscillation (NAO)/Arctic Oscillation (AO). They suggested large-scale pressure patterns include a high over Eurasia, which favors cold winters in Europe and northeastern Eurasia. However, several other factors generate a large interannual variability and often mask the effects of sea ice decline. In addition, the small sample of years with a large sea ice loss makes it difficult to distinguish the effects directly attributable to sea ice conditions.

Screen (2017) performed simulations of atmospheric circulation changes to regional and pan-Arctic ice loss with the Met Office Unified Model (6.6.3). The percentage of the area 30–90° N over which significant responses were found in 500-hPa geopotential heights were as follows:

December–January–February: 28.8 percent for the Sea of Okhotsk and 12.3 percent for the Barents–Kara seas
March–April–May: 14.9 percent for the Barents–Kara seas and 13.7 percent for the Beaufort–Chukchi seas
June–July–August: 17.7 percent for the Bering Sea, 15.3 percent for the Beaufort–Chukchi seas, and 13.8 percent for the Greenland Sea
September–October–November: all values ≤8.2 percent

A unique effect occurs with sea ice loss in the Barents–Kara seas. This loss drives a weakening of the stratospheric polar vortex, followed months later by a tropospheric circulation response that resembles the negative NAO. Screen (2017) also found that whereas regional ice losses mainly cause cooling over the high- and mid-latitude continents, pan-Arctic ice losses cause continental warming.

The effects of different Arctic processes on winter atmospheric circulation on weekly to monthly time scales have been studied by Kretschmer et al. (2016) for 1979–2014. Barents and Kara sea ice concentrations are important external drivers of the mid-latitude circulation, influencing winter AO via tropospheric mechanisms and through processes involving the stratosphere. Eurasian snow cover also has a causal effect on sea level pressure in Asia.

9.4 Snowfall

Rising temperatures will have major effects on the rain/snow ratio and the duration of the snow cover season, but snowfall amounts in winter will be affected by the warming-induced increases in atmospheric vapor content. Simulations performed with twenty-four coupled atmosphere–ocean general circulation models (GCMs) from the CMIP5 projections of northern hemisphere daily snowfall events under the RCP 8.5 emissions scenario were analyzed for 2021–2050 and 2071–2100 and compared to the historical period of 1971–2000 by Danco et al. (2016). Large portions of the northern hemisphere, including much of Canada, Tibet, northern Scandinavia, northern Siberia, and Greenland, are projected to experience increases in average daily snowfall and event frequency in midwinter. In warmer months, the regions with increased snowfall are predicted to become fewer in number and to be limited to northern Canada, northern Siberia, and Greenland. These increases result from increases in the water-holding capacity of the atmosphere by 7 percent per 1 °C of temperature rise and from enhanced water vapor transport from the tropics. A reinforcement of the early-period tendencies is predicted to occur in the later period, suggesting that the continuing increase in temperature during the twenty-first century will be the primary driver of the changes in snowfall intensity and frequency.

9.5 Permafrost Thawing

Permafrost is deep over wide areas of northern high latitudes; only the upper few meters are considered to be susceptible to thaw. Several GCM experiments have explicitly addressed the thaw and retreat of permafrost conditions in the northern hemisphere. Lawrence et al. (2008) simulated near-surface soil temperature in the Community Land Model (CLM). The model projection of near-surface permafrost degradation was assessed after explicitly accounting for the thermal and hydrologic properties of soil organic matter and for a soil column of 50 m or more. The rate of near-surface permafrost degradation, in response to strong simulated Arctic warming (+7.5 °C) over Arctic land areas from 1900 to 2100 (A1B greenhouse gas emissions scenario), is slower in the improved version of CLM, particularly during the early twenty-first century. Even at the depressed rate, however, the warming is enough to drive near-surface permafrost extent sharply down by 2100 (Figure 9.4). In the high-emission scenario, the area with permafrost in the upper 3-m soil layer shrinks from 10 million km^2 to slightly more than 2.5 million km^2 by the year 2050 and decreases further to about 1 million km^2 by 2100. In the low-emission scenario, the upper permafrost area shrinks to about

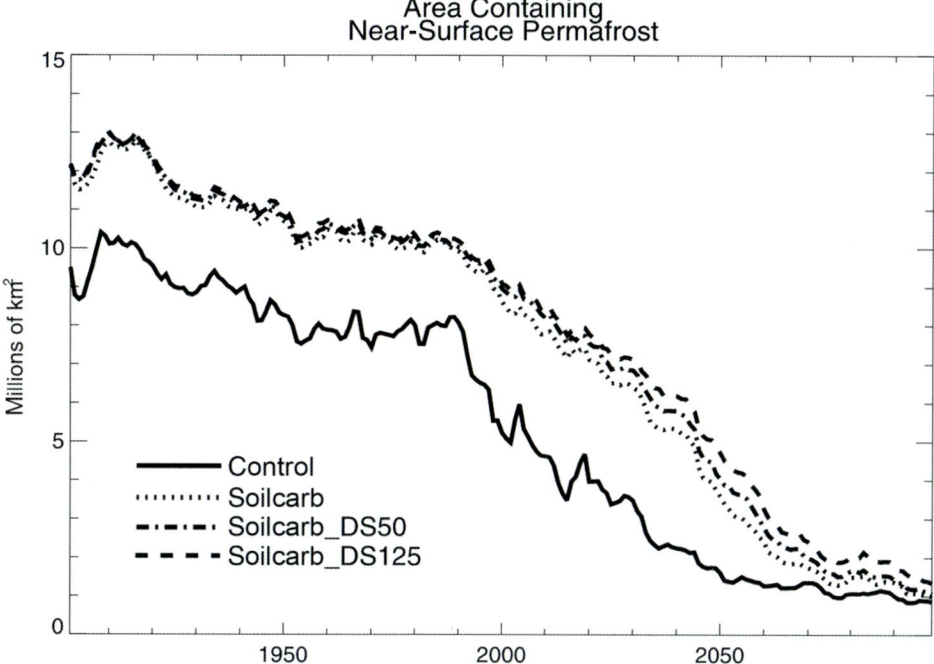

Figure 9.4 Time series of total area containing near-surface permafrost (north of 45° N and excluding ground underneath glaciers) for selected experiments.
Source: Lawrence et al. 2008, 9, figure 5.

3.75 million km² by 2100. There are complex feedbacks involving snow cover, shrub coverage, and Arctic sea ice extent. For the CMIP5 models, Slater and Lawrence (2013) estimate a reduction of the 2080–2099 near-surface continuous and discontinuous permafrost area of 37 percent for RCP 2.6 and 81 percent for RCP 8.5, compared to the 1986–2005 diagnosed near-surface permafrost area.

Biskaborn et al. (2015) used fifteen climate models to determine, for RCP 4.5 and 8.5 scenarios, the projected temperature increases at Global Terrestrial Network for Permafrost (GTN-P) and Circumpolar Active Layer Monitoring (CALM) sites for 2070–2099 over 1970–1999. For RCP 4.5, most bore holes and CALM sites are located in relatively narrow zones of less extreme projected temperature change (3–6 °C for Thermal State of Permafrost [TSP] and 2–5 °C for CALM). The high-emission scenario RCP 8.5 projects a more extreme temperature increase for larger areas and more GTN-P monitoring sites are located in zones of up to a 10 °C potential temperature rise.

A key concern with permafrost thawing is the potential for massive release of frozen carbon, especially in the form of methane (CH_4). Cooper et al. (2017) have reported fluxes of more than 20 g CH_4 m^{-2} yr^{-1} from thawing peatlands in northern Canada. However, less than 10 percent of this methane was derived from previously frozen carbon; instead, fluxes were driven by anaerobic decomposition of recent organic carbon inputs. These findings suggest that thaw-induced changes in surface wetland area may determine methane release from northern peatlands.

Thermokarst formation has been assessed by Olefeld et al. (2016) for the circum-Arctic area. The distribution of thermokarst landscapes, defined as landscapes composed of current thermokarst landforms and areas susceptible to future thermokarst development, was mapped from a set of six spatial circumpolar data layers. These layers described key landscape characteristics: (1) permafrost zonation (isolated, sporadic, discontinuous or continuous); (2) ground ice content (less than 10, 10–20, or more than 20 percent); (3) sedimentary overburden thickness (thin or thick); (4) terrestrial ecoregion (boreal or tundra); (5) topographic ruggedness (flat, undulating, hilly or mountainous/rugged); and (6) landscape coverage of permafrost peat soils (less than 10, 10–30, or more than 30 percent). Spanning 3.6×10^6 km^2, thermokarst landscapes are estimated to cover approximately 20 percent of the northern permafrost region, with approximately equal contributions from three landscape types – wetland, lake, and hillslope –where characteristic thermokarst landforms occur.

Wetland thermokarst landscapes include thermokarst bogs, fens, and shore fens, whose development is dependent on hydrological landscape position. Development causes transition from boreal forest or tundra dry shrub ecosystems into sedge or *Sphagnum* moss wetland ecosystems with near-surface water tables. Wetland thermokarst landforms are typically 0.5–10 ha in size, but can reach sizes up to 100 ha. Thermokarst development leads to 1–3 m of land settlement (Olefeld et al. 2016). There is very high wetland thermokarst landscape coverage in the extensive boreal peatland regions of the West Siberian Lowlands, the Hudson Bay Lowlands, and the Mackenzie River valley.

Lake thermokarst landforms include deep, shallow, and glacial thermokarst lakes' thermokarst lake basins; alas basins; and thaw sinks. Land settlement varies between 1 and 20 m, with potentially substantial lateral soil movement. Very high regional coverage of lake thermokarst landscapes is found in lowland tundra regions, including the Yukon delta, the Alaska north slope, and the coastal regions along the Kara Sea, Laptev Sea, and East Siberian Sea.

Hillslope thermokarst landscapes include active layer detachment slides, retrogressive thaw slumps, thermal erosion gullies, beaded streams, and thermokarst water tracks. They are most likely to be found in undulating and hilly topography. These thermokarst landforms are generally smaller than landforms of the other thermokarst landscapes, but can sometimes cover 10 ha.

Vulnerability to thermokarst development is likely to increase this century due to both climate change and the associated higher frequencies of disturbances such as wildfire and floods.

9.6 Tundra Vegetation

The expansion of tundra vegetation is expected to continue as a result of Arctic warming trends. The Normalized Difference Vegetation Index (NDVI) trend for 91 percent of the non-water, non-snow land area of Canada and Alaska was determined by Ju and Masek (2016) using the peak-summer Landsat surface reflectance data for 1984–2012. Their analysis indicated that 29.4 and 2.9 percent of the land area of Canada and Alaska showed statistically significant positive (greening) and negative (browning) trends, respectively. The greening occurred primarily in the tundra of western Alaska, along the north coast of Canada, and in northeastern Canada; the most intensive and extensive greening occurred in Quebec and Labrador. The browning occurred mostly in the boreal forests of eastern Alaska.

Shrub growth appears to be affecting the ability of caribou herds to obtain food because the shrubs are non-edible – one factor leading to the decline in caribou populations (Fauchald et al. 2017). Climatic warming is also enhancing the northward spread of parasites that severely impact ungulate populations.

9.7 Glacier and Ice Cap Recession

During the last few decades, glacier recession (ongoing since the end of the Little Ice Age) has accelerated in most parts of the world, a trend that is projected to continue. Using dynamic ice-flow models, the response of twelve glaciers and ice caps to global warming over the twenty-first century has been analyzed by Oerlemans et al. (1998). The ice bodies studied ranged in area from 3 to 1,400 km^2 and were located in Europe, Iceland, New Zealand, and Antarctica. The researchers found that if warming is restricted to 0.018 °C yr^{-1} and precipitation increases by 10 percent per 1 °C of warming, ice losses will be restricted to 10–20 percent of their 1990 values. However, with a warming rate of 0.048 °C yr^{-1} (4.8 °C century^{-1}) and no increase in precipitation, little glacier ice would remain in 2100. Projected glacier ice losses in Central Asia by the end of this century were discussed in Chapter 8.

Gordon et al. (2017) have investigated the future of the Barnes ice cap in Baffin Island, which is the last remnant of the Laurentide ice sheet. It has been sustained during the Holocene by its topography but no longer has an accumulation zone.

Modeling for RCP scenarios 2.6–8.5 suggests that this ice cap will disappear between 150 and 530 years from now, despite the near-minimum levels of summer insolation. Gordon et al. note that only during the warmest interglacials MIS 5e and 11c was the Barnes ice cap as small as it is at present, but its projected disappearance points to the exceptional nature of the current greenhouse gas forcing.

9.8 The Ice Sheets, Sea Level Rise, and Coastal Erosion

Polar temperatures over the last several million years have, at times, been slightly warmer than today, yet global mean sea level was 6–9 m higher as recently as the Last Interglacial (130,000–115,000 years ago) and possibly higher during the Pliocene epoch (about 3 MYA). In both cases, the West Antarctic ice sheet has been implicated as the primary contributor, hinting at its future vulnerability. DeConto and Pollard (2016) developed a model that couples ice sheet and climate dynamics – including processes linking atmospheric warming with hydrofracturing of buttressing ice shelves and structural collapse of marine-terminating ice cliffs – that is calibrated against Pliocene and Last Interglacial sea-level estimates. When this model was applied to future GHG scenarios, Antarctica was shown to have the potential to contribute more than 1 m of sea-level rise by 2100 and more than 15 m by 2500, if emissions continue unabated. In this model, atmospheric warming is predicted to soon become the dominant driver of ice loss, but prolonged ocean warming will delay its recovery for thousands of years. Global sea levels are projected to rise approximately 1 m by 2100 as a result of ocean thermal expansion and ice melt (glaciers, Greenland, and the West Antarctic ice sheet [WAIS]).

Nauels et al. (2017) employ a sea level model that emulates global-mean long-term process-based model projections for all major sea level components. They estimate a 0.35–0.56 m rise in seal level will occur by 2100 based on the RCP 2.6 scenario, 0.45–0.67 m based on RCP 4.5, 0.46–0.71 m based on RCP 6.0, and 0.65–0.97 m based on RCP 8.5.

The consequences will be felt primarily in densely populated coastal areas of mid- and tropical latitudes (Bangladesh, Thailand, Florida, and so on). However, in the Arctic, coastal effects will be substantial. Since 2000, the coastal cliffs that predominantly comprise permafrost in silts have retreated by an average of 1 m per year, but locally by as much as 20–30 m. These losses are especially apparent in the Laptev, East Siberian, and Beaufort seas.

On a 60-km section of the Beaufort Sea coast of Alaska, east of Barrow, Jones et al. (2009) reported that mean annual erosion rates increased from 6.8 m yr^{-1}

over the 1955–1979 period, to 8.7 m yr^{-1} during 1979–2002, to 13.6 m yr^{-1} during 2002–2007. They also observed that spatial patterns of erosion became more uniform across shoreline types with different degrees of ice richness. Further, during the remainder of the 2007 ice-free season, 25 m of erosion occurred locally in the absence of a westerly storm event. Sea-level rise of 2.61 ± 0.47 mm yr^{-1} has also been documented along Arctic coastlines from 1954 to 2007, with an abrupt increase noted between 2000 and 2007 (Richter-Menge 2008).

9.9 Hydrological Cycle Effects

Large increases in precipitation are expected to occur in polar regions in the future, for two reasons. First, both observations and climate models indicate that the warming rate has been and will continue to be the highest there, and warmer air can hold more water vapor. Second, the warming will reduce the extent of sea ice, thereby allowing more evaporation from open water. Accordingly, runoff is expected to increase as well as the area of wetlands and lakes.

Rawlins et al. (2010) investigated trends in observed and modeled precipitation over the Arctic. For precipitation time series (1950–1999) from nine CMIP3 GCMs, the linear trends were all positive, ranging from 0.12 to 0.63 mm yr^{-1}, with a multiple-model mean trend of 0.37 mm yr^{-1}. Three observational data sets showed little evidence of any trend, though these data were uncorrected for gauge undercatch. However, a circumpolar increase of 12 percent has occurred for heavy precipitation events since 1950 for the region north of 50° N, with most of the increase taking place in Eurasia. From 1950 through 2004, annual pan-Arctic discharge exhibits a significant, positive trend of 0.23 mm yr^{-1} (5.3 km^3 yr^{-1}) (see also Section 5.5).

A decline in lake abundance and area has been noted throughout the region of discontinuous, sporadic, and isolated permafrost of Siberia (Smith et al. 2005). Between 1973 and 1997–1998, the total number of lakes greater than 40 ha in size decreased by 11 percent, from 10,882 to 9,712. Most shrank to sizes less than 40 ha. Total regional lake surface area decreased by 93,000 ha (approximately 6 percent); 125 lakes vanished completely. In contrast, increases in lake area and number occurred across the continuous permafrost zone over the same time span. Total lake area increased by 13,300 ha (12 percent), and the number of lakes increased from 1,148 in 1973 to 1,197 by 1997–1998 (4 percent).

According to Anisimov and Vaughan (2007), projected increases of runoff by 2080 in the Arctic are generally in the range of 10–30 percent. There are also links between this phenomenon and permafrost thawing, which contributes to enhanced runoff and expanded wetlands.

9.10 Future Polar Research and Development

In September 2016, US President Obama hosted a twenty-five-nation ministerial meeting to advance international research (www.whitehouse.gov/the-press-office/2016/09/28/fact-sheet-united-states-hosts-first-ever-arctic-science-ministerial). The four themes of this ministerial and the participants' joint statement were as follows:

1. Arctic science challenges and their regional and global implications
2. Strengthening and integrating Arctic observations and data sharing
3. Applying expanded scientific understanding of the Arctic to build regional resilience and to shape global responses
4. Empowering citizens through science, technology, engineering, and mathematics (STEM) education leveraging Arctic science

Highlights of the meeting included the European Union's announcement of a new five-year project (2016–2021), coordinated by Norway, to develop an Integrated Arctic Observing System (INTAROS). The European Union will also initiate two new projects to elucidate the impact of the changing Arctic on the weather and climate of the northern hemisphere.

In 2007, the US Inter-agency Arctic Research Policy Committee (IARPC) directed NOAA and the National Science Foundation (NSF) to develop an Arctic Observing Network (AON) as part of the Study of Environmental Arctic Change (SEARCH) program to provide a lasting legacy from the International Polar Year. AON includes physical, biological, and human observations and is part of NOAA's Arctic Program (www.arctic.noaa.gov/).

In 2011, the Arctic Council established the Sustaining Arctic Observing Networks (SAON) under the Arctic Monitoring and Assessment Program (AMAP) Working Group. Its purpose is to support and strengthen the development of multinational engagement for sustained and coordinated pan-Arctic observing and data sharing systems. It operates through the Arctic Data Committee (ADC) and the Committee on Observations and Networks (CON). Details can be found at www.arcticobserving.org/.

In May 2017, the Year of Polar Prediction (YOPP) was launched; it will continue to mid-2019. YOPP will include three Special Observing Periods (SOPs):

- February 1 to March 31, 2018 in the Arctic
- July 1 to September 30, 2018 in the Arctic
- November 16, 2018 to February 15, 2019, in the Antarctic

The purpose of the SOPs is to enhance the routine observations in an attempt to close the gaps in the conventional Arctic and Antarctic observing systems for an extended period of time (www.polarprediction.net/). The aim is to enhance the current observation network by (1) more frequent observations from existing platforms and/or (2) adding observations in regions where the observation network is not sufficiently dense. This will include buoy networks and Argo floats.

In autumn 2019, the Multidisciplinary Drifting Observatory for the Study of Arctic Climate (MOSAiC) will become the first year-round expedition into the central Arctic (www.mosaicobservatory.org/index.html). The Polarstern station will be frozen into the ice at $84°$ N, $120°$ E and will drift southwestward into the Greenland Sea. The program will be led by the Alfred Wegener Institute, Helmholtz Centre for Polar and Marine Research (AWI), the Arctic and Antarctic Research Institute (AARI), and the University of Colorado, Cooperative Institute for Research in Environmental Sciences (CIRES).

In terms of Antarctic study, the Scientific Committee on Antarctic Research (SCAR) is an interdisciplinary committee of the International Council for Science (ICSU), established in 1957. It has Science Groups on geosciences, life sciences, and physical sciences that initiate, develop, and coordinate interdisciplinary, international research in the Antarctic and Southern Ocean. SCAR also has a Scientific Research Program on Antarctic Climate Change in the 21st Century (AntClim[21]) and another on Past Antarctic Ice Sheet dynamics (PAIS), as well as a Standing Committee on Antarctic Data Management. In 2016, it issued the SCAR Strategic Plan 2017–2022 (http://doi.org/10.5281/zenodo.229139). Its goals include enhancing the research capacity of SCAR countries and increasing public awareness of the Antarctic.

SUMMARY

The rates of increase in atmospheric carbon dioxide concentration and in global air temperature since 1750 are about an order of magnitude greater than the corresponding rates in the Paleocene–Eocene Thermal Maximum. Four greenhouse gas (GG) representative concentration pathways (RCPs) for the twenty-first century have been defined by the IPCC as radiative forcings relative to pre-industrial levels and are widely used in modeling studies. The CO_2 concentration in RCP 8.5 (W m^2) reaches 936 ppm by 2100, more than double the present level of 400 ppm. The sensitivity of global temperature to CO_2 doubling has been found to depend on the background climate: It is approximately 1.8 K during cold phases and approximately 4.9 K during warm phases.

The effects of GHG and the Atlantic Multidecadal Oscillation have been shown to account for 86 percent of Arctic temperature variance from 1900 to 2012. About half of the 1985–2012 warming of the Arctic seems attributable to anthropogenic causes. Arctic warming of 3 °C during 1913–2012 was offset by aerosol cooling of 1.8 °C. The Antarctic spring stratospheric ozone hole was first reported in 1985. Adoption of the Montreal Protocol of 1987 gradually slowed the growth of this hole, but full recovery will not occur before 2050. There is less ozone depletion in the spring Arctic stratosphere, where temperatures are higher.

The Arctic Ocean in July–October will be most reliably ice-free by the middle of the twenty-first century in the Eurasian Arctic. Various analytical approaches suggest a nearly ice-free Arctic summer will emerge between 2020 and 2040. Internal variability alone adds a two-decade uncertainty to these predictions. There is a 40 percent likelihood of an ice-free Arctic summer occurring with global warming of 2 °C. September sea ice decline is linearly related to atmospheric CO_2 concentration. Based on this sensitivity, September will be ice-free before 2050. The surface albedo feedback will reach a peak in 2100, associated with spring sea ice cover. Around 2150, winters will become too mild to permit sea ice formation.

By mid-century, the frequency of navigable periods for ocean-going vessels is expected to double. By the late twenty-first century, ice-strengthened vessels will be able to transit the Arctic for 10–12 months out of the year.

Mid-latitude cold air outbreaks in winter do not become more frequent or longer-lasting as a result of sea ice loss, but they do become less severe. This kind of winter cooling is also associated with an increase in temperature variability. The small number of years with large sea ice loss makes statistical analysis difficult. Sea ice decline and increased Eurasian snow cover favor negative NAO/AO patterns that promote cold winters in Europe and northeastern Eurasia. Sea ice loss in the Sea of Okhotsk affects 29 percent of the area 30–90° N for winter, or 500 hPa at the geopotential height. The corresponding percentage for the Bering Sea in summer is almost 18 percent.

CMIP5 projections of daily snowfall under RCP 8.5 suggest increased winter snowfall will occur over large parts of the northern hemisphere in the mid- to late twenty-first century. In warmer months, the increases will be limited to northern Canada, Greenland, and northern Siberia.

GCM experiments indicate that by 2100 the near-surface permafrost area will have shrunk from 10 million km^2 to just 1 million km^2 for the high-emission scenario. Thermokarst landscapes are estimated to cover approximately 20 percent of the northern permafrost region. Wetland thermokarst has experienced land settlement of 1–20 m. Vulnerability to thermokarst development will increase this century due to climate change and more frequent floods and wildfires.

Tundra vegetation in Canada and Alaska, based on 1984–2012 NDVI trends, has demonstrated significant greening in Quebec–Labrador, the north coast of Canada, and western Alaska. Conversely, tundra browning has affected boreal forest in eastern Alaska.

Based on dynamic ice flow modeling, with twenty-first century warming of 4.8 °C, little glacier ice is projected to remain in 2100. The Barnes ice cap in Baffin Island is projected to disappear within 150–530 years.

The Antarctic ice sheet has the potential to cause a sea level rise of 1 m by 2100 if the emission of GHGs is not reduced. Ocean thermal expansion and land ice melt will lead to coastal impacts not only in the mid-latitudes and the tropics, but also in the Arctic due to coastal erosion and cliff retreat in ice-rich silts. Coastal recession rates in the Beaufort Sea coast of Alaska have already reached 14 m yr^{-1}.

Precipitation in the Arctic is predicted to increase in the twenty-first century due to the capacity of warmer air to hold more water; moreover, the loss of sea ice will increase evaporation. The multiple-model mean trend for 1950–1999 was 0.37 mm yr^{-1}. Lake abundance and area decreased (increased) during 1973–1998 in the discontinuous and sporadic (continuous) permafrost regions of Siberia. By 2080, runoff is projected to increase by 10–30 percent in the Arctic.

Research initiatives in the Arctic and Antarctic include the Year of Polar Prediction (YOPP). The Strategic Plan of the Scientific Committee for Antarctic Research (SCAR) for 2017–2022 advocates enhancing the research capacity of SCAR countries.

QUESTIONS

1. What causes the stratospheric ozone hole in austral spring?
2. Describe the projections for Arctic sea ice up to 2100.
3. What are the relationships between Arctic sea ice loss and mid-latitude climate anomalies?
4. How will the cryosphere in the Arctic have changed by 2100?
5. Discuss the projected changes in Arctic vegetation and the landscape during this century and their causes.
6. What are the consequences of sea level rise in the Arctic?
7. Discuss the causes and impacts of changes in the Antarctic ice sheet on global sea level.
8. What are the projected changes in the Arctic hydrological cycle in this century?

References

Andry, O., R. Bintanja, and W. Hazeleger. 2017. "Time-Dependent Variations in the Arctic's Surface Albedo Feedback and the Link to Seasonality in Sea Ice." *Journal of Climate* 30: 393–410.

Anisimov, O. A., and D. G. Vaughan. 2007. "Polar Regions." In *Climate Change 2007: Working Group II: Impacts, Adaptation and Vulnerability*, edited by M. L. Parry et al. Cambridge: Cambridge University Press.

Ayarzagüena, B., and J. A. Screen. 2016. "Taking the Chill Off: Future Arctic Sea-Ice Loss Reduces Severity of Cold Air Outbreaks in Midlatitudes." *Geophysical Research Letters*. doi: 10.1002/2016GL068092.

Barnhart, K. R., et al. 2015. "Mapping the Future Expansion of Arctic Open Water." *Nature Climate Change* 6: 280–5.

Biskaborn, B. K., et al. 2015. "The New Database of the Global Terrestrial Network for Permafrost (GTN-P)." *Earth System Science Data* 7: 245–59.

Chylek, P., et al. 2014. "Isolating the Anthropogenic Component of Arctic Warming." *Geophysical Research Letters* 41: 3569–78.

Cohen, J. 2016. "An Observational Analysis: Tropical Relative to Arctic Influence on Midlatitude Weather in the Era of Arctic Amplification." *Geophysical Research Letters* 43: 5287–94.

Cooper, M. D. A., et al. 2017. "Limited Contribution of Permafrost Carbon to Methane Release from Thawing Peatlands." *Nature Climate Change* 7: 507–11.

Crutzen, P. 1973. "A Discussion of the Chemistry of Some Minor Constituents in the Stratosphere and Troposphere." *Pure and Applied Geophysics* 106: 1385–99.

Danco, J. F., et al. 2016. "Effects of a Warming Climate on Daily Snowfall Events in the Northern Hemisphere." *Journal of Climate* 29: 6295–318.

DeConto, R. M., and D. Pollard. 2016. "Contribution of Antarctica to Past and Future Sea-Level Rise." *Nature* 531: 591–602.

Farman, J. C., B. G. Gardiner, and J. D. Shanklin. 1985. "Larfe Losses of Total Ozone in Antarctica Reveal Seasonal ClO_x/NO_x Interactions." *Nature* 315: 207–10.

Fauchald, P., et al. 2017. "Arctic Greening from Warming Promotes Declines in Caribou Populations." *Science Advances* 3: 1601365.

Friedrich, T., et al. 2016. "Nonlinear Climate Sensitivity and Its Implications for Future Greenhouse Warming." *Science Advances* 2(11): e1501923.

Gearheard, S. F., et al. 2013. *The Meaning of Ice: People and Sea Ice in Three Arctic Communities*. Lebanon, NH: University Press of New England.

Glikson, A. 2016. "Cenozoic Mean Greenhouse Gases and Temperature Changes with Reference to the Anthropocene." *Global Change Biology* 22: 3843–58.

Gordon, A., et al. 2017. "The Projected Demise of Barnes Ice Cap: Evidence of an Unusually Warm 21st Century Arctic." *Geophysical Research Letters* 44. doi: 10.1002/2016GL072394.

Intergovernmental Panel on Climate Change (IPCC). 2013. *Climate Change: The Physical Science Basis. Contribution of Working Group I to the Fifth Assessment Report of the Intergovernmental Panel on Climate Change*. Cambridge: Cambridge University Press.

Jahn, A., et al. 2016. "How Predictable Is the Timing of a Summer Ice-Free Arctic?" *Geophysical Research Letters*. doi: 10.1002/2016GL070067View.

Jones, B. M., et al. 2009. "Increase in the Rate and Uniformity of Coastline Erosion in Arctic Alaska." *Geophysical Research Letters* 36: L03503.

Ju, J., and J. G. Masek. 2016. "The Vegetation Greenness Trend in Canada and US Alaska from 1984–2012 Landsat Data." *Remote Sensing of the Environment* 176: 1–16.

Kretschmer, M., et al. 2016. "Using Causal Effect Networks to Analyze Different Arctic Drivers of Midlatitude Winter Circulation." *Journal of Climate* 29: 4069–81.

Laliberté, F., S. E. L. Howell, and P .J. Kushner. 2016. "Regional Variability of a Projected Sea Ice-Free Arctic during the Summer Months." *Geophysical Research Letters* 43: 256–63.

Lawrence, D. M., et al. 2008. "Sensitivity of a Model Projection of Near-Surface Permafrost Degradation to Soil Column Depth and Representation of Soil Organic Matter." *Journal of Geophysical Research* 113: F02011.

Letterly, A., J. Key, and Y.-H. Liu. 2016. "The Influence of Winter Cloud on Summer Sea Ice in the Arctic, 1983–2013." *Geophysical Research Letters*. doi: 10.1002/ 2015JD024316.

McElroy, M. B., et al. 1986. "Reductions of Antarctic Ozone due to Synergistic Interactions of Chlorine and Bromine." *Nature* 321: 729–32.

Melia, N., K. Haines, and E. Hawkins. 2016. "Sea-Ice Decline and 21st Century Trans-Arctic Shipping Routes." *Geophysical Research Letters* 43(18): 9720–8.

Molina, M. J., and F. S. Rowland. 1974. "Stratospheric Sink for Chlorofluoromethanes: Chlorine Atom-Catalysed Destruction of Ozone." *Nature* 249(5460): 810–2.

Moss, R., et al. 2008. *Towards New Scenarios for Analysis of Emissions, Climate Change, Impacts, and Response Strategies: Technical Summary*. Geneva: Intergovernmental Panel on Climate Change.

Najab, M. R., F. W. Zwiers, and N. P. Gillett. 2015. "Attribution of Arctic Temperature Change to Greenhouse-Gas and Aerosol Influences." *Nature Climate Change* 5: 246–9.

Nauels, A., et al. 2017. "Synthesizing Long-Term Sea Level Rise Projections: The MAGICC Sea Level Model v2.0." *Geoscientific Model Development* 10: 2495–524.

Notz, D., and J. Stroeve. 2016. "Observed Arctic Sea-Ice Loss Directly Follows Anthropogenic CO_2 Emission." *Science*. doi: 10.1126/science.aag2345.

Oerlemans, J., et al. 1998. "Modelling the Response of Glaciers to Climate Warming." *Climate Dynamics* 14: 267–74.

Olefeld, D., et al. 2016. "Circumpolar Distribution and Carbon Storage of Thermokarst Landscapes." *Nature Communications* 7. doi: 10.1038/ncomms13043.

Onarheim, Å. M. 2017. "Toward an Ice-Free Barents Sea." *Geophysical Research Letters*. doi: 10.1002/2017GL074304.

Overland, J. E., and M.-Y. Wang. 2013. "When Will the Summer Arctic Be Nearly Sea Ice Free." *Geophysical Research Letters* 40: 2097–101.

Perlwitz, J., M. Hoerling, and R. Dole. 2015. "Arctic Tropospheric Warming: Causes and Linkages to Lower Latitudes." *Journal of Climate* 28: 2154–67.

Pizzolato, L., et al. 2016. "The Influence of Declining Sea Ice on Shipping Activity in the Canadian Arctic." *Geophysical Research Letters* 43: 12146–54.

Previdi, M., and L. M. Polyani. 2017. "Impact of the Montreal Protocol on Antarctic Surface Mass Balance and Implications for Global Sea Level Rise." *Journal of Climate* 30. doi: 10.1175/JCLI-D-17-0027.1.

Rawlins, M. A., et al. 2010. "Analysis of the Arctic System for Freshwater Cycle Intensification: Observations and Expectations." *Journal of Climate* 23: 5715–37.

Richter-Menge, J. 2008. *Arctic Report Card*. Silver Spring, MD: National Oceanic and Atmospheric Administration. www.arctic.noaa.gov/reportcard.

Screen, J. A. 2017. "Simulated Atmospheric Response to Regional and Pan-Arctic Sea Ice Loss." *Journal of Climate* 30: 3945–62.

Screen, J. A., and D. Williamson. 2017. "Ice-Free Arctic at 1.5° C?" *Nature Climate Change* 7: 230–1.

Slater, A. G., and D. M. Lawrence. 2013. "Diagnosing Present and Future Permafrost from Climate Models." *Journal of Climate* 25: 5608–23.

Smith, L. C., et al. 2005. "Disappearing Arctic Lakes." *Science* 308: 1429.

Solomon, S. 1999. "Stratospheric Ozone Depletion: A Review of Concepts and History." *Reviews of Geophysics* 37: 275–316.

Vihma, T. 2014. "Effects of Arctic Sea Ice Decline on Weather and Climate: A Review." *Surveys in Geophysics* 35: 1175–214.

Walsh, J. E. 2008. "Climate of the Arctic Marine Environment." *Ecological Applications* 18 (2 suppl): S3–22.

Appendix A: Polar Institutes

Alfred Wegener Institute for Polar and Marine Research (AWI), Bremerhaven, Germany. www.awi.de/

Antarctic Climate & Ecosystems Cooperative Research Centre, Hobart, Tasmania. www.acecrc.org.au

Antarctic Research Centre, Victoria University of Wellington, New Zealand. www.victoria.ac.nz/

Arctic and Antarctic Research Institute (AARI), St. Petersburg, Russia. www.aari.ru/index_en.html

Australian Antarctic Division, Kingston, Tasmania. www.antarctica.gov.aun

Austrian Polar Research Institute, University of Vienna, Austria. www.polarresearch.at/

British Antarctic Survey (BEDMAP2) https://www.antarctica.ac.uk//bas_research/data/access/bedmap/

Bulgarian Antarctic Institute, Sofia, Bulgaria. www.bai-bg.net/

Byrd Polar and Climate Research Center, Ohio State University, Columbus, OH, United States. https://bpcrc.osu.edu/

Centre for Antarctic Studies and Research, University of Canterbury, Christchurch, New Zealand. www.anta.canterbury.ac.nz/

Institute of Arctic and Alpine Research (INSTAAR), University of Colorado, Boulder, CO, United States. https://instaar.colorado.edu/

Institute for Marine and Antarctic Studies (IMAS), University of Tasmania, Hobart, Australia. www.imas.utas.edu.au/

National Institute of Polar Research. www.nipr.ac.jp/english/outline/index.html

International Centre for Terrestrial Antarctic Research (ICTAR), University of Waikato, New Zealand. www.waikato.ac.nz/research/units/ictar.shtml

International Arctic Research Center (IARC), University of Alaska, Fairbanks, AK, United States. https://uaf-iarc.org/

Korean Polar Research Institute, Incheon, Korea. https://www.uarctic.org/member-profiles/non-arctic/28245/korea-polar-research-institute

National Snow and Ice Data Center (NSIDC), CIRES, University of Colorado, Boulder, CO, United States. http://nsidc.org/

Northwest Institute of Eco-Environment and Resources (NIEER), CAS (merger of five former organizations including the Cold and Arid Regions Engineering and Environmental Research Center [CAREERI, CAS]). http://english.nieer.cas.cn/

Polar Research Institute of China. www.polar.org.cn/en/

Scott Polar Research Institute (SPRI), University of Cambridge, Cambridge, United Kingdom. www.spri.cam.ac.uk/

Swiss Polar Institute, Lausanne, Switzerland. http://polar.epfl.ch/

University Centre Svalbard, Longyearbyen, Svalbard, Norway. www.unis.no/

Glossary

Ablation   Processes that remove material from a snow or ice surface by vaporization and melt.

Active layer   The upper layer of permafrost that thaws seasonally to a depth of between 0.5 and 3–4 m, depending on latitude and soil conditions.

Albedo   The reflectivity of a surface to incoming solar radiation.

AMSR-E   Advanced Scanning Microwave Radiometer – Earth Observing System.

Antarctic Circumpolar Current (ACC)   A massive current system in the Southern Ocean that circles Antarctica.

Anthropogene   The name recently applied to the period of time when human activity began to dominate global environmental processes.

Anticyclone   A high pressure cell.

Aphelion   The farthest location of the Earth from the Sun in its seasonal orbit (approximately July 4).

Arctic dipole anomaly   Opposite pressure anomalies over the Canadian Arctic archipelago and northern Greenland, and over the Kara and Laptev seas.

Arctic Oscillation   Opposing atmospheric pressures in the northern middle and high latitudes. The oscillation exhibits a "negative phase" with high pressure over the polar region and low pressure at mid-latitudes (about 45° N), and a "positive phase" in which the pattern is reversed.

Atlantic Meridional Overturning Circulation (AMOC)   Atlantic thermohaline circulation that involves wind-driven surface currents (the Gulf Stream) traveling poleward from the tropical Atlantic Ocean, cooling en route, and eventually sinking at high latitudes, forming North Atlantic Deep Water. This dense water then flows into the ocean basin.

AVHRR   Advanced Very High Resolution Radiometer on NOAA polar-orbiting satellites.

Baroclinic   The depth-dependent part of the flow.

Barotropic fluid   A fluid whose density is only a function of pressure.

Benthic   The bottom layer of a water body.

Biome   A community of plants and animals that have common characteristics relative to the environment in which they exist.

Blocking   The interruption of the westerly wind belt by an anticyclone that is more or less stationary.

Blue ice area   An area of bare ice in Antarctica that results from net ablation when low snowfall is accompanied by sublimation.

Boreal forest   Forest characterized by coniferous trees in northern high latitudes.

Boron isotopes ^{10}B and ^{11}B; stable isotopes that make up about 20 and 80 percent, respectively, of natural boron.

Chlorofluorocarbons (CFCs)   Halogenated paraffin hydrocarbons that contain only carbon, chlorine, and fluorine.

CO$_1$ fertilization   The increase of carbon dioxide in the atmosphere owing to an increased rate of photosynthesis in plants.

Conductivity   The property of conducting heat, electricity, or sound.

Continentality   An increased annual and daily range of temperatures that occurs over land compared with water.

Continental shelves   Underwater landmass that extends from a continent, resulting in an area of relatively shallow water.

Continuous permafrost   Terrain where more than 90 percent is underlain by permafrost.

Cosmogenic nuclide (CN) dating   Surface exposure dating that is applied to samples taken from bedrock or boulders chosen for the information they provide on deglaciation. The preferred mineral is quartz, in which four nuclides (three of which are radioactive) are produced through exposure to cosmic rays. Beryllium-10 has been by far the most reliable nuclide.

Coupled Model Intercomparison Project   A project that compares the results of coupled climate model experiments.

Cryodepology   The study of frozen ground and frost action.

Cryosphere   Originally derived from the Greek word *kryos*, for "cold"; the frozen parts of the Earth's surface, including sea ice, lake ice, river ice, snow cover, glaciers, ice caps and ice sheets, and frozen ground, which includes permafrost.

Cryoturbation   Soil movement due to frost action.

Cyclogenesis   The development of a cyclone system.

Cyclone   A low pressure area.

Cyclonic vorticity   Cyclonic rotation of air (defined as positive in the northern hemisphere).

Dansgaard–Oeschger Oscillation   A climatic fluctuation averaging about 1,500 years, recurring in the late glacial period.

DEM   Digital elevation model; a model that specifies the topography of a terrain by digital numbers representative of the mean elevations of cells of certain resolutions.

Desert pavement   A surface covered with closely packed, interlocking angular or rounded rock fragments of pebble and cobble size.

Diatoms   A major group of algae belonging to the phytoplankton.

Discontinuous permafrost   Terrain underlain by 50–90 percent permafrost.

DMSP   Defense Meteorological Satellite Program, operated by the US military.

Eccentricity   The degree of departure from spherical in the Earth's orbit about the Sun.

ECMWF   European Centre for Medium-Range Weather Forecasts.

Ekman pumping   Downwelling in the ocean as a result of surface convergence.

El Niño–Southern Oscillation (ENSO)   Irregular, periodic variations in winds and sea surface temperatures over the tropical eastern Pacific Ocean, affecting much of the tropics and subtropics. The warming phase is known as El Niño and the cooling phase as La Niña.

Energy budget   The sum of all energy components, including solar and infrared radiation and sensible, latent, and conductive heat fluxes.

Ensemble (of models)   A collection of models whose results are averaged.

Eolian silt   Windblown material in the 20–50 μm size range.

Epilimnion   The warm surface layer of a stratified lake.

Equilibrium-line altitude (ELA)   The average elevational boundary between the accumulation and ablation zones on a glacier or ice sheet.

Equinoctial maximum   A maximum value near the equinoxes.

ERS-1 and ERS-2   Earth Resources Satellite 1 and 2, operated by the European Space Agency.

Eustatic sea level   The global sea level determined by either changes in the volume of water in the world's oceans or net changes in the volume of the ocean basins (as opposed to vertical land movements).

Feedback   A process in which "information" about the past or the present influences the same phenomenon in the present or future. A positive (negative) feedback amplifies (dampens) the response. A negative feedback tends to be self-regulating, whereas a positive feedback enhances the original effect. Feedbacks are a major feature of the climate system.

Ferrel cell   A mean meridional circulation in the atmosphere at middle latitudes that is thermally indirect; named after American meteorologist William Ferrel.

First-year ice   Sea ice that forms in the autumn and lasts through the following winter–spring.

Foraminifera   Single-celled marine organisms (protists) with shells (tests).

Frazil   Needle-shaped ice crystals formed in turbulent water.

Freezing degree-day   Determined by accumulating the value of mean daily air temperature below the freezing threshold value (0 °C).

Geomorphology   The scientific study of the origin and evolution of topographic features formed by physical, chemical, or biological processes operating at or near the Earth's surface.

Geopotential height   A height, adjusted for latitudinal variations of gravity, that is used to indicate the altitude of pressure levels in the atmosphere.

GHz   Gigahertz (10^9 Hz).

Glacial cycle   An interval comprising a glacial period (about 90 percent of the time) and an interglacial (10 percent of the time).

Glacial drift   Gravel, sand, or clay that is transported and deposited by a glacier or by glacial meltwater.

Glacial loading   The effect of an ice body in depressing the Earth's crust.

Glacial trimline   A clear line on the side of a valley formed by a glacier; it marks the most recent highest extent of the glacier in the valley.

Gravity Recovery and Climate Experiment (GRACE)   NASA mission comprising two satellites in a tandem, near-polar orbit. Using data from this project, it is possible to calculate, for a particular region, the change in mass that would have been necessary to cause the observed change in the gravity field.

Greenhouse effect   The effect in the atmosphere of warming due to the trapping of infrared radiation by greenhouse gases.

Greenhouse gas   A gas such as water vapor, carbon dioxide, or methane that absorbs infrared radiation in the atmosphere.

Grounding line   The position where a marine-terminating glacier starts to float.

Ground-penetrating radar   A geophysical method that uses radar (microwave) pulses to image the subsurface.

Hadley cell   A thermally direct meridional circulation in the tropical atmosphere; named after British scientist George Hadley.

Halocline   A steep vertical gradient of salinity.

Heinrich events   Irregular, massive discharges of icebergs from glacial-age ice sheets around the North Atlantic that transport sediments across the ocean.

Hummock   A small knoll or mound on the land surface, typically less than 15 m high.

Hydrolaccolith   A mound of earth-covered ice formed by frost heave in the Arctic and sub-Arctic.

Hypolimnion   The cooler, bottom layer of a stratified lake.

Iceberg rafting   The transport of rocks by icebergs.

Ice-rafted debris (IRD) Rock fragments transported by icebergs and deposited in layers across the oceans.

Igneous Related to rocks of volcanic origin.

Inversion A reversal of the usual decrease of temperature with height in the atmosphere.

Insolation A contraction of incoming solar radiation.

Interferometric Synthetic Aperture Radar (InSAR) Use of two or more Synthetic Aperture Radar (SAR) images of an area to identify surface movements through time.

Interglacial A geological interval of warmer global average temperature lasting thousands of years that separates consecutive glacial periods within an ice age.

Interstadials Periods of temporary retreat of ice during a glacial stage.

Isohaline Equal salinity.

Isostatic adjustment (rebound) Glacial isostatic adjustment, or postglacial rebound; the response of the solid Earth to the changing surface load brought about by the waxing and waning of ice sheets.

Isotherm A line of equal temperature.

Jokulhlaup The sudden and rapid draining of a glacier-dammed lake or of water impounded within a glacier.

Katabatic winds Cold air drainage winds moving down a slope, especially at night.

Lagrangian analysis Analysis of currents following their motion.

Landfast ice Sea ice that is attached to the land, in shallow water, by bottom freezing or extending outward between grounded ice ridges.

Last Glacial Maximum (LGM) The last glacial phase of the Pleistocene, which culminated circa 21 ka.

Latent heat The amount of energy released or absorbed by a substance during a change of state that occurs without changing its temperature; occurs during a phase transition such as evaporation/condensation or melting/freezing.

Lead A linear fracture in sea ice.

Leaf Area Index (LAI) A dimensionless quantity that characterizes plant canopies; defined as the one-sided green leaf area per unit ground surface area.

Lidar Light detection and ranging.

Limnology The study of lakes.

Little Ice Age A cold interval, particularly around the North Atlantic and in Europe, during AD 1550–1850.

Madden–Julian Oscillation   An eastward-propagating oscillation in the tropics with a time scale of 30–90 days.

MAGT   Mean annual ground temperature.

Marine Isotope Stage (MIS)   A numbering system for glacial and interglacial stages based on oxygen isotopes in marine fauna.

Mass balance   The mass (e.g., of snow and ice) that enters a system must, by conservation of mass, either leave the system or accumulate within the system.

Meridional   In the direction of latitude.

Methane clathrate   A large amount of methane that becomes trapped within a crystal structure of water and forms a solid similar to ice.

Microwave   Electromagnetic waves ranging from 1 mm to 30 cm in wavelength.

Mires   A wetland terrain covered by swampy or boggy ground; also called quagmire or peatland.

MODIS   Moderate Resolution Imaging Spectroradiometer; a system operated on NASA's Terra (launched December 1999) and AQUA (launched May 2002). It has thirty-six bands from 0.4 to 14.3 μm, with spatial resolution from 250 to 1000 m.

Monsoon   A large-scale, seasonal wind reversal.

Moraine   Accumulation of unconsolidated glacial debris that was carried by a glaciers and then deposited laterally or at the glacier terminus.

Moulin   A shaft that forms at the surface of an ice sheet and transports supraglacial streams into the ice.

Multiyear ice   Sea ice that survives more than two summers.

Net radiation   The amount of solar and infrared radiation that is received by a surface, taking into account all wavelengths and both incoming and outgoing radiation.

Naled   A Russian term for river icing.

Naryn   A Russian term for a ground icing from a spring.

NDVI   Normalized Difference Vegetation Index.

Neoglacial   A cold interval in the late Holocene when glaciers readvanced, most prominently in the Little Ice Age.

North Atlantic Oscillation (NAO)   A pressure oscillation, which in its positive phase has a deep Icelandic low and a strong Azores high.

Northern Annular Mode (NAM)   A pressure oscillation, which in its positive mode has low pressure in high latitudes and high pressure in mid-latitudes (and vice versa for negative mode). Also call the Arctic Oscillation (AO).

Obliquity   The axial tile of the Earth, currently 23.4°.

Optically simulated luminescence (OSL)   A method of dating based on the buildup of a luminescence signal in quartz grains that are shielded from sunlight through burial.

Orographic uplift   Uplift of an air current by mountains.

Oxygen isotope ratio   The ratio of ^{18}O to ^{16}O.

Pacific Decadal Oscillation (PDO)   A multidecadal oscillation in sea surface temperature between the eastern and western North Pacific Ocean.

Pacific–North America (PNA) pattern   A teleconnection pattern involving, in the positive phase, above-average heights in the vicinity of Hawaii and over the intermountain region of North America, and below-average heights located south of the Aleutian Islands and over the southeastern United States.

Palsa   A low, often oval, frost-heave mound occurring in polar and subpolar climates with discontinuous permafrost. It contains permanent ice lenses, consists of an ice core with overlying soil, and often occurs in groups.

Passive microwave radiation   Electromagnetic radiation emitted naturally by objects in the microwave range.

Patterned ground   Surface material arranged in polygons, circles, or stripes through freeze–thaw processes.

Penultimate glaciation   The glacial period that preceded the final one before the Holocene.

Periglacial   An environment that is adjacent to ice bodies; more generally applied to cold climate processes.

Perihelion   The Earth's position when it is closest to the Sun (approximately January 2).

Permafrost   Perennially frozen ground; ground that remains frozen for at least two summers.

Permittivity   The measure of a material's ability to resist an electric field.

Piedmont glacier   A steep valley glacier that spills out into relatively flat land, where it spread outs into bulb-like lobes.

Pingo   A large mound of earth-covered ice found in the Arctic and sub-Arctic, which can be up to 50 m high.

Planetary waves   A long wave in the atmospheric circulation; there are between two and five around a hemisphere.

Planktonic   The near surface of ocean layers.

Polar amplification   The amplification of temperature trends in the polar regions as a result of various feedback processes.

Polar front   The boundary between cold polar air and warm tropical air; in the ocean, a corresponding boundary between water masses.

Polar vortex   A persistent large-scale cyclonic circulation pattern in the middle and upper troposphere and the stratosphere, centered generally in the polar regions of each hemisphere. In the Arctic, the vortex is asymmetric and typically features a trough over eastern North America.

Polynya   A Russian term for an irregular open water area that is usually adjacent to the coast or between landfast and pack ice, but occasionally forms in an ice-covered ocean.

Polythermal glacier   A thermally complex glacier with both warm and cold ice.

Practical salinity units (psu)   The conductivity ratio of a sea water sample to a standard solution. It is almost equal to parts per thousand (‰), which is approximately the number of grams of salt per kilogram of solution.

Precession of the equinox   The shift in timing of the vernal equinox from perihelion to aphelion over a 21,000-year cycle.

Proglacial lake   A lake formed immediately in front of a glacier that is bordered by a moraine.

Proxy record   A biological, chemical, or physical characteristic that indicates past environments or climate.

Pycnocline   A layer of strong, vertical density gradient within a body of water.

Radar   Radio direction and ranging.

Radiative forcing   The difference between the energy received by the Earth and that radiated back to space. It can be calculated for individual greenhouse gases and aerosols, among other things.

Reanalysis   The production of multiyear, global, state-of-the-art, gridded representations of atmospheric states, generated by a constant numerical model and constant data assimilation system.

Rock glacier   A glacier that consists of either angular rock debris frozen in interstitial ice, a former "true" glacier overlain by a layer of talus, or something in between. Rock glaciers may extend outward and downslope from talus cones, glaciers, or terminal moraines.

Rossby wave train   A sequence of Rossby (planetary) waves in the atmosphere.

Salinity   The measure of all the salts dissolved in water.

Sastrugi   Linear ridges in the snow surface produced by wind action.

Sensible heat   Thermal energy that is added to or removed from the air by conduction and convection.

SMMR   Scanning Multifrequency Microwave Radiometer, 1978–1987. The five dual-polarized (horizontal, vertical) frequencies ranged from 6.6 to 37.0 GHz; the devices were flown on NASA's Nimbus 7 satellite from October 25, 1978, through August 20, 1994.

Solifluction   Downslope flow of water-saturated soil down a steep slope.

Solsistial maxima   Maxima at the time of the solstices.

Sonar   Sound detection and ranging.

Southern Annular Mode (SAM)   A pressure oscillation, which in its positive phase has low pressure in high southern latitudes and high pressure in midlatitudes (and vice versa in its negative mode). Also called the Antarctic Oscillation (AAO).

Southern Oscillation (SO)   An east–west oscillation in the atmosphere over the equatorial Pacific Ocean; the atmospheric part of the coupled El Niño–Southern Oscillation (ENSO) phenomenon.

Spectroradiometer   An instrument used to measure the spectral power distributions of solar radiation.

SSM/I   Special Sensor Microwave Imager, which was flown on Defense Meteorological Satellite Program (DMSP) satellites (USA) from late 1987 to the present. It has four frequencies from 19.35 to 85.5 GHz that operate in horizontal and vertical polarizations, except at 22.35 GHz, which is only vertical.

Stalagmites   Limestone columns in caves formed by deposition of calcium carbonate.

Steppe tundra   A plant community with codominance of both steppe and tundra species that is found in cold, dry climates.

Sublimation   The direct transition from solid to vapor phase, bypassing the intermediate liquid stage.

SWE   Snow water equivalent; the amount of liquid water contained in a volume of snow.

Synoptic meteorology   The study of synoptic-scale (cyclones) phenomena.

Talik   An unfrozen layer within permafrost.

Teleconnection pattern   Climate anomalies that are related to each other at large distances.

Thermal expansion   Expansion of the ocean volume due to rising water temperatures.

Thermal gradient   A gradient of temperature.

Thermocline   A steep temperature gradient in a body of water.

Thermohaline circulation   A circulation in the ocean driven by contrasts in temperature and salinity (and therefore density differences).

Thermokarst   A land surface form that results from the melting of ground ice in a permafrost region.

Third Pole   Geographical term used to designate high-mountain Central Asia (Tibet plateau and mountain ranges surrounding it).

Till   Unsorted glacial sediment deposited by a glacier or ice sheet.

Varve   Laminated sediment formed by the deposition in a lake of coarse-grained sand and silt in summer and fine-grained material in winter. The annual paired layers can be used for dating.

Vernal equinox   The date in spring when the sun is directly overhead at noon on the equator.

Younger Dryas   A cold event (circa 12.9–11.6 ka) when glacial conditions briefly returned during the postglacial warming.

Zonal   In the direction of longitude.

Index